STRUCTURED PROGRAMMING:
PL/I with PL/C

STRUCTURED PROGRAMMING: PL/I with PL/C

Kathi Davis
and
Lyle Domina

NORTHERN ILLINOIS UNIVERSITY

Holt, Rinehart and Winston
New York   Chicago   San Francisco
Philadelphia   Montreal   Toronto
London   Sydney   Tokyo

Requests for permission to make copies of any part
of the work should be mailed to: Permissions, Holt,
Rinehart and Winston, Inc., 111 Fifth Avenue, New
York, New York 10003.

Library of Congress Cataloging-in-Publication Data

Davis, Kathi.
    Structured programming.

    Includes index.
    1. PL/I (Computer program language)   2. PL/C
(Computer program language)   3. Structured programming.
I. Domina, Lyle.   II. Title.
QA76.73.P25D66   1987   005.13'3   87-12473

ISBN 0-03-003723-9

8 9 0   039   9 8 7 6 5 4 3 2 1
Printed in the United States of America
Published simultaneously in Canada

Holt, Rinehart and Winston, Inc.
The Dryden Press
Saunders College Publishing

We dedicate this book to our families,
for their love and patience.

PREFACE

Since structured programming is nearly universally accepted today, *Structured Programming: PL/I with PL/C* begins with the elements that create a structured program, and this concern is carried throughout the book. It is also widely recognized that beginning programmers have as many problems with the logic of a program as they do with PL/I syntax. Chapter 1, therefore, stresses the importance of algorithm development prior to coding and provides the necessary tools (flowcharts and pseudocode), and Chapter 6 discusses the subject in more detail. Because there is a movement away from flowcharts toward pseudocode, most of the complete programs in the text are accompanied by their logic expressed in pseudocode. For those who prefer them, flowcharts for some of the complete programs are gathered together in Appendix A.

For those with no previous computer background, Chapter 2 opens with a discussion of terminal-mainframe communications, a brief introduction to computer memory, and a discussion of the distinction between compilation and execution. Those with some background in computing can easily skip over Chapter 1 and/or the first section of Chapter 2.

In the presentation of the PL/I programming language, we have stressed getting quickly into programming before interest is lost. So as not to confuse the programmer unnecessarily, we present only those aspects of our topic that are required at the time.

Many programming language textbooks rather gloss over debugging and documentation; however, this text offers some early guidance on both topics. A discussion of syntax errors immediately follows the introduction of the basic PL/I language elements (Chapter 2), and a section of Chapter 6 (including an example program) is devoted to a discussion of both syntax and logic errors. By the same token, we emphasize documentation regularly throughout the text, not only by discussing it but more important, by illustrating it in terms of providing documentation for each complete program.

Given the generally business applications orientation of the book, the material is presented in an order that allows it to be taught and learned with no skipping around. We have successfully used this order in our PL/I classes.

Each chapter ends with a series of exercises, which include several programming assignments. The answers to selections from the exercises are given in Appendix B. Appendix D contains a summary of all PL/I built-in functions.

All example programs are written in PL/I, using the Optimizing Compiler, Version 1, Release 3.1. The limitations of PL/C are noted wherever necessary, and its differences from PL/I are summarized in Appendix C.

Finally, the text is suitable for a highly intensive single-semester course for beginning or experienced programmers. It is also suitable for a less intensive two-semester course for beginners.

An instructor's manual to accompany *Structured Programming: PL/I with PL/C* is available. Please contact

CBS Educational and Professional Publishing

A Division of CBS Inc.

383 Madison Avenue

New York, New York 10017

We wish to acknowledge and thank those who reviewed the text in manuscript; they spared us a great deal of embarrassment. Any remaining infelicities and errors are our responsibility, not theirs.

CONTENTS

C H A P T E R 1

ALGORITHMS,

PSEUDOCODE,

AND FLOWCHARTS

In this first chapter we introduce you to some concepts that are preliminary to actually coding a PL/I program. Computer experts debate at length the question, "Does the computer solve problems?" The question, like many others, is a semantic one; that is, what is meant by "solve"? If the word means to formulate a solution to a problem, clearly computers using languages like PL/I are incapable of solving problems. If "solve" means doing the calculations once the problem has been formulated, computers just as clearly do solve problems. The purpose of this chapter, at any rate, is to introduce you to some methods by which problems can be formulated in such a manner that the instructions necessary to direct the computer in its computations can easily be arrived at; in other words, the logic for solving the problem will have been determined.

Algorithms

The first step in solving any problem is to define the problem. This perhaps strikes you as a statement of the obvious, but any teacher of programming languages can give you example after example of students who attempted to write programs before they had even defined the problem. The aggregate hours of wasted labor that resulted is staggering. Any problem can be said to consist of three parts: information that is given, results that are desired, and actions that must be taken to get from the given information to the desired results. The first and second parts of the problem definition usually pose only the difficulty of being so obvious that they are taken for granted. We strongly urge you to begin your solution of a problem by formulating in your mind or (preferably) on paper a clear statement of what information is given and what results are desired. Keep in mind that a crucial aspect of both (and especially of desired results) includes the unit(s) in which the information is (or is to be) expressed, and we are convinced you will find it useful to provide unit labels at all times. Are the desired results, for example, to be expressed in time units (days, months, years), in money units (dollars, cents), or in money units per time unit (dollars per month)? Until these questions have been clearly resolved, it is foolish to begin work on the third part of the problem definition, the actions necessary to transform the given information into the desired results.

Once the first and second parts of the definition of a problem have been clearly formulated, however, you can begin on the third, which is usually the most difficult of the three. It is in determining the actions necessary to produce the desired results that algorithms come into play. An **algorithm** is a set of unambiguous instructions to solve a problem in a finite number of steps. Any human activity can be stated in algorithmic format. For example, let's see what an algorithm for preparing for class in the morning might look like.

1. Shut off the alarm.

2. Get out of bed.

3. Shower.

4. Dress and complete personal grooming.

5. Make the bed.

6. Gather up books and go to breakfast.

It is ironic to note that the term *ambiguous* is itself ambiguous in this context.

In the algorithm, for example, what do we mean by "personal grooming"? The term can be clarified somewhat by rephrasing step 4 as follows:

4. If male, shave, apply lotion, and comb hair.

Otherwise, apply makeup and comb hair

Even though this revision helps to clarify it, the algorithm still raises a good many questions. How long should I stay in the shower? What shall I wear? What books should I gather up? These questions, in turn, raise two more general questions: how much detail should be included in an algorithm? and since we are interested here in solving problems with a computer, what kinds of algorithms (ultimately, what kinds of problems) can be solved by the computer?

No algorithmic answer can be given to the question of how much detail to include in an algorithm, but two general principles will help. First of all, always keep in mind that creating an algorithm is seldom, if ever, an end in itself; instead, it is a means to the orderly and efficient solution to a problem. The second principle follows from the first, and one way of stating it is to say, "Include as much detail as necessary to allow you to solve the problem in an orderly and efficient manner." You, as a beginner at computer programming, will undoubtedly find it desirable to include more detail than would someone who has many years of experience. Suppose, on the other hand, that you are an accomplished (even if amateur) photographer or automobile mechanic. In these cases, the algorithms for changing spark plugs or shooting a snapshot will hardly enter your conscious mind. In fact, most of us have known people so expert that they make poor teachers because they are hardly conscious of the steps in their process. At this point, the guiding principle for how much detail to include might be: "How much detail will be necessary to allow someone else to get the desired results from the algorithm?" Since computer programs in business and industry might have a life span of 20 years or more (but may need periodic updating), it is crucial that a clear statement of the algorithm be maintained. For you as a student, it is probably best if you include enough detail that another person at your level of expertise can understand and reproduce the program that results from your algorithm.

The second question—what kinds of problems are amenable to computer solution—is more easily answered: any problem that either the computer or the programmer can quantify (i.e., assign numeric values to the variables in the problem). The computer is built in such a way that each letter of the alphabet, for example, is assigned a numeric value, with A having the lowest value and Z the highest. Thus the computer can easily alphabetize a list of words. What about the problem of who is the greater composer, Beethoven or Mozart? There is nothing in this problem that the computer can quantify; therefore, if the computer is to be used in solving the problem, a human being will have to provide the quantification. Someone will have to decide what elements in a musical composition lead to "greatness" and the relative importance of each of the elements. Finally, someone will have to provide a means by which the computer can count the number of incidences of each element in the music of each composer. Then, and only then, can the computer aid in the solution of the problem.

In light of the foregoing discussion we can begin to see how an algorithm developed in a computer context not only must not be, but also cannot be, ambiguous. It is possible, of course, to be ambiguous with numbers, but the computer is built with "rules" to resolve these ambiguities, although in the process it might produce a result other than the individual intended. Given the problem 20 / 2 + 8, a human being might perceive an ambiguity. Is 20 / 2 = 10, and then 10 + 8 = 18 intended; or is 2 + 8 = 10, and then 20/10 = 2 intended? For a computer, however, there is no ambiguity; it will resolve the problem with an answer of 18. As the math student would probably guess, the computer can be presented with the problem of 20 / (2 + 8), in which case it will return an answer of 2.

With the problem of ambiguity behind us, let us turn back to algorithms. Suppose your checkbook contains five entries since you last brought it up to date. An algorithm to update your checkbook would contain the following six steps.

1. Get balance.

2. The first entry is a check, so subtract its amount from the (previous) balance to get the (new) balance.

3. The second entry is also a check, so subtract its amount from the (previous) balance to get the (new) balance. *Note*: The new balance of step 2 has become the previous balance of step 3; there is actually only one balance.

4. The third entry is a deposit, so add its amount to the (previous) balance to get the (new) balance.

5. The fourth entry is a check, so subtract its amount from the (previous) balance to get the (new) balance.

6. The fifth entry is a check, so subtract its amount from the (previous) balance to get the (new) balance.

With this basic understanding of the algorithm as a set of instructions for transforming the given information into the desired results, we can now go on to discuss two common methods of expressing algorithms for problems to be solved by the computer. In the process we will introduce more complex algorithms.

1.2

Pseudocode

The algorithm just presented will clearly solve the problem of updating a checkbook, but it is rather wordy; and if the problem were more complex, a good deal more would be necessary. To avoid the wordiness of the example above, there are two widely used methods of expressing algorithms: **pseudocode** and **flowcharts**. Pseudocode is similar to computer code, but no computer can understand it (hence "pseudo," or "false"). It is closer to English than are computer languages. There are no universally (or even widely) accepted standards for writing pseudocode, and we present those standards we prefer. The advantage of pseudocode is that, using language, it is close to the way people are used to expressing solutions to problems; therefore, we begin by discussing pseudocode, which will be followed by a discussion of flowcharting.

Notice that in the algorithm for updating a checkbook no decisions are required. In pointing out that the computer can alphabetize a list of names, however, we implied that the computer can make decisions (which comes first, ANDERSON or ANDERSEN?) Taking this computer ability into account, we can improve on the algorithm as follows (now presented in pseudocode, although pseudocode statements are not normally numbered).

```
START
   1.   Get balance
   2.   IF (first entry is a check)
   3.       Balance ← balance - check amount
   4.   ENDIF
   5.   IF (second entry is a check)
   6.       Balance ← balance - check amount
   7.   ENDIF
   8.   IF (third entry is a deposit)
   9.       Balance ← balance + deposit amount
  10.   ENDIF
  11.   IF (fourth entry is a check)
  12.       Balance ← balance - check amount
  13.   ENDIF
```

```
14.  IF (fifth entry is a check)
15.      Balance ← balance - check amount
16.  ENDIF
STOP
```

First, a few observations on the pseudocode:

1. Notice that the condition on which the decision is made is placed within parentheses.

2. Notice that the key words such as START, STOP, IF, and ENDIF are printed in uppercase. These are simply conventions that we are used to and that we believe produce a clear and easily readable pseudocode.

3. The arrow (←) means "move the results of the operation on the right side of the arrow to the area designated by the name on the left of the arrow."

4. "+" and "−" are, of course, the standard mathematical symbols for addition and subtraction.

5. The purpose of the ENDIF is to clearly indicate how many statements (computer actions) are dependent upon the IF condition.

6. The indentation serves the same purpose of showing the subordination of statements to IF conditions.

7. Now notice that we have allowed the computer to make the decision that it is capable of making.

8. Finally, notice that this algorithm is both quite wordy and absolutely rigid; that is, if the entries are not in the specified order of check, check, deposit, check, check, the algorithm will not work because no instruction has been provided for the alternative case.

This rigidity can be somewhat overcome by the following version of the algorithm (now presented with no line numbers).

```
START
   Get previous balance
      IF (the entry is a check)
         Balance ← balance - check amount
      ELSE [i.e., otherwise]
         Balance ← balance + deposit amount
      ENDIF
```

```
IF (the entry is a check)
    Balance ← balance - check amount
ELSE
    Balance ← balance + deposit amount
ENDIF
IF (the entry is a check)
    Balance ← balance - check amount
ELSE
    Balance ← balance + deposit amount
ENDIF
IF (the entry is a check)
    Balance ← balance - check amount
ELSE
    Balance ← balance + deposit amount
ENDIF
IF (the entry is a check)
    Balance ← balance - check amount
ELSE
    Balance ← balance + check amount
ENDIF
STOP
```

Because we have provided an alternative for each step, our algorithm is now more flexible; that is, any of the five entries can now be either a check or a deposit, but the algorithm still demands exactly five entries; there is no provision for either more or fewer entries. Given this limitation, the algorithm is still too rigid to function as a general algorithm for bringing a checkbook up to date. Before we attempt to overcome this limitation, however, let's consider what we have done so far. In the first (nonpseudocode) attempt we used a structure called **sequence**. Sequence is simply carrying out one instruction after another in the order in which they appear (i.e., top to bottom), with no decisions involved. In the second attempt we introduced the **decision** structure in its first format. In this case the computer is told to make a decision and to act on the basis of the decision; in effect we tell the computer to ask a TRUE-FALSE question and to act if the answer is true, but to do nothing if the answer is false. This structure is commonly referred to as the **IF-THEN** structure. In our third attempt we provide for an action in the case of a false as well as a true condition. This structure is often referred to as the **IF-THEN-ELSE** structure. The generic term covering both the IF-THEN and the IF-THEN-ELSE is sometimes called the **selection** structure, as well as the decision structure.

The next structure to consider is **iteration** (also called **repetition** or

DO WHILE). Iteration is the iterated or repeated execution of one or more instructions. The effect of iteration is to create a **loop**, and the process is often referred to as "looping." In order to present iteration in its simplest form, we must change our problem slightly: We are going to assume that all five checkbook entries are checks; there are no deposits. Our algorithm will now look like this:

```
START
    Counter ← 0
    DO WHILE (Counter < 5)
        Balance ← balance − check amount
        Counter ← Counter + 1
    ENDDO
STOP
```

The ENDDO serves the same purpose as the ENDIF; it shows us how many statements are to be iterated. It is important to understand that iteration also inherently involves decision. Notice that the area referred to as Counter is initially set to a value of 0 (i.e., **initialized**). This is done to signify the fact that no entries (checks or deposits) have been processed. Each time the loop is executed, an entry is processed and counted, and a 1 is added to the current value of Counter, thereby increasing its value by 1. Each time the DO WHILE is encountered, a comparison is made between the current value of Counter and the value 5 (the number of entries to be processed), and on the basis of that comparison a decision is made whether or not to execute the statements within the loop another time. On successive iterations Counter will have values of 1, 2, 3, 4, and 5. After five entries have been processed and counted, Counter will have the value of 5, the condition (Counter < 5) will no longer be true, and the decision will be to ignore the statements within the loop.

The most obvious benefit of this format over the previous attempts is the drastic reduction in the number of pseudocode statements required. Some reduction is accounted for by the absence of a decision structure, the rest by the introduction of the iteration structure.

The next step is to generalize the algorithm a bit more. In the iteration example above the algorithm is good for any number of entries (i.e., by changing the constant with which Counter is compared), but we must know in advance how many entries there will be. Since advance knowledge of how many entries there will be is often not available, we will now take advantage of the computer's ability to detect the end of the input data.

```
START
   Get balance
   Get entry
   DO WHILE (more entries)
      Balance ← balance - check amount
      Get entry
   ENDDO
STOP
```

In our final step we will combine iteration with decision and tie the algorithm more closely to an actual computer process. To accomplish the first, we will alter our problem back to the situation in which any entry can be either a check or a deposit. We will assume that each input entry contains a code that identifies the type of transaction. As for tying the algorithm to the computer process, it is necessary to point out that the data the program manipulates ordinarily lies outside the confines of the program itself. To be manipulated it must be brought into designated areas within the program; to indicate this process we will use the pseudocode verb **Read**. Also, the program is of little value unless we can know the results of the computations. The process by which a program makes known these results (in this case by sending them to a printer) we will designate with pseudocode verb **Print**. Thus our algorithm in its final form is as follows:

```
START
   Read balance
   Read first entry
   DO WHILE (more entries)
      IF (transaction code = check)
         Balance ← balance - check amount
      ELSE
         Balance ← balance + deposit amount
      ENDIF
      Print balance
      Read next entry
   ENDDO
STOP
```

There are a number of points in this version of the algorithm worth noting. Within a single program all three structures have been used; in fact, the decision structure lies *within* the iteration structure. It has been proved that any problem a computer can solve can be solved using only the three structures of sequence, decision, and iteration. The three structures, as we will see at various points later in the text, can be combined in any manner and to any level of resulting complexity that the problem demands. Furthermore, an experienced programmer can write code that the computer understands in any computer language he or she is familiar with directly from the pseudocode. Before the computer code is written, however, the programmer should **debug** (i.e., check for errors in the logic of) the pseudocode. To debug pseudocode (also called "desk check") one must take hypothetical data and trace its path through the pseudocode. This process might need to be undertaken several times because all the possibilities must be explored. Debugging our example, however, is relatively simple. We first assume an input entry that is a check and make sure its amount is subtracted from the balance. Since we want the balance printed after each transaction, we must be sure that printing will occur. Next, we must be certain that the next entry will be read in. Finally, we must assume a deposit entry and trace its route through the algorithm in the same manner as we did with the check entry.

One feature of the latest version of our algorithm that might puzzle the beginner is the two "read entry" statements—one outside the loop and the other at the bottom of the loop. Structured programming is explained later in the book; for now suffice it to say that structured programs are obviously best produced from structured pseudocode or flowcharts. Structured algorithms, moreover, are first of all restricted to the three structures of sequence, decision, and iteration. In addition, the decision (Is the DO WHILE condition still true?) must be made at the top of the loop. It therefore follows that an entry must be read before we ask, "Are there any more?" (The first time we are in effect asking, "Are there any at all?") The entry that has been read must now be processed before another entry is read. After the entry is processed and an attempt has been made to read another entry, we return to the top of the loop and again check the DO WHILE condition, remembering that if the DO WHILE condition is false the statements between it and the ENDDO will be skipped over.

1.3

Flowchart Symbols

The second major method of expressing algorithms is through flowcharts. A **flowchart** is a visual representation of the logic necessary to solve a problem. The advantage of flowcharts over pseudocode is that for many people, once they get used to them, logic in flowchart form is easier to follow than it is in pseudocode. Because flowcharts present in effect a schematic diagram of the flow of data and the action taken at each point, sample data can be literally traced through the flowchart. Like pseudocode, a flowchart uses only the three structures of sequence, decision, and iteration. Also like pseudocode, there is no single standard for drawing flowcharts, although there is more standardization here than in the case of pseudocode.

The symbols used in flowcharting each represent an action on the part of the computer, and the nature of the action is represented by the shape of the symbol. These symbols are standard and consist of the following:

1. The "terminal interrupt" symbol which in practice signals the start and stop of program execution.

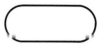

2. The input/output (I/O) symbol, which corresponds to a "read" or "print" in pseudocode.

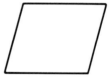

3. The process symbol, which indicates an action that occurs entirely within the computer. It corresponds to such pseudocode statements as balance ← balance − check.

4. The decision symbol, which corresponds to the pseudocode statements IF and DO WHILE.

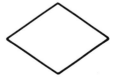

**Structured
Flowcharts**

In structured flowcharting, as in structured pseudocoding, only the three constructs of sequence, decision, and iteration are permitted. The symbols given above are combined in various ways to create the three structures. Although the symbols themselves are standard, the precise manner in which they are combined is not. We believe that the following formats shown in Figure 1-1 through Figure 1-4 produce flowcharts that are the easiest to read.

• SEQUENCE.   In sequential processing (Fig. 1–1), the first symbol could be an input/output symbol. The chain of steps (boxes) can be as long as necessary.

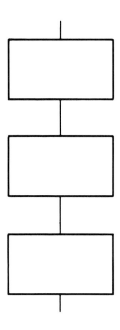

**FIGURE 1–1
Sequential processing.**

• DECISION (IF-THEN VERSION).   In Figure 1–2 it can be seen that an action is taken only if the condition is true. The **true** path could lead to a sequence of actions, to another decision, to an iteration, or to some complex combination of any of the three.

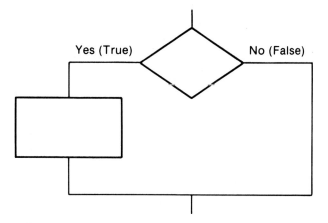

FIGURE 1-2
An IF/THEN structure with no ELSE.

• DECISION (IF-THEN-ELSE VERSION). In this case, one action is taken if the condition is false; another if it is true (see Fig. 1–3). Either path could lead to a complex structure consisting of any combination of sequence, decision, and iteration.

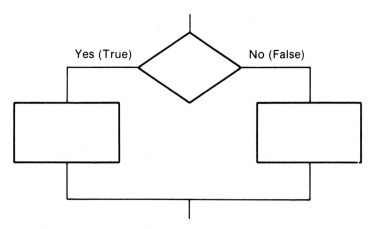

FIGURE 1–3
An IF/THEN/ELSE structure.

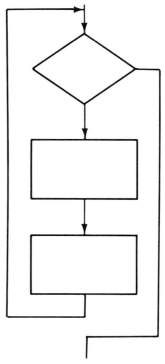

FIGURE 1–4
The iteration (loop structure).

• ITERATION.   This is the only instance in which the flow is not from top to bottom; note that in this case an arrowhead is used to make clear the flow (see Fig. 1–4). In the absence of arrowheads, flow is always assumed to be from top to bottom. By bringing the **false** path back in under the last sequence box, a nearly vertical line is maintained. We believe this enhances the readability of the flowchart. The actions along the **true** path could once again be a complex structure of sequence, decision, and iteration.

In addition to the standards illustrated in the preceding flowcharts, we recommend that neither on-page nor off-page connectors be used. Creating structured flowcharts makes on-page connectors unnecessary. As for off-page connectors, we believe that larger sheets of paper, combined with the judicious use of subroutines is preferable to the confusion caused by off page connectors.

It should now be easy to determine what a flowchart counterpart to our final pseudocode attempt of the checkbook update would look like. We repeat the pseudocode here for your convenience and once again provide numbers that correspond to the numbers also included in the flowchart shown in Figure 1–5.

```
(1)   START
(2)       Read balance
(3)       Read first entry
(4)       DO WHILE (more entries)
(5)           IF (it is a check)
(6)               Balance ← balance – check-amount
              ELSE
(7)               Balance ← balance + deposit-amount
              ENDIF
(8)           Print balance
(9)           Read next entry
          ENDDO
(10)  STOP
```

Notice that the coming together of the two paths of the IF-THEN-ELSE corresponds to the pseudocode ENDIF, and the branch upwards of the flow of execution corresponds to the pseudocode ENDDO. Finally, note that the action represented by each symbol is indicated within the symbol.

1.5

Summary

To conclude the chapter let's review what we have presented concerning problem solving.

1. The first two steps are to use the information given and the results desired to define the problem.
2. The third step is to determine the actions to be performed on the information given in order to produce the desired results.
3. Step 3 is best accomplished by the use of algorithms expressed either as pseudocode or as a flowchart.
4. Because our ultimate goal is to solve the problem with the aid of the computer, and since this is best accomplished by writing structured programs, you should create structured pseudocode and flowcharts that can be readily translated into structured computer programs; the three structures of sequence, decision, and iteration are the only structures needed (or permitted) to produce a structured flowchart or pseudocode.

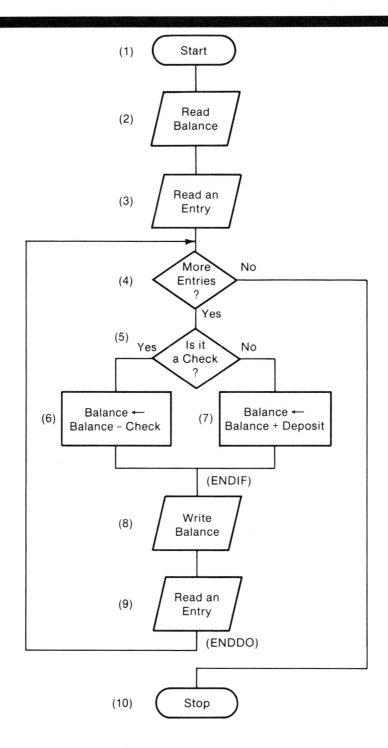

FIGURE 1–5
An example of a complete flowchart.

5. No one should attempt to code a program until steps 1 to 3 above have been completed and carefully checked. Once you have defined the problem and stated the logic of its solution as either pseudocode or a flowchart, you are ready to begin translating your solution in the specific form demanded by PL/I. This is the subject of Chapter 2.

1.6

Exercises

1. What value is printed in each of the algorithms presented in the following flowcharts?
   a. Use Figure 1–6.
      Input: 19 4 22 5 2 6 9 11 15 30 3 21 45 7 34 13 12 40 8 27
   b. Use Figure 1–7.
      Input: −3 8 12 −1 14 0 9 19 23 -7 27 4 −13 22 6
   c. Use Figure 1–8.
      Input: 16 10 15 99 74 11 55 14 −16 62

2. Write the pseudocode represented by the flowcharts in Exercise 1.

3. Draw the flowchart corresponding to each of the following algorithms.

   a.
   ```
   START
       Read X,Y
       Print X,Y
       IF (X < Y)
           Z ← Y - X
       ELSE
           Z ← X - Y
       ENDIF
       Print Z
   END
   ```

   b.
   ```
   START
       Read N
       CTR ← 0
       PCTR ← 0
       DO WHILE (CTR < N)
           Read X
           IF (X < 0)
               PCTR ← PCTR + 1
           ENDIF
           CTR ← CTR + 1
       ENDDO
       Print CTR, PCTR
   END
   ```

   c.
   ```
   START
       Read A, B
       Print A, B
       C ← 2 · B + A - 3
       Print C
   END
   ```

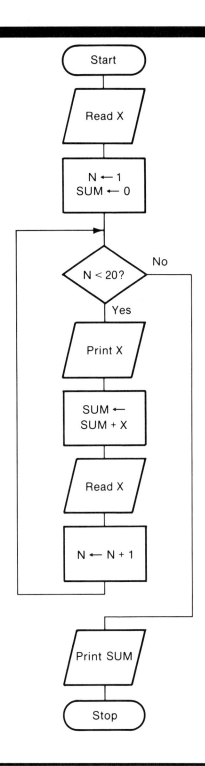

FIGURE 1-6

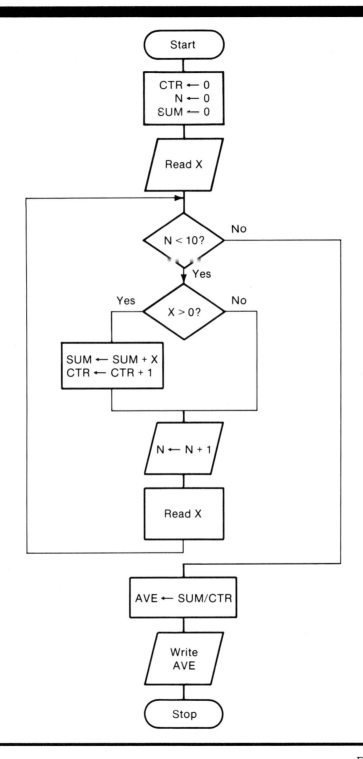

FIGURE 1–7

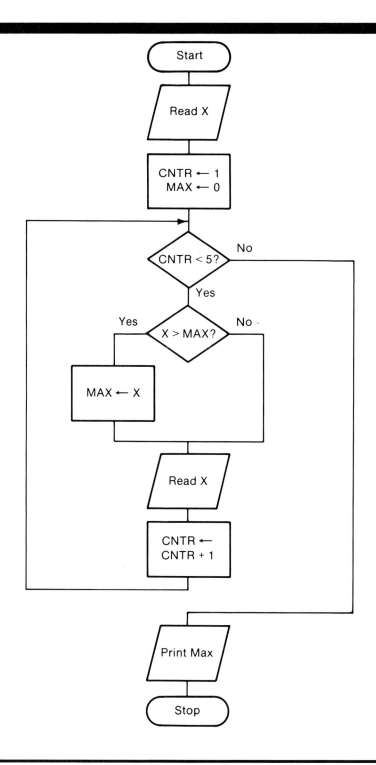

FIGURE 1-8

4. Write the pseudocode and draw the flowchart for each of the following programs.

    a.   Read, print and calculate the sum of 100 numbers. Print the sum when finished.

    b.   Read ten numbers, and compute the average of those numbers that are greater than zero. Print the average.

    c.   Read five numbers, and print the maximum value.

C H A P T E R 2

INTRODUCTION

TO PL/I

This chapter is concerned with giving enough information to enable you to submit simple programs in PL/I and see them execute successfully. The first step is to gain some idea of how we communicate with a computer; then we look at the basic format and syntax of a PL/I program and the fundamental statements it contains.

2.1

Computer Fundamentals

Programmers using high-level languages such as PL/I actually are required to know very little about the internal workings of the computer. On the other hand, situations do arise when such knowledge is extremely useful, if not absolutely essential. In this section we present a brief overview of the sequence of events beginning with the submission of a program (**job**) to the computer through the generated output on the printed page. We will present technical details only to the extent necessary to allow you to understand the discussion. A conceptual understanding of this material, moreover, will enable you to grasp more readily some of the details of PL/I.

2.1.1

Programmer-Computer Communications

Historically, programs were submitted to the computer via punched cards. The programmer first typed the program on a card punch machine. Each key caused one or more holes to be punched in the card at specific locations, and one card generally represented one line of type. When this process was complete, the programmer took the deck of punched cards to the card reader. The card reader rapidly "read" the holes punched in each card and transferred the contents to the computer's memory. Card readers often had printers attached to them and were capable of having the contents of the cards (the **source code**) printed. Meanwhile, as soon as possible the computer would begin processing the program, eventually producing a listing of the executed program, including output, if any, and/or error and warning messages. The major drawback of the punched cards as a means of job submission was that errors were quite difficult to spot on the cards. In many cases the programmer had to execute ("run") the program before even the simplest error such as a spelling error was detected. The improperly punched card(s) had then to be located in the deck, corrected on a new card, the new card inserted in its proper place in the deck, and the program run again.

The cathode-ray tube (CRT) terminal that has largely replaced the card punch machine in recent years represents a considerable advance. Although the details vary somewhat from one system to another, the basic process is similar to this: As you hit a key on the keyboard, the character appears on the television-like CRT screen. When a special key such as "RETURN" or "SEND" is pressed, the data on the screen is transferred to the computer's memory over a telephone line or special cable connecting the two devices. This means that you can type one or more lines (depending on the system), inspect the data for errors, make any necessary corrections, and then send it to the computer. If you discover an error after you have sent the data, you can call it back to the terminal, where it will be displayed on the screen. You can then make the needed corrections and send the revised version to the computer to replace the original version. After you have entered the entire program, you can issue a command to cause the computer to write the program to a magnetic disk from which it can be recovered at any time. When you feel the program is ready for execution, you can issue a command that will cause the computer to read the program from the magnetic disk back into memory. One variation of this command will cause the program to execute but will "hold" it from going to the printer. In this case the executed program also constitutes a data set and is likewise written to magnetic disk.

You should keep in mind that you now have two data sets: the source code you typed and the executed program. You can now issue a command to bring the executed program to your CRT screen and check it for error or warning messages and output. If you want a printed copy, on the other hand, the second variation of the "execute" command will cause the executed program to be sent to the printer, or alternatively, you might type a command to send a copy of the executed program data set to the printer.

In order to submit a program from a CRT terminal, however, you must learn a **text editing language**, such as WYLBUR, SUPERWYLBUR, or INTERACT. It is these languages that contain the commands that allow for the process described above, and you will need to learn the one that is installed at your school or shop. These languages are, for the most part, easily learned. An hour or two of practice will teach you enough to begin submitting programs.

A question often raised by those with no experience with computers is, "How does the computer know I am submitting a PL/I program rather than a COBOL or FORTRAN program?" The answer lies in a series of coded statements that must be placed before the beginning of the PL/I program. These statements are collectively known as **Job Control Language** (JCL), and they not only identify the language the program is written in, but they also give the computer other information it requires to process the program. The exact JCL statements required vary, however, from one installation to another; you will be informed by your teacher or you will have to inquire in your computer center as to exactly what JCL is required for your system.

2.1.2

Computer
Memory

It would be quite easy to write a book the length of this one on the composition and arrangement of computer memory. What follows is a highly simplified explanation, stressing those details that will be most helpful to you as a PL/I programmer.

Anyone working in any depth with large mainframe computers must sooner or later become familiar with three number systems. The first is the **decimal** (base 10) system, with which you are already familiar. The second, which is used by the computer itself, is the **binary** (base 2) system, which consists entirely of the digits 0 and 1. The third is the **hexadecimal** (base 16), which is used for two closely related reasons: (1) Large numbers quickly become unwieldy when represented in the binary system, and therefore (2) many computers will display binary numbers in hexadecimal because four-digit binary numbers can be represented by one hexadecimal digit, a correspondence that does not exist between binary and decimal numbers.

The hexadecimal system uses the decimal digits 0–9, plus the letters A, B, C, D, E, and F. The following chart lists base 10 numbers from 1 to 15 and their counterparts in binary and hexadecimal.

Decimal	Binary	Hexadecimal
0	0	0
1	1	1
2	10	2
3	11	3
4	100	4
5	101	5
6	110	6
7	111	7
8	1000	8
9	1001	9
10	1010	A
11	1011	B
12	1100	C
13	1101	D
14	1110	E
15	1111	F

As you have perhaps guessed, the next numbers in the hexadecimal system are 10, 11, 12, 13, 14, 15, 16, 17, 18, 19, 1A, 1B, 1C, 1D, 1E, 1F, 20.

Computer memory is composed of millions of **bistable devices**. A bistable device is one that, first of all, has two states; these two states can be expressed by any pair of opposing terms such as on-off, negatively charged–positively charged, or absent-present. Second, the device is stable; whichever state it is in, it remains in until some force causes it to change states. A common electrical wall switch is a familiar example of a bistable device. In the computer world the bistable device is referred to as a **bit** (BInary digiT), and its two states are designated by the binary digits 0 (off) and 1 (on). The bits, moreover, are grouped together into units called bytes, most commonly with eight bits composing one byte. (Thus, we begin to see the significance of one hexadecimal digit representing four binary digits; a byte can be expressed with two hexadecimal digits.)

The computer is built in such a way that each byte is assigned a numerical **address**. A commonly used analogy is that of house numbers in a city or mail boxes in a post office. Given an address in memory, the computer can locate that byte and store data there or retrieve the data already present at the address. The **byte**, then, *is the smallest addressable unit of computer memory*. Through various means, including statements written in a program, a predetermined number of adjacent bytes can be combined into a unit known as a **field**. The address of a field is always the address of its leftmost byte. There are three fields of specific sizes that have been given special names. A field composed of two adjacent bytes and whose address (i.e., address of its leftmost byte) is evenly divisible by 2 is called a **half-word**. A field of four adjacent bytes starting at an address evenly divisible by 4 is a **fullword** or simply a **word**. A **doubleword** is a field of eight adjacent bytes whose address is evenly divisible by 8. All of this is depicted graphically in Figure 2-1.

Modern computer languages are constructed in such a way that the computer is given the responsibility for determining and keeping track of addresses; ordinarily the programmer need not even know that such a thing as an address exists.

To the computer the address of a particular field might look like this:

001010101100000111110010

(which the computer would print out as 2AC1F2). Instead of forcing the programmer to remember either the binary or hexadecimal representation of the address, PL/I allows the user to represent addresses with **symbolic names**. A symbolic name is an arbitrary set of characters (usually suggesting the significance of the contents of that memory location) used to represent an address. Symbolic names can be either variables or labels; here we will be concerned only with variables. A PL/I programmer might ask PL/I to create a field which he or she will designate with the name AVERAGE (presumably the field will contain the average of some set of numbers). PL/I will locate a portion of memory not currently in use; it will equate that address with the word AVERAGE, and every time it finds AVERAGE in a PL/I statement within the program, it will substitute the address for it. Therefore, the only thing the programmer has to remember is the symbolic name or variable (including its spelling).

byte0	byte1	byte2	byte3	byte4	byte5	byte6	byte7
HALFWORD		HALFWORD		HALFWORD		HALFWORD	
FULLWORD				FULLWORD			
DOUBLEWORD							

FIGURE 2-1
Bytes, halfwords, fullwords, and doublewords.

The Compiler

Now that we have seen how jobs are submitted and how memory is organized, we can consider the two stages that are involved in the processing of a PL/I program. The first stage is **compilation**. The **compiler** is a program that resides on a magnetic disk that is permanently available to the computer. When a certain one of the Job Control Language statements is encountered, it causes the compiler to be read into memory. The compiler then accepts the PL/I program as input data and processes it just as the PL/I program will later process its input. The compiler has three major functions. First of all, it scans the PL/I code for syntactical errors, errors in the "grammatic" formation of the statements (these are described in Section 2.2). When the compiler encounters such an error, it generates a warning or error message. The compiler's second function is to translate the PL/I code into machine code (sequences of 0's and 1's), including the assignment of numerical addresses to all labels and variables encountered in the program. This translation is necessary because the computer does not understand PL/I, only 0's and 1's. Its third major function is to generate a copy of the PL/I code exactly as it was typed and to include with it any warning and/or error messages that have been generated.

The second stage in the processing of a PL/I program is that of **execution**. During this stage, the computer actually carries out the translated instructions within the PL/I program. Once the translation and syntactical error checking processes have been completed, the compiler must determine whether or not the translated code can be successfully executed. In general, if the compiler has found only those syntactical errors that result in warning messages being generated, it will send the translated code for execution. If, on the other hand, the syntactical errors have generated error messages, the compiler normally will conclude that execution could not succeed. It will therefore withhold the program from execution, generate the printed PL/I code and messages and terminate processing. This ability of the compiler to assess the likelihood of successful execution and to withhold those apparently doomed to failure increases considerably the efficient use of the computer's resources.

2.2

Example Program— PL/I Format

In order to avoid a highly inefficient use of facilities, a programmer ordinarily codes the program on paper, transferring the handwritten code to the computer by whichever means is available. Coding sheets, similar to the one illustrated in Figure 2–2, are available for this purpose. Some students find them useful for the first few programs they code, but generally they abandon them rather quickly. Other students find them unnecessary from the begin-

CSD-27

COMPUTER CENTER
Northern Illinois University

PROJECT NO.＿＿＿＿＿＿＿   **CODED BY**＿＿＿＿＿＿＿   **DATE**＿＿＿＿＿＿＿

REMARKS (TYPE OF CARD, INTERPRETING, LISTING, ETC.)＿＿＿＿＿＿＿＿＿＿

```
/* THE PROGRAM READS 2 VALUES, ASSIGNS 2 VALUE

CH2EX1:    PROCEDURE OPTIONS (MAIN);

DECLARE NUM1;
DECLARE NUM2;
DECLARE NUM3;
DECLARE NUM4;

     GET LIST (NUM1, NUM2);

     NUM3 = 15;
     NUM4 = 79;

     PUT LIST (NUM1, NUM2, NUM3, NUM4);
END CH2EX1;
```

FIGURE 2–2
Complete handwritten PL/I program.

48 49 50 51 52 53 54 55 56 57 58 59 60 61 62 63 64 65 66 67 68 69 70 71 72 73 74 75 76 77 78 80

```
S , | A N D | P R I N T S | T H E M | * /
```

48 49 50 51 52 53 54 55 56 57 58 59 60 61 62 63 64 65 66 67 68 69 70 71 72 73 74 75 76 77 78 80

ning. We have reproduced the handwritten code in Figure 2–2 because periodic reference to it, as well as to the following results of executing the program, will make the subsequent discussion more clear.

When the code from the coding sheet was entered into the computer and executed, it produced the program identifed as CH2EX1, which is shown in Figure 2–3.

2.2.1

PROCEDURE and END Statements

As we noted above, each program must be preceded by certain Job Control Language statements to cause it to be compiled and executed. Since these statements vary from one installation to another, we have not reproduced them here. The **statement** is the primary unit of PL/I; it is the code between two semicolons, except for the first statement, which is not preceded by a semicolon. The program itself begins with the PROCEDURE statement. The numbers 1–10 on the far left under the column heading "STMT" are provided by the compiler. If the compiler issues warning, error, or information messages, it uses these numbers as a means of identifying the PL/I statement being referenced.

Although compilers do vary, it is often the case that the first column in PL/I is reserved for a carriage control character. Leaving this column blank causes the printer to advance one line before printing a line (i.e., single spacing); other carriage control characters will be introduced later. The blank lines in the program above were produced simply by leaving a blank line when the program was typed in. PL/I statements and comments, then, can begin in column 2 or any succeeding column up to and including column 72. Starting a statement in column 1 will produce an error; that portion of a statement or comment extending into column 73 and beyond will usually be ignored and therefore will also often produce an error.

The word CH2EX1 in the PROCEDURE statement is referred to as a **label**; the PROCEDURE statement must have a label, which must consist of from 1 to 7 characters. The characters in a label must be composed of **alphabetic** characters (A–Z), **national** characters (@, #, $) and **numeric** characters (0–9); the first character in the label, however, must be alphabetic or national. *Within* a label (i.e., with alphabetic, national, or numeric characters preceding and following it) the break (_) (underline) character can also be used (some compilers also allow the break character as the last character in a label). The final rule in creating labels is that no blanks can be embedded within the label. The table on page 32 illustrates legal and illegal labels.

```
PL/I OPTIMIZING COMPILER  VERSION 1 RELEASE 5.1  TIME: 14.53.39  DATE: 1 MAR 87   PAGE   1
OPTIONS SPECIFIED
GOSTMT;


PL/I OPTIMIZING COMPILER  /* READS 2 VALUES, ASSIGNS 2 VALUES, AND PRINTS THEM */ PAGE   2
                    SOURCE LISTING
    STMT

        /* READS 2 VALUES, ASSIGNS 2 VALUES, AND PRINTS THEM */

    1  CH2EX1:  PROCEDURE OPTIONS (MAIN);

    2  DECLARE NUM1;
    3  DECLARE NUM2;
    4  DECLARE NUM3;
    5  DECLARE NUM4;

    6     GET LIST (NUM1, NUM2);

    7     NUM3 = 15;
    8     NUM4 = 79;

    9     PUT LIST (NUM1, NUM2, NUM3, NUM4);
   10  END CH2EX1;


PL/I OPTIMIZING COMPILER  /* READS 2 VALUES, ASSIGNS 2 VALUES, AND PRINTS THEM */ PAGE   3
COMPILER DIAGNOSTIC MESSAGES
ERROR ID L   STMT    MESSAGE DESCRIPTION

COMPILER INFORMATORY MESSAGES

IEL0533I I        NO 'DECLARE' STATEMENT(S) FOR 'SYSPRINT'.

END OF COMPILER DIAGNOSTIC MESSAGES


    19              87              15              79
```

FIGURE 2–3
The complete CH2EX1 PL/I program compiled.

Legal Labels	Illegal Labels	
PROG1:	PROGRAM#1:	(too long; produces a warning)
PROG_1:	1_PROG:	(starts with a number)
#1_PROG:	_PROG1:	(starts with a break character)
	PROG1_:	(ends with a break character)
	PROG 1:	(embedded blank)

The label *must be followed by a colon.*

The word PROCEDURE is a PL/I **keyword**. A keyword is a word that in certain contexts has special meaning to PL/I. OPTIONS (MAIN) tells PL/I that what follows is an independent program. Later you will see PRO-CEDURE statements without OPTIONS (MAIN), but for the time being you can assume that each program must begin with **label**: PROCEDURE OP-TIONS (MAIN);. Note that the statement ends with a semicolon; *all PL/I statements end with a semicolon.* The PROCEDURE statement is represented in a flowchart by the top terminal interrupt and in pseudocode by START or the PROCEDURE statement label.

The final PL/I statement (STMT 10) is END CH2EX1;. The END is required, but the CH2EX1 is optional. When it is used (and we recommend that it always be used), it is the same word as the label you have chosen for the PROCEDURE statement. If the label is not used on the END statement, the syntax is simply END;. Every PROCEDURE statement must have a matching END statement. This END statement, which serves to mark the physical end of the PL/I code, is represented in a flowchart by the bottom terminal interrupt and in pseudocode by STOP.

2.2.2

A First Look at the DECLARE Statement

You will recall from our earlier discussion that the compiler assigns a numeric address to each variable found in the program. The function of the DECLARE statement is to provide the compiler with information about the variables you wish to use in the program so that storage can be reserved in memory for the values that will be assigned to each variable. Since DECLARE statements have no bearing on the logic of the program, they are not represented in flowcharts or pseudocode. The rules for forming variable names are the same as those for labels except that variable names can be up to 31 characters long.

As you will discover in Chapter 3, the DECLARE statement can become quite complex. As a matter of fact, in CH2EX1 the DECLARE statement is not strictly required at all. The compiler would pick out the variables NUM1 and NUM2 as it scanned the program. Thus, variables can be declared either **implicitly** or **explicitly**. We feel it is good programming practice to declare *all* variables explicitly. It is, first of all, a courtesy to a reader of your program to introduce the "cast of characters" before the play begins, and declaring all variables in one location within the program will make it easier for the reader to spot them. We think it will save you time, too. If for example, you have put the program away for a time, you will probably have forgotten what variable names you used. If you have not explicitly declared them, you will have to scan the various statements in the program to locate them. For reasons that will become clear in Chapter 3, we further recommend that all variables start with one of the letters I–N until the material in Chapter 3 has been mastered.

2.2.3

The Assignment Statement

Statements (STMTs) 7 and 8 of CH2EX1 are assignment statements. The "equal" sign found in these two statements does not function in a PL/I program in the way you are accustomed to in mathematics. Here it means "assign the value of the **expression** on the right of the assignment symbol to the variable on the left of the assignment symbol. In the case of CH2EX1, the expression is merely an integer constant; it could be, however, a complex arithmetic expression composed of variables with or without constants intermixed. For example, one might code:

```
BIG = SMALL * 3 + LITTLE / .002 - TINY;
```

In this case the computer would **evaluate** the expression by carrying out the necessary computations according to a predetermined set of rules until it arrived at a single value. This value would then be assigned to (i.e., placed in) the memory address designated by BIG. The assignment statement, then, can be used simply to move data from one place to another (BIG = SMALL;) or to collect the results of the evaluation of a more complex arithmetic expression such as the one illustrated above.

The verb "move" is firmly entrenched in the language of computer programmers (it is in fact a keyword in some programming languages), and it is not likely to disappear. In some respects it is an unfortunate choice, however, because it can be misleading in its implications. Using "move" to describe the action caused by the assignment statement BIG = SMALL; is to imply that the contents of SMALL are transferred to BIG and, since the contents were "moved," are no longer present in SMALL. Such, however, is not the case: the execution of an assignment statement *has no effect on the contents of the variable(s) and constant(s) on the right side of the assignment symbol.* "Copy" is the verb that more accurately describes the result of executing the assignment statement. Thus, when BIG = SMALL; is executed, the contents of both will be identical. Of course, the previous contents in the variable to the left of the assignment symbol are lost (think of a figure written on a blackboard; the figure is erased and a new one written in its place), but the contents of the variable(s) on the right side of the assignment symbol remain unaffected. The assignment statement is represented in a flowchart by a process symbol and in pseudocode by the arrow (←).

2.2.4

A First Look at I/O Statements

Statement 6 in Figure 2–3 is an example of an **input** statement, and its purpose is to bring data from some location outside the program to the location within the program designated by the variables in the input statement. One input value is placed in each variable. The input statement consists of the keywords GET LIST, followed by one or more variables enclosed in parentheses. If a single variable is named, the format is, for example,

```
GET LIST (NUM1);
```

As the CH2EX1 GET LIST statement illustrates, a series of variables must be separated by commas.

Turning to the very last line of the example program listing of Figure 2–3, you will notice:

19	87	15	79

This line represents the **output** of the program, the data that the statements within the program read (GET LIST) or caused to be generated (assignment statements) and printed. The specific statement that caused the output to be printed is the PUT LIST; statement. PUT LIST prints the value contained in the variable at the time the statement is encountered by the computer. The programmer has little or no control over the placement of the output on the printed line. The printed values are placed in fields that begin in columns 1, 25, 49, 73, and 97 if there are five variables in the program. If there are six or more variables, the first five will appear on one line and the rest in the same columns of the succeeding line(s) until all the variables have been printed. The print line for PL/I normally contains 120 print positions; it is possible to extend the print line, in which case the sixth variable begins in column 121. Certain restrictions or controls can be placed on the PUT LIST statement; these will be introduced in Chapter 3. The PUT LIST statement is represented in a flowchart by an I/O symbol and in pseudocode by the word "Print."

Comments are used in a PL/I program primarily to explain various aspects of the program that would not be immediately obvious to someone reading it. Comments are recognized by PL/I by the fact that they are enclosed by a /* in front of the comment and by a */ following the comment. The computer prints these **delimiters** (special symbols that mark boundaries; the semicolon is the PL/I statement delimiter) and the material between them exactly as it encounters them, and their presence has no effect whatever on the execution of the program. Comments can be placed nearly anywhere in a PL/I program. They might be placed, for example, following a PL/I statement:

```
NUM1 = 15; /* ASSIGN VALUE OF 15 TO NUM1 */
```

Another possibility is to intersperse the comments among the PL/I statements but on a separate line:

```
/* ASSIGN VALUE OF 15 TO NUM1 */
NUM1 = 15;
/* ASSIGN VALUE OF 79 TO NUM2 */
NUM2 = 79;
```

Notice, however, that for anyone who is at all familiar with PL/I code the comments give no information that cannot be gained from the code itself; this is especially true when meaningful names are chosen for the variables. The more common practice in regard to comments, then, is to place them in a group at the top of the program. The contents of the comments vary according to the situation and also according to personal or installation preference. One possibility is to describe briefly the function of the program and the content and format of the input and output records. Perhaps you have noticed that CH2EX1 contains one line of comment placed above the beginning of the program. It is a good programming practice to place a single line of comment in this position because it will automatically be printed at the top of each page of source code.

Syntax Errors—
A First Look

At this point it might be worthwhile to summarize the sources of problems commonly faced by beginning programmers:

1. *Spelling:* Keywords must be spelled exactly as given; to type OPTION instead of OPTIONS will produce one or more error messages. With variables the common problem is spelling what is intended to be the same variable in two different ways. Remember that PL/I will set aside a memory area for each spelling and might not warn you of the fact. Therefore, the value you expect to find in one variable might in fact have been assigned to the variable created by the alternative spelling.

2. *Punctuation:* The failure to place a colon after a label will produce an error. The failure to place a semicolon after a PL/I statement can lead to erroneous results, an error, or a warning. PL/I is "trained" to make certain assumptions and to proceed on the basis of the assumptions. When this occurs, PL/I produces a warning message in the source code to alert the programmer that an assumption was made. Keep in mind that the assumption that PL/I makes might not be the one you intended; in either case, good programmers do not consider their programs "debugged" (error free) until such warning messages have been eliminated by making the necessary changes in the source code. Also bear in mind that the change suggested by PL/I might not be the correct one; it is only a guess on the compiler's part. It is up to you to decide whether that guess is correct or that a different change is needed. If PL/I does not have enough information to make an assumption, an error condition is created. In this case PL/I generates an error message describing (or at least hinting at) the error, and processing stops.

3. *No END statement:* Normally PL/I inserts an END following the last line of code, and processing continues. PL/I will produce a warning message, however, and this message should be eliminated by inserting the END statement and resubmitting the program.

4. *Format errors:* The most common formatting error is starting a comment or a statement in column 1; this produces an error in either case. A second common format error is that of forgetting the beginning or ending delimiter (or both) on a comment. If the beginning delimiter is omitted, PL/I will attempt to read the comment as a statement, fail, and stop processing. If the ending delimiter is omitted, PL/I will attempt to read statements following the comment as part of the comment. This, too, typically creates an error condition.

We have been discussing **syntax errors,** errors in the format or syntax of the statement. You will quickly discover an entirely different set of errors called **logic errors.** PL/I cannot detect most logic errors; often the only clue that a logic error exists is that the output simply is not correct. Less often, the logic error will cause execution to cease as soon as the error is detected. The important point to remember is that you must begin at once to establish the habit of checking a representative sample of the output for correctness. You should also get in the habit of scanning the entire printed listing. Programmers have wasted untold hours searching for errors because messages that would have aided them were not printed exactly where the programmer expected them.

2.2.7

The PL/I
Character Set

A character set is simply the set of characters that is recognizable to the computer language being used. PL/I accepts two character sets: a 48-character set and a 60-character set. Since most terminals and card punch machines, as well as most printers, now have 60-character capacity, only it will be described here. The PL/I character set, then, consists of the following:

1. The three national symbols: $ # @

2. The 26 uppercase letters of the English alphabet

3. The ten decimal digits: 0–9

4. Twenty-one special characters; some of these in certain contexts, others in all contexts, have special meaning for PL/I. The semicolon, for example, is the PL/I statement delimiter.

	Space	
+	Plus sign	(add)
−	Minus sign	(subtract)
*	Asterisk	(multiply)
/	Slash	(divide)
=	"Equal" symbol	(assignment)
>	"Greater than" symbol	(is greater than)
<	"Less than" symbol	(is less than)
^	NOT symbol	(is not)
&	Ampersand	(logical AND)
|	Stroke	(logical OR)
(	Left parenthesis	

)	Right parenthesis
'	Single quotation mark or apostrophe
,	Comma
.	Period
;	Semicolon
:	Colon
%	Percent
_	Break or underline character
?	Question mark

You will not find the question mark in any PL/I program because no meaning has been attached to it as yet. The NOT symbol ("^") is also represented by "⌐" in PL/I and will be printed by some printers. We will use the "^" throughout the book. You may assume that the two symbols are interchangeable.

Certain of the special characters can be combined: NOT EQUAL is formed by combining ^ and = (^=); ^< produces NOT LESS THAN, and ^> produces NOT GREATER THAN. Equal (=) can likewise be combined with < (<=) and > (>=) to produce LESS THAN OR EQUAL TO and GREATER THAN OR EQUAL TO. When characters are combined in this manner, no space can intervene between the two characters.

2.3

PL/I vs. PL/C— An Introduction

PL/I (Programming Language One) is a major high-level computer programming language. A **high-level language** is one that produces several lines of machine code for each line of code in the high level language program. PL/I was preceded in its development by two other major high level languages. The first to be developed was FORTRAN (**FOR**mula **TRAN**slator). FORTRAN was developed primarily to facilitate the coding of programs needed by scientists and mathematicians to solve their problems. Consequently, it was given the capacity to deal easily and efficiently with numbers, especially the very large numbers so often dealt with by scientists, and to provide for easy control of precision. In spite of its origins as a scientific language, however, it was soon adopted by many computer centers in the business world.

Programmers working in business applications soon became dissatisfied with the difficulties they encountered in adapting FORTRAN to their needs. Hence a second major language, COBOL (**CO**mmon **B**usiness **O**riented **L**anguage), was developed. COBOL was developed with the needs of the business community in mind just as the developers of FORTRAN had the scientific community in mind. COBOL quickly took over as the major programming language of the business community.

With two major high-level programming languages available, computer specialists—who were aware of certain weaknesses in both FORTRAN and COBOL—raised the possibility of combining the best features of each of the two languages into a single high-level programming language, suitable alike to the needs of the scientific community and the business community; it was to be *the* programming language. The result is PL/I, and, although it has not become *the* programming language, it has succeeded, in our estimation, in combining the best features of FORTRAN and COBOL and adding other features not available to either of the two earlier languages. It is particularly noteworthy that as new versions of FORTRAN appear, they incorporate more and more of the features originally developed in PL/I. In addition to being a dual-purpose language, PL/I has the advantage over COBOL (and until recently FORTRAN also) of being a much more structured programming language. Still another advantage of PL/I is that it is **modular**; that is, a subset of the language can be used by someone who has no knowledge of many other features available.

As PL/I began to be taught in college computer science courses, it quickly became apparent that beginning PL/I students had no use for many aspects of PL/I. At the same time, the rather obscure error messages printed by PL/I made it difficult for beginners to debug their programs. As a result of this realization, a student version of PL/I, called PL/C, was developed at Cornell University. The relationship of PL/I to PL/C, then, is this: Anything that is acceptable to the PL/C compiler is equally acceptable without modification to the PL/I compiler. It is for this reason that, with a few exceptions, we make no reference to PL/C in this text. Keep in mind, however, that many programs written for a PL/I compiler are not acceptable to a PL/C compiler because they contain certain elements of the language not contained in the PL/C subset.

At the same time, PL/C provides clearer error messages than does PL/I. Several aspects of PL/C that make programming easier for the student also make the programs inefficient. The inefficiency is relatively insignificant, however, in terms of the short programs written by students and executed only once after they are debugged. For the massive programs written in industry and executed perhaps daily for many years, the inefficiency of PL/C would be significant indeed.

Later on we will go into much more detail regarding differences between PL/I and PL/C. This brief introduction is designed primarily to satisfy the natural curiosity aroused by the title of this book.

2.4

Summary

In this chapter we have given you some idea of how to communicate with the computer from a remote terminal. Keeping it clearly in mind that all data is stored in the computer as binary numbers and that symbolic names represent numeric addresses within the computer's memory will help you to understand many concepts you will encounter throughout this book.

The syntax of PL/I statements and of the overall program, perhaps bewildering now, will soon become second nature to you. Two details that are especially troublesome to beginners are (1) forgetting to leave column 1 blank, and (2) omitting or misplacing the semicolon.

In regard to the DECLARE statement, we urge you to declare all variables explicitly and to check carefully the spelling each time you reference a variable. Remember: A misspelling causes PL/I to set up a new variable (implicitly declared), perhaps with no warning to you that it has done so. The error messages produced because the declared variables did not contain the expected values are fairly clear in indicating the nature of the problem.

The one aspect of programming programmers dislike the most is that of providing adequate documentation; yet no program can be said to be completely finished without such documentation, and the comment statement is the means by which documentation is entered into the program.

Finally, let us remind you that the PL/I character set is composed of the complete list of characters recognized by the PL/I compiler, and the list numbers 60 characters.

We closed the chapter with a brief sketch of the development of high-level languages with the purpose of acquainting you with how PL/I fits into the "big picture." We also gave you some indication of the relationship of PL/C to PL/I. We (and, we trust, you also) are now ready to consider the intricacies of numeric data in PL/I.

2.5

Exercises

1. How many halfwords are in two doublewords? How many bytes? How many bits?

2. What number system is used by the computer?

3. Why are hexadecimal numbers used in computing?

4. Briefly explain the concept of *addresses* and their relation to symbolic names.

5. List and describe the two stages that are involved in the processing of a PL/1 program.

6. Identify the invalid labels and give the reason(s) why they are not valid. Assume the labels are on PROCEDURE statements.
   a. ABC_DEF_GHI_JKL_MNO_PQR_STU_VWX_YZ :
   b. @PRICE :
   c. SUM_UP :
   d. 260_PL1 :
   e. CODE 3 :
   f. TOTALS :

7. Identify the invalid variable names and give the reason(s) why they are not valid.
   a. ALPHA
   b. ALPHA_BETA
   c. ALPHA-DELTA
   d. $123_45
   e. THIS_IS_THE_NUMBER_OF_STUDENTS_HERE
   f. NUM STUDENTS
   g. #_STUDENTS
   h. _#_
   i. 87_ENROLL
   j. Q

8. List the PL/I delimiter for each of the following:
   a. label
   b. word
   c. statement
   d. comment

9. What is the purpose of including comments in a PL/I program?

10. Explain why the word "move" is misleading when applied to the action of the assignment statement.

11. Code the following assignment statements.
    a. Put 79 into the variable named AVERAGE.
    b. Put the contents of STUDENT_NUM into STUDENT_ID.
    c. Put the contents of COURSE into CLASS.
    d. Put the present year (e.g., 1987) into YEAR.

12. Code the following I/O statements.
    a. Read a value into a variable named X.
    b. Read a value into each of the variables named GRD_1, GRD_2, and GRD_3.
    c. Print the values in GRD_1, GRD_2, and GRD_3 above.
    d. Print the value in X above.

13. Write a PL/I program in which you assign a value to each of three variables. Print all three variables with one output statement. Then print each variable with a separate output statement.

C H A P T E R 3

MORE

PL/I

BASICS

In this chapter we explain, first of all, some of the details of manipulating numeric data. The discussion includes the details of the DECLARE statement for numeric and character data, implicit declarations, and defaults. We also consider the means by which arithmetic operations are carried out and introduce PL/I built-in functions that act on numeric data. Arithmetic operations are illustrated in the program below. We suggest you scan the example program, identified as CH3EX1 and shown in Figure 3–1, then go on to the discussion. Return to examine pertinent statements in CH3EX1 when the discussion directs your attention to them. The logic of CH3EX1 is extremely simple, but we include the pseudocode to begin familiarizing you with this means of expressing the logic of a program. Notice that the name of the program replaces the START we used in Chapter 1 as the first word in the pseudocode.

3.1

**Example Program—
Numeric Data**

The pseudocode for CH3EX1 is as follows:

```
CH3EX1
    Print fixed diameters, pi
    Radius ← diameter1 / 2.0
    Print radius
    Area1 ← radius * radius * pi
    Radius ← diameter2 / 2.0
    Print radius
    Area2 ← radius * radius * pi
    Print area1, area2
    Sum ← area1 + area2
    Print sum
    Read diameter1, diameter2
    Print diameter1, diameter2, pi
    Area1 ← (diameter1 / 2) **2 * pi
    Area2 ← (diameter2 / 2) **2 * pi
    Print area 1, area 2
    Difference ← area1 − area2
    Print difference
STOP
```

```
PL/I OPTIMIZING COMPILER  VERSION 1 RELEASE 5.1  TIME: 15.30.02  DATE: 12 FEB 87  PAGE   1
OPTIONS SPECIFIED
GOSTMT;


PL/I OPTIMIZING COMPILER  /* ILLUSTRATES FIXED VS. FLOAT DECIMAL NUMERIC DATA */ PAGE   2
              SOURCE LISTING
    STMT

        /* ILLUSTRATES FIXED VS. FLOAT DECIMAL NUMERIC DATA */

    1  CH3EX1:  PROCEDURE OPTIONS (MAIN);
```

FIGURE 3–1
CH3EX1 (FIXED vs. FLOAT DECIMAL numeric data).

FIGURE 3–1 *(continued)*

```
/**********************************************************************/
/*                                                                  */
/*    FUNCTION:   TO COMPUTE THE AREA OF TWO CIRCLES WITH THE        */
/*                DIAMETER AS THE GIVEN DATA.  THE AREA IS           */
/*                COMPUTED TWICE.                                    */
/*                (1) USING FIXED DECIMAL ATTRIBUTES.  IN THIS CASE  */
/*                    THE SUM OF THE TWO AREAS IS FOUND.  THE        */
/*                    DIAMETER VALUES ARE SUPPLIED BY THE INITIAL    */
/*                    ATTRIBUTE                                      */
/*                (2) USING FLOAT DECIMAL ATTRIBUTES.  IN THIS CASE  */
/*                    THE DIFFERENCE BETWEEN THE TWO AREAS IS        */
/*                    FOUND                                          */
/*                                                                  */
/*    INPUT:      A SINGLE CARD CONTAINING THE DIAMETERS            */
/*                                                                  */
/*    OUTPUT:     THE ORIGINAL DATA VALUES, THE CALCULATED RADII     */
/*                (FOR FIXED DECIMAL), THE AREAS, THE SUM, AND       */
/*                THE DIFFERENCE                                     */
/*                                                                  */
/**********************************************************************/

2    DCL DIAMETER1_FXD         FIXED DECIMAL (7,4)  INITIAL (18.74),
         DIAMETER2_FXD         FIXED DECIMAL (7,4)  INITIAL (31.44),
         PI_FXD                FIXED DECIMAL (5,4)  INITIAL (3.1416),
         RADIUS_FXD            FIXED DECIMAL (7,4),
         AREA1_FXD             FIXED DECIMAL (9,4),
         AREA2_FXD             FIXED DECIMAL (9,4),
         SUM_FXD               FIXED DECIMAL (9,4),

         DIAMETER1_FLT         FLOAT DECIMAL (7),
         DIAMETER2_FLT         FLOAT DECIMAL (7),
         PI_FLT                FLOAT DECIMAL (6)    INITIAL (3.1416),
         AREA1_FLT             FLOAT DECIMAL (7),
         AREA2_FLT             FLOAT DECIMAL (7),
         DIFFERENCE_FLT        FLOAT DECIMAL (7);

3    PUT LIST (DIAMETER1_FXD, DIAMETER2_FXD, PI_FXD);
4    RADIUS_FXD = DIAMETER1_FXD / 2.0;
5    PUT LIST (RADIUS_FXD);
6    AREA1_FXD = RADIUS_FXD * RADIUS_FXD * PI_FXD;
7    RADIUS_FXD = DIAMETER2_FXD / 2.0;
```

```
     8      PUT LIST (RADIUS_FXD);
     9      AREA2_FXD = RADIUS_FXD * RADIUS_FXD * PI_FXD;
    10      PUT LIST (AREA1_FXD, AREA2_FXD);
    11      SUM_FXD = AREA1_FXD + AREA2_FXD;
    12      PUT LIST (SUM_FXD);
```

```
    13      GET LIST (DIAMETER1_FLT, DIAMETER2_FLT);
    14      PUT LIST (DIAMETER1_FLT, DIAMETER2_FLT, PI_FLT);
    15      AREA1_FLT = (DIAMETER1_FLT / 2.0) **2 * PI_FLT;
    16      AREA2_FLT = (DIAMETER2_FLT / 2.0) **2 * PI_FLT;
    17      PUT LIST (AREA1_FLT, AREA2_FLT);
    18      DIFFERENCE_FLT = AREA2_FLT - AREA1_FLT;
    19      PUT LIST (DIFFERENCE_FLT);
    20  END CH3EX1;
```

COMPILER INFORMATORY MESSAGES

IEL0533I I   NO 'DECLARE' STATEMENT(S) FOR 'SYSPRINT','SYSIN'.

END OF COMPILER DIAGNOSTIC MESSAGES

18.7400	31.4400	3.1416	9.3700	15.7200
275.8227	776.3471	1052.1698	1.873999E+01	3.143999E+01
3.14159E+00	2.758227E+02	7.763470E+02	5.005243E+02	

3.1.1

The DECLARE	When we introduced the DECLARE statement in Chapter 2, we made the
Statement—	following points:
A Second Look	

1. The function of the DECLARE statement is to provide information about the variables so that memory for the value to be associated with the variable can be reserved.

2. Variables can be explicitly or implicitly declared.

3. The DECLARE statement can become quite complex and confusing.

4. We recommend that *all* variables be declared.

In the following sections we present the details of fully declaring variables to contain numeric data. You will discover that variables can be declared with a full description of the numeric data, a partial description, or no description at all. Remember that the compiler uses defaults for any portion of the description which you do not explicitly describe. These defaults are perhaps the most confusing aspect of data declarations; therefore, it is worthwhile to again recommend not only that all variables be declared, but also that a full description be given for each one.

You will no doubt recall that in the example program of Chapter 2 our declarations looked like this:

```
DECLARE NUM1 ;
DECLARE NUM2 ;
```

The declaration for CH3EX1 (see Fig. 3–1) looks quite different, and the difference we are concerned with here is that CH3EX1 has a single instance of the keyword DECLARE. As you can see, PL/I accepts either syntax. Note, however, that if DECLARE is repeated for each variable, then each is a separate statement and ends with a semicolon. If, on the other hand, DECLARE is used only once, the declaration of all the variables is a single statement. In this case a *comma* is placed after the declaration of each variable except the last; it, of course, must be delimited as usual with the semicolon. Still another variation is to declare the variables in several statements, but not a single statement for each variable. One might, for example, declare all variables of a single data type in one statement. Had we followed this scheme in CH3EX1, the results would look like this:

```
DCL   DIAMETER1_FXD FIXED DECIMAL (7,4) INITIAL (18.74),
      DIAMETER2_FXD FIXED DECIMAL (7,4) INITIAL (31.44),
      PI_FXD        FIXED DECIMAL (5,4) INITIAL (3.1416),
      RADIUS_FXD    FIXED DECIMAL (7,4),
      AREA1_FXD     FIXED DECIMAL (9,4),
      AREA2_FXD     FIXED DECIMAL (9,4),
      SUM_FXD       FIXED DECIMAL (9,4);

DCL   DIAMETER1_FLT   FLOAT DECIMAL (7),
      DIAMETER2_FLT   FLOAT DECIMAL (7),
```

```
PI_FLT            FLOAT DECIMAL (6) INITIAL (3.1416),
RADIUS_FLT        FLOAT DECIMAL (7),
AREA1_FLT         FLOAT DECIMAL (7),
AREA2_FLT         FLOAT DECIMAL (7),
DIFFERENCE_FLT    FLOAT DECIMAL (7);
```

In any case, DECLARE can be abbreviated DCL.

The DECLARE statements can appear anywhere within the PL/I program. They could be scattered, for example, at various points throughout the program, or they could all be grouped together somewhere in the middle of it. It is common practice, however, to group them all together at the beginning or end of the program. It is our practice to group them at the beginning.

Just as the DECLARE can be placed anywhere within the program, so also can the variables be declared in any order; here, too, practice varies. One possibility is to list the variables in alphabetical order. A second possibility (and the one we favor) is to group the variables according to the types of data they contain.

3.1.2

Numeric Variables and Attributes

The data type or data format of the data to be stored in a variable is described to PL/I through terms called **attributes**, and the specific words used are keywords. In this discussion we are going to present four classes of attributes and briefly mention a fifth.

The first attribute is the **scale** attribute, and the available options are FIXED and FLOAT. When a variable is declared as a fixed-point variable, the numbers are printed in the form you are most used to. As its name implies, the decimal point is *fixed* in this representation; given a series of numbers, the decimal point is always aligned (a number with no decimal point contains an implied decimal to the right of the rightmost digit).

125

12.5

1.25

.125

When a variable is declared to have a floating-point attribute, on the other hand, it is printed in what you might know as "scientific notation." A glance at the variable PI_FLT in CH3EX1 will confirm that the value 3.1416 prints as 3.14159E+00. A floating-point number always consists of two parts: the

mantissa, which is the fixed-point decimal portion of the number (3.14159 in our example); and the **exponent**, which consists of the letter E, a sign, and an integer. The integer specifies a power of 10. In other words, it indicates the number of digits the decimal point should be shifted to the right if the sign is positive or to the left if the sign is negative. In the case of PI_FLT the decimal point is not to be shifted at all. DIFFERENCE_FLT in its floating-point representation is 5.005244E+02; the E+02 indicates that the decimal point is to be shifted two digits to the right, giving a fixed-point representation of 500.5244. The following chart of the variables from CH3EX1 shows the relationship between fixed- and floating-point representations.

Fixed-Point	Floating-Point
18.7400	1.873999E+01
331.4400	3.143999E+01
275.8227	2.758227E+02
776.3471	7.763471E+02

The following chart makes the changes in the fixed-point scale necessary to illustrate a negative exponent in the floating-point scale.

Fixed-Point	Floating-Point
.187400	1.873999E−01
.314400	3.143999E−01
.02758227	2.758227E−02
.07763471	7.763471E−02

Keep in mind that it is of course possible to have a floating-point number with a negative mantissa, which can have either a positive or a negative exponent.

−776.3471	−7.763471E+02
−.07763471	−7.763471E−02

No doubt you are wondering about the minor discrepancy between PI_FXD and PI_FLT, DIAMETER1_FXD and DIAMETER1_FLT, and DIAMETER2_FXD and DIAMETER2_FLT. The answer lies in the fact that floating-point numbers are stored in the computer in coded arithmetic form, which for our purposes is simply binary representation. As you know, the decimal fraction 1/3 cannot be expressed with absolute precision in the decimal system. If we divide 1 by 3, the answer is .3. We should now be able to add .3 + .3 + .3 and regain our original 1. No matter how many decimal fraction digits are attached to .3 (i.e., .33, .333, etc.), however, we will never get back to 1. Exactly the same problem exists in the binary system; some fractions cannot be expressed with absolute precision in a finite number of binary digits.

Numbers are generally given a scale of FLOAT only when the program is dealing with very large or very small numbers, which occurs primarily in scientific calculations. A floating-point variable can express a value range of approximately 10^{-78} to 10^{75}, while a fixed-point variable is restricted to the range of 15 decimal digits. A second use of floating-point decimal is when a given precision is desired and the location of the decimal point is unknown. The scale of FIXED is generally assigned in the case of business or general purpose applications programs.

The second attribute is **base**, and again we are given two options: DECIMAL and BINARY. Both bases print out in exactly the same form, but there are significant differences between them, as follows:

1. DECIMAL allows for a maximum value of 999,999,999,999,999 (15 digits), while the minimum and maximum values for BINARY are −2,147,483,648 and +2,147,483,647.

2. Instructions (i.e., statements) involving BINARY based numbers ordinarily execute faster than those involving DECIMAL based numbers.

3. As we saw with floating-point numbers, many decimal fractions cannot be represented precisely in binary.

Because instructions involving BINARY based numbers execute faster, variables that contain whole numbers and are involved in a very large number of computations are often given the BINARY attribute. A prime example is a variable used as a counter. Suppose that a college with an enrollment of approximately 10,000 students wanted to determine the mean grade point average of all its students. In order to do so, the computer must know exactly how many individual grade point averages have been processed. One solution is to declare a variable, give it an initial value of 0, and add 1 to it each time an individual grade point average is processed. Such a variable would be an excellent candidate for the use of the BINARY base attribute.

At this point we have four possible combinations for our DECLARE statement:

```
DECLARE
VARIABLE_1    FIXED DECIMAL,
VARIABLE_2    FIXED BINARY,
VARIABLE_3    FLOAT DECIMAL,
VARIABLE_4    FLOAT BINARY;
```

The order in which these two attributes are listed in the DECLARE statement is not fixed; PL/I accepts any of the following:

```
DECLARE
    VARIABLE_1    DECIMAL FIXED,
    VARIABLE_2    BINARY FIXED;
```

and so forth.

The next attribute to be considered is **precision**. Its function is to specify the maximum number of digits the variable is to contain. In the case of FIXED scale the precision attribute can also indicate the number of digits to the right of the *implied* decimal point. The precision is listed following either the base or the scale attribute. It is placed within parentheses, and if the number of digits to the right of the implied decimal point is specified, the two are separated by a comma. Efficiency is improved if the precision of FIXED DECIMAL variables are declared as an odd number (3, 5, 7, etc.) of digits. A variable of precision (4), for example, takes up as much space as one with a precision (5); in addition, PL/I will have to generate extra code to make precision (4) the correct size. Some examples are given in the following paragraphs.

1. The declaration

```
EX_VARIABLE1    FIXED DECIMAL (7);
```

says that the largest number to be placed in EX_VARIABLE1 will contain no more than seven digits and that the implied decimal point is to be to the right of the rightmost digit. This in turn means that if the value currently in EX_VARIABLE1 is printed, no decimal point will print. Note also that there is no requirement that a seven-digit number ever be placed in the variable; we have merely reserved space for that many digits.

2. In the declaration

```
EX_VARIABLE2    FIXED DECIMAL (7,4);
```

we are specifying that the variable can contain up to seven digits; of those seven digits, four will be assigned to the fractional digits. We have spoken of the *implied* decimal point, which means that no decimal point is actually stored in the variable. In determining the precision to declare in a particular situation, the decimal point is not counted. The computer keeps track of the position specified for the decimal point, but no decimal point is actually present in memory; however, when the value in the variable is printed, the decimal point is printed. The following assignment statements, with the value as it would print indicated as comments, will help to illustrate precision involving fractions. Notice that any places to the right of the decimal point as specified in the declaration that are not used for significant digits are padded with zeros.

```
EX_VARIABLE2 = 123.4567;    /* PRINTS AS 123.4567   */
EX_VARIABLE2 = 234.567;     /* PRINTS AS 234.5670   */
EX_VARIABLE2 = 34.56;       /* PRINTS AS  34.5600   */
EX_VARIABLE2 = .5678;       /* PRINTS AS .5678      */
EX_VARIABLE2 = 123.;        /* PRINTS AS 123.0000   */
```

3. These examples illustrate the truncation that will occur when the value in one variable is assigned to another that has a different precision.

```
DECLARE NUM_1  FIXED DECIMAL (8,5);
NUM_1 = 123.45678;
EX_VARIABLE2 = NUM_1;            /* PRINTS AS 123.4567   */
                                 /* NOTE TRUNCATION      */

DECLARE NUM_2  FIXED DECIMAL (10,3);
NUM_2 = 1234567.890;
EX_VARIABLE2 = NUM_2;            /* PRINTS AS 567.8900   */
                                 /* NOTE TRUNCATION      */
```

4. Finally, the following two declarations produce exactly the same result:

```
EX_VARIABLE1        FIXED DECIMAL (7);
EX_VARIABLE1        FIXED DECIMAL (7,0);
```

Either causes the whole number to be printed with no decimal point.

In declaring the precision of a FIXED BINARY variable, it is important to keep in mind that you are specifying the maximum number of *binary* digits (bits) the variable can accommodate, not the number of decimal digits. One of two precisions is ordinarily declared for a FIXED BINARY variable. If the maximum *decimal* value to be stored in the binary variable is not greater than 32,767 or less than −32,768, a precision of 15 is used because any precision up to and including 15 requires a halfword of storage (two consecutive bytes starting at an address evenly divisible by 2). If the maximum decimal value is greater than 32,767 or less than −32,768, a precision of 31 is specified, again because any precision of 16 to 31 requires a fullword of storage (four consecutive bytes starting at an address evenly divisible by 4); 31 is the largest precision it is possible to declare. FIXED BINARY values, furthermore, can be declared to have a fractional portion exactly as fixed-point decimals can. This, however, is a seldomly used feature because, as we noted earlier, many decimal fractions have no exact binary counterparts.

Variables used to maintain a count, such as a count of the number of records read, are very heavily involved in arithmetic operations. At the same time, they do not involve a fraction; therefore, such fields are commonly declared as BINARY, and we highly recommend the practice; this is the main use of the fixed-point binary field in general data processing.

When declaring the precision of floating-point variables (either binary or decimal), only the total number of significant digits is declared. In Figure 3–1 you will notice that PI_FLT is given a precision of 6. The following declaration would result in an error.

```
DECLARE PI_FLT    FLOAT DECIMAL (6,2);    /* ERROR */
```

The fourth attribute is the INITIAL attribute, and its use is to assign a beginning or initial value to a variable. Notice in Figure 3–1 that all the fixed-point decimal variables, as well as PI_FLT, make use of the INITIAL attribute. The initial value is placed within parentheses, and decimal numbers in fixed-decimal format can be used to specify the initial value, even if the variable has a base of BINARY and/or a scale of FLOAT. With the type of variables we are concerned with for the time being (i.e., automatic variables), it is important to keep in mind that values assigned by the INITIAL attribute are assigned when the compiled program is loaded into memory; consequently, the value is assigned only once. If a new value is assigned to the variable through an assignment statement, the initial value is lost; the only way to regain it is to reassign it to the variable with an assignment statement. The keyword can be abbreviated INIT. There is more to be said on the subject; other aspects of the INITIAL attribute are introduced later.

The fifth attribute, which we do not discuss in detail, is the MODE attribute. The options are REAL and COMPLEX. If this attribute is omitted (i.e., not coded), it is assigned a MODE of REAL. Since only mathematics and science use complex numbers, we do not go into any detail here. Unless some of your variables are complex, there is no need to code this attribute (an exception to our general rule that all attributes should be explicitly declared).

There are other possible attributes that can be declared. Some, such as BIT, CHARACTER, VARYING, BUILTIN, STATIC, AUTOMATIC, and FILE, are introduced in later chapters; others are beyond the scope of this text.

3.1.3

Character Variables and Attributes

We introduce the DECLARE statement for character variables in this section, but a more detailed description of the power that PL/I has with respect to character variables is delayed until Chapter 8.

A character variable is a variable that can contain any of the 256 characters recognized by the computer, and it is called a **character string**. Character strings are declared using the CHARACTER attribute, along with a length attribute (as opposed to the precision attribute of numeric data) and the optional INITIAL attribute. PL/I recognizes CHAR as a valid abbreviation of CHARACTER.

```
DECLARE NAME CHARACTER (20) INITIAL ('BOB SMITH');
DECLARE NAME2 CHAR     (8)  INITIAL ('BOB SMITH');
```

Given the declaration above of NAME, the variable can contain up to 20 characters and has been initialized to 'BOB SMITH'. You should note three things about the use of character variables. First, a blank anywhere within the character string is part of the data and is included in the length of the string. Second, if the initial value is shorter than the declared length of the character variable, the value is *left-justified* and padded on the right with blanks. And third, if the initial value is longer than the declared length, the value is *truncated* on the right. The contents of the character variables NAME and NAME2 are:

NAME: | B | O | B | | S | M | I | T | H | | | | | | | | | | | |

NAME2: | B | O | B | | S | M | I | T |

One rule that must be remembered when you are using character strings is that when an apostrophe is to be included in the string, you must use two single quotation marks (with no space between them).

```
DECLARE TITLE CHARACTER (26) INITIAL ('ROSEMARY''S BABY');
```

| R | O | S | E | M | A | R | Y | ' | S | | B | A | B | Y | | | | | | | | | | | |

```
DECLARE NAME CHARACTER (15) INITIAL ('JIM O''CONNELL');
```

| J | I | M | | O | ' | C | O | N | N | E | L | L | | |

The following DECLARE statement illustrates how you can initialize a character variable to all blanks:

```
DECLARE CITY CHARACTER (17) INITIAL (' ');
```

The declaration works in this manner: A blank is inserted into the leftmost position of CITY, and because the initial value is less than the declared length, the rest of the variable is padded with blanks, thereby achieving an entirely blank variable.

When character variables are used in an assignment statement, the same rules (regarding padding, truncation, and the apostrophe) apply. The following are examples of character variables used in assignment statements:

```
DECLARE NAME CHARACTER (15) INITIAL (' ');
```

Initial value of NAME:

| | | | | | | | | | | | | | | |

```
NAME = 'TOM JOHNSON';
```

| T | O | M | | J | O | H | N | S | O | N | | | | |

```
NAME = 'EBENEEZER SCROOGE';
```

| E | B | E | N | E | E | Z | E | R | | S | C | R | O | O |

```
NAME = 'TOM O''HARA';
```

| T | O | M | | O | ' | H | A | R | A | | | | | |

PL/I has many built-in functions that can be used for the manipulation of character strings. We can put two strings together side by side, divide a string into substrings, and verify the contents of strings. All of these capabilities are described in Chapter 8.

3.1.4

Implicit Declarations and Default Values

As we noted in Chapter 2 in connection with the example program CH2EX1, we were not required to declare those variables at all. PL/I assigns default attributes to any variables for which no attributes are declared.

If a variable is implicitly declared (i.e., not listed in a DECLARE statement or listed with only a variable name), the following rules apply.

1. Variables beginning with any letter I–N inclusive (think of **IN**teger) are used for whole numbers only. In IBM machines such variables are the same as if you explicitly declared FIXED BINARY (15).

2. Variables beginning with the letters A–H, O–Z inclusive, or @, #, or $ are the same as if they were explicitly declared to be DECIMAL FLOAT (6).

Keep in mind, however, that when you explicitly declare a field, it can begin with any alphabetic or national character; an explicit declaration overrides the rules that apply for implicit declaration. Note, too, that the rules above leave no beginning characters for the implicit declaration of variables to hold character data. *Character variables must be explicitly declared.*

A variable can also be partially declared; that is, only one or two of the attributes can be coded. If only the base is declared, the scale defaults to FLOAT, as illustrated in the following pairs of declarations

```
DCL D DECIMAL;  = DCL D DECIMAL FLOAT (6);
DCL B BINARY;   = DCL B BINARY FLOAT (21);
```

If only the scale is declared, the base defaults to DECIMAL, as follows:

```
DCL FX FIXED;  = DCL FX FIXED DECIMAL (5,0);
DCL FL FLOAT;  = DCL FL FLOAT DECIMAL (6);
```

If both the base and the scale are explicitly declared, the precision is that given in the examples above. The precision, you recall, cannot be declared without a base and/or scale. The following chart summarizes the defaults for the given explicitly declared attributes. Explicitly declaring only a variable name is the same as "None" because the variable name is not an attribute. Therefore, a declared variable name with no attributes serves only to provide documentation.

Declared Attributes	Default Attributes
None; starts with I-N	FIXED BINARY (15)
None; starts with A-H, O-Z, or National	FLOAT DECIMAL (6)
FIXED	DECIMAL (5,0)
FLOAT	DECIMAL (6)
DECIMAL	FLOAT (6)
BINARY	FLOAT (21)
FIXED DECIMAL	(5,0)
FLOAT DECIMAL	(6)
FIXED BINARY	(15,0)
FLOAT BINARY	(21)

3.1.5

Arithmetic Operators and Expressions

Arithmetic operations are carried out using the assignment statement. Recall from Chapter 2 that the assignment statement is one way to place a value in a storage location. In our first example (CH2EX1) we simply assigned a numeric literal to two variables and read values into the other two. In CH3EX1, on the other hand, we actually carried out some arithmetic operations, which require the use of arithmetic operators. **Arithmetic operators** are special symbols that tell PL/I what arithmetic operation to carry out; there are five such operators.

Operator	Operation
+	Addition
−	Subtraction
*	Multiplication
/	Division
**	Exponentiation

We pointed out in Chapter 1 that a computer cannot tolerate ambiguity, and by way of illustration we asked if the expression 20 / 2 + 8 was intended to equal 18 or 2. We then pointed out that for the computer there is no ambiguity—that it resolves the expression to equal 18. The reason there is no ambiguity lies in the fact that the computer has definite rules for the order in which arithmetic operations are to be carried out. The order is:

1. Parentheses from innermost to outermost.

2. Exponentiation from right to left. Thus, given the expression W * X ** 3 ** 2, these steps occur:

   a. 3 is first raised to the power of 2 (i.e., squared).

   b. X is then raised to the power of 9 (i.e., multiplied times itself eight times).

   c. The resulting value of X is multiplied by W. Notice that the result is the same as if the expression had been coded as W * (X** (3 ** 2)).

3. Multiplication and division from left to right. Given the expression W / X * Y / Z, these steps occur:

   a. W is divided by X.

   b. The result of step a is then multiplied by Y.

   c. The result of step b is then divided by Z.

   (((W / X) * Y) / Z) illustrates again the use of parentheses and again makes no difference in the order in which the operations are carried out.

4. Addition and subtraction from left to right. Given the expression A + B − C + D, the computer takes the following actions:

   a. B is added to A.

   b. C is subtracted from the result of step a.

   c. D is added to the result of step b.

   Adding parentheses to clarify the order of operations but not change them gives us (((A +B) - C) + D).

To illustrate rules 2–4 in combination, we can use the expression A + B * C ** D / E − F, which results in the following order of evaluation:

   a. C is raised to the power of D.

   b. B is multiplied by the result of step a.

   c. The result of step b is divided by E.

   d. A is added to the result of step c.

   f. F is subtracted from the result of step d.

Before we continue, we need to remind you, first of all, that multiplication and division are carried out in the order encountered from left to right, and the same is true for addition and subtraction. Only exponentiation proceeds from right to left. Secondly, we need to point out that spaces can be placed between the variables (or literals) and the arithmetic operators or not as you please: A+B is the same as A + B. Remember, however, that the two asterisks signifying exponentiation are a *single* operator; therefore, no space can be placed between them.

```
VALID:     A**B
VALID:     A ** B
INVALID:   A* *B
INVALID:   A * * B
```

Obviously, there are times when the situation demands that the expression not be evaluated according to the rules we have presented; how can we inform the computer that the expression 20 − 8 / 2, for example, is supposed to evaluate to 6 instead of 16? The answer, of course, lies in rule 1: The order of evaluation can be changed through the use of parentheses. To repeat, the parentheses rule is that the expression(s) in parentheses is evaluated first, from the innermost to the outermost set of parentheses. Therefore, while the expression 20 − 8 / 2 evaluates to 16, the expression (20 − 8) / 2 evaluates to 6. The expression ((8 + 9) − 5) / 3 evaluates to 4: (8 + 9 = 17; 17 − 5 = 12; 12 / 3 = 4). Finally, the expression (8 + 8) * (10 − 4) evaluates to 96: (8 + 8 = 16; 10 − 4 = 6; 16 * 6 = 96).

As their use in the illustration of rules 2–4 suggest, parentheses can be (and often are) used for purposes of documentation where their presence has no effect on the order of evaluation. To give one more example, we might code 20 − (8 / 2) simply to make it clear to a human reader that the expression is intended to evaluate to 16; as a matter of fact, we encourage you to make use of parentheses for purely documentary purposes.

As a final note on arithmetic, we need to point out that the variables in an arithmetic operation must have the same scale and base. If they have been declared with differing attributes, the value of one variable must be converted to the scale and/or base of the other. The rules governing these conversions are found in Appendix E rather than here in the text because we believe that it is far better to avoid this situation. In business data processing in particular, it is almost never necessary to involve variables of different scales and bases in an arithmetic operation.

3.2

Numeric BUILTIN Functions

Certain arithmetic and mathematical processes are carried out many times in the course of a single program, as well as in program after program. To handle these recurring operations the concept of functions was developed. A **function** is a block of code that performs a single process, that returns a single result, and that can be invoked from anywhere in a program simply by coding the function name and providing the data to be acted upon. The advantages of using functions should be immediately obvious.

1. Functions reduce the number of lines to be coded.

2. As the number of lines to be coded is reduced, the likelihood of error is correspondingly reduced.

3. By using functions, all programmers use the same formulas, thereby guaranteeing uniformity of results.

PL/I has gone a step farther by providing built-in functions. A **built-in function** is a function that is a part of the language itself. Only built-in functions (and only a few of them) are discussed here.

Although it is not always necessary to declare a built-in function explicitly, it is a good practice to do so. The syntax of the declaration is

```
DECLARE function-name BUILTIN;
```

and a specific example is

```
DECLARE SQRT BUILTIN;
```

The syntax of the statement that invokes ("calls") a built-in function is

```
Variable = function name (arguments);
```

"Variable" is the name representing the location at which you want the computer to place the results of the function's computations. The equal sign, of course, serves the same purpose here as it does in any assignment statement. The function name is a PL/I keyword naming the function to be invoked. An **argument** is an expression, and the simplest form of an expression is a variable or a constant. The number of arguments permitted varies from one built-in function to another. Although the assignment statement is the general form used in invoking a built-in function, it is not always required; this exception is discussed later. Also, you are later introduced to functions that have no arguments.

Let us now turn to some specific built-in functions. The following DECLARE statement is used to illustrate each built-in function, and the comment on each line shows the value returned by that execution of the function.

```
DECLARE N1 FIXED DECIMAL (7,4) INIT (-119.8472),
        N2 FIXED DECIMAL (7,4) INIT (84.1919),
        N3 FIXED DECIMAL (7,4) INIT (48.1948),
        N4 FIXED DECIMAL (7,4) INIT (231),
        N5 FIXED DECIMAL (3,1) INIT (17),
        NX FIXED DECIMAL (7,4),
        NB FIXED BINARY  (15);
```

3.2.1

ABS

The ABS function has the syntax of ABS (x); x is the argument, and we see that in this case the function handles only a single argument. The purpose of ABS is to return the absolute value of x.

```
NX = ABS (N1);              /* NX = 119.8472 */
NX = ABS (N2);              /* NX =  84.1919 */
NX = ABS (N1 + N3);         /* NX =  71.6524 */
```

The final example illustrates that the expression forming the argument can contain arithmetic operators; in this case, the arithmetic operations are performed first; the function is then called to find the absolute value of the result of the arithmetic operation.

3.2.2

CEIL and FLOOR

The CEIL built-in function has the syntax of CEIL (x), and so once again the function is restricted to a single argument. Its purpose is to return the smallest integer (whole number) value equal to or larger than the argument.

```
NX = CEIL (N1);             /* NX = -119.0000 */
NX = CEIL (N2);             /* NX =   85.0000 */
NX = CEIL (N4);             /* NX =  231.0000 */
```

FLOOR has the same syntax as CEIL, but its purpose is to return the largest integer value less than or equal to the argument.

```
NX = FLOOR (N1);                    /* NX = -120.0000 */
NX = FLOOR (N2);                    /* NX -   84.0000 */
NX = FLOOR (N4);                    /* NX =  231.0000 */
```

3.2.3

MAX and MIN

The MAX function has the syntax of MAX (x1, x2, . . ., xn). Here we see a function that can have a varying number of arguments, ranging from 1 to 64. MAX returns the largest value from the arguments provided.

```
NX = MAX (N1, N3);                  /* NX =   84.1919 */
NX = MAX (N1, N3, N4 - 76.25);      /* NX =  154.7500 */
NX = MAX (N1, N2, N3, N4);          /* NX =  231.0000 */
```

MIN is identical to MAX except that it returns the smallest value from the argument list.

```
NX = MIN (N1, N2);                      /* NX = -119.8472 */
NX = MIN (N1, -125.0, -105.7550);       /* NX = -125.0000 */
NX = MIN (N2, N3, N4, N1 + 999.9);      /* NX =   48.1948 */
```

Note that when multiple arguments are used, they are separated by commas (with or without a space following the comma).

3.2.4

ROUND and TRUNC

The syntax of the ROUND built-in function is ROUND (x, y), where x is an expression and y is the digit at which it is to be rounded. ROUND is one of several built-in functions that must be supplied with exactly two arguments. If y is positive it refers to the digit to the right of the decimal point; in this case, a 5 is added to the y+1 digit (if the y+1 digit is 5 or greater it will cause the carry of a 1), and the y+1 and following digits are set to zero. If y is negative, it refers to the digit to the left of the decimal point. A 5 is added to the y digit, and the y and following digits are set to zero.

```
NX = ROUND (N1, 3);                 /* NX = -119.8470 */
NX = ROUND (N1, 2);                 /* NX = -119.8500 */
NX = ROUND (N1, 1);                 /* NX = -119.8000 */
```

```
NX = ROUND (N1, 0);            /* NX = -120.0000 */
NX = ROUND (N1, -1);           /* NX = -120.0000 */
NX = ROUND (N1, -2);           /* NX = -100.0000 */
NX = ROUND (N2, 3);            /* NX =   84.1920 */
NX = ROUND (N2, 2);            /* NX =   84.1900 */
NX = ROUND (N2 / 4.0, 2);      /* NX =   21.0500 */
```

The ROUND function can also be used to round floating-point numbers. In this case, the second argument, although it must be present, is ignored. Instead, the rightmost bit of the mantissa is set to 1 if it is currently 0; if it is currently 1, no change is made. The ROUND function is not often used with floating-point numbers because in some types of output the rounding is done automatically by PL/I. When ROUND is used, the floating-point number is normally rounded just before output.

The syntax of the TRUNC function is TRUNC (x), and it returns the value of the argument with the fractional portion set to zero (i.e., it truncates the fraction or "remainder").

```
NX = TRUNC (N1);               /* NX = -119.0000 */
NX = TRUNC (N2 / 4.0);         /* NX =   21.0000 */
```

3.2.5

MOD

The syntax of the MOD function is MOD (x, y). In this case, both x and y are expressions. MOD returns the value that must be subtracted from x in order to make it perfectly divisible by y; in other words, it returns the remainder expressed as an integer, not as a fraction.

```
NX = MOD (N4, 3);              /* NX =  0       */
NX = MOD (N4, 4);              /* NX =  3.0000 */
NX = MOD (N4, N5);             /* NX = 10.0000 */
```

3.2.6

SIGN

The SIGN built-in function has the syntax of SIGN (x), and it returns a binary value that indicates the sign of the argument.

If $x > 0$ (positive) the binary value is $+1$

$x = 0$ (zero) 0

$x < 0$ (negative) -1

The SIGN function is used primarily as a basis on which to make decisions;

the variable containing the returned binary value is checked, and the computer is instructed according to that result.

```
NB = SIGN (N1);                    /* NB = -1 */
NB = SIGN (N2);                    /* NB - +1 */
NB = SIGN (N4 - 231);              /* NB =  0 */
```

3.2.7

SQRT

The built-in functions we have looked at thus far are all **arithmetic** functions; SQRT, however, is a **mathematical** built-in function. Its syntax is SQRT (x), and it returns the square root of the argument. The value of the argument must be greater than zero.

```
NX = SQRT (N4);                    /* NX = 15.1980 */
```

Since PL/I is designed as both a mathematics-science language and as a business-oriented language, it contains many built-in functions. Several of these are introduced at various points in the book; all of them are described in Appendix C.

3.3

Example Program—Numeric Built-In Functions

The CH3EX2 program (see Fig. 3–2) introduces two features of the text that will remain consistent throughout. Each chapter, first of all, will end with a complete program that illustrates those aspects of PL/I introduced in that chapter. These programs, moreover, will all deal with some aspect of the same problem: a student file. Thus CH3EX2 reads in the name and identification number of a single student. It then reads in three test scores for that student and finds their average; it prints the student's name, identification number, total points, a label, and the average of the three scores. The average is recalculated, this time using the ROUND built-in function, and once again a label and the average are printed. CH4EX1, at the end of Chapter 4, implements exactly the same algorithm as CH3EX2, only with several control statements to illustrate the degree of formatting that list-directed I/O is capable of. The example program in Section 5.2.4 of Chapter 5 (CH5EX1) illustrates the various aspects of the IF statement that are presented in the first portion of the chapter. CH5EX2, which appears in Section 5.3.1 of Chapter 5, illustrates the DO WHILE statement. They, too, deal with some aspect of a student file, and this common problem is carried throughout the remaining example programs. Finally, at the end of Chapter 5 we present a program that combines the features of CH5EX1 and CH5EX2.

CH3EX2, like CH3EX1 and all future programs, is accompanied by pseudo-code to aid you in understanding the logic of the program.

Before we present CH3EX2, however, we need to call your attention to some of its features. Notice, first of all, that the variables ACCUMULATOR and COUNTER are initialized to 0. Looking down at the executable statements you see that the first one that mentions ACCUMULATOR is Statement 14.

```
ACCUMULATOR = ACCUMULATOR + TEST_SCORES;
```

This tells us, of course, that the contents of ACCUMULATOR and the contents of TEST_SCORE are to be added together, with the results placed in ACCUMULATOR. This format is standard for those instances in which a running total is to be maintained. The problem, however, lies in the fact that the computer does not clear out its memory after a program has run. Therefore, at the beginning of execution there can be any kind of data in the memory location designated by ACCUMULATOR, left over from a previous program; such leftover data is referred to as **garbage**. Garbage in a variable can pose either one of two problems.

1. If the garbage happens to be (in this case) a valid fixed-decimal, the contents of TEST_SCORE will be added to it, and immediately the results will be incorrect.

2. The more likely possibility is that the garbage will not be a valid fixed-decimal (i.e., the garbage will not have whatever attributes are declared for the current variable).

In this second case, when the computer attempts to carry out addition using the garbage as one of the operands, it will be unable to do so, and the program will **abend** (come to an **ab**normal **end** or termination). Even if the space occupied by ACCUMULATOR happened to be clear, moreover, no valid number would be present to add with TEST_SCORE. The INITIAL attribute assures us that the variable contains data with the declared attributes, and the first TEST_SCORE must be added to 0 to achieve the correct results, hence INIT (0). The situation is precisely the same for COUNTER, and so it, too, is initialized to 0. Incalculable amounts of programmer time has been spent as a result of the failure to initialize a field that is first used in an addition statement.

The second feature of CH3EX2 you should notice is that instead of declaring a separate variable for each of the three test scores, we read all of them into the same variable (TEST_SCORES). The use of a single variable is made possible by the fact that the first test score is processed (in this case, added to ACCUMULATOR) before the second test score is read in, which, of course, wipes out the first one. The technique used here is a simple but excellent illustration of the need to think through and plan carefully (a plan usually expressed as a flowchart or pseudocode) before you begin to code the program. The CH3EX2 pseudocode is as follows:

```
CH3EX2
    Read student name
    Read student number
    Read a test score
    Accumulator ← accumulator + test score
    Counter ← counter + 1
    Read a test score
    Accumulator ← accumulator + test score
    Counter ← counter + 1
    Read a test score
    Accumulator ← accumulator + test score
    Counter ← counter + 1
    Compute the average
    Print student name, student number, accumulator, label,
        average
    Compute the average (rounded)
    Print label, average
STOP
```

```
PL/I OPTIMIZING COMPILER  VERSION 1 RELEASE 5.1  TIME: 15.30.17  DATE: 12 FEB 87 PAGE   1
OPTIONS SPECIFIED
GOSTMT;




PL/I OPTIMIZING COMPILER        /* CHAPTER 3 EXAMPLE PROGRAM 2 */              PAGE   2
                SOURCE LISTING
    STMT


        /* CHAPTER 3 EXAMPLE PROGRAM 2 */

    1  CH3EX2:  PROCEDURE OPTIONS (MAIN);
```

FIGURE 3–2
CH3EX2 (list-directed I/O).

FIGURE 3–2 *(continued)*

```
/************************************************************************/
/*                                                                    */
/*    FUNCTION:   THIS PROGRAM PRODUCES A REPORT ON A STUDENT          */
/*                GIVING HER AVERAGE ON THE TESTS IN THE CLASS         */
/*                                                                    */
/*    INPUT:      A SINGLE RECORD CONTAINING A STUDENT'S              */
/*                IDENTIFICATION NUMBER AND THREE RECORDS, EACH        */
/*                CONTAINING A TEST SCORE                             */
/*                                                                    */
/*    OUTUT:      THE STUDENT'S IDENTIFICATION NUMBER AND HER         */
/*                AVERAGE TEST SCORE,FIRST WITHOUT USING THE ROUND,    */
/*                THEN USING IT TO COMPUTE THE AVERAGE SCORE           */
/*                                                                    */
/*                                                                    */
/************************************************************************/

2      DCL STUDENT_NAME      CHARACTER     (20);

3      DCL STUDENT_NUMBER    FIXED DECIMAL (9);
4      DCL TEST_SCORES       FIXED DECIMAL (3);
5      DCL AVERAGE           FIXED DECIMAL (5,2);

6      DCL LABEL_1           CHARACTER     (21)     INIT
             ('AVERAGE WITHOUT ROUND');
7      DCL LABEL_2           CHARACTER     (18)     INIT
             ('AVERAGE WITH ROUND');
8      DCL ACCUMULATOR       FIXED BINARY  (15)     INIT (0);
9      DCL COUNTER           FIXED BINARY  (15)     INIT (0);

10     DCL ROUND             BUILTIN;

11     GET LIST (STUDENT_NAME);
12     GET LIST (STUDENT_NUMBER);
13     GET LIST (TEST_SCORES);
14     ACCUMULATOR = ACCUMULATOR + TEST_SCORES;
15     COUNTER = COUNTER + 1;

16     GET LIST (TEST_SCORES);
17     ACCUMULATOR = ACCUMULATOR + TEST_SCORES;
18     COUNTER = COUNTER + 1;
```

```
19    GET LIST (TEST_SCORES);
20    ACCUMULATOR = ACCUMULATOR + TEST_SCORES;
21    COUNTER = COUNTER + 1;

22    AVERAGE = ACCUMULATOR / COUNTER;
23    PUT LIST (STUDENT_NAME, STUDENT_NUMBER, ACCUMULATOR,
```

PL/I OPTIMIZING COMPILER   /* CHAPTER 3 EXAMPLE PROGRAM 2 */   PAGE   3
   STMT

              LABEL_1, AVERAGE);

```
24    AVERAGE = ROUND (ACCUMULATOR / COUNTER, 2);
25    PUT LIST (LABEL_2, AVERAGE);
26 END CH3EX2;
```

PL/I OPTIMIZING COMPILER   /* CHAPTER 3 EXAMPLE PROGRAM 2 */   PAGE   4
COMPILER DIAGNOSTIC MESSAGES
ERROR ID L   STMT   MESSAGE DESCRIPTION

COMPILER INFORMATORY MESSAGES

IEL0533I I   NO 'DECLARE' STATEMENT(S) FOR 'SYSIN','SYSPRINT'.

END OF COMPILER DIAGNOSTIC MESSAGES

BETTY JONES   123456789   253   AVERAGE WITHOUT ROUND   84.33
AVERAGE WITH ROUND   84.25

3.4

Summary

The major topics introduced in this chapter include handling numeric data and using numeric built-in functions.

Data can be declared explicitly or implicitly; that is, if you do not provide variable names and attributes, PL/I will assume defaults for them. Because the default attributes are quite confusing and because declarations provide good documentation, we urge you to declare explicitly *all* variables and all their attributes. The keyword DECLARE can be appended to each variable declaration, in which case the statement is ended with a semicolon, or it can be coded once to apply to all or a number of variables; in this case each declaration except the last in the series is followed by a comma. The last variable declaration in the series, of course, is followed by the usual delimiting semicolon.

Numeric data in PL/I has four main attributes. The scale attribute has the options of FIXED and FLOAT, with the second being in the form commonly called scientific notation. The second attribute is base, and it offers the options of DECIMAL and BINARY. We recommend that all variables heavily involved in arithmetic operations, such as counters, be given the base attribute of BINARY. The third attribute is precision. Used with fixed-point scale, it specifies the maximum number of digits the variable is capable of containing; it also specifies the number of digits to be placed to the right of the decimal point. A precision of (7) is the same as a precision of (7,0). Used with floating-point scale, only one number is allowed to specify precision, and it specifies the number of significant digits to be included. The precision attribute is not coded without either the scale or base attribute, or both. The fourth attribute is INITIAL, which is used to assign a beginning or initial value to a variable. Remember, however, that this value is assigned only once.

To allow us to perform arithmetic operations, PL/I provides arithmetic operators, and there are strict rules controlling the order in which arithmetic operations are carried out.

1. Parentheses from innermost to outermost.

2. Exponentiation from right to left.

3. Multiplication and division from left to right.

4. Addition and subtraction from left to right.

Parentheses can (and often should) be included even when they have no effect on the order of operations.

The last major topic of this chapter is numeric built-in functions, blocks of prewritten code invisible to the user that carry out a single function, return a single value, and are easily invoked from anywhere in a PL/I program. The value returned logically replaces the function name. To invoke a function, you give the name of the function and provide the specified number of arguments. The function, the number of arguments, and the significance of the value returned is presented in the following table.

Chapter 3 also contains example programs; the second (CH3EX2) introduces the student file, which is the basis for all following program examples.

In the next chapter we expand upon the list-directed reading and printing of data. You will discover possibilities for greater flexibility in formatting the printed data.

Function/Argument	Returns
ABS(x)	Absolute value of the argument
CEIL(x)	Smallest integer value equal to or larger than the argument
FLOOR(x)	Largest integer value less than or equal to the argument
MAX(x1,x2,. . .xn) [n <= 64]	Largest value from the argument list
MIN(x1,x2,. . .xn) [n <= 64]	Smallest value from the argument list
ROUND(x,y)	The argument (x) rounded at the position signified by (y)
TRUNC(x)	The whole number with the fraction set to 0
MOD(x,y)	The remainder resulting from the division of x by y, expressed as an integer
SIGN(x)	A binary value: positive value: x = +1 value = 0; x = 0 negative value: x = -1
SQRT(x)	The square root of x (x must be greater than 0)

3.5

Exercises

1. Write a DECLARE statement for each of the following values, giving the attributes revealed in the value. For numeric data, assume decimal.
   - a. 123
   - b. 797632
   - c. 123.123
   - d. 875.08
   - e. 2.652398E+03
   - f. 84.40361E+04
   - g. 16.97531E-01
   - h. 505.4927E-02
   - i. 'DECLARATION'
   - j. 'PL/I'

2. Add the INITIAL attribute to each declaration of Exercise 1, using the values given.

3. Show the internal storage of each of the following.
   - a. DECLARE ALPHA_1 CHARACTER (6) INITIAL ('A');
   - b. DECLARE ALPHA_2 CHARACTER (4) INITIAL ('PL/I');
   - c. DECLARE ALPHA_3 CHARACTER (10) INITIAL ('BROWN COW');
   - d. DECLARE ALPHA_4 CHARACTER (3) INITIAL ('SNOW');

4. Give the remaining attributes for each of the following declarations.
   a.   `DECLARE IN_VALUE;`
   b.   `DECLARE K_SUM;`
   c.   `DECLARE C_NUM DECIMAL;`
   d.   `DECLARE D_NUM BINARY;`
   e.   `DECLARE E_NUM FIXED;`
   f.   `DECLARE F_NUM FLOAT;`

5. Give the results of evaluating the following expressions.
   a.   4 ** 3
   b.   8 * 2 / 4
   c.   19 − 16 + 34
   d.   3 + 9 / 3
   e.   50 − 5 * 9
   f.   (3 + 9) / 3
   g.   (50 − 5) * 9
   h.   (2 + 9 / 3) + 18
   i.   (6 * 6) + 36
   j.   (81 / 9) − 8
   k.   ((106 − 12 * 8) + 10)

6. Invoke the appropriate built-in function to find the answer to the following questions.
   a.   Which contains the smallest value—N1, N2, N3, or N4?
   b.   When stripped of its sign, what is the value of QX?
   c.   What is the whole number portion of the value in K3?
   d.   What is the remainder (as an integer) when BJ is divided by KD?
   e.   What is the square root of ROOT?
   f.   Is DD a positive or negative number?

7. Write a program to read in two numbers, each in the form of nnnn. Use the appropriate built-in function to find which is greater and find its square root. Print each of the values (two numbers, greater number, and square root).

8. Write a program that reads in the name of an article purchased, its cost (in dollars and cents), and the tax rate of the item. Your program should print the input and the total cost of the item.

9. Write a program that reads in and print three values. Using the appropriate built-in function find and print the maximum of the three values and the square root of each.

10. Write a program that reads in the name of a professional football player (20 characters), the team he plays for (10 characters), the number of yards gained by each of rushing, pass receiving, and returning punts. Print the input and the total yards gained.

C H A P T E R 4

READING AND

WRITING USING

LIST I/O

In this chapter, after introducing the concept of files, we expand on the list-directed method of I/O that we touched on briefly in Chapter 2. You will learn some of the common problems beginners encounter in using LIST I/O and what to look for to correct them. We then explain what can be done in the way of formatting list-directed output. We close the chapter with a version of the example program CH3EX2, now modified to illustrate some output formatting possibilities.

4.1
Introduction to I/O in PL/I

Input is the term used to designate data that is brought into the program from the outside. This data might be found on magnetic tape or magnetic disk, or it might be stored with the program itself; this last is the most common case for student programs. **Output** is the term used to designate data sent from the program to the outside. The destination of the output, like the origin of the input, might be magnetic tape or magnetic disk. Other possible destinations are a terminal monitor and microfiche, but the most common destination for the output from student programs is the printer. It is important to keep in mind, moreover, that the same data might be input at one point in the program and output at another point. The status of the data is determined from the point of view of the program: Data coming in is input; data leaving is output. **I/O** is the common abbreviation for the two terms when they are used together, as they often are.

4.1.1

Files in PL/I
(SYSIN and SYSPRINT)

The following is the input to CH4EX1, the program found at the end of the chapter. We are using that input here to illustrate three terms—field, record, and file—that are necessary for you to understand before we proceed.

```
'BETTY JONES'
123456789
91
75
87
```

A **field** is a single data item; it is the unit that normally is placed in a single variable. Although the concept of a record is somewhat misleading for stream I/O, nevertheless both input and output is usually spoken of as a collection of records. A **record**, then, is one or more fields treated as a unit for purposes of I/O. Although it is not technically correct, we can think of a record as consisting of the field or list of fields referenced in a single GET or PUT statement. Particularly in student programs, a record is most commonly the field or group of fields placed on a single line in either the input or output. Thus we can think of each line of the input above as a record, as well as each line of the output from CH4EX1 at the end of the chapter. In the input to CH4EX1, moreover, each record consists of a single field.

A **file** is a collection of related records. It is possible (though rarely the case) for a file to consist of a single record. Ordinarily each record in the file consists of the same fields. The input to CH4EX1, viewed collectively, is a file; the output is a second file. If you are familiar with computer literature, you have probably come across the term "data set." A **data set** is synonymous with a file.

Your PL/I program, of course, must know the name of the file you want to access in an I/O (GET or PUT) statement; therefore, within an individual program each file must be given a unique name. PL/I provides a means by which the program can be informed of the file intended for use with each I/O statement. We noted above, however, that for student programs in particular the input usually is stored with the program and output is directed to the printer. As long as these files are treated as stream files, PL/I supplies default file names: SYSIN for the input file and SYSPRINT for the output file to the printer. Therefore, you need not concern yourself with file names at this point. The output from the programs you have run has probably contained a message along the lines of

```
NO FILE SPECIFIED.   SYSIN/SYSPRINT   ASSUMED.
```

This is PL/I telling you that it has made the assumption that SYSIN and SYSPRINT are the files you wished to access.

4.1.2

Introduction to LIST I/O

Stream I/O is so named because the data in the file (either input or output) is treated by PL/I as a continuous stream of data. A GET LIST statement causes PL/I to read a data item for each variable named in the GET LIST statement. A subsequent GET LIST statement causes the first variable to receive the next input data item, regardless of whether it is the middle of a line or the beginning of a new line. To understand this concept it might be helpful to think of all of the input data items, no matter how many items and no matter how many lines they are *actually* on, as being a single record. The first GET, then, moves down the list of input data items, sending them one by one to all variables named in that GET. We might think of PL/I "checking off" each input data item as it is read in, thereby "knowing" which values have been read in and which haven't. When the next GET is encountered, PL/I remembers where in the input data item list it left off. It picks up with the next item and again sends items until all variables in that GET have received values. This process continues until either no more GET statements are encountered or there are no more input data items to send.

When PL/I executes the first GET LIST statement, it begins at the start of the file and scans, byte by byte, until it encounters a nonblank character. It considers this nonblank character to be the start of a data item, and it then scans until it encounters either a blank or a comma. The comma, blank, or the two together delimit a data item in the input file. If you remember that wherever a single blank is valid, any number of blanks are valid, you will see that the following are equally valid input formats.

```
1066,1142,1492,1941
1066 1142 1492 1941
```

```
1066       1142 , 1492
    1941
1066
1142
1492
1941
```

Since CHARACTER data items often contain embedded spaces, PL/I
must have some other means of knowing how much data is to be considered
a single field, and this is accomplished by enclosing CHARACTER data in
single quotation marks (apostrophes):

```
'LYLE DOMINA' 'KATHI DAVIS' '1066' '@>•#&'
```

The data items are still separated by a comma and/or one or more blanks,
but the quotation marks tell PL/I not to consider an embedded blank as a
delimiter. In list-directed input, as we have seen, a single input statement
reads input data items until it has supplied a value for each variable in the
input variable list. The first input value read is placed in the first variable in
the variable list, the second input value in the second variable, and so forth.
Consequently, the variables in the variable list must correspond in order
with the order of the input values; otherwise, the variables will contain
values you do not expect, and the program will produce (at best) erroneous
results.

In list-directed output, tab positions are set at columns 25, 49, 73, and
97. When a PUT LIST statement is encountered, the contents of the first
variable will be placed left-justified in column 1, the contents of the second
in column 25, and so forth until the five print fields are full. If a PUT LIST
statement names two variables, their values will print beginning in columns
1 and 25, and again PL/I will remember where it left off. If the next PUT LIST
statement also names two variables, their values will begin in columns 49
and 73. If the third PUT LIST statement names three variables, the value in
the first variable will print beginning in column 97; PL/I will then automatically cause the printer to advance one line (single space), and the values
of the second and third variables will print beginning in columns 1 and 25. If
the value being printed consists of 25 characters or more (i.e., extends to or
past the next tab position), that field is considered as used. As an example
suppose the following:

```
ALPHA = 'ABCDEFGHIJKLMNOPQRSTUVWXYZ';
BETA = 'NEXT';
PUT LIST (ALPHA, BETA);
```

The contents of ALPHA will occupy columns 1–26, and NEXT will begin in column 49.

Knowing the alignment of list-directed output on column 1 and the tab positions, students are often mystified by the extra blanks that appear in front of a FIXED DECIMAL value when it prints. The solution to the mystery lies in the fact that PL/I first moves the value into an intermediate field that is three columns wider than the declared width of the variable. Thus, a variable declared as FIXED DECIMAL (5) would be moved to a field eight columns wide. The extra three columns are to allow for the possibility of a minus sign, a decimal or binary point, and a leading zero in front of the point. In addition, leading zeros are replaced by blanks, except for the zero in front of the point. The contents of the intermediate field (including the leading blanks) are then moved into the output field. The following examples show the contents of the intermediate fields

Precision	Value	Intermediate Field Contents
(5,0)	1978	'bbbb1978'
(5,0)	−11662	'bb-11662'

We include the single quotation marks around the intermediate field contents to emphasize the fact that the contents are now character data.

Finally, list-directed output allows for character and numeric literals, as well as expressions to appear in the output variable list; we explain and illustrate these in Section 4.3.

4.2

The GET LIST Statement

The list-directed input statement has the syntax of

```
GET LIST (variable list);
```

Variable list can be a single variable, or it can consist of a series of variables, each separated from the next by a comma (and optionally one or more blanks). CH4EX1 reads only one data item at a time, and so it has only a single variable (STUDENT_NAME, STUDENT_NUMBER, TEST_SCORES) in each variable list. Given a different program logic, the GET LIST statement might have been:

```
GET LIST (SCORE1, SCORE2, SCORE3, SCORE4, SCORE5);
```

The exact same effect could also have been achieved by coding

```
GET LIST (SCORE1);
GET LIST (SCORE2);
GET LIST (SCORE3);
GET LIST (SCORE4);
GET LIST (SCORE5);
```

It is extremely important to remember that list-directed input reads in values until all variables in a given GET have received a value. As we noted above, the input for CH4EX1 was entered one data item per line. It could have been entered in almost any manner, as long as the data on a given line did not exceed 80 columns. SYSIN, the default file, is described to PL/I as having 80-byte (column) records. The first record above consists of 11 bytes of character data, 2 bytes for the quotation marks, and 67 bytes of blanks. The second record consists of 9 bytes of numeric data and 71 bytes of blanks; the remaining records have 2 bytes of data and 78 bytes of blanks. The input data, then, might have been entered on a single line:

```
'BETTY JONES' 123456789 90 75 82 89 79
```

The results would be the same; the first GET LIST would cause BETTY JONES to be placed in STUDENT_NAME; the second GET LIST would cause STUDENT_NUMBER to receive the value of 123456789. Each subsequent GET LIST would cause a new value to be placed in TEST_SCORES. The input data also could have been entered in any manner between the two extremes of one data item per line and all the data on one line.

```
'BETTY JONES'     123456789 91
75 87
```

4.2.1

Things that Can Go Wrong

Beginning students of PL/I are introduced to list-directed I/O early because it is easier to use than edit-directed I/O, another of the types of stream I/O. Therefore, the number of things that can go wrong are minimized; nevertheless, problems do arise. One common error results from the failure to keep in mind that a GET LIST causes data items to be read in until each variable in the variable list has received a value. If this fact is not kept in mind, it is likely that at least some of your variables will receive the wrong values. As long as the data is numeric, the program will proceed with execution to a normal termination; no error messages are printed because PL/I has no way of knowing that the variables received improper values. The only way you have of knowing anything went wrong is by hand checking the printed results, and this situation demonstrates the importance of making such a check.

When the input consists of both numeric and character data, the results of failing to correlate variables with input data items becomes more dramatic. Consider the following sequence of events: PL/I reads character data and discovers that the variable this data item is to be placed in is declared to be a numeric field. PL/I then attempts to convert the character data to numeric data but cannot. Not knowing what it is supposed to do in such a situation, it abends with a message probably including the following:

```
'CONVERSION' CONDITION RAISED
CONVERSION FROM CHARACTER TO ARITHMETIC ON INPUT OR OUTPUT
```

The most common practice in entering input data, therefore, is to place on one line those values that will be read by a single GET LIST statement; this practice helps to avoid mistakes in correlating the input values with the proper variable that might otherwise occur.

A second thing that commonly goes wrong when using list-directed input is mistakes in entering the input data. You must keep in mind that the comma and/or blank(s) are the delimiters in the input data that mark the end of a data item. If, for example, you typed in 9075 82 where you intended 90 75 82, then issued a

```
GET LIST (SCORE1, SCORE2, SCORE3);
```

PL/I would put 9075 (or as much as would fit) in SCORE1, 82 in SCORE2, etc. Suppose now that you have a five-digit number and that for the sake of readability you enter it as 52,346. On a GET LIST statement PL/I would put 52 in the first variable and 346 in the second. When your program abends with a 'CONVERSION' CONDITION RAISED message, or when the results are wrong, one of the first things you want to check is your input data. Have you made a mistake in entering it?

Still another error arising from input mistakes involving list-directed input is that of having fewer items in the input data than there are variables in (all of) the GET LIST statements. When this situation arises, PL/I will give you a message similar to the following:

```
IBM1811 'ONCODE'=0070 'ENDFILE' CONDITION RAISED ('ONFILE'=SYSIN)
          IN STATEMENT n AT OFFSET +nnnnnn IN PROCEDURE WITH ENTRY
          label
```

PL/C will notify you with a rather chilling message that might include the following:

```
END OF FILE REACHED
ABOVE ERROR IS FATAL.  PROGRAM IS STOPPED.
```

When you see this message (or something similar to it), you again need to check your input for mistakes; if there is no error in the input data items, check carefully the correlation between the variables in the various GET LIST statements and the input data.

Finally, some compilers require that any options (SKIP, PAGE, COPY, presented in the following several sections) be placed in front of LIST. Note that in the following sections we consistently place LIST as the last word before the variable list:

```
GET options LIST (variable list);
```

Valid GET LIST Statements

The basic syntax of the list-directed input statement is, as we noted earlier,

```
GET LIST (variable list);
```

The variable list can contain a single variable, or it can contain several; in the second case, the variables are separated by a comma (and optionally one or more blanks).

```
GET  LIST (TEST_SCORE);              /* ONE VARIABLE */
GET  LIST (SCORE1, SCORE2, SCORE3);  /* SEVERAL VARIABLES */
```

The two examples illustrate all that is required on a GET LIST statement if the input file has the default name of SYSIN. PL/I, however, does provide several other options on the GET LIST statement.

The first such option is COPY, and it has the effect of causing input data to be "echoed" to the printer. In other words, the input data is printed in exactly the form in which it appears in the input file. The format in which the output from COPY will appear depends on the compiler used: in some cases it is in a variable-name = value format; in other cases each input value will begin in column 1, with no indication of the variable included. The primary use of the COPY option, of course, is for debugging. If you suspect an error in the input data, you can easily insert a COPY into the appropriate GET LIST statements to see your input; once the input data has been corrected, you should remove the COPY from the code.

The second option offered on the GET LIST statement is the SKIP. Including the SKIP option on a GET LIST statement causes a skip to the next record (line) of the input file before values are read into variables. Suppose, for example, that each record in the input file contains a student identification number and that student's score on each of two tests and two programming assignments. The input, then, looks like this:

```
12345 85 72 88 81
12435 91 95 97 93
12534 77 72 81 75
```

Suppose also that the teacher wants the average of only the tests for each student. If she codes only

```
GET LIST (STUDENT_ID, TEST1, TEST2);
```

for each student, STUDENT_ID contains 12345, TEST1 contains 85, and TEST2 contains 72 after execution of the first GET LIST. So far so good; however, when the next GET LIST executes, STUDENT_ID contains 88, TEST1 contains 81, and TEST2 contains 12435 (or as much of it as fits). This problem can be overcome by coding instead

```
GET SKIP LIST (STUDENT_ID, TEST1, TEST2);
```

for each student after the first one. Now, since the SKIP causes position to be set at the beginning of the next record, the variables will contain the correct values for each student. Entire records can be skipped by coding SKIP (n), where n is any integer or any expression that evaluates to an integer (SKIP is the equivalent of SKIP (1)). Thus, one might code

```
GET SKIP (5) LIST (TEST1, TEST2);
GET SKIP (NUM2 - NUM1) LIST (TEST1, TEST2);
```

Obviously, several entire lines of input are skipped when, for some reason, only a sample of the input is desired. The COPY and SKIP options can also be combined in a single GET statement:

```
GET SKIP COPY LIST (variable list);
```

To summarize, the minimum list-directed input statement is

```
GET LIST (variable list);
```

Everything else is optional, to be used when the occasion suggests or demands it. In this case, to be on the safe side, code

```
GET options LIST (variable list);
```

4.3

The PUT LIST Statement

The list-directed output statement has the syntax of

```
PUT options LIST (variables, literals, expressions);
```

The PUT LIST statement, as the syntax statement illustrates, can contain any combination of variables, literals, and expressions. A **literal** (also called a **constant**) is simply any self-defining term and can be either of two types. A **character literal** (also called a character string) is composed of any one or more characters in the PL/I character set. PL/I recognizes a character literal by the fact that it is enclosed with single quotation marks (apostrophes). A **numeric literal** is any number between 0 and 9 or any combination of numbers; it is *not* enclosed by quotation marks. Thus, '921' is a character literal, and 921 is a numeric literal. The distinction is important where numbers are involved because although arithmetic operations can be carried out on both, to perform an arithmetic operation on a character literal composed of numbers is very inefficient and not at all recommended.

4.3.1

Formatting the Output with PUT LIST

There are a number of common uses for character literals in the output variable list. The following statement illustrates two such uses.

```
PUT LIST (' ', ' ', 'STUDENT GRADE REPORT');
```

When this statement executes, the first literal, a blank enclosed in quotation marks (remember a blank *is* a character) has the effect of causing the printer to print blanks in the field starting in column 1. In list-directed output the blank literal is the only means we have to exert control over the columns the data is printed in. The second literal, of course, causes the printer to print blanks in column 25. The third literal is printed just as it is listed above except that the quotation marks are stripped away; the delimiting quotation marks are always stripped away when a character literal is printed. The effect of executing the PUT LIST above, then, is to print a roughly centered report title. A title that appears at the top of each page of a report is referred to as **page header**. The effect of the two blank literals, then, is to cause the page header to be roughly centered on the page. In real programming situations, the programmer is normally given exact output specifications, which require more editing facilities than list-directed output provides for. In the absence of detailed specifications, however, you should always strive to make your output *easily readable and attractive*. This means, for example, that the output should be roughly centered on the page, not huddled on the lefthand side of the page. Normally a report is given a title that is descriptive of the report's function. **Column headers** are also often included to indicate the nature of the data in the columns.

To illustrate the use of column headers, suppose a program is designed to read in test scores for many students, that for each student the input consists of a student identification number and two test scores, and that the average of the test scores for each student is to be computed and printed along with the test scores. The PUT LIST statement to print the column headers might contain this:

```
PUT LIST (' ', 'STUDENT-ID', 'TEST1', 'TEST2', 'AVERAGE');
```

The statement causes blanks to be printed in the field starting at column 1 and the column headers to be printed in starting columns 25, 49, 73, and 97. Once again the data is roughly centered on the printed page. To align the data in the columns with the column headers, it is necessary that the PUT LIST statement for each student have this format.

```
PUT LIST (' ', STUDENT_ID, SCORE1, SCORE2, AVG);
```

Note carefully that in the first PUT statement, 'STUDENT-ID' (with quotation marks) is a literal, while in the second PUT statement STUDENT_ID (with no quotation marks) is a variable.

There are other refinements we could make in formatting our report. List-directed output gives us several options for controlling the vertical formatting of the output. We briefly introduced the first of them, SKIP, in conjunction with the GET LIST statement. SKIP is used much more commonly for output, however. For example, we might want to underline our column headers; we can accomplish this by coding these two lines:

```
PUT LIST (' ', 'STUDENT-ID', 'TEST 1', 'TEST 2', 'AVERAGE');
PUT SKIP (0) LIST (' ', '_____', '_____', '_____',
                   '_____');
```

SKIP (0) specifies that the data specified in the variable list should print with no advance; in other words, the printer first prints the column headers, then with no advance of the paper, it returns and prints the underlines. The printed results look like this:

```
STUDENT-ID    TEST 1    TEST 2    AVERAGE
_____    _____    _____    _____
```

The following points are pertinent to the use of SKIP.

1. An integer literal, or any expression that evaluates to an integer, can be placed in parentheses following the keyword.

2. If the integer is less than 1 (i.e., 0 or negative), the data will print on the same line.

3. The default integer is 1; SKIP is equivalent to SKIP (1).

4. If the integer is 1 or greater, it specifies the line, *relative to the current line*, to skip *to*; it does not specify the number of lines to skip. For example, SKIP (2) leaves *one* blank line between the two printed lines (double spacing); SKIP (3) leaves two blanks lines between the printed lines.

We have seen how we can force the printer to print on a relative line through the use of SKIP. SKIP (3), for example, causes the printer to skip to the third line relative to the current line before it starts to print. We can also request the printer to print on an absolute line through the LINE option. If we want to print a page header on line 5 of the first page, we code

```
PUT LINE (5) LIST (' ', ' ', 'A REPORT');
```

If we want the next line printed on line 12, we code

```
PUT LINE (12) LIST (' ', STUDENT_ID, TEST_SCORE);
```

LINE can also be combined with PAGE:

```
PUT PAGE LINE (10) LIST (' ', 'THIS IS PAGE 2');
```

or

```
PUT LINE (10) PAGE LIST (' ', 'THIS IS PAGE 2');
```

The order in which the PAGE and LINE options are coded in the same statement is of no significance. Regardless of their order in the statement, the PAGE option is executed first, followed by the LINE option. This means that the printer goes to a new page, then to the designated line (here 10) and prints. Note, however, that SKIP cannot be combined with PAGE.

PL/I has a line counter that keeps track of the current line. If you specify an absolute line number less than the current line, PL/I will go to a new page, but it will print on line 1 rather than on the line specified. Suppose, for example, that you are currently printing on line 10. You then code:

```
PUT LINE (9) LIST ('I WANT LINE 9 OF A NEW PAGE');
```

PL/I will skip to a new page and print the literal on the first line. There is also a default value for the number of lines to be printed on a page (blank lines also count as lines); when the line counter value equals the default value or when a SKIP or LINE option causes the line counter to advance to a value equal to or greater than the default value, PL/I forces a new page. If the LINE (without PAGE) or SKIP option forces a new page, the values in the variable list will again print on the first line of the new page. If you want a line other than the first on the new page, you must explicitly force the new page with the PAGE option.

Still another formatting refinement provided by PL/I is the use of SKIP, LINE, and PAGE with no list. To use this feature simply code:

```
PUT SKIP; or PUT SKIP (n);
PUT LINE; or PUT LINE (n);
PUT PAGE;
```

The value of these possibilities will become apparent as you learn to write more sophisticated and complex programs.

A summary of the options available to list-directed output is presented in the following section.

4.3.2

Valid PUT LIST Statements

The following examples are intended to illustrate and summarize the various possibilities of the list-directed output statement. The comment on each statement indicates the key point being illustrated. The syntax of the statement is

```
PUT LIST (variable list);
PUT LIST (VAR);                /* ONE VARIABLE */
PUT LIST (VAR1, VAR2, VAR3):   /* SEVERAL VARIABLES */
```

```
            PUT LIST (' ', ' ', 'LITERALS');    /* LITERALS */
            PUT LIST ((BASE * .5) * ALT);       /* EXPRESSION */
            PUT LIST (ROUND (NUM_1 / NUM_2, 2);
                                                /* BUILT-IN FUNCTION */
            PUT SKIP LIST (VAR);                /* SKIP TO NEXT LINE */
            PUT SKIP (3) LIST (VAR);            /* TRIPLE SPACE */
            PUT SKIP (0) LIST ('_____');        /* DO NOT ADVANCE PAPER */
            PUT LINE (10) LIST ('ON LINE 10');
                                                /* PRINT ON ABSOLUTE LINE  10 */
            PUT PAGE LIST ('PAGE HEADER');      /* FORCE A NEW PAGE */
            PUT LINE (8) PAGE LIST ('PAGE HEADER');
                                                /* NEW PAGE AND ABSOLUTE LINE 8 */
            PUT SKIP;                           /* CREATE BLANK LINE--NO PRINT */
            PUT SKIP (2);                       /* CREATE TWO BLANK LINES--NO PRINT */
            PUT PAGE;                           /* NEW PAGE--NO PRINT */
            PUT LINE (5);                       /* ABSOLUTE LINE 5--NO PRINT */
```

The examples just presented should give you, first of all, a good idea of the various ways in which the options available to the PUT LIST statement can be combined; at the same time, you should now see the degree of flexibility provided by the PUT LIST options for the formatting of output.

4.4

Example Program—
LIST I/O

So that you can get an idea of formatting in list-directed I/O, we have repeated CH3EX2 in example program CH4EX1 (see Fig. 4–1), simply adding some control options (LINE and SKIP). As you might suspect, there is no difference whatsoever in the logic of the two programs; therefore, the pseudocode is not repeated here.

```
PL/I OPTIMIZING COMPILER  VERSION 1 RELEASE 5.1  TIME: 15.30.54  DATE: 12 FEB 87 PAGE   1
OPTIONS SPECIFIED
60STMT;


PL/I OPTIMIZING COMPILER        /* CH4EX1 - USING CONTROL OPTIONS */           PAGE    2
                    SOURCE LISTING
      STMT


        /* CH4EX1 - USING CONTROL OPTIONS */
```

FIGURE 4–1
CH4EX1 (using control options in list-directed I/O).

```
1   CH4EX1:  PROCEDURE OPTIONS (MAIN);

    /********************************************************************/
    /*                                                                  */
    /*   FUNCTION:   THIS PROGRAM PRODUCES A REPORT ON A STUDENT        */
    /*               GIVING HER AVERAGE ON THE TESTS IN THE CLASS       */
    /*                                                                  */
    /*   INPUT:      A SINGLE RECORD CONTAINING A STUDENT'S             */
    /*               IDENTIFICATION NUMBER AND THREE RECORDS, EACH      */
    /*               CONTAINING A TEST SCORE                            */
    /*                                                                  */
    /*   OUTUT:      THE STUDENT'S IDENTIFICATION NUMBER AND HER        */
    /*               AVERAGE TEST SCORE,FIRST WITHOUT USING THE ROUND,  */
    /*               THEN USING IT TO COMPUTE THE AVERAGE SCORE         */
    /*                                                                  */
    /*                                                                  */
    /********************************************************************/

2       DCL STUDENT_NAME      CHARACTER      (20);

3       DCL STUDENT_NUMBER    FIXED DECIMAL (9);
4       DCL TEST_SCORES       FIXED DECIMAL (3);
5       DCL AVERAGE           FIXED DECIMAL (5,2);

6       DCL LABEL_1           CHARACTER      (21)      INIT
              ('AVERAGE WITHOUT ROUND');
7       DCL LABEL_2           CHARACTER      (18)      INIT
              ('AVERAGE WITH ROUND');
8       DCL ACCUMULATOR       FIXED BINARY  (15)     INIT (0);
9       DCL COUNTER           FIXED BINARY  (15)     INIT (0);

10      DCL ROUND             BUILTIN;

11      GET LIST (STUDENT_NAME);
12      GET LIST (STUDENT_NUMBER);
13      GET LIST (TEST_SCORES);
14      ACCUMULATOR = ACCUMULATOR + TEST_SCORES;
15      COUNTER = COUNTER + 1;

16      GET LIST (TEST_SCORES);
17      ACCUMULATOR = ACCUMULATOR + TEST_SCORES;
18      COUNTER = COUNTER + 1;
```

FIGURE 4–1 *(continued)*

```
19      GET LIST (TEST_SCORES);
20      ACCUMULATOR = ACCUMULATOR + TEST_SCORES;
21      COUNTER = COUNTER + 1;

22      AVERAGE = ACCUMULATOR / COUNTER;
23      PUT LINE (5) LIST (STUDENT_NAME, STUDENT_NUMBER, ACCUMULATOR,
```

```
PL/I OPTIMIZING COMPILER          /* CH4EX1 - USING CONTROL OPTIONS */          PAGE   3
  STMT
                LABEL_1, AVERAGE);

24      AVERAGE = ROUND (ACCUMULATOR / COUNTER, 2);
25      PUT SKIP (4) LIST (' ', ' ', ' ', LABEL_2, AVERAGE);
26   END CH4EX1;
```

```
PL/I OPTIMIZING COMPILER          /* CH4EX1 - USING CONTROL OPTIONS */          PAGE   4
COMPILER DIAGNOSTIC MESSAGES
ERROR ID L   STMT    MESSAGE DESCRIPTION

COMPILER INFORMATORY MESSAGES

IEL0533I I          NO 'DECLARE' STATEMENT(S) FOR 'SYSIN','SYSPRINT'.

END OF COMPILER DIAGNOSTIC MESSAGES

BETTY JONES              123456789              253        AVERAGE WITHOUT ROUND    84.33

                                                          AVERAGE WITH ROUND       84.25
```

4.5

Summary

This chapter has introduced you to input–output operations in PL/I, although much remains to be said because we have restricted ourselves to stream I/O, and within stream I/O to its list-directed form. To lay the necessary groundwork we introduced the concept of a file, which is a collection of similar and related records; a record, in turn, is a collection of related fields. Although we follow common practice in referring to "records" in connection with stream I/O, we urge you to keep in mind that in reality the entire input file is a single record, and the same is true of the output file. Thus, the only way to control the number of items read in by a given PUT is through the number of variable names in the variable list. Also on the subject of files, we pointed out that PL/I provides default file names: SYSIN for the input file and SYSPRINT for the output file to the printer.

The data items in the input file being processed with list-directed input must be separated by a comma and/or one or more blanks. PL/I scans until it encounters a nonblank. It then scans until it finds a comma or a blank character. What is in between, it considers to be a data item; therefore, the data cannot contain an embedded blank or comma. The exception to this rule is character data, which must be enclosed in single quotation marks; PL/I treats blanks and commas within quotation marks as ordinary characters. The data can be placed in the file in any manner, so long as these restrictions are met.

PL/I reads data into each variable in the the variable list and remembers where it left off. Unless modified by a SKIP option, at a subsequent GET LIST it will pick up where it left off for the previous GET LIST. The relationship between variables in the list and data items in the input file is positional—the first data item is placed in the first variable in the list, the second data item in the second variable, and so on.

For list-directed output, tab positions are established at columns 25, 49, 73, and 97, logically dividing the print page into five fields (the first starting in column 1). When a PUT LIST is encountered, the value in the first variable is printed left-justified in column 1, the next in column 25, and so on. If necessary, PL/I will direct the printer to return to column 1 of the following line and continue printing. One or more blank literals can be coded in the PUT LIST statement to provide a measure of horizontal formatting. Vertical formatting can be attained by the SKIP option, which designates the line, relative to the current line, to be skipped to. LINE specifies an absolute line on which the data is to be printed, and PAGE causes the printer to skip to a new page before printing. LINE can be combined with PAGE (in either order), and the combination will force a new page and a skip to the specified line before printing occurs.

Having mastered the rudiments of I/O in PL/I, you are now ready to proceed in Chapter 5 to an introduction to the manner in which PL/I implements the decision and iteration structures.

4.6

Exercises

1. Define the terms *field*, *record*, and *file*. Give an example of what kind of information would be held in each.

2. Show the values of the variables after the execution of all the GET LIST instructions.

```
GET LIST (A,B,C);
GET LIST (D,E,B);
GET LIST (C,B);
GET SKIP LIST (E,A,C);
Data:   1,2,3,4,5,6,7
        8,9,10,11,12
        13,14,15,16
```

3. What would be printed with the following PUT LIST instructions?

```
PUT LIST ('ABC',25,'23+2+4',23+2+4);
PUT LIST ('PRICE','AREA');
PUT SKIP LIST ('ALL DONE');
```

4. What would be printed after the execution of the following instructions?

```
GET LIST (VALUE,COUNT,A);
GET LIST (COUNT,VALUE);
PUT LIST (VALUE,COUNT,'NOTHING','A',A);
GET LIST (A,VALUE,COUNT);
PUT LIST ('VALUE IS:',VALUE);
PUT LIST (COUNT,'IS THE COUNT');
GET SKIP LIST (COUNT);
GET LIST (VALUE,A,COUNT);
PUT SKIP LIST (COUNT+A,VALUE,A);
Data:   10,15,85,100,200
        0,65,78,2
        6,79,54,9
```

5. Provide input data for the following GET LIST statements.

```
DCL A_1    CHAR     (5);
DCL NUM_1 FIXED DEC (3);
DCL NUM_2 FIXED DEC (4,2);
GET LIST (A_1, NUM_1, NUM_2);
```

6. Write a PL/1 program to print out an address label with your name, address, and telephone number. The address label should be similar to the following.

RUTH A. BAKER

1125 S. ELM

KIRKLAND, IL 60185

815-344-7986

7. Write a PL/1 program to perform the following algorithm. Do not forget to declare all variables.

```
START
    Write a heading.
    Read 3 numbers of precision (3,1).
    Triple-space and print the 3 numbers labeled and starting
        in columns 25, 49, and 73
    Calculate the sum of the 3 numbers.
    Double-space and print the sum with a label beginning in
        column 49
END
```

8. Redo any of Exercises 8, 9, or 10 in Chapter 3 using GET LIST and PUT LIST. Use labels to identify all values printed.

C H A P T E R 5

CONDITIONAL

EXECUTION

AND LOOPING

In this chapter we introduce several elements that allow you to write some fairly complex programs. The first step is to become familiar with the comparison and logical operators that are the basis for all decisions the computer is able to make. We then present various aspects of the decision structures of IF-THEN and IF-THEN-ELSE. In order to process large amounts of data the programmer must be able to direct the computer to execute the same lines of code over and over; this is the iteration structure implemented by the DO WHILE statement, which is the subject of Section 5.3. We close Chapter 5 by explaining the PL/I facility that allows the program to know when the last input record has been read, and we introduce a new data type called bit data. Although there are numerous combinations of program structures that will be presented in later chapters, you will have been presented with all the basics introduced in Chapter 1 by the time you finish Chapter 5.

5.1

Comparison and Logical Operators

One capability the computer must have in order to allow for programs that are at all complicated is that of making decisions. Since all data is stored in the computer ultimately in the form of numbers, it follows that any decision a computer makes must be based on a comparison between the numbers at two separate locations. Because only comparisons of numbers are possible, it is clear that every decision must be made on the basis of whether the expression is *true* or *false*. An expression that can evaluate to only true or false is called a **logical expression**. Of course, it is possible to restate the logical expression in the form of a question, in which case the answer must be either *yes* or *no*. Consider the following:

Expression/Question	Answer
5 is less than 7	True
Is 9 greater than 0?	Yes
91 is equal to 95	False
Is 28 not equal to 28?	No
45 is not less than 45	True
Is 45 not greater than 45?	Yes

The words (less than, greater than, etc.) used to state the relationships between the numbers are called **comparison operators**. The numbers themselves are the **operands** of the comparison operators. The following chart illustrates the comparison operators available to PL/I and how they are coded.

Symbol	Meaning
<	Less than
¬<	Not less than
=	Equal to
¬=	Not equal to
<=	Less than or equal to
>	Greater than
¬>	Not greater than
>=	Greater than or equal to

A note of caution: In each case in the chart above in which two symbols appear together, the two symbols are considered to be one by PL/I; therefore, as we saw earlier with the exponentiation symbol, no space can intervene between the two symbols.

In addition to the comparison operators, PL/I has three **logical operators**, and each has a special symbol in PL/I. The logical operators, moreover, have several uses, but here we are concerned only with their use in the decision structures. The first, NOT, uses the symbol ^, and its purpose is to allow you to state the converse (i.e., opposite) of a logical expression containing less than, equal to, or greater than, as illustrated in the table above. The condition on which PL/I makes a decision can be based on a single factor (e.g., A < B), or it can be based on several factors. The function of the other two logical operators is to allow several conditions to be tested, one after another, in a single PL/I statement.

The next logical operator, then, is OR, signified to PL/I by the symbol |. When two conditions are joined together in a single statement by OR, *the statement is true if any one or more of the conditions are true.*

17 < 20 | 30 = 30	condition 1 true; condition 2 true; statement true
17 > 20 | 30 = 30	condition 1 false; condition 2 true; statement true
17 < 20 | 30 = 17	condition 1 true; condition 2 false; statement true
17 > 20 | 30 = 17	condition 1 false; condition 2 false; statement false

These examples illustrate all possible combinations of two conditions. Note that in three of the four combinations the statement is true; in only one is it false.

The last logical operator is AND (&); it, too, is used to join two or more conditions in a single statement. In the case of conditions joined by AND, however, *all of the conditions must be true for the statement to be true.*

17 < 20 & 30 = 30	condition 1 true; condition 2 true; statement true
17 > 20 & 30 = 30	condition 1 false; condition 2 true; statement false
17 < 20 & 30 = 17	condition 1 true; condition 2 false; statement false
17 > 20 & 30 = 17	condition 1 false; condition 2 false; statement false

With all possible combinations again included, we see that two conditions joined by AND produce three false statements and one true statement.

The above material is usually condensed into a **truth table**. In this case A and B each stand for a condition and S stands for "statement." T and F, of course, stand for "true" and "false" respectively. Finally, you should be aware that truth tables are often presented with a 1 replacing the T and 0 replacing the F.

	A or B			A and B	
A	B	S	A	B	S
F	F	F	F	F	F
F	T	T	F	T	F
T	F	T	T	F	F
T	T	T	T	T	T

5.2

The Decision Structures— IF Statements

It is easily understandable that you might have forgotten the IF-THEN and IF-THEN-ELSE decision structures as they were presented in Chapter 1; therefore, we repeat them here in Figure 5–1. These structures require the computer to make a decision and to take some action depending on the outcome of the decision. In the IF-THEN structure, no action is taken on the False (No) path—only on the True (Yes) path. In the IF-THEN-ELSE, actions are taken on both paths.

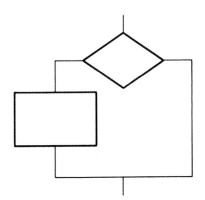 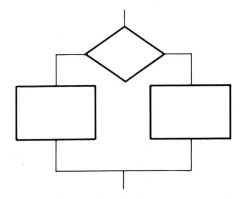

FIGURE 5–1
IF-THEN construct (left) and IF-THEN-ELSE construct (right).

The IF Statement
with Simple Conditions

The terms "simple" and "compound" as used here are analogous to their use in grammar. A simple condition, like a simple sentence, states one proposition; a compound condition, like a compound sentence, states two or more propositions. The syntax for the IF statement with a simple condition is

```
IF condition THEN
    imperative statement;
```

As a matter of fact, the statement following the IF-THEN need not be an imperative (e.g., GET LIST or A = B + C); it might also be a conditional (IF) statement. In this second case, however, the result is a nested IF, which we introduce in a later chapter; therefore, for the time being we can pretend that the IF-THEN is followed by an imperative. There are no instances of the IF-THEN with a simple condition in CH5EX1, but suppose our program is intended to produce a list of those students who have failed the course; suppose, furthermore, that the student's average has been computed and the result is in FINAL_AVG. At this point we might very well code

```
IF FINAL_AVG < 68.0 THEN
    PUT SKIP (2) LIST (' ', STUDENT_ID, FINAL_AVG);
```

Two items should be noted:

1. Even though two statements are coded, the *single* semicolon is placed after the imperative. PL/I, of course, demands this placement of the semicolon.

2. Putting the statement on two lines is a matter of making the program more readable and its logic easier to follow. Note the indentation of the imperative to show its relation to the conditional.

The syntax of the IF-THEN-ELSE with a simple condition is

```
IF condition THEN
    imperative;
ELSE
    imperative;
```

Notice again the placement of the semicolons. Notice also the alignment of the ELSE with the IF and that of the two imperatives; this alignment is not a

PL/I requirement, but again it makes the logic of the program easier to follow.

CH5EX1 (in Statements 14–17) provides examples of the IF-THEN-ELSE with simple conditions. Let us remind you once more that the THEN imperative is executed when the condition is true; the ELSE imperative is executed when the condition is false. Under no circumstances will both be executed in a single pass through the IF-THEN-ELSE code.

5.2.2

The IF Statement With Compound Conditions

In the case of the IF statement with simple conditions, a single condition is checked, and the statement executed depends on the outcome of the comparison. Often, however, the situation requires that the statement to be executed depend on the outcome of more than one comparison. In some cases we want to execute one statement if two or more conditions are all true and perhaps another if any one condition is false. In other cases we want to execute one statement if any one condition is true and perhaps another if all the conditions are false. Although there are other methods of handling such a situation, the IF with compound conditions is usually the most convenient. The syntax of the statement is:

```
IF condition-1 logical-operator condition-2 THEN
    imperative statement;
```

where "logical-operator" is either & (AND) or | (OR). Any number of conditions can be put together using these two logical operators. Remember, however, that no matter how many conditions are included, for that statement to be true all of the conditions must be true if all of the logical operators are AND and that only one need be true to make the statement true if all conditions are joined by OR.

It is also quite possible to have a statement containing a pair of conditions joined by OR and another pair joined by AND; the two pairs, of course, can be joined by either AND or OR. In general, the AND is evaluated first, then the OR. As with any expression, however, this order can be altered through the use of parentheses, and it is a good idea to use parentheses to group the conditions in such a way that they will be evaluated in the order you intend; if nothing else, this practice aids another person in determining what was intended.

```
IF (A = B & C > D) | (E <= D & G ^= H) THEN
    I = I + 1;
```

The parentheses in this example do not change the order of evaluation, but they do make the expression clear to the reader.

DO Groups in an IF Statement

The IF statements we have examined so far lead to the execution of a single statement in each case when the condition is true and, when ELSE is included, another single statement when the condition is false. Often times, though, it is necessary to execute more than one statement following either one or both of the IF and the ELSE. PL/I allows for this situation by means of a do-group. A **do-group** is simply one or more statements whose beginning is marked by the keyword DO and whose ending is marked by the keyword END; in other words, the do-group delimits the number of statements that are controlled by the IF and by the ELSE when it is included. Statements 20 through 27 of CH5EX1 provide one example; another is given here.

```
IF CODE = 1 THEN
    DO;
        BALANCE = BALANCE + TRANSACTION;
        DEPOSIT_COUNT = DEPOSIT_COUNT + 1;
    END;
ELSE
    DO;
        BALANCE = BALANCE - TRANSACTION;
        WTHDRW_COUNT = WTHDRW_COUNT + 1;
    END;
```

As always with examples, you should pay special attention to the placement of semicolons: With the do-group we see a semicolon after each imperative statement, including DO and END.

A number of variations on this example are possible. For example, PL/I will accept a single statement within a do-group, but because the DO and END are redundant in this case it is pointless to go to the trouble of including them. You can also code a do-group following the IF and a single statement following the corresponding ELSE, or vice versa.

```
IF CODE = 1 THEN
    DO;
        BALANCE = BALANCE + TRANSACTION;
        DEPOSIT_COUNT = DEPOSIT_COUNT + 1;
    END;
ELSE
    BALANCE = BALANCE - TRANSACTION;
```

At this point you might wonder how PL/I knows that the END you coded as part of a do-group is not the END that signals the end of the program. The answer is that PL/I keeps track of each DO that it encounters, and it interprets the first END encountered after the DO as belonging to the DO. It is not unusual for beginning (and even experienced) users of PL/I to lose track and provide one less DO than PL/I requires. The results of omitting an END vary according to the situation, but the result is never desirable. In Section 5.3.3 we propose one aid in avoiding this problem.

5.2.1

Example Program—
IF-THEN-ELSE

CH5EA1 illustrates much of our discussion of the decision structure. The program reads in a single record listing a student identification number, his average program grade, and his average test grade (see Fig. 5.?) The program is designed to print a two-line page header and the disposition of the student as passing or failing. The IF-THEN structure is not illustrated in the program, but the IF-THEN-ELSE with a simple condition and a single statement following each of the IF-THEN and the ELSE is illustrated in Statements 14–17. The NOTES in the comments box tell us that in order to pass the course the student must have a minimum average of 68 percent for his programs and for his tests; he must also have a minimum overall average of 68 percent. This logic is implemented in Statements 20–27, which also illustrate IF-THEN-ELSE using do-groups and controlled by a compound condition. Note that if the student has an average of 68 percent or higher on both his programs and his tests, he of necessity also has an overall average of 68 percent or higher; therefore, no comparison of his overall average against the minimum need be made. All of these matters of logic are reflected in the following pseudocode for the program.

```
START
    Print page headers
    Read student name
    Print label, student name
    Read averages
    Print labels, averages
    IF (test average < 68)
        Print 'Fail' message
    ELSE
        Print 'Pass" message
    ENDIF
    IF (program average < 68)
        Print 'Fail' message
```

```
                          ELSE
                              Print 'Pass' message
                          ENDIF
                          Final average ← (test average + program average) / 2.0
                          Print label, final average
                          IF (test average and program average >= 68)
                              Print 'Pass' message
                              Print 'Enroll' message
                          ELSE
                              Print 'Fail' message
                              Print 'See advisor' message
                  STOP
```

PL/I OPTIMIZING COMPILER VERSION 1 RELEASE 5.1 TIME: 15.42.07 DATE: 12 FEB 87 PAGE   1
OPTIONS SPECIFIED
GOSTMT;

PL/I OPTIMIZING COMPILER   /* CH5EX1 - USING IF-THEN-ELSE */   PAGE   2
                        SOURCE LISTING
     STMT
          /* CH5EX1 - USING IF-THEN-ELSE */

       1 CH5EX1: PROCEDURE OPTIONS (MAIN);
          /***/
          /*   */
          /*   FUNCTION: THIS PROGRAM PRODUCES A REPORT ON A STUDENT   */
          /*   REGARDING THE OUTCOME OF HIS CLASS   */
          /*   */
          /*   INPUT: A SINGLE RECORD CONTAINING A STUDENT'S   */
          /*   NAME AND HIS AVERAGE FOR EACH OF (1) TESTS AND   */
          /*   (2) PROGRAMMING ASSIGNMENTS   */
          /*   */
          /*   OUTPUT: THE INPUT AND APPROPRIATE MESSAGES   */
          /*   */
          /*   NOTES: 1. THE MINIMUM PASSING GRADE IS 68%   */
```

**FIGURE 5–2**
**CH5EX1 (the IF-THEN-ELSE structure).**
```

```
      /*              2.   THE STUDENT MUST HAVE A PASSING AVERAGE ON       */
      /*                   BOTH THE EXAMS AND THE PROGRAMS IN ORDER TO       */
      /*                   PASS THE COURSE                                   */
      /*                                                                     */
      /***********************************************************************/
   2    DCL STUDENT_NAME     CHARACTER      (20);

   3    DCL TEST_AVG        FIXED DECIMAL (3);
   4    DCL PGM_AVG         FIXED DECIMAL (3);
   5    DCL FINAL_AVG       FIXED DECIMAL (5,2);

   6    DCL ROUND           BUILTIN;

   7    PUT LIST (' ', 'GRADE REPORT', ' ', 'SPRING 1987');
   8    PUT SKIP (2) LIST (' ', ' ', 'COMPUTER SCIENCE 101');

   9    GET LIST (STUDENT_NAME);
  10    PUT SKIP (3) LIST (' ', ' ', 'STUDENT NAME', STUDENT_NAME);

  11    GET LIST (TEST_AVG, PGM_AVG);
  12    PUT SKIP (3) LIST (' ', ' ', 'YOUR TEST AVERAGE IS', TEST_AVG);
  13    PUT SKIP (2) LIST (' ', ' ', 'YOUR PROGRAM AVERAGE IS', PGM_AVG);

  14    IF TEST_AVG < 68 THEN
          PUT SKIP (2) LIST (' ', ' ', '*** YOU FAILED TESTS ***');
  15    ELSE
          PUT SKIP (2) LIST (' ', ' ', 'YOU PASSED TESTS');

  16    IF PGM_AVG < 68 THEN
          PUT SKIP (2) LIST (' ', ' ', '*** YOU FAILED PROGRAMS ***');
  17    ELSE
          PUT SKIP (2) LIST (' ', ' ', 'YOU PASSED PROGRAMS');
  18    FINAL_AVG = ROUND ((TEST_AVG + PGM_AVG) / 2.0,2);
```

```
   STMT
   19    PUT SKIP (3) LIST (' ', ' ', 'YOUR FINAL AVERAGE IS', FINAL_AVG);

   20    IF TEST_AVG >= 68 & PGM_AVG >= 68 THEN
          DO;
   21       PUT SKIP (4) LIST (' ', ' ', 'YOU HAVE PASSED CSCI 101');
   22       PUT SKIP (2) LIST (' ', ' ', 'YOU MAY ENROLL IN CSCI 102');
```

FIGURE 5-2 *(continued)*

```
23        END;
24     ELSE
         DO;
25          PUT SKIP (4) LIST (' ', ' ', 'YOU HAVE FAILED CSCI 101');
26          PUT SKIP (2) LIST (' ', ' ', 'SEE YOUR ADVISOR AT ONCE');
27        END;
28  END CH5EX1;
```

```
PL/I OPTIMIZING COMPILER        /* CH5EX1 - USING IF-THEN-ELSE */            PAGE   4
COMPILER DIAGNOSTIC MESSAGES
ERROR ID L   STMT   MESSAGE DESCRIPTION

COMPILER INFORMATORY MESSAGES

IEL0533I I         NO 'DECLARE' STATEMENT(S) FOR 'SYSPRINT','SYSIN'.

END OF COMPILER DIAGNOSTIC MESSAGES
```

```
              GRADE REPORT                                SPRING 1987

                           COMPUTER SCIENCE 101

                           STUDENT NAME          HILL, HARRY

                           YOUR TEST AVERAGE IS       93

                           YOUR PROGRAM AVERAGE IS    84

                           YOU PASSED TESTS

                           YOU PASSED PROGRAMS


                           YOUR FINAL AVERAGE IS    88.50


                           YOU HAVE PASSED CSCI 101

                           YOU MAY ENROLL IN CSCI 102
```

5.3

The Iteration Structure— DO WHILE

The purpose of the example program CH4EX1 (see Fig. 4–1) in Chapter 4, you will recall, is to read in a student identification number and three scores for that student and to compute and print the student's average. The program consists of 15 statements following the DECLARE to process the data for one student. It follows, therefore, that it would require 525 executable statements to process the data for the 35 students in one class, and 26,250 statements to process the data for each student in each of the 50 sections of classes offered by a single department. Clearly such a situation is impractical. In this section we introduce you to a PL/I facility (DO WHILE) through which large amounts of similar input data can be handled without a corresponding increase in the number of lines of code required.

5.3.1

The DO WHILE Statement

You have been introduced to two of the three structures that are permissible in structured programming—sequence and decision. The third, iteration, is the subject of this section, but first for your convenience we repeat in Figure 5–3 the flowchart of the iteration structure. Iteration is, of course, the repeated execution of one or more statements, and its syntax is

```
DO WHILE (logical expression);
          .
          .
          .
END;
```

As with IF-THEN, the logical expression of a DO WHILE can contain simple or compound conditions. Two things to keep in mind with regard to the syntax are, first, that there must be a space between DO and WHILE and, second, that PL/I must also have some way of knowing how many statements following the DO WHILE are to be executed repeatedly—that is, the point in the program at which it is to branch back to the DO WHILE statement. The END statement marks the bottom of the loop, just as it marks the end of a do-group and the end of a program. Everything between the DO WHILE and corresponding END are said to be in the **range** of the DO WHILE or loop.

The function of the logical expression is to control the number of iterations of the statements within the loop; as long as the logical expression evaluates to true the statements in the range of the DO WHILE will be repeatedly executed. It follows, therefore, that if the logical expression is

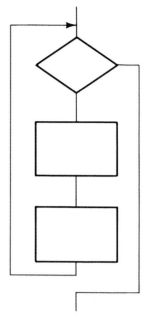

FIGURE 5–3
The iteration (loop) structure.

true at the outset, at least one of its operands must be modified in order to terminate the iteration. It is almost certain that sooner or later you will forget to modify an operand and that you will find yourself in an **infinite loop**, repeatedly executing the same set of instructions until your program has used its allotted amount of computer time or until you have printed the number of lines allowed for your program. At either of these points your program is automatically terminated by the operating system (a set of programs that controls the overall operation of the computer resources).

One way of modifying a variable used in the logical expression is that illustrated in example program CH5EX2 in the following section. In this method the first input record contains a number specifying how many data records are to follow (a record such as this is referred to as a **header record**). This value is read into a variable (NUM_RECS); meanwhile, another variable (REC_COUNT) has been initialized (usually to 0 or 1). These two variables are then used as the operands of the DO WHILE condition. Somewhere within the loop (i.e., within the range of the DO WHILE) the second variable (REC_COUNT) is incremented by the value 1. Thus the value of REC_COUNT changes with each iteration, while the value of NUM_RECS remains constant. It follows, then, that eventually the logical expression will become false; when this occurs, PL/I will skip over all statements in the range of the DO WHILE (i.e., it will "drop" or "fall" through the loop) and execute the statement immediately following the END statement. In CH5EX2 we chose to initialize REC_COUNT to 1 and use <= as the logical operator; the same results could be obtained by initializing REC_COUNT to 0 and using only < as the logical operator. Why wouldn't it be simpler, though, just to code

```
DO WHILE (REC_COUNT <= 10);
```

since you have to know how many records are in the input file in order to create the header record? Coding literals within the program is known as **hard coding,** and if this program is to be executed only once there is no reason not to do it. In industry, however, programs are executed repeatedly over a period of years. It seems a bit unreasonable to assume that every time the program executes it will have exactly ten input records. By putting the number of input data records on a header and using that value to control the DO WHILE iterations, CH5EX2 will perform equally well for one input record or 99 input records. The only reason the maximum is 99 is that we declared that NUM_RECS would contain only two digits.

In the case of CH5EX2 notice that the original value in NUM_RECS is available at the end of the program. The program, however, has no use for that value; therefore, instead of using REC_COUNT and adding the value 1 to it each time through the loop, we could have chosen instead simply to decrement NUM_RECS each time through the loop. Using this logic we would have to make three changes in the present version of CH5EX2.

1. Eliminate the declaration of REC_COUNT.

2. Change the DO WHILE to read

```
DO WHILE (NUM_RECS > 0);
```

3. Change Statement 18 (REC_COUNT = REC_COUNT + 1;) to

```
NUM_RECS = NUM_RECS - 1;
```

In creating the logic for your DO WHILE loop there are two things you must understand. First of all, the computer evaluates the DO WHILE logical expression when it encounters the DO WHILE statement (i.e., at the top of the loop). This in turn means that if the logical expression is false the first time it is evaluated, the statements within the range of the DO WHILE will not get executed at all. The second thing to remember is that once the loop is entered, the computer attempts to execute every statement through to the END statement. In Section 5.4.3 we introduce a new feature to the DO WHILE that alters the logic of the loop, but for now you should code a loop that carries out the three following steps in the order in which they are listed.

1. Read a record.

2. Process (i.e., do arithmetic, print, etc.) that record.

3. Modify one of the operands of the logical expression in a manner consistent with the logic of the DO WHILE statement.

5.3.2

Example Program—
DO WHILE

Most of the details of CH5EX2 (shown in Fig. 5–4) that you need to pay special attention to have been pointed out in the preceding section. You should note, however, that Statements 13, 20, 22, 26, and 30 each start on one line and are continued on the next. Most PL/I compilers stop processing a statement when column 73 is reached (that is, statements must be found in columns 2–72 inclusive).

In the following example assume that PUT starts in column 2 and that the right-hand margin of the text is column 72. In that case the continuation of a literal looks like this:

```
PUT SKIP (3) LIST (' ', 'NOW IS THE TIME FOR ALL GOOD MEN TO CO
ME TO THE AID OF THE HARRIED PROGRAMMER');
```

If it is necessary (as it often is) to continue a statement within the data list, you will want to end the line with the comma following a variable or literal. It is not necessary to end the line with the comma, but doing so improves readability. If a literal is continued on a second line, however, one of two things is likely to happen. It might be that your compiler will flag the continuation as an error, or it might insert blanks in the literal to represent any blank columns between the column with which you discontinued your literal and column 72, and on the next line any blank columns between column 2 and the column in which you continue your literal. Clearly, the safest course is to break the line following a comma that separates data items.

You have also no doubt noticed the use of labels on each DO and DO WHILE statement. This is the technique mentioned in the previous section that we recommend for reducing the confusion over which END corresponds with which DO or DO WHILE. Not only does the technique reduce the confusion for the programmer, but also it makes it considerably easier for someone else to determine quickly the range of a particular DO or DO WHILE. Unlike the label on the PROCEDURE statement, which must consist of seven or fewer characters, the label on other PL/I statements can be as many as 31 characters long. Looking at Statements 24, 28, and 29 (the END statements) of CH5EX2, you notice the syntax is END label;, just as it is for the END statement that corresponds to the PROCEDURE statement with which the program begins. The use of labels, however, does not change the basic rule that an END statement will be taken by PL/I to correspond with the closest previous DO or DO WHILE that has not already been assigned an END statement.

What we recommend, then, and what we practice in our examples is the following:

1. Columns 2–10 are reserved for labels and the required trailing colon.

2. Column 11 is left blank.

3. Statements begin in column 12, except for those that are indented to show continuation or relationships.

4. If you choose a label longer than eight characters, begin the DO or DO WHILE in column 12 of the following line, again unless it is indented to show relationships.

```
LABEL_TOO_LONG:
          DO WHILE (. . .);
                   .
                   .
                   .
          END LABEL_TOO_LONG;
```

The CH5EX2 pseudocode is as follows:

```
CH5EX2
    Print page, column headers
    Read number of records
    DO WHILE (record count < number of records)
        Read a record
        Record count ← record count + 1
        Final average ← (test average + program average) / 2.0
        Print name, averages
```

```
                     IF (test average < 68.0 OR program average < 68.0 OR final
                         average < 68.0)
                        Print 'Fail' message
                        Fail count ← fail count + 1
                     ELSE
                        Print Pass' message
                        Pass count ← pass count + 1
                     ENDIF
                  ENDDO
                  Print label, pass total
                  Print label, fail total
               STOP
```

PL/I OPTIMIZING COMPILER VERSION 1 RELEASE 5.1 TIME: 15.43.15 DATE: 12 FEB 87 PAGE   1
OPTIONS SPECIFIED
GOSTMT;

PL/I OPTIMIZING COMPILER   /* CH5EX2 - USING DO WHILE */   PAGE   2
                   SOURCE LISTING
    STMT

          /* CH5EX2 - USING DO WHILE */

     1 CH5EX2:   PROCEDURE OPTIONS (MAIN);

        /**/
        /*   */
        /*   FUNCTION: THIS PROGRAM PRODUCES A REPORT ON THE STUDENTS' */
        /*   STATUS IN A SINGLE CLASS   */
        /*   */
        /*   INPUT: 1. A HEADER RECORD CONTAINING THE NUMBER OF   */
        /*   STUDENTS IN THE CLASS   */
        /*   2. FOR EACH STUDENT A RECORD CONTAINING THE   */
        /*   STUDENT'S NAME, TEST AVERAGE, AND PROGRAM   */
        /*   AVERAGE   */
        /*   */
        /*   OUTPUT: THE INPUT AND AN APPROPRIATE MESSAGE   */
        /*   */

FIGURE 5–4
CH5EX2 (the DO WHILE statement).

```
/*   NOTES:    1.  THE MINIMUM PASSING GRADE IS 68%           */
/*             2.  THE STUDENT MUST HAVE A PASSING AVERAGE ON  */
/*                 BOTH THE EXAMS AND THE PROGRAMS IN ORDER TO */
/*                 PASS THE COURSE                             */
/*                                                             */
/***************************************************************/

2     DCL STUDENT_NAME    CHARACTER    (20);

3     DCL TEST_AVG        FIXED DECIMAL (3);
4     DCL PGM_AVG         FIXED DECIMAL (3);
5     DCL FINAL_AVG       FIXED DECIMAL (5,2);

6     DCL NUM_RECS        FIXED BINARY  (15);
7     DCL REC_COUNT       FIXED BINARY  (15)      INIT (1);
8     DCL PASS_COUNT      FIXED BINARY  (15)      INIT (0);
9     DCL FAIL_COUNT      FIXED BINARY  (15)      INIT (0);

10    DCL ROUND           BUILTIN;

11          PUT LIST (' ', 'GRADE REPORT', ' ', 'SPRING 1987');
12          PUT SKIP (2) LIST (' ', ' ', 'COMPUTER SCIENCE 101');

13          PUT SKIP (3) LIST ('STUDENT NAME', 'TEST AVERAGE',
                'PROGRAM AVERAGE',  'FINAL AVERAGE', 'MESSAGE');
14          PUT SKIP;

15          GET LIST (NUM_RECS);
16 LOOP:    DO WHILE (REC_COUNT <= NUM_RECS);
17              GET LIST (STUDENT_NAME, TEST_AVG, PGM_AVG);
18              REC_COUNT = REC_COUNT + 1;
19              FINAL_AVG = ROUND ((TEST_AVG + PGM_AVG) / 2.0, 2);
20              PUT SKIP LIST (STUDENT_NAME, TEST_AVG, PGM_AVG,
```

```
                        FINAL_AVG);

21              IF TEST_AVG < 68.0 ; PGM_AVG < 68.0 ; FINAL_AVG < 68.0
                    THEN
    COND1:          DO;
```

FIGURE 5–4 *(continued)*

```
22                      PUT SKIP (0) LIST (' ',' ',' ',' ',
                                        'FAILED CSCI 101');
23                      FAIL_COUNT = FAIL_COUNT + 1;
24                    END COND1;

25              ELSE
    COND2:        DO;
26                    PUT SKIP (0) LIST (' ',' ',' ',' ',
                                   'PASSED CSCI 101');
27                    PASS_COUNT = PASS_COUNT + 1;
28                  END COND2;

29            END LOOP;

30          PUT SKIP (3) LIST (' ',' ','NUMBER PASSING IS',
                           PASS_COUNT);
31          PUT SKIP LIST (' ',' ','NUMBER FAILING IS', FAIL_COUNT);
32        END CH5EX2;
```

```
PL/I OPTIMIZING COMPILER          /* CHAPTER 5 EXAMPLE PROGRAM 2 USING DO WHILE */              PAGE   4
COMPILER DIAGNOSTIC MESSAGES
ERROR ID L   STMT    MESSAGE DESCRIPTION

COMPILER INFORMATORY MESSAGES

IEL0533I I        NO 'DECLARE' STATEMENT(S) FOR 'SYSPRINT','SYSIN'.

END OF COMPILER DIAGNOSTIC MESSAGES
```

```
                    GRADE REPORT                          SPRING 1987

                              COMPUTER SCIENCE 101
```

STUDENT NAME	TEST AVERAGE	PROGRAM AVERAGE	FINAL AVERAGE	MESSAGE
BAKER, BOB	93	84	88.50	PASSED CSCI 101
CARLSON, CATHY	65	80	72.50	FAILED CSCI 101
DEVON, DEBBIE	81	84	82.50	PASSED CSCI 101

Note: This printout has been reduced slightly for design purposes.)

FARGO, CARL	79	61	70.00	FAILED CSCI 101
JAMES, JENNY	86	95	90.50	PASSED CSCI 101
LLOYD, LINDA	77	74	75.50	PASSED CSCI 101
MASON, TERRY	95	98	96.50	PASSED CSCI 101
OLSEN, BEVERLY	66	50	58.00	FAILED CSCI 101
PEARSON, PAULA	68	68	68.00	PASSED CSCI 101
WILSON, BILL	97	92	94.50	PASSED CSCI 101

```
                        NUMBER PASSING IS        7
                        NUMBER FAILING IS        3
```

5.4

Flags and Switches

As you have seen, the computer makes its decisions based on a logical expression that can evaluate only to true or false. The proposition formed by the logical expression can also be stated in the form of a question, in which case the only possible answer is either yes or no. At times it might appear as if the computer is responding to a proposition with multiple possible evaluations, but when the actions actually taken by the computer are examined in detail, it is discovered that the computer breaks these seemingly complex decisions down into a series of propositions, each of which evaluates to true or false. Consider, for example, the logical expression $W < X$ & $Y > Z$, assuming that both parts of the expression are true. $W < X$ is evaluated, found to be true, and that result is saved. $Y > Z$ is evaluated, found to be true, and the result ANDed with the first result. The results of this comparison is also found to be true, and the entire expression is then considered true. The point is that the computer didn't ever work with a proposition having more than two possible answers.

A variable designed to contain one of only two possible values, that value set by the evaluation of a condition, is called a **flag** or **switch**; the terms are interchangeable. A flag or switch value can be declared in any way you please. You might, for example, declare a FIXED DECIMAL (9) variable and use nine zeros (or any other nine-digit value) to represent a condition evaluating to true and nine ones (or other value) to represent a condition evaluating to false; to do so, though, would be rather extravagant. There is a better way.

5.4.1

BIT Variables as Flags and Switches

Programmers regularly make use of the binary nature of a computer's operation by declaring variables that will contain one of only two possible values. You will recall from our discussion of computer memory that the smallest unit capable of holding data is the bit; you will also recall that the two possible states a bit can be in are represented by a 0 and a 1. The bit, therefore, is ideal for use as a flag or switch, and PL/I allows you to declare a variable to be a bit; for example,

```
DECLARE END_FLAG     BIT (1);
```

It is quite common to initialize a variable with the BIT attribute to the value representing a true condition. PL/I assumes a binary 0 to represent the false condition and a binary 1 to represent the true; therefore, it is more efficient to follow this pattern. A declaration of END_FLAG with the INITIAL attribute is coded as follows:

```
DECLARE END_FLAG     BIT (1)    INIT ('1'B);
```

You have already become familiar with decimal numeric literals and character literals (though not in the INITIAL attribute). Bit literals (also called **bit strings**) are designated by placing the binary value within single quotation marks and following the closing quotation mark with a B. In case you have forgotten, one difference between a binary and a decimal 0 is that the binary 0 can be represented by a single bit, while it requires four or eight bits to represent a decimal 0, depending on its internal representation. Finally, you will gain a clearer understanding of the use of flags and switches as you proceed to the end of the chapter.

5.4.2

Conditions and the ON Statement

During the execution of a program certain conditions can arise that make it impossible for processing to continue. When such a situation arises, a **program interrupt** occurs. Two possibilities exist for handling program interrupts. First, if the programmer has not provided instructions for the contingency, PL/I will direct the computer to stop processing, write an error message, and return control to the computer's operating system. These three steps (and others not detailed here) constitute the abend we spoke of earlier. The second possibility is for the programmer to provide the necessary instructions through the ON statement.

The syntax of the ON statement is

```
ON condition imperative;
```

where ON condition is a PL/I keyword such as ENDFILE, ENDPAGE, CON-
VERSION, or SIZE. The imperative tells the computer what to do when the
error condition arises; often this direction is to change the value in a flag or
switch, which in turn will control the order of the processing of the PL/I
code. When an IF or an ELSE is followed by more than a single imperative,
you will recall, a do-group is required. By the same token, if you want to
follow the ON condition with several statements you must code a BEGIN
block, which is described in the following section (5.4.3). In that section we
also discuss ENDFILE , which is the only condition we are concerned with at
this time; a more complete discussion of conditions is presented in Section
6.5.1 of Chapter 6.

5.4.3

GET LIST
Using ENDFILE

One program interrupt condition that arises in almost any program is the
end-of-file condition. End-of-file occurs when, after the last record in the
input file has been read, another GET is encountered in the program code.
The GET causes the computer to *attempt* to read another record; it cannot do
so, of course, and so not knowing what else to do, it ceases processing,
writes an error message, and returns control to the operating system.

In the example programs presented thus far, we have prevented the
end-of-file condition from arising by two closely related techniques, each of
which presumes an intimate knowledge of the input data. The first technique involves the sequence structure, wherein we simply code enough
consecutive GET statements to read all the input data, and no more. To use
this technique we obviously must be aware of exactly how many data items
(or records) appear in the input stream. The second technique gives us more
flexibility because we use the iteration structure with a header to the input
telling us how many data records are to follow in this particular run. We
have now removed the straitjacket from the input data, but we still must
know the exact number of data records for the run. Imagine now a large
credit card operation, in which dozens of employees all day long have been
creating input records consisting of purchases made by customers. Who is
going to count these records and how long will it take? Clearly, what is
needed is a facility whereby the programmer can tell the computer what to
do when end-of-file occurs. In other words, our previous techniques were
designed to prevent the end-of-file condition from arising. Now we want to
let it arise and still remain in control when it does, and we do so with the ON
ENDFILE statement.

The syntax of the ON ENDFILE statement is

```
ON ENDFILE (file name) imperative;
```

File name, of course, is the name of the file on which the ENDFILE condition will occur, that is, an input file. In terms of the default file names of SYSIN for the input file and SYSPRINT for the output file, the only file on which ENDFILE can occur is SYSIN. Therefore, for the time being we can specify the statement as

```
ON ENDFILE (SYSIN) imperative;
```

The imperative is most commonly used to assign a value to a flag; this flag, in turn, is used to control the iteration of the loop.

```
DECLARE MORE_RECS BIT (1) INIT ('1'B);
ON ENDFILE (SYSIN) MORE_RECS = '0'B;
DO WHILE (MORE_RECS = '1'B);
```

When the ON ENDFILE statement executes, MORE_RECS is assigned the value of '0'B, and the DO WHILE condition is no longer true. To make the program more readable, however, programmers often initialize three bit-variables, instead of the one variable used in the example above. The next example program, CH5EX3, for instance, contains the following among its DECLARE statements:

```
DCL   MORE_RECS    BIT (1);
DCL      YES       BIT (1)   INIT ('1'B);
DCL      NO        BIT (1)   INIT ('0'B);
```

(YES and NO are indented, by the way, to show that they represent the two possible states of MORE_RECS.)

Later in the program we find these statements:

```
        MORE_RECS = YES;
        ON ENDFILE (SYSIN) MORE_RECS = NO;

              .
              .

              .
        GET LIST (STUDENT_ID, TEST_AVG, PGM_AVG);
LOOP:   DO WHILE (MORE_RECS):
              .
              .

              .
        GET LIST (STUDENT_ID, TEST_AVG, PGM_AVG);
        END LOOP;
```

When the DO WHILE is first encountered in execution, the value in MORE_RECS is equal to the value in YES. When end-of-file occurs, the ON ENDFILE statement is executed, and control returns to the statement following the one which raised the condition (that is, in this case ON ENDFILE is raised by the GET—Statement 30—and control returns to the END at the bottom of the loop). The END statement returns control to the DO WHILE; now MORE-RECS has the value of '0'B, having received it from NO. The DO WHILE condition is no longer true, and control passes to the statement following the end of the loop. As long as you provide variable names that suggest in and of themselves the status of the file, this technique of declaring three bit-variables makes your program considerably more readable than is the case with the use of the binary 0 and 1.

With any of the conditions, what we have presented here is adequate so long as the ON condition is followed by only a single imperative. It is not uncommon, however, that you want to follow the ON condition with several imperatives; the begin block facility allows for this situation. A **begin block** is a series of one or more PL/I statements functioning as an ON-unit and delimited at the beginning by the keyword BEGIN and at the end by the keyword END; it is analogous to the do-group that can follow an IF or an ELSE. Here is an example of a begin block.

```
                ON ENDFILE (SYSIN)
    B_BLOCK:        BEGIN;
                    PUT SKIP (3) LIST (' ', ' ',
                        'NUMBER PASSING IS', PASS_COUNT);
                    PUT SKIP LIST (' ', ' ',
                        'NUMBER FAILING IS', FAIL_COUNT);
                    END B_BLOCK;
```

You should, as usual, pay careful attention to the placement of the semicolons in the example. Also, since CH5EX3 does not use a begin block, it might be instructive to compare this example with Statements 13, 32, and 33 from that program; they carry out the same tasks as those in the BEGIN block do in the example above.

A final consideration of the ON ENDFILE statement is its placement within the program. It should be placed before the first GET. It is always possible that, through some accident or oversight, the input file contains no data. The first GET will then raise the ENDFILE condition. If the ON ENDFILE statement has not been encountered prior to the execution of the GET, the result of the ENDFILE being raised is an abend.

**Example Program—
ON ENDFILE**

One aspect of using the ON ENDFILE statement remains to be discussed: its impact on the logic of the program. To illustrate this impact we have chosen to repeat CH5EX2 in CH5EX3, adding ON ENDFILE and making no other changes except those dictated by its use. You can therefore see what changes in logic are rendered necessary by ON ENDFILE through a comparison of the two programs.

The first difference in the two programs is in the DECLARE statement. Whereas CH5EX2 must declare a variable (NUM_RECS) to receive the value from the header, and a second variable (REC_COUNT) to be incremented, CH5EX3 (see Fig. 5–5) requires neither variable because both are replaced by the ON ENDFILE statement. CH5EX3, on the other hand, requires at least one flag variable; we have chosen to code three (MORE_RECS, YES, and NO) for the reasons we explained in Section 5.4.3. A second difference, of course, is that CH5EX3 contains an ON ENDFILE statement (Statement 13) which is not found in CH5EX2. The condition on the DO WHILE is different in each program, also, and CH5EX2 must increment REC_COUNT (Statement 18) within the loop.

By far the most significant difference between the two programs, however, is the placement of the GET statements. CH5EX2 has one GET (Statement 15) to read in the value on the header record; it then has a single GET (Statement 17) to read in the data records, and it is the *first* statement in the loop. CH5EX3, on the contrary, has two GET statements to read in data records (remember, with ON ENDFILE the header record is no longer necessary and has been removed). The first of these (Statement 18) is outside the loop, while the second (Statement 30) is the *last* statement in the loop. If you think about the logic of each program for a moment, perhaps you will determine why there is a discrepancy in the placement of the GET statements.

The clue to the puzzle lies in the fact, mentioned in the preceding section (5.4.3), that in CH5EX2 we want to *prevent* end-of-file from occurring, whereas in CH5EX3 we want to *allow* end-of-file to occur. Notice carefully what happens inside the CH5EX2 loop: we read a record, process it, then loop back to read another record. In conjunction with this, recall that once a loop has been entered the computer must attempt to execute each statement within the loop. As soon as a record is read, REC_COUNT is incremented, and it is REC_COUNT that controls the iterations of the loop. When the last record is read, REC_COUNT is incremented, and the record is processed. Control then passes to the DO WHILE; its condition is found to be false, and control falls through the loop. In other words, it is REC_COUNT that prevents end-of-file from occurring.

Now consider the quite different situation of CH5EX3. The first record is read outside the loop; the placement of this **initial read** serves two purposes:

1. If the file contains no data, this GET will trigger the ON ENDFILE statement and the loop will not be entered.

2. Within the loop it allows us to process a record, then read another.

Within its loop CH5EX3 processes, then reads. What is so significant, you ask, about this reversal of order? Consider that in CH5EX3 the iterations of the loop are being controlled by the ON ENDFILE statement. Remember also that ON ENDFILE executes when an attempt is made to read a record which does not exist in the file. Suppose for a moment that the initial GET in CH5EX3 is omitted and that the GET in the loop is the first statement in the loop. Everything is fine until there is an attempt to read a record that isn't there; at this point the ON ENDFILE is executed; control then passes to the statement following the GET. In short, since the loop has already been entered, the computer must attempt to process that record which doesn't exist.

When CH5EX3 was modified by removing the initial GET and placing the remaining GET as the first statement in the loop, this is what happened. The computer attempted to read a record; since no record was available, ON ENDFILE was executed, and control returned to the statement following the GET. Also since no record was read, the values in STUDENT_ID, TEST_ AVG, and PGM-AVG did not change. Processing continued to a normal termination. A careless programmer (one who does not check the output carefully) would assume that everything was fine. The programmer who does check the output carefully would discover, however, that student 14235 appears twice in the output and that the sum of PASS_COUNT and FAIL_COUNT equals 11 instead of the expected 10. When we are attempting to prevent end-of-file with faulty logic, end-of-file occurs; the computer writes an error message, and returns control to the operating system. Thus, we have only to look for an error message in the output to recognize that something is wrong. When ON ENDFILE is used, however, its imperative is executed, and processing continues. Since we want end-of-file to occur, we do not direct the computer to write an error message. Instead we check our output very carefully. You need to understand that it is standard practice in programming to test a new program with known data so that the output can be readily checked against the correct results.

Let us now contrast the sequence of events just outlined with those that take place within the loop when the initial GET is coded, and the GET

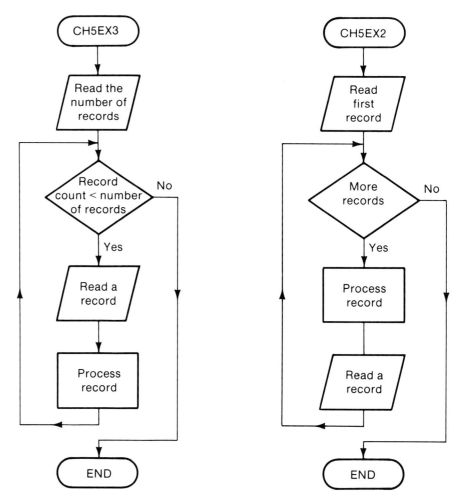

FIGURE 5–6
The DO WHILE vs. the ON ENDFILE.

within the loop is the last statement in the loop. Now a record is processed and another is read. Eventually the GET tries to read a nonexistent record, the ON ENDFILE statement is executed, control returns to the statement following the GET, which is END LOOP; the END statement passes control to the DO WHILE, the condition is checked, found to be false, and control passes to the statement following the END statement (Statement 32). In this scenario there has been no attempt to process a nonexistent record.

The moral should be perfectly clear by now. When ON ENDFILE is controlling the iterations of the loop, you must code an initial GET outside the loop. The second GET must be the last statement in the loop. When iterations of the loop are controlled by a header value, however, you must not code an initial read, and the GET must be the first statement within the loop; otherwise, you will either fail to process the final record, or you will raise the end-of-file condition and abend.

The two flowcharts shown in Figure 5–6 illustrate this point. Note that they are "skeleton" flowcharts in that they deal only with the portion of CH5EX2 and CH5EX3 we have been discussing. Also, in each case the process symbol marked "Process record" might require more than one symbol in a fully developed flowchart. The flowchart on the right is controlled by ON ENDFILE, whereas the one on the left is controlled by the header value.

The pseudocode for CH5EX3 is as follows:

```
CH5EX3
    Print page, column headers
    Read first record
    DO WHILE (more records)
        Final average ← (test average + program average) / 2.0
        Print a detail line
        IF (test average OR program average OR final average
            < 68.0)
            Print 'Fail' message
            Fail count ← fail count + 1
        ELSE
            Print 'Pass' message
            Pass count ← pass count + 1
        ENDIF
        Read next record
    ENDDO
    Print label, pass count
    Print label, fail count
STOP
```

PL/I OPTIMIZING COMPILER VERSION 1 RELEASE 5.1 TIME: 15.44.20 DATE: 12 FEB 87 PAGE   1
OPTIONS SPECIFIED
60STMT;

FIGURE 5–5
CH5EX3 (ON ENDFILE).

FIGURE 5–5 *(continued)*

```
PL/I OPTIMIZING COMPILER              /* CH5EX3 - USING ON ENDFILE */           PAGE   2
                  SOURCE LISTING
     STMT


              /* CH5EX3 - USING ON ENDFILE */


     1  CH5EX3:   PROCEDURE OPTIONS (MAIN);


        /*******************************************************************/
        /*                                                               */
        /*   FUNCTION:   THIS PROGRAM PRODUCES A REPORT ON THE STUDENTS'  */
        /*               STATUS IN A SINGLE CLASS                         */
        /*                                                               */
        /*   INPUT:   1.  A HEADER RECORD CONTAINING THE NUMBER OF        */
        /*               STUDENTS IN THE CLASS                           */
        /*            2.  FOR EACH STUDENT A RECORD CONTAINING THE        */
        /*               STUDENT'S IDENTIFICATION NUMBER, TEST            */
        /*                AVERAGE, AND PROGRAM AVERAGE                    */
        /*                                                               */
        /*   OUTUT:     THE INPUT AND AN APROPRIATE MESSAGE              */
        /*                                                               */
        /*   NOTES:   1.  THE MINIMUM PASSING GRADE IS 68%               */
        /*            2.  THE STUDENT MUST HAVE A PASSING AVERAGE ON      */
        /*               BOTH THE EXAMS AND THE PROGRAMS IN ORDER TO      */
        /*               PASS THE COURSE                                 */
        /*                                                               */
        /*******************************************************************/

     2        DCL     STUDENT_NAME    CHARACTER    (20);

     3        DCL     TEST_AVG        FIXED DECIMAL (3);
     4        DCL     PGM_AVG         FIXED DECIMAL (3);
     5        DCL     FINAL_AVG       FIXED DECIMAL (5,2);

     6        DCL     PASS_COUNT      FIXED BINARY  (15)    INIT (0);
     7        DCL     FAIL_COUNT      FIXED BINARY  (15)    INIT (0);

     8        DCL     MORE_RECS       BIT          (1);
     9        DCL     YES             BIT          (1)     INIT ('1'B);
    10        DCL     NO              BIT          (1)     INIT ('0'B);
```

```
11              DCL       ROUND       BUILTIN;

12          MORE_RECS = YES;
13          ON ENDFILE (SYSIN) MORE_RECS = NO;
14          PUT LIST (' ', 'GRADE REPORT', ' ', 'SPRING 1987');
15          PUT SKIP (2) LIST (' ', ' ', 'COMPUTER SCIENCE 101');

16          PUT SKIP (3) LIST ('NAME', 'TEST AVERAGE',
                 'PROGRAM AVERAGE',  'FINAL AVERAGE', 'MESSAGE');

17          PUT SKIP;

18          GET LIST (STUDENT_NAME, TEST_AVG, PGM_AVG);

19 LOOP:    DO WHILE (MORE_RECS);
```

```
  STMT
   20              FINAL_AVG = ROUND ((TEST_AVG + PGM_AVG) / 2.0, 2);
   21              PUT SKIP LIST (STUDENT_NAME, TEST_AVG, PGM_AVG,
                              FINAL_AVG);

   22              IF TEST_AVG < 68.0 ! PGM_AVG < 68.0 ! FINAL_AVG < 68.0
                   THEN
      COND1:          DO;
   23                     PUT SKIP (0) LIST (' ', ' ', ' ', ' ',
                                               'FAILED CSCI 101');
   24                     FAIL_COUNT = FAIL_COUNT + 1;
   25                  END COND1;

   26              ELSE
      COND2:          DO;
   27                     PUT SKIP (0) LIST (' ', ' ', ' ', ' ',
                                     'PASSED CSCI 101');
   28                     PASS_COUNT = PASS_COUNT + 1;
   29                  END COND2;

   30              GET LIST (STUDENT_NAME, TEST_AVG, PGM_AVG);
   31          END LOOP;

   32          PUT SKIP (3) LIST (' ', ' ', 'NUMBER PASSING IS',
                              PASS_COUNT);
```

FIGURE 5–5 *(continued)*

```
33            PUT SKIP LIST (' ', ' ', 'NUMBER FAILING IS', FAIL_COUNT);
34          END CH5EX3;


PL/I OPTIMIZING COMPILER              /* CH5EX3 - USING ON ENDFILE */         PAGE   4
COMPILER DIAGNOSTIC MESSAGES
ERROR ID L   STMT    MESSAGE DESCRIPTION


COMPILER INFORMATORY MESSAGES

IEL0533I I          NO 'DECLARE' STATEMENT(S) FOR 'SYSIN','SYSPRINT'.


END OF COMPILER DIAGNOSTIC MESSAGES
```

5.6

Summary

You have now learned to implement each of the three structures allowed to structured programming. The sequence structure is simplest, of course, consisting only of a series of statements to be executed in chronological order. The decision structures are implemented by the IF-THEN and IF-THEN-ELSE statements, with or without do-groups following either or both. The iteration structure is implemented by the DO WHILE statement, which, like a do-group, uses an END statement as the ending delimiter.

The ON statement can, in certain cases, allow the programmer to retain control when a program interrupt occurs, particularly when the interrupt is caused by end-of-file. In most cases it is much more efficient to allow end-of-file to occur and retain control than it is to prevent the occurrence of end-of-file. One common method for retaining control when a program interrupt occurs is to declare and initialize a variable with the BIT attribute and have the program interrupt change the value in the variable (called a flag or switch). If the ON condition is followed by more than one statement, a begin block, delimited by BEGIN and END, must be used.

The logic of a program that uses ON ENDFILE to control the iterations of the loop differs significantly from one whose loop iterations are controlled by a counter because in the first case we are allowing end-of-file, and in the second we are preventing it. The first must contain a GET before the loop and another at the bottom of the loop; the second must contain a GET at the top of the loop.

Although there is much to be added to your stock of PL/I knowledge, you are now in a position to write programs of considerable sophistication. Therefore, it is time to pause and consider how we might do more than just get correct results from our programs, how we can make them easily readable and modifiable, as well as accurate and reliable. This is the subject of Chapter 6.

5.7

Exercises

1. Which of the following logical expressions evaluate to true and which to false? Given W = 20; X = 5; Y = 2; Z = 0.
   a. W = 85 OR X = 5 OR Z = -9
   b. (X + Y) = 7 AND X < Z AND W NOT = Z
   c. Z = W OR Z = 0 AND 2 = 2 OR X = (W - 5)
   d. 20 < = 18 AND Y = Z OR X NOT < Y AND (Z · W) = 0

2. Code the logical expressions from Exercise 1 into PL/I code. (Use the IF portion of the IF-THEN statement.)

3. Show what will print in each of the following:
   a. ```
       A = 4;
       DO WHILE (A >= 0);
           PUT LIST (A);
           A = A - 1;
       END;
       ```
   b.  ```
       B = 0;
       DO WHILE (A < 12);
           PUT LIST (A);
           A = A + 3);
       END;
       ```
   c. ```
       C = 0; Q = 10;
       DO WHILE (C < Q OR Q > 6);
           PUT LIST (A, Q);
           C = C + 2;
           Q = Q - 1;
       END;
       ```
   d.  ```
       D = 20; Q = 1;
       DO WHILE (D > Q AND Q < 10);
           PUT LIST (D, Q);
           D = D - 5;
           Q = Q + 2;
       END;
       ```

4. Write the pseudocode and a PL/I program to read a file of integers, the first of which is a header that tells how many follow. Calculate (a) the sum of the odd

integers and (b) the sum of the even integers. Print the two sums when you are finished. (*Hint:* Use a DO WHILE loop.)

5. a. Write the pseudocode and a PL/1 program that will
   (1) Read an unknown number of three-digit integers.
   (2) Print the numbers (one per line).
   (3) Print the smallest and largest values (labeled).
   (4) Print an error message if the largest value is more than 100 greater than the smallest value.
   b. Redo part a using a header count record to determine how many numbers are to be read.

6. Write the pseudocode and a PL/1 program to read a set of records, each containing the following:

```
PART_NAME      CHARACTER (20)
PART_NUMBER    FIXED DEC (2)
PRICE          FIXED DEC (5,2)
PART_QUANTITY  FIXED DEC (3)
```

Compute the total cost for each item ordered, and print the input and total cost. Include column headers, and check for the possibility of a "no data" error.

7. Write the pseudocode and a program to analyze and report on donations to the Red Cross. Keep a running total of donations of less than $100 and a separate running total for donations of $100 or more. The input will consist of a donor name (20 characters) and a donation (nnnn.nn). Print each input record, the total of donations under $100, the total of donations of $100 or more, and the total of all donations. The three total lines should be appropriately labeled.

8. Write the pseudocode and a program to determine and report on running yards versus passing yards through the first three games of the NFL season. The input consists of the team name and for each team its running and passing yardage for each of three games. Report the input for each team and the difference between running and passing yardage (i.e., subtract the lesser from the greater). Also report the total of each type of yardage for all teams and the difference in yards between the two.

CHAPTER 6

PROGRAM FORMAT;

SUBROUTINES;

DEBUGGING

Up to this point you have been concerned with very little except writing a program that produces the desired results without producing any error messages. If you think only in terms of the classroom situation, you need few goals beyond the one of obtaining the desired results because your program is executed only once after it is debugged. The assumption of this book, however, is that you are studying PL/I with the aim of becoming a professional programmer, or, at the very least, writing for your own use programs that will have a life span of months or years. In either of these cases the requirements of the program change quite drastically.

Two facts regarding the vast majority of programs written by professional programmers can be stated as follows:

1. They have a life span of up to 20 years, while nationally a programmer remains on his or her first job for about two years.

2. **Systems** (a large number of interacting programs) are the norm in industry.

The first fact implies that someone other than the programmer who wrote the original code will have to maintain it for up to 18 years of its 20-year life span. As for the second fact, systems are usually written by teams, with each member of the team working on one or more programs that make up the system. Obviously, since they interact, all of these programs must fit together to form a coherent whole, with no contradictions between any two programs. It is clear, then, that requirements of these programs are more than just producing the desired results when run by themselves.

For those who intend to write programs for their own use only, one major consideration applies: It is amazing how much you can forget over a period of a few weeks about the details of a program that gave you so much trouble in debugging that you were convinced its details were permanently etched into your brain. The purpose of this chapter, then, is to suggest some programming goals beyond simply getting the desired results. To meet these goals we suggest some programming techniques, some features of structured programming, subroutine procedures, and some debugging aids.

6.1

Programming Techniques

In the following sections we discuss some of the techniques that have been devised over the years to aid in the development of successful programs. Historically, computer time has cost more than programmer time, and during this period programmers spent a good deal of time creating programs that would solve the problem in the fewest lines of code, use up the least amount of the computer's central processing unit time (i.e., run most efficiently), and require the least amount of storage. These programs were often marvels of efficiency, as long as "efficiency" was measured according to the three criteria just listed, but the programs were nearly impossible to maintain by anyone other than their author. As mainframe computers have become faster (at the time of this writing some computers are capable of executing two billion instructions per second) and cheaper, programmer time has come to be more expensive than central processing unit time. Consequently, the goal of "efficiency" has given way to "cost efficiency," the amount of programmer time required to develop a new system and the amount of time required to maintain it over its life span. This chapter, then, is written to reflect this relatively new state of affairs.

A "cost efficient" program today, given the current state of computer programming in a "real world" situation, must meet four goals: accuracy, reliability, readability, and maintainability. We discuss each of these in turn, attempting to show you the importance of each. We then (in the following sections) suggest some techniques for accomplishing these goals. You will discover, moreover, that the four goals are closely interrelated, as are, though to a lesser extent, the techniques for achieving the goals.

• ACCURACY.   By accuracy, first of all, we mean that the program must produce the desired results. To give a trivial example, a program is of no value whatsoever if it is expected to add 5 to 7 and if it then produces a result of 13. Accuracy, as we use the term here, goes well beyond simply achieving the correct result from an arithmetic operation. A program, for example, might produce totally accurate results in terms of what the programmer *thinks* is the desired result, only to learn that the **end-user** (the person, for example, for whom the report is being produced) doesn't really want that information at all. Such a situation can arise from a lack of communication between user and programmer; it might result from the user not really knowing initially what she or he does want, or it might result from the programmer's failure to analyze the problem carefully. In any case, it is a problem of analysis. Analysis, of course, is not the thrust of this book, but at the same time no programmer is totally free of its demands.

Even though a program produces exactly the data that is desired, it cannot be said to be accurate unless it is formatted on the printed page exactly as desired. In some cases you are expected simply to produce an attractive and easily readable report. In other cases you will be given a very rigid set of specifications for the output format. In such instances your program is not accurate unless these specifications are met *exactly*. One way to understand the importance of rigid formatting specifications is to assume that the output will be entered on a preprinted form (a program to write checks, for example). We also believe that no programming course is complete unless the student is given at least one assignment in which rigid and fairly complex formatting specifications are supplied, but this cannot occur until edit-directed output has been introduced.

• RELIABILITY.   Reliability might be thought of as accuracy extended over the lifetime of the program. It is axiomatic in computer programming that one can never say with complete assurance that any nontrivial program is ever completely debugged. Since the combinations of data are almost endless, it is impossible to test the program completely. The likelihood of achieving reliability, however, is enhanced first of all by top-down development (introduced in Section 6.2) and the closely related modular programming techniques. You will learn modular programming in Section 6.2 as you study subroutine procedures and eventually external procedures (presented in Chapter 10).

Because a program cannot be fully tested is no reason not to test it at all, however. Testing is, as a matter of fact, an extremely important phase in the cycle of program development. Thoroughly tested programs in fact often live out their lifetimes without ever encountering that once-in-a-billion combination of data that it is unprepared to handle. In an ideal testing situation, you would provide data that would cause each flow-of-execution path to be taken; supply data that would cause each of the IF-THEN paths to be taken; and supply data (this would require more than one testing run) that would thoroughly test the condition of all DO WHILE loops, including their never being entered at all. This ideal can best be approached if each program is tested when it is written. The programs you have written thus far have required very little testing. In Chapter 7, however, we introduce nested IF-THEN and DO WHILE statements; they increase the complexity of the programs dramatically, and correspondingly increase the complexity of the testing required.

• READABILITY.   The goal of producing a program (i.e., a listing of the source code, not the output produced by the program) that is easily readable perhaps could be included with the final goal of maintainability because it is obvious that if a program is difficult to read, it will be difficult to maintain. We have chosen to make it a separate goal, however, because it is not uncommon that a program is read for reasons other than maintaining it. The following three sections provide our detailed recommendations for creating readable code. For now, we simply list some of the qualities of a program that make it readable.

1. You should be able to obtain a sense of the function of a variable from its name.

2. You should be able to locate quickly the declaration of any variable and be able to determine its attributes at a glance.

3. You should be able to follow the logic of the program easily; blocks of code that perform specific functions should be separated from surrounding lines by one or more blank lines. You should also be able to determine at a glance which lines or blocks of code are dependent upon

a condition (either IF-THEN-ELSE or DO WHILE) and which lines are continuations of preceding lines.

We have already commented on how these qualities can be achieved, and we will have more to say about them, but one feature that is absolutely essential to creating a readable program is *consistency* in coding.

• MAINTAINABILITY. We have used the terms "maintain," "maintenance," and "maintainability" without giving you a clear sense of what we mean by the concept. In addition to correcting bugs that arise, maintenance is the process of changing the code of a program to reflect the changes that are bound to occur over the life of the program. One example of the changes that might require maintenance is if the person using the output decides that he or she wants a different format and/or more, less, or different information included in the output. A second example is this: The United States Postal Service is currently encouraging businesses to switch from a five-digit zip code to a nine-digit (plus hyphen) zip code. Any company deciding to make the switch will have to consider the cost of updating (i.e., maintaining) all its programs that include zip codes. A third example of why maintenance is required is that the installation of new hardware might require modifications to existing programs.

Maintainability originates in a well-thought-out program design and is enhanced by writing structured code that meets the test for readability. Another technique to make maintenance easier is to avoid using literals (including numeric constants) in the program, even though we do not rule them out entirely. Three factors might be considered in determining whether or not to hard-code (use a literal).

1. How likely is it that the literal value will change?

2. In how many lines of code does the literal appear?

3. How obvious is the significance of the literal in its context?

The number of days in a week, for example, seems unlikely to change in the next 20 years, and if the value 7 appears in only one or two lines of code, it might just as well be hard-coded if the context makes clear what the 7 stands for. DAYS_IN_WEEK (declared with an INITIAL attribute of 7), however, will more certainly make the significance clearer than will 7. If the value is found in a relatively large number of lines of code, it becomes a more certain candidate for avoiding hard-coding. If the literal value is likely to change, it is foolish to hard-code it, and if you answer the three questions above with "very likely," "many lines," and "not very obvious," you have definitely slipped into the realm of poor programming practice if you choose to hard-code it with a literal. Finally, keep in mind that if you have used good practices in your declaration, the initialized variable will be easier to find there than in any one line of code within the program.

Many professionals consider scandalous the national ratio of maintenance time to development time. Good programming practices in the development stage can significantly reduce the maintenance time.

6.1.2

The DECLARE Format

We have already suggested some techniques for coding the DECLARE statement to make it easy to locate variables. We review them here and go on to suggest others.

1. Use meaningful names for all variables. A "meaningful" name is one which suggests the function or significance of the variable. Surely STUDENT_ID accomplishes this much more nearly than X. Because you often have to repeat a variable many times in the source code, you will find yourself deciding to "over abbreviate" variables to save typing. Resist this natural impulse; again STUDENT_ID is much clearer than SID or S_ID.

2. Use one variable for one purpose. Often a variable will no longer be needed within a program. It is then tempting to use that same variable for another purpose in order to avoid coding a second variable. Such a practice proves confusing to anyone else, and can in fact confuse you in your attempts to debug the program.

3. Explicitly declare all variables and all of their attributes, even if the attributes you want are the defaults. For the beginner the default attributes are quite confusing, and you should not depend on the reader having them memorized. In other words, the list of attributes should serve the purpose of documentation, if nothing else.

4. Be consistent in the placement of the DECLARE statements within your program. You might recall that these statements can appear anywhere in your program after the PROCEDURE statement and before the final END statement. They should be, at a minimum, gathered together in one place, and the two favored places are immediately after the PROCEDURE statement and immediately preceding the final END statement. Many programmers prefer the second, but we believe the first position is better because it allows the reader to scan through the variables in the DECLARE before encountering them in the code. We do stress, however, that consistency is more important than the choice between the two locations.

5. List only one variable per line and start each variable name and each attribute in the same column on each succeeding line. All example programs illustrate this principle, and so you can check any of them for examples.

6. There is one more technique for declaring variables that we have not yet introduced. If handled properly it fulfills both the programmer's desire to reduce the amount of typing required and the reader's desire to be able to spot a declaration quickly. **Factoring** is the technique by which two or more variables sharing the same scale, base, and precision are grouped together within parentheses. Each variable within the parentheses has the attributes specified outside the parentheses. An example:

```
DECLARE (TOTAL_PGM_PTS,
         TOTAL_TEST_PTS,
         PGM_POINT_TOTAL,
         TEST_POINT_TOTAL)  FIXED DECIMAL (5);
```

Each of the variables listed will have the attributes specified following the closing parenthesis. So long as each variable is listed on a separate line, moreover, this technique is quite satisfactory, although you might wish to reconsider in light of our next point.

7. The question of the order in which the variables should be listed is one that has no single satisfactory answer. There are four schools of thought on the matter: the first is that the variables be listed in the order in which they are encountered in the program; the second is that they be listed in alphabetical order; the third is that the variables be grouped according to what they describe. All of the variables in the input record, for example, are grouped together. The fourth school of thought is that the variables be grouped together according to scale and base. The advantages of the first three methods are obvious. What attracts us to the fourth method, however, is that it allows comments to be inserted as a header identifying each group.

```
DECLARE
              /*** FIXED DECIMAL VARIABLES ***/
       TOTAL_PGM_POINTS   FIXED DECIMAL (5),
       TOTAL_TEST_POINTS  FIXED DECIMAL (5),
              /*** FIXED BINARY VARIABLES ***/
       SUBSCRIPT          FIXED BINARY  (15),
       INDEX              FIXED BINARY  (15);
```

Notice, too, that this is the only method that really allows for factoring. All example programs in the text make use of this method (though the comment is not necessarily included).

Let us remind you once more, however, that consistency is more important than the various choices of format.

*Indentation
and Spacing*

PL/I is, of course, a free-form computer language. This means that programs can be written in almost any format, the only restrictions (for most compilers) being that column 1 is reserved for a special purpose and that the compiler will not read past column 72. Several statements (delimited by the semicolon) can be placed on one line, or a single statement can be interrupted at almost any point and continued on another line. PL/I treats a large number of consecutive blank columns the same as it does a single blank column. It should be obvious, however, that coding no more than a single statement on a line makes a program much more readable. In addition to placing no more than one statement on a line, indentation should be used to show the lines of code that are subordinates of conditions (IF-THEN-ELSE or DO WHILE). We recommend that the IF and its corresponding ELSE begin in the same column. If either or both contain a DO, we recommend that its corresponding END begin in the same column as the DO, and we make the same recommendation in regard to the DO WHILE and its corresponding END. We also recommend the use of labels with the DO of either the do-group or the DO WHILE.

One question that might be occurring to you right now is how many columns to indent. We can only reply by suggesting that the indentation should be noticeable at a glance: this seems to rule out a one-column indentation. On the other hand, as you will quickly learn when you code several levels of nesting, through excessive indentation you can quickly get the left margin so far to the right that you have little room left on the line for code. In our examples we indent three columns to show logical relationships, but more important than the number of columns you indent is that you be *consistent* in your indentation.

A use of indentation, aside from that of showing logical relationships, is to show statement continuation. You might want to choose a second indenting factor for continuation to clearly distinguish between it and indentation to show logical relationships. If you choose a three-column indentation for logical relationships, for example, you might choose five for continuations. If a statement must be continued on a third line, it would normally begin in the same column as the line above it.

Indentation, then, is one way in which readability is enhanced, and another is through spacing—leaving blank lines. We have already recommended to you that blocks of code that perform a single function be set off by blank lines. Sometimes, however, each line of code performs a function distinctly different from the one above and below it. In this case it is better to consider a grouping, such as "initialization," and create blank lines accordingly. Under no circumstances is it necessary to double-space the entire program or even four or five consecutive lines. In the matter of spacing, even more than indentation, you must finally rely on your own good judgment.

We have suggested that blank lines be created simply by inserting blank lines in the source code, and this is certainly a satisfactory method; however, there is another technique that provides more flexibility. We have also told you that column 1 of the PL/I source code must be left blank, but that is not quite true. As a matter of fact, column 1 is reserved for a **carriage control character**, which controls the vertical spacing of the printed listing of the source code. (It has no effect on the spacing of the output from the program.) The acceptable characters that can be placed in column 1 and the results produced are:

blank	single-space before printing
0	double-space before printing
+	do not advance carriage before printing
1	skip to a new page before printing

Instead of inserting a blank line, therefore, you can put a 0 in column 1 of the line you wish to have double spaced. You can also force a new page by placing a 1 in the first column.

```
1EXAMPLE:  PROCEDURE OPTIONS (MAIN);
0DECLARE TEST_SCORES FIXED DECIMAL (3) INIT (0);
```

When the source code is printed, the PROCEDURE statement will be at the top of the page, and the DECLARE statement will be double-spaced below it. When you begin to code subroutine procedures, you might want to start each procedure on a new page, and the carriage control character allows for that. The + character is used primarily for underlining, and since the underline character (_) has special meaning to PL/I, items in a program cannot be underlined; therefore, it is unlikely you will ever use the + carriage control character in formatting your source code. Of the goals we set forth for good programming, indentation and spacing contribute primarily to readability. Comments, too, enhance readability, but perhaps they are most helpful in maintenance.

Comments

As with other matters regarding program format, the practice of what should be included in comments and where they should be placed varies a good deal. At one extreme are those who feel that PL/I is an almost entirely self-documenting language and consequently that few or no comments need be included. At the other extreme are those who feel that detailed pseudo-code and detailed descriptions of file layouts should be included in a large comment section at the head of the program and that the code itself must be liberally sprinkled with comments. It is sometimes the case that the documentation of source code is separated from the code and kept in a hard-copy file. When this is true, much more is required in comments, but we are here concerned with comments that remain a part of the source code.

We take a position between the two extremes just outlined. We believe that the majority of the comments should appear together at the head of the program; only on occasion do we find it necessary to intersperse the code with comments, and when we do find it necessary, we place that comment on a line by itself rather than attaching it to the end of a statement. We have found the headings of FUNCTION, INPUT, OUTPUT, and NOTES are satisfactory for the gathered comments. Under the heading of FUNCTION simply state in a sentence or two what the program does.

```
/* FUNCTION:  TO COMPUTE STUDENTS' FINAL AVERAGES          */
```

If your program has more than one major function, you might provide a numbered list:

```
/* FUNCTION:  1.  TO COMPUTE STUDENTS' FINAL AVERAGES      */
/*            2.  TO SORT GRADES IN DESCENDING SEQUENCE    */
/*            3.  TO SORT SOCIAL SECURITY NUMBERS AND       */
/*                GRADES IN ASCENDING ORDER ON SSN KEY     */
```

Under the heading of INPUT we recommend that you give a general description of the input record contents but not go into detail:

```
/* INPUT:     FOR EACH STUDENT A RECORD CONTAINING         */
/*            1.  SOCIAL SECURITY NUMBER                    */
/*            2.  FIVE PROGRAM GRADES                       */
/*            3.  TWO TEST GRADES                           */
```

For OUTPUT we suggest you briefly describe the contents of the report(s) produced.

```
/* OUTPUT:    THREE REPORTS                               */
/*            1.  FOR EACH STUDENT, THE INPUT AND THE      */
/*                COMPUTED FINAL AVERAGE                   */
/*            2.  THE SORTED GRADES                        */
/*            3.  THE SORTED SSN'S WITH INCLUDED FINAL      */
/*                AVERAGE                                  */
```

Many programs will not require a NOTES heading, and it should not be included unless necessary. Continuing the example above, however, you might find it useful to point out how the final average was computed.

```
/* NOTES:     THE FINAL AVERAGE IS DETERMINED BY          */
/*            (PROGRAMMING AVERAGE + TEST AVERAGE) / 2.0   */
```

We have three further points to make on comments.

1. You should supply blank lines at appropriate places within the gathered comments. A blank comment line is produced by coding

```
/*                                                        */
```

2. We cannot provide examples to cover all situations; you will have to exercise your judgment.

3. Consistency is more important than the exact format you choose.

6.2

Structured Programming

We have been insisting on structured programming throughout the book, but so far we have considered only one of its requirements—the allowable structures. After introducing the concept of top-down development, we review these structures, then proceed to discuss two more requirements of structured programming: top-driven loops and the one entrance/one exit construction.

6.2.1

Top-Down Development

Let us note at the outset that there is little you can do to actually implement top-down development in your programs until you become familiar with subroutine procedures. Nevertheless, it is important that you be introduced to it early and begin to think in its terms. We believe this approach will aid you in developing even those programs that do not formally implement top-down development.

Top-down development is a structured means of analyzing a problem to see what steps are required to solve it. Such an analysis will be very general at the first level, consisting of a single step that names the overall function of the program. This first level is broken down at the next level into several somewhat more specific steps, and the process continues until you are prepared to draw a detailed flowchart or write detailed pseudocode. (The process of breaking a large function into several smaller, more detailed functions, by the way, is called **decomposition**.) At some point in the process, however, instead of creating more steps you begin to see how the various steps are implemented (which, of course, is still movement from the general to the specific).

Let us illustrate this process by supposing you have been asked to write a program that will read an unknown number of input records, each containing a student's social security number and that student's semester average expressed as a percentage. Your program is to compute the class average and keep a count of the number of students having grades of 95 percent or higher and a separate count of those having a grade of 68 percent or lower. Your report is to list the social security number and grade of all the students in the class. In addition, you are to report the class average, the number of students in each of the two categories, and the total number of students processed. Although each of you would provide a top-down development that varied in its details, it might in general look like the following:

Step 1: Create a program

Step 2: The program must

     a. Read in data

     b. Do the computations

     c. Print data for each student

     d. Print the totals

At this point it looks as if we have determined the major steps required of the program. Now we need to discover what is involved in each of the steps.

Step 3: The program must

    a. Read in data

        (1) Social security number

        (2) Average grade

    b. Do the computations

        (1) Count the number of students processed

        (2) Count the number of students with

            (a) Grades $>= 95$

            (b) Grades $<= 68$

        (3) Sum the average grade for determining the class average

    c. Print data for each student

        (1) Student's social security number

        (2) Student's average grade

    d. Print the totals

        (1) Number of students processed

        (2) Number of students with

            (a) Grades $>= 95$

            (b) Grades $<= 68$

        (3) The class average

At this point it would appear that you are ready to draw a flowchart or write pseudocode, but there is one more aspect to consider: "bosses" and "subordinates." Top-down development is implemented in a manner that closely resembles the typical business organization. There is one "boss" at the top that controls all the subordinates, and this boss makes all the major decisions. Some of the subordinates might also be bosses, controlling "workers" at a still lower level. These subordinates make decisions for the workers under them. The actual work (doing the computations, moving data around, etc.) is done by these workers at the lowest level. It is important to remember that here (as is usually the case in a business organization) no one tells someone else at the same or a higher level what to do. All control flows from the top down. How a system of bosses and subordinates is set up is the subject of Section 6.2.2, which follows.

Although we have presented a relatively simple problem, we trust that we have illustrated the manner in which top-down development can supply a discipline and coherence to your problem solving approach.

<table>
<tr><td>*6.2.2*</td><td></td></tr>
<tr><td>*Allowable*
Structures</td><td>Structured programming allows three—and only three—structures and PL/I contains keywords to implement each of them. Sequence structure is implemented simply by coding a series of imperative statements; decision structure is implemented by an IF-THEN or IF-THEN-ELSE, with or without do-groups; and iteration structure is implemented by a DO WHILE. You are surely aware of the structures and their implementations by now, and we bring them up here only to prepare you for the shocking revelation that while PL/I *allows* you to write structured programs, it does not *force* you to do so. In other words, PL/I contains keywords that allow you to write unstructured programs. If you ever get into a situation in which you are required to read (and perhaps maintain) a significant number of PL/I programs, you will discover that unstructured PL/I programs do indeed exist, and you will need to know and understand the statements you find in them (although you, of course, would never use them). We explain two such statements as we consider top-driven loops and the one entrance/one exit construction.</td></tr>
</table>

We have stressed the fact that the DO WHILE condition is checked when the DO WHILE statement is encountered, which, of course, is at the top of the loop. A DO WHILE loop is therefore a top-driven loop because the term means simply that the condition is checked at the top of the loop. The primary corollary is that if the condition is false when the DO WHILE is first encountered, the loop is never entered. Some PL/I compilers, however, provide a second means of creating an iteration structure, and it is DO UNTIL.

In syntax the DO UNTIL is essentially the same as the DO WHILE in that the keyword is followed by a condition, and its range is delimited by an END. There are significant differences, however. First of all, unlike the DO WHILE condition, the condition in a DO UNTIL statement is not tested when the DO UNTIL statement is encountered. Even though the condition is coded as part of the DO UNTIL statement, it is not tested until the END statement delimiting the range of the loop is encountered. This in turn means that the statements within the loop are always executed at least once. It also means, of course, that the DO UNTIL statement does not create a top-driven loop; therefore, DO UNTIL falls outside the restrictions imposed by structured programming. The second significant difference is that the statements within the range of the DO UNTIL are executed as long as the condition is *false*, unlike those in the DO WHILE, which are executed as long as the condition is true. Notice what happens in the following skeleton example, assuming an empty input file:

```
EOF = FALSE;
ON ENDFILE (SYSIN) EOF = TRUE;
GET LIST (A, B, C);
DO UNTIL EOF = TRUE;
        .
    GET LIST (A, B, C);
```

The initial GET causes the ON ENDFILE condition to be raised, and the DO UNTIL condition becomes true. Because the condition is not checked until the bottom of the loop, however, PL/I must attempt to process a nonexistent record. If the statements in the loop contain arithmetic operations on the input data, it is likely that an abend will occur because, with no record read in, the variables will contain garbage. In other words, always use top-driven loops.

The third requirement of structured programming is that the program itself and any loops within it must have one entry point and one exit point. Even though it is not finally true, we will pretend for the time being that a program can have only one entry point. RETURN is a PL/I keyword that allows for multiple exits. When a RETURN is encountered, control returns to the calling program, and if the word is encountered in a main procedure, control returns to the operating system. The reason that RETURN has not been introduced is that if it is absent, PL/I assumes a RETURN when it encounters the END statement which signals the physical end of the program. It is possible to code more than one RETURN statement, thereby violating the principle of one exit.

Another statement available in PL/I and several other high-level languages that creates even more problems is the GO TO (or GOTO) statement. With a few exceptions the GO TO allows you to branch to any statement in the program, provided the statement has a label on it. One of the exceptions is that you cannot branch into the range of a DO WHILE (although you can branch to the DO WHILE itself); thus, the one entrance is preserved. Because a GO TO can be coded anywhere, including in the middle of the DO WHILE range, it effectively destroys the one-exit concept of structured programming. The early version of FORTRAN, in particular, was developed before structured programming techniques were available, and so GO TO was required. It was carried over into later languages, such as PL/I, which have no need of it. It is the greatest single enemy of structured programming; in fact, structured programming is sometimes called GO TO-less programming.

Most data processing installations today insist on structured programming, and you are well advised to cultivate its habits assiduously. At the same time, unstructured programs still abound at most installations, and you need to be aware of that fact.

6.3

Internal Subroutine Procedures

In addition to MAIN procedures, PL/I allows two types of procedures to be used. These are **external** and **internal** procedures. External procedures are known to the system by name, in the manner of a MAIN procedure. They are self-contained units of code, each beginning with a PROCEDURE statement and ending with an END statement. Internal procedures are separate units of code contained within an external procedure. We present external procedures in Chapter 10. We discuss internal procedures in the following sections. For the sake of simplicity, we use the term **subroutine** to denote internal subroutine procedures. An external procedure that is invoked by another program is called a **subprogram.**

We continue the example shown previously by saying the 'main' procedure is the procedure that invokes ("calls") the subroutines. Below is a diagram showing the flow of execution, that is, what code is executed at what point. When a subroutine is called, the code within that subroutine is executed. When execution is completed within the subroutine, control returns to the instruction in the main procedure that immediately follows the call. As you are studying the diagram, follow the lines. They will show you the flow of execution.

Main procedure:

Do While (there are more students to process)

    Call Read Data.   Read data subroutine:
                                Read:
                                   Social security number.
                                   Average grade.
                             End Read data.

    Call Do Computations.   Do Computations subroutine:
                                Count the number of
                                   students processed.

                                                        Count the number of
                                                            students with
                                                            Grades >= 95
                                                            and Grades <= 68.
                                                        Sum the average grade
                                                            for determining the
                                                            class average grade.
                                                    End Do Computations.

                    Call Print Data.   Print Data Subroutine:
                                                        Print:
                                                            Student's social
                                                                security number
                                                            Student's average grade.
                                                    End Print data.

            Enddo.

                    Call Print the Totals.   Print the Totals Subroutine:
                                                        Print:
                                                            Number of students
                                                                processed
                                                            Number of students
                                                                with Grades >= 95
                                                                Grades <= 68
                                                            The class average
                                                    End Print the Totals.

            End Main.

        At this point you are ready to write more thorough pseudocode for each
of the five separate procedures (don't forget the main) that will comprise the
entire program. The next step then is to write the actual PL/I code. In the
following sections we cover how to implement subroutine procedures
within PL/I.

6.3.1

The CALL Statement   The main procedure invokes a subroutine via a **CALL** statement. The CALL
and Subroutine Format   statement passes control to the subroutine named.

                CALL   READDATA;
                CALL   DOCALCULATIONS;
                CALL   PRINTDATA;

A subroutine begins with a PROCEDURE statement similar to the one that you have been using for the main procedures of the programs you have written so far. It is distinguished from a main procedure by not containing the keywords OPTIONS (MAIN). The naming conventions of the subroutine follow the PL/I internal naming conventions. The subroutine names can be up to 31 characters in length. The END statement is used as the last line within a procedure. A RETURN statement can be coded to cause normal termination of the subroutine. If you omit the RETURN statement, the END statement terminates the subroutine. An outline of a subroutine procedure might look like:

```
PRINTDATA:  PROCEDURE;
        .
        .
        .
     RETURN;
END PRINTDATA;
```

Below we show the PL/I outline of the example we have been using in the discussion of subroutine procedures.

```
MAIN: PROCEDURE OPTIONS (MAIN);
   CALL READ_A_STUDENT;
   DO WHILE (END_FLAG = MORE_RECS);
           .
           .                           The code in the
           .                           subroutines is
      CALL DO_THE_CALCULATIONS;        executed when
           .                           the routine
           .                           is invoked.
      CALL PRINT_INDIVIDUAL_STUDENT;
      CALL READ_A_STUDENT;
   END;
   CALL PRINT_TOTALS;
   READ_A_STUDENT: PROCEDURE;
           .
           .
           .
        RETURN;
   END READ_A_STUDENT;
   DO_THE_CALCULATIONS: PROCEDURE;
           .
```

```
        .
        .
      RETURN;
  └→  END DO_THE_CALCULATIONS;
  ┌→  PRINT_INDIVIDUAL_STUDENT: PROCEDURE;

        .
        .
        .

      RETURN;
  └→  END PRINT_INDIVIDUAL_STUDENT;
  ┌→  PRINT_TOTALS: PROCEDURE;

        .
        .

      RETURN;
  └→  END PRINT_TOTALS;
  └→  END MAIN;
```

6.3.2

The Active Procedure

In PL/I a procedure becomes active when it is invoked and stays active until it is terminated. A procedure is invoked either by being executed from the system, as with the main, or by being called from another procedure. There are several ways in which a procedure can be terminated, some of which we covered above. The other means of terminating a procedure we will not cover in this book. (Please refer to your PL/I Language Reference Manual for details that are beyond the scope of this book.) The following is a list of the methods that can used in PL/I to terminate a procedure (these pertain to both main procedures and subroutine procedures).

1. The RETURN statement terminates a procedure by switching the program execution back to the calling procedure at the instruction immediately following the CALL statement that was issued. If used at the end of the main procedure, control is passed back to the system.

2. The END statement has the same effect as the RETURN statement and is used to terminate a procedure when the RETURN statement is missing.

3. A STOP or EXIT statement causes the subroutine invoked to terminate. They, however, also cause the entire program, main procedure and all, to terminate. These statements are usually used for unusual error conditions where processing is no longer feasible. This is called an "abnormal termination" condition.

The main procedure in a PL/I program always remains active until the entire program is terminated. Other procedures invoked by the main procedure remain active, along with the main, until they are terminated.

Example Program—
Internal Procedures

CH6EX1 is the complete program that we have been discussing within this chapter. CH6EX1 shows the use of internal procedures all invoked from the main procedure. As you are reading the example please keep in mind the topics of structured programming that we covered in this chapter. Notice that the main program is the "boss" of all the subroutines and that all the subroutines are "subordinates." We decomposed the program into the basic functions that it performs and made each a subroutine. CH6EX1 has only one level of subordinates. Please review the following pseudocode before looking at the actual program, given in Figure 6−1.

```
MAIN:
     CALL READ_A_STUDENT
     DO WHILE (there are more records to process)
          CALL DO_THE_CALCULATIONS
          CALL PRINT_INDIVIDUAL_STUDENT
          CALL READ_A_STUDENT
     ENDDO
     CALL PRINT_TOTALS
READ_A_STUDENT:
     Read the student's Social Security Number and the
          Average Grade
END READ_A_STUDENT.
DO_THE_CALCULATIONS:
     Increment the count of students processed
     IF (Student's Average Grade >= 95)
         Increment the count for Grade >= 95
     ENDIF
     IF (Student's Average Grade <= 68)
         Increment the count for Grade <= 68
     ENDIF
     Add this Student's Average Grade to the sum of
          all the students average grades
END DO_THE_CALCULATIONS.
PRINT_INDIVIDUAL_STUDENT:
     IF (the Line Count > 50)
          Increment the Page Count
          Print the Headings
```

```
                              Print the column Headings
                              Initialize the Line Count to 5
                   ENDIF
                   Print the Student's Social Security Number and
                              the Average Grade
                   Increment the Line Count by 2
           END PRINT_INDIVIDUAL_STUDENT.
           PRINT_TOTALS:
                   Print the Page Heading on a New Page
                   Increment the Page Count
                   Print the line containing the Page Count
                   Print the total number of students processed
                   Print the number of students with average grade >= 95
                   Print the number of students with average grade <= 68
                   Print the class average
           END PRINT_TOTALS.
           END MAIN.
```

PL/I OPTIMIZING COMPILER VERSION 1 RELEASE 5.1 TIME: 16.02.07 DATE: 12 FEB 87 PAGE   1
OPTIONS SPECIFIED
60STMT;

PL/I OPTIMIZING COMPILER   /* CH6EX1 - USING INTERNAL PROCEDURES */   PAGE   2
                  SOURCE LISTING
    STMT

```
     /* CH6EX1 - USING INTERNAL PROCEDURES */
   1 CH6EX1:  PROCEDURE OPTIONS (MAIN);
     /***************************************************************/
     /*                                                           */
     /* FUNCTION:  TO PRODUCE A STUDENT GRADE REPORT SHOWING THE   */
     /*            FINAL AVERAGE GRADE FOR EACH STUDENT AND A TOTALS */
     /*            PAGE.                                          */
     /*                                                           */
     /* INPUT:     ONE CARD FOR EACH STUDENT CONTAINING           */
     /*            STUDENT'S SOCIAL SECURITY NUMBER               */
     /*            STUDENT'S FINAL AVERAGE GRADE                  */
     /*                                                           */
```

FIGURE 6-1
CH6EX1 (internal subroutines).

FIGURE 6–1 *(continued)*

```
/*  OUTPUT:    A REPORT LISTING                                   */
/*                    EACH STUDENT'S SOCIAL SECURITY NUMBER       */
/*                    THE FINAL AVERAGE GRADE FOR EACH STUDENT    */
/*                    A TOTALS PAGE LISTING                       */
/*                    TOTAL NUMBER OF STUDENTS PROCESSED          */
/*                    TOTAL NUMBER OF STUDENTS WITH GRADES >= 95  */
/*                    TOTAL NUMBER OF STUDENTS WITH GRADES <= 68  */
/*                    THE AVERAGE OF ALL THE GRADES               */
/*                                                                */
/******************************************************************/



2  DECLARE
       SOC_SEC_NUM                    CHARACTER (9),
       STUDENTS_AVERAGE_GRADE         FIXED DEC (3),
       GRADES_GREATER_95              FIXED DEC (3) INIT (0),
       GRADES_LESS_68                 FIXED DEC (3) INIT (0),
       TOTAL_STUDENTS_PROCESSED       FIXED DEC (3) INIT (0),
       TOTAL_OF_ALL_GRADES            FIXED DEC (6) INIT (0),
       PAGE_COUNT                     FIXED DEC (6) INIT (0),
       LINE_COUNT                     FIXED DEC (2) INIT (55);

3  DECLARE
       SYSIN FILE INPUT STREAM,
       SYSPRINT FILE OUTPUT STREAM;

4  DECLARE
       END_FLAG                       BIT (1) INIT ('1'B),
          MORE_RECS                   BIT (1) INIT ('1'B),
          FINISHED                    BIT (1) INIT ('0'B);

5  ON ENDFILE (SYSIN) END_FLAG = FINISHED;

6  CALL READ_A_STUDENT;

7  DO WHILE (END_FLAG = MORE_RECS);
8     CALL DO_THE_CALCULATIONS;
9     CALL PRINT_INDIVIDUAL_STUDENT;
10    CALL READ_A_STUDENT;
11 END;
```

    12 CALL PRINT_TOTALS;
    13 RETURN;

    14 READ_A_STUDENT: PROCEDURE;

        /**/
        /*   READ_A_STUDENT   */
        /*   */
        /* FUNCTION: TO READ THE RECORD OF ONE STUDENT   */
        /*   */
        /* GLOBAL VARIABLES: SOC_SEC_NUM   */
        /*   STUDENTS_AVERAGE_GRADE   */
        /*   */
        /* INPUT: SOC_SEC_NUM AND STUDENTS_AVERAGE_GRADE   */
        /*   */
        /* OUTPUT: NONE   */
        /*   */
        /**/

    15 GET LIST (SOC_SEC_NUM, STUDENTS_AVERAGE_GRADE);

    16 RETURN;

    17 END READ_A_STUDENT;

    18 DO_THE_CALCULATIONS: PROCEDURE;

        /**/
        /*   */
        /*   DO_THE_CALCULATIONS   */
        /*   */
        /* FUNCTION: TO PERFORM THE CALCULATIONS NECESSARY TO   */
        /*   FIND THE TOTAL NUMBER OF STUDENTS   */

FIGURE 6-1 *(continued)*

```
/*               FIND THE NUMBER OF STUDENTS WITH GRADE >= 95   */
/*               FIND THE NUMBER OF STUDENTS WITH GRADE <= 68   */
/*               TOTAL THE STUDENTS AVERAGE GRADE FOR LATER USE */
/*                                                              */
/*  GLOBAL VARIABLES:   TOTAL_STUDENTS_PROCESSED                */
/*                      STUDENTS_AVERAGE_GRADE                   */
/*                      GRADES_GREATER_95                        */
/*                      GRADES_LESS_68                           */
/*                      TOTAL_OF_ALL_GRADES                      */
/*                                                              */
/*  INPUT:    NONE                                              */
/*                                                              */
/*  OUTPUT:   NONE                                              */
/*                                                              */
/***************************************************************/
```

19 TOTAL_STUDENTS_PROCESSED = TOTAL_STUDENTS_PROCESSED + 1;

20 IF STUDENTS_AVERAGE_GRADE >= 95 THEN
        GRADES_GREATER_95 = GRADES_GREATER_95 + 1;

21 IF STUDENTS_AVERAGE_GRADE <= 68 THEN
        GRADES_LESS_68 = GRADES_LESS_68 + 1;

22 TOTAL_OF_ALL_GRADES = TOTAL_OF_ALL_GRADES + STUDENTS_AVERAGE_GRADE;

23 RETURN;

24 END DO_THE_CALCULATIONS;

PL/I OPTIMIZING COMPILER   /* CH6EX1 - USING INTERNAL PROCEDURES */   PAGE   6
    STMT
    25 PRINT_INDIVIDUAL_STUDENT: PROCEDURE;

    /***/
    /*   */
    /*   PRINT_INDIVIDUAL_STUDENT   */
    /*   */
    /* FUNCTION: TO PRINT A LINE FOR EACH STUDENT AS THEY ARE */
    /*   PROCESSED.   */
```
```

```
      /*                                                               */
      /*  GLOBAL VARIABLES:   LINE_COUNT                               */
      /*                      PAGE_COUNT                               */
      /*                      SOC_SEC_NUM                              */
      /*                      STUDENTS_AVERAGE_GRADE                   */
      /*                                                               */
      /*  INPUT:      NONE                                             */
      /*                                                               */
      /*  OUTPUT:     THE STUDENT'S SSN AND AVERAGE GRADE ARE PRINTED  */
      /*                                                               */
      /*****************************************************************/

  26  IF LINE_COUNT > 50 THEN
         DO;
  27        PAGE_COUNT = PAGE_COUNT + 1;

  28        PUT PAGE LIST (' ', ' ', 'STUDENT AVERAGE GRADE LIST');
  29        PUT SKIP LIST ('PAGE NUMBER: ', PAGE_COUNT);

  30        PUT SKIP(2) LIST (' ', 'STUDENT SSN', ' ', 'STUDENT GRADE');
  31        PUT SKIP LIST (' ');

  32        LINE_COUNT = 5;
  33     END;

  34  PUT SKIP LIST (' ',SOC_SEC_NUM, ' ', STUDENTS_AVERAGE_GRADE);
  35  LINE_COUNT = LINE_COUNT + 2;

  36  RETURN;
  37  END PRINT_INDIVIDUAL_STUDENT;
```

PL/I OPTIMIZING COMPILER   /* CH6EX1 - USING INTERNAL PROCEDURES */   PAGE   7
   STMT
    38 PRINT_TOTALS: PROCEDURE;

```
      /*****************************************************************/
      /*                                                               */
      /*                    PRINT_TOTALS                               */
      /*                                                               */
      /*  FUNCTION:  TO PRINT THE TOTALS OF THE REPORT ON STUDENT GRADES*/
      /*                                                               */
```

FIGURE 6–1 *(continued)*

```
       /*  GLOBAL VARIABLES:  PAGE_COUNT                                 */
       /*                     TOTAL_STUDENTS_PROCESSED                   */
       /*                     GRADES_GREATER_95                          */
       /*                     GRADES_LESS_68                             */
       /*                                                                */
       /*  INPUT:    NONE                                                */
       /*                                                                */
       /*  OUTPUT:   TOTAL NUMBER OF STUDENTS PROCESSED                  */
       /*            TOTAL NUMBER OF STUDENTS WITH GRADE >= 95           */
       /*            TOTAL NUMBER OF STUDENTS WITH GRADE <= 68           */
       /*            THE AVERAGE GRADE FOR THE ENTIRE CLASSS             */
       /*                                                                */
       /*****************************************************************/

   39  PUT PAGE LIST (' ', ' ', 'TOTALS FOR STUDENT GRADE REPORT');
   40  PAGE_COUNT = PAGE_COUNT + 1;
   41  PUT SKIP LIST ('PAGE NUMBER: ', PAGE_COUNT);

   42  PUT SKIP(2) LIST ('TOTAL NUMBER OF STUDENTS PROCESSED = ',
           TOTAL_STUDENTS_PROCESSED);

   43  PUT SKIP(2) LIST ('TOTAL NUMBER OF STUDENTS WITH GRADES >= 95 = ',
           GRADES_GREATER_95);

   44  PUT SKIP(2) LIST ('TOTAL NUMBER OF STUDENTS WITH GRADES <= 68 = ',
           GRADES_LESS_68);

   45  PUT SKIP(2) LIST ('THE CLASS AVERAGE = ',
           TOTAL_OF_ALL_GRADES / TOTAL_STUDENTS_PROCESSED);

   46  RETURN;
   47  END PRINT_TOTALS;
   48  END CH6EX1;


PL/I OPTIMIZING COMPILER      /* CH6EX1 - USING INTERNAL PROCEDURES */      PAGE  8
NO MESSAGES PRODUCED FOR THIS COMPILATION
```

```
                              STUDENT AVERAGE GRADE LIST
PAGE NUMBER:                  1

            STUDENT SSN                                STUDENT GRADE


            345678675                                       82
            454323445                                       93
            234556677                                       99
            234455665                                       63
            345654654                                       23
            435576543                                       75
            233453228                                       98
            355554577                                       95
            345667745                                       68
            455654543                                       75
            254564656                                       99
            353343435                                       64
            435436544                                       88
            122344556                                       34
            464765746                                      100
            123445456                                       87
            232343435                                       62
            324344545                                       87
            223543556                                       67
            987421344                                       34
            335567677                                       99
            233445545                                       65


                            TOTALS FOR STUDENT GRADE REPORT
PAGE NUMBER:                  2

TOTAL NUMBER OF STUDENTS PROCESSED =              23

TOTAL NUMBER OF STUDENTS WITH GRADES >= 95 =      6

TOTAL NUMBER OF STUDENTS WITH GRADES <= 68 =     10

THE CLASS AVERAGE =        74.391304347
```

6.4

**Nesting
Subroutines**

Although all internal subroutines are located within a main procedure, they can also be "nested" within each other. The nesting of subroutines limits the "scope" of the subroutine, as we demonstrate in this section.

A **nested subroutine** is one which is located within another subroutine. For example, in CH6EX1 all the subroutines are located within the MAIN procedure. We define the term "located within" to mean that the source code for the subroutine appears after the PROCEDURE statement and before the END statement of another routine. In the code below, we show that subroutine READLINE is located within READRTN (demonstrating the nesting of subroutines).

```
MAIN:  PROCEDURE OPTIONS (MAIN);
       CALL READRTN;

              .
              .
              .

READRTN:  PROCEDURE;
       CALL READLINE;

              .
              .
              .

READLINE:  PROCEDURE;

              .
              .
              .

END READLINE;
END READRTN;
END MAIN;
```

There are two questions that we need to answer pertaining to the scope of nested subroutines. The first is: "What other subroutines can be called from the nested subroutine?" The second is: "What variables does the nested subroutine have access to?" We answer each question separately in the following two sections.

6.4.1

*Scope of
Subroutines*

Nested subroutines can call (have access to) any other subroutine that

1. is nested within itself at the first level down.

2. is nested at the same level as any subroutine that is currently active.

Let us illustrate these two points by showing you pictures of nested subroutines.

```
MAIN:   PROCEDURE OPTIONS (MAIN);
            .
            .
            .
SUBA:   PROCEDURE;
            .
            .
SUBA_1:   PROCEDURE;
            .
            .
END SUBA_1;
SUBA_2:   PROCEDURE;
    .
    .
    ¡
SUBA_2_1:   PROCEDURE;
        .
        .
END SUBA_2_1;
END SUBA_2;
END SUBA;
SUBB:   PROCEDURE;
        .
        .
        .
SUBB_1:   PROCEDURE;
        .
        .
END SUBB_1;
END SUBB;
END MAIN;
```

We think the easiest way to understand the scope of each subroutine is to look at the above example while studying Figure 6–2.

```
PROCEDURE       CAN CALL THESE PROCEDURES
---------       --- ---- ----- ----------
MAIN            SUBA and SUBB
SUBA            SUBB, SUBA_1, and SUBA_2,
SUBA_1          SUBA_2 and SUBB
SUBA_2          SUBA_1, SUBA_2_1, and SUBB
SUBA_2_1        SUBA_1 and SUBB
SUBB            SUBA and SUBB_1
SUBB_1          SUBA
```

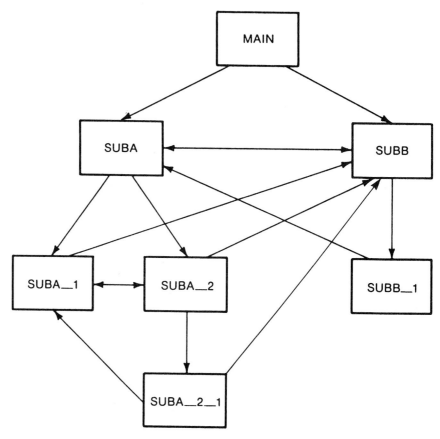

FIGURE 6-2
Valid calls from each module.

We revisit the idea of active subroutines here in order to assist you in understanding the scope of the subroutines that can be invoked by other subroutines. When the MAIN procedure above is called from the system, it becomes active and stays active until it is terminated. When the MAIN procedure calls SUBA, SUBA becomes an active procedure as well as the MAIN. Now when SUBA calls SUBA-1, we have three active procedures. When SUBA_1 terminates, there are only two active procedures again. The maximum depth of nested procedures is limited to 50 in PL/I (using the optimizing compiler). Examine the following sequence of CALL statements and see whether you can determine the active procedures after every CALL statement and every RETURN statement.

```
(1)      MAIN:
(2)          CALL SUBA;
(3)          CALL SUBB;
(4)          RETURN;
(5)      SUBA·
(6)          CALL SUBA_1;
(7)          CALL SUBA_2;
(8)          RETURN;
(9)      SUBA_1:
(10)     END SUBA_1;
(11)     SUBA_2:
(12)         CALL SUBA_2_1;
(13)         CALL SUBB;
(14)         RETURN;
(15)     SUBA_2_1;
(16)         RETURN;
(17)     END SUBA_2_1;
(18)     END SUBA_2;
(19)     END SUBA;
(20)     SUBB:
(21)       CALL SUBB_1;
(22)       RETURN;
(23)     SUBB_1:
(24)       RETURN;
(25)     END SUBB_1;
(26)     END SUBB;
(27)     END MAIN;
```

Here we give you the correct results of the active procedures after each statement above is executed. We use the statement number to refer to the statement. Remember we are presenting the active procedures *after* the statement is executed and in the order the statements are executed.

```
(1)  MAIN
    (2)  MAIN, SUBA
        (6)  MAIN, SUBA, SUBA_1
            (10) MAIN, SUBA
        (7)  MAIN, SUBA, SUBA_2
            (12)  MAIN, SUBA, SUBA_2, SUBA_2_1
                (16) MAIN, SUBA, SUBA_2
            (13)  MAIN, SUBA, SUBA_2, SUBB
```

```
                             (21) MAIN, SUBA, SUBA_2, SUBB, SUBB_1
                                 (24) MAIN, SUBA, SUBA_2, SUBB
                             (22) MAIN, SUBA, SUBA_2
                      (14)   MAIN, SUBA
                (8)   MAIN
           (3)   MAIN, SUBB
                (21) MAIN, SUBB, SUBB_1
                    (24) MAIN, SUBB
                (22) MAIN
           (4) none now active
```

We hope that you achieved the correct answer for the active procedures. If you did not, please go back through this chapter until you completely understand the concept of active procedures before you continue.

6.4.2

Global and Local Variables and the Scope of Variables

When you are coding subroutines, you will find that you need two types of variables. Variables that are used by more than one of the subroutines and are declared within a main procedure and outside any subroutine are considered **global** variables. **Local** variables are variables that are declared within a subroutine itself. They can be used only by that subroutine.

The scope of variables that a subroutine can access differs slightly from the scope of the procedures that a nested subroutine can call. In the example above, we demonstrated the scope of the CALL statement. In this section we will answer the second question that we posed in the beginning of the section on nested subroutines—that being the scope of the variables.

Please examine the example above as you study the following chart.

```
PROCEDURE        CAN ACCESS VARIABLES DECLARED IN:
---------        --- ------ --------- -------- --
MAIN             MAIN
SUBA             MAIN, SUBA
SUBA_1           MAIN, SUBA, SUBA_1
SUBA_2           MAIN, SUBA, SUBA_2
SUBA_2_1         MAIN, SUBA, SUBA_2, SUBA_2_1
SUBB             MAIN, SUBB
SUBB_1           MAIN, SUBB, SUBB_1
```

We can state the scope-of-variables rule generally as follows: "A subroutine can access the variables that have been declared within the same PROCEDURE and END statement pairs that the subroutine itself resides." To illustrate this to yourself, draw lines from each PROCEDURE statement above to its matching END statement. A subroutine can access any variable declared within the code residing within the lines you have drawn.

STATIC vs.
AUTOMATIC Storage

In this section we discuss two types of storage classes that are included in PL/I. (There are other types, which we will present in Chapter 15.) PL/I uses different types of storage classes to distinguish between the different ways in which a variable can be used. For example, one variable might be needed during the entire active period of a procedure and then reinitialized the next time the procedure becomes active. Another variable, however, might be needed to contain the same value when a procedure becomes active as it had the last time the procedure was terminated (also called **deactivated**).

6.4.3.1.   AUTOMATIC Storage Allocation
Each time a routine is entered (either from the system or via a CALL statement), the variables that are declared within that routine are allocated storage. That is, memory is obtained of the length and format specified in the DECLARE statement and initialized as requested. When a routine is exited, the storage is returned to the system so that it can be used by another routine. We illustrate this idea using the example program shown in CH6EX2 (see Fig. 6–3). Please study CH6EX2 before going on. The first storage allocation takes place when the MAIN program is entered from the system:

Storage for MAIN Program code
Storage for READ_A_STUDENT code
Storage for PROCESS_STUDENT code
Storage for PRINT_STUDENT code
Storage for PRINT_HEADINGS code
Storage for COMPUTE_HONOR_POINTS code
Storage for PRINT_GPA code
Storage allocation for variables DECLARED in MAIN STUDENT_SSN, COURSE_ID, COURSE_CREDIT_HOURS, COURSE_GRADE, TOTAL_HONOR_POINTS, TOTAL_CREDIT_HOURS, TOTAL_STUDENTS_PROCESSED, END_FLAG, MORE_RECS, and FINISHED

As each subroutine is called from the main program, storage is allocated for the variables declared within that routine. As you will notice, this storage is obtained following the storage that was allocated for the MAIN program. Since the READ_A_STUDENT subroutine does not have any variables declared, no new storage allocation would take place upon calling this subroutine. However, if you look below you will see that after PROCESS_STUDENT is invoked the storage allocation changes.

Storage for MAIN Program code
Storage for READ_A_STUDENT code
Storage for PROCESS_STUDENT code
Storage for PRINT_STUDENT code
Storage for PRINT_HEADINGS code
Storage for COMPUTE_HONOR_POINTS code
Storage for PRINT_GPA code
Storage allocation for variables DECLARED in MAIN STUDENT_SSN, COURSE_ID, COURSE_CREDIT_HOURS, COURSE_GRADE, TOTAL_HONOR_POINTS, TOTAL_CREDIT_HOURS, TOTAL_STUDENTS_PROCESSED, END_FLAG, MORE_RECS, and FINISHED
Storage allocation for variables DECLARED in PROCESS_STUDENT OLD_STUDENT_SSN

When PROCESS_STUDENT is exited and deactivated, the storage allocated when PROCESS_STUDENT was entered is "deallocated," given back to the system. Deallocation, then, destroys the contents of the variables because they are reallocated and reinitialized the next time the subroutine is invoked. Thus after PROCESS_STUDENT is exited, the storage allocation for the example program looks the same as it did when the main program was first entered. As you know, COMPUTE_HONOR_POINTS is called from PROCESS_STUDENT next in the execution of the CH6EX2 example program. The storage allocation would thus be as follows:

Storage for MAIN Program code
Storage for READ_A_STUDENT code
Storage for PROCESS_STUDENT code
Storage for PRINT_STUDENT code
Storage for PRINT_HEADINGS code
Storage for COMPUTE_HONOR_POINTS code
Storage for PRINT_GPA code
Storage allocation for variables DECLARED in MAIN STUDENT_SSN, COURSE_ID, COURSE_CREDIT_HOURS, COURSE_GRADE, TOTAL_HONOR_POINTS, TOTAL_CREDIT_HOURS, TOTAL_STUDENTS_PROCESSED, END_FLAG, MORE_RECS, and FINISHED
Storage allocation for variables DECLARED in PROCESS_STUDENT OLD_STUDENT_SSN
Storage allocation for variables DECLARED in COMPUTE_HONOR_POINTS HONOR_POINTS

The storage allocations that we have been discussing are called "dynamic storage allocations." The variables that are allocated memory during the dynamic storage allocations are called "automatic" variables. Automatic is the default attribute for any variables declared. The storage class of a variable is assigned within the DECLARE statement as are the other attributes. However, since automatic is the default storage class, we do not need to code it within the DECLARE statement.

6.4.3.2. STATIC Storage Allocation

As you will soon discover when running PL/I programs using subroutines, there will be times when you will not want the variables within a subroutine deallocated upon exit. For example, in the PRINT_STUDENT of CH6EX2 we use a variable called LINE_COUNT. We want this variable to be initialized once and only once, because it must "remember" the count of the number of lines printed. It is to keep track of the lines printed, and when more than 50 have been printed, cause a new page to be started. As you can see, if this variable was declared (perhaps implicitly) as an automatic variable, storage would be allocated every time PRINT_STUDENT was entered, and LINE_COUNT would be reinitialized to 1.

The STATIC storage class of variables allows us to do exactly what we wish to do concerning LINE_COUNT. When a variable is declared as STATIC, storage is allocated for it prior to the execution of the program. The static variable remains active until the program terminates. During this storage allocation, the static variable is initialized. Thus, a static variable is initialized *only* prior to the execution of the entire program. Therefore, the static variable is allocated and initialized prior to the first time the subroutine in which it is declared is called. The storage for STATIC variables is *not* deallocated when a subroutine is exited. Therefore, the value within a static variable remains the same upon entering a subroutine as it was when the subroutine was previously exited. The storage allocation is shown below as it would appear after PRINT_STUDENT from CH6EX2 was called for the first time.

Storage for MAIN Program code
Storage for READ_A_STUDENT code
Storage for PROCESS_STUDENT code
Storage for PRINT_STUDENT code
Storage for PRINT_HEADINGS code
Storage for COMPUTE_HONOR_POINTS code
Storage for PRINT_GPA code
Storage for any STATIC variables within the program.   LINE_COUNT, and PAGE-COUNT

Storage allocation for variables DECLARED in MAIN
  STUDENT_SSN, COURSE_ID, COURSE_CREDIT_HOURS,
    COURSE_GRADE, TOTAL_HONOR_POINTS,
    TOTAL_CREDIT_HOURS,
    TOTAL_STUDENTS_PROCESSED,
    END_FLAG, MORE_RECS, and FINISHED

Storage allocation for variables DECLARED in
  PROCESS_STUDENT
    OLD_STUDENT_SSN

Here are some examples of how to DECLARE variables as STATIC:

```
DCL  PAGE_COUNTER   FIXED DEC (5) INIT (1) STATIC;
DCL  LINE_COUNTER   FIXED DEC (2) INIT (0) STATIC;
DCL  LAST_NAME      CHAR (25) STATIC;
```

Notice that the STATIC attribute is placed in the DECLARE statement. As you know, it could be placed anywhere within the attribute list of the variable; however, we recommend that you put the STATIC attribute at the end so that it is very noticeable. This way you can tell immediately whether or not a variable is to be allocated as static or automatic.

6.4.4

Example Program—
Nesting Procedures

In CH6EX2 we demonstrate the use of nested subroutines. In this example we show the scope of the subroutines and the scope of the procedures. CH6EX2 is a program that reads in student information and produces a report. For each student in the input stream, the student's social security number, a course identification, the credit hours of the course, and the grade the student received in the course is read on a single line. The program prints this information and calculates and prints the student's grade point average. Therefore, there might be more than one line for each student. The grade point average is printed after all the course information for a single student has been processed and printed.

The total number of students processed is printed on the last page of the report so that we can check to make sure all the students were read in and processed. No other totals are printed.

CH6EX2 shows you how we suggest subroutines be documented. You will notice that two subtopics have been added: (1) **GLOBAL VARIABLES** and (2) **SUBROUTINES CALLED.** These areas have been added in order to ensure complete documentation. It is necessary to document the global variables that are used within subroutines because when changes are made to the global variables, the subroutines must be checked to see whether they also need changing. When we state explicitly which subroutines are called, a programmer who is changing a subroutine can examine the procedures that call it. These documentation standards assist in making modifications to programs easier. However, the documentation is only an assistance if it is strictly followed. If not adhered to or if inconsistent, the documentation really does no good and actually becomes a hindrance to maintenance of programs.

Below is the pseudocode for CH6EX2.

```
CH6EX2:
    CALL READ_A_STUDENT for the first student
    DO WHILE (there are more records to process)
        CALL PROCESS_STUDENT
    CALL PRINT_GPA
    ENDDO
READ_A_STUDENT:
    Read the student's social security number, course id,
        course credit hours, and course grade
END READ_A_STUDENT.
PROCESS_STUDENT:
    Save the student's social security number as a check
        to see whether a new student has been read
    Initialize the total honor points field to zero
    Initialize the total credit hours field to zero
    DO WHILE (a new student has not been read)
        CALL COMPUTE_HONOR_POINTS
        CALL PRINT_STUDENT to print the course information
        Call READ_A_STUDENT for the next card
    ENDDO
PRINT_STUDENT:
    IF (the line count > 50)
        CALL PRINT_HEADINGS
    ENDIF
    Print the student's social security number,
        the course id, the course's credit hours,
        and the course grade for this student
    Increment the line count by 1
PRINT_HEADINGS:
    Print the first line on a new page
    Increment the page count by 1
    Print the page count line
    Print the subheadings
    Initialize the line count to 5
END PRINT_HEADINGS.
END PRINT_STUDENT.
END PROCESS_STUDENT.
COMPUTE_HONOR_POINTS:
    Increment the total number of students processed by 1
    Find the honor points for the student:
        honor points <- course credit hours * course grade
    Add these honor points to the total honor points for
        the student
```

```
                    Add the credit hours to the total credit hours for
                        the student
                END COMPUTE_HONOR_POINTS.
                PRINT_GPA:
                    Compute the student's grade point average (GPA):
                        Total honor points / total credit hours
                    Print the total credit hours
                    Print the total honor points
                    Print the student's GPA just calculated
                END PRINT_GPA.
                END CH6EX2.
```

PL/I OPTIMIZING COMPILER VERSION 1 RELEASE 5.1 TIME: 16.02.33 DATE: 12 FEB 87 PAGE   1
OPTIONS SPECIFIED
GOSTMT;

PL/I OPTIMIZING COMPILER   /*   CH6EX2 - NESTED PROCEDURE */   PAGE   2
                    SOURCE LISTING
    STMT

```
      /*   CH6EX2 - NESTED PROCEDURES */
   1  CH6EX2:  PROCEDURE OPTIONS (MAIN);
      /*******************************************************************/
      /*                                                             */
      /*  FUNCTION:  TO PRINT A REPORT OF STUDENTS COURSE GRADES AND GPA */
      /*                                                             */
      /*  INPUT:     ONE CARD FOR EACH STUDENT CONTAINING            */
      /*             STUDENT'S SOCIAL SECURITY NUMBER                */
      /*             THE COURSE ID                                   */
      /*             CREDIT HOURS OF THE COURSE                      */
      /*             COURSE GRADE AS FOLLOWS                         */
      /*                 4 FOR AN A                                  */
      /*                 3 FOR A  B                                  */
      /*                 2 FOR A  C                                  */
      /*                 1 FOR A  D                                  */
      /*                 0 FOR AN F                                  */
      /*                                                             */
```

FIGURE 6–3
CH6EX2 (nested internal subroutines).

FIGURE 6–3 *(continued)*

```
     /*  OUTPUT:    A REPORT LISTING                                */
     /*                    EACH STUDENTS SOCIAL SECURITY NUMBER      */
     /*                    THE COURSE ID FOR EACH COURSE             */
     /*                    THE CREDIT HOURS FOR EACH COURSE          */
     /*                    THE GRADE AS LISTED ABOVE                 */
     /*                    AFTER ALL THE COURSES FOR A STUDENT ARE PRINTED */
     /*                        THE STUDENTS GPA IS PRINTED           */
     /*                                                              */
     /****************************************************************/

  2  DECLARE
          STUDENT_SSN                   CHAR (9),
          COURSE_ID                     CHAR (8),
          COURSE_CREDIT_HOURS           FIXED DEC (3),
          COURSE_GRADE                  FIXED DEC (1),

          TOTAL_HONOR_POINTS            FIXED DEC (7) INIT (0),
          TOTAL_CREDIT_HOURS            FIXED DEC (7) INIT (0),
          TOTAL_STUDENTS_PROCESSED      FIXED DEC (3) INIT (0);
  3  DECLARE
          SYSIN FILE INPUT STREAM,
          SYSPRINT FILE OUTPUT STREAM;


  4  DECLARE
          END_FLAG                      BIT (1) INIT ('1'B),
              MORE_RECS                 BIT (1) INIT ('1'B),
              FINISHED                  BIT (1) INIT ('0'B);
  5  ON ENDFILE (SYSIN) END_FLAG = FINISHED;

  6  CALL READ_A_STUDENT;

  7  DO WHILE (END_FLAG = MORE_RECS);
```

PL/I OPTIMIZING COMPILER   /*   CH6EX2 - NESTED PROCEDURES */   PAGE   3
    STMT

```
    8       CALL PROCESS_STUDENT;

    9       CALL PRINT_GPA;
```

```
    10  END;

    11  PUT SKIP (2) LIST ('TOTAL NUMBER OF RECORDS PROCESSED = ',
                   TOTAL_STUDENTS_PROCESSED);

    12  RETURN;
```

```
    13  READ_A_STUDENT:   PROCEDURE;

        /*********************************************************************/
        /*                                                                 */
        /*                   READ_A_STUDENT                                 */
        /*                                                                 */
        /*  FUNCTION:  TO READ THE INFORMATION FOR A SINGLE STUDENT.       */
        /*                                                                 */
        /*  GLOBAL VARIABLES:    STUDENT_SSN                               */
        /*                       COURSE_ID                                 */
        /*                       COURSE_CREDIT_HOURS                       */
        /*                       COURSE_GRADE                              */
        /*                                                                 */
        /*  SUBROUTINES CALLED: NONE                                       */
        /*                                                                 */
        /*  INPUT:    STUDENT_SSN, COURSE_ID, COURSE_CREDIT_HOURS, AND     */
        /*            COURSE_GRADE (IN NUMBER AS 0 : F, 1 : D, 2 : C,      */
        /*                 3 : B, AND 4 : A)                               */
        /*                                                                 */
        /*  OUTPUT:   NONE                                                 */
        /*                                                                 */
        /*********************************************************************/

    14  GET LIST (STUDENT_SSN, COURSE_ID, COURSE_CREDIT_HOURS, COURSE_GRADE);

    15  RETURN;
    16  END READ_A_STUDENT;
```

FIGURE 6–3 *(continued)*

```
17  PROCESS_STUDENT:    PROCEDURE;


    /*********************************************************************/
    /*                                                                 */
    /*                    PROCESS_STUDENT                              */
    /*                                                                 */
    /*            FILE.  A REPORT IS PRODUCED SHOWING THE COURSES IN   */
    /*            WHICH THE STUDENT WAS ENROLLED ALONG WITH THE GRADE  */
    /*            AND CREDIT HOURS GIVEN.  THE PURPOSE IS TO CALCULATE */
    /*            THE GPA FOR THE STUDENT AFTER PRINTING ALL THE       */
    /*            COURSES.                                             */
    /*                                                                 */
    /*  GLOBAL VARIABLES:    STUDENT_SSN                               */
    /*                       COURSE_ID                                 */
    /*                       COURSE_CREDIT_HOURS                       */
    /*                       COURSE_GRADE                              */
    /*                       TOTAL_HONOR_POINTS                        */
    /*                       TOTAL_CREDIT_HOURS                        */
    /*                       ENDFLAG                                   */
    /*                                                                 */
    /*  SUBROUTINES CALLED: COMPUTE_HONOR_POINTS                       */
    /*                      PRINT_STUDENT                              */
    /*                      READ_A_STUDENT                             */
    /*                                                                 */
    /*  INPUT:   EACH STUDENTS INFORMATION IS READ IN ONE COURSE AT    */
    /*           A TIME BY CALLING READ_A_STUDENT.                     */
    /*                                                                 */
    /*  OUTPUT:  EACH STUDENT'S COURSE INFORMATION IS PRINTED          */
    /*                                                                 */
    /*********************************************************************/

18  DECLARE
        OLD_STUDENT_SSN             CHAR (9);

19  OLD_STUDENT_SSN = STUDENT_SSN;
20  TOTAL_HONOR_POINTS = 0;
21  TOTAL_CREDIT_HOURS = 0;

22  DO WHILE (OLD_STUDENT_SSN = STUDENT_SSN & END_FLAG = MORE_RECS);
```

```
23    CALL COMPUTE_HONOR_POINTS;

24    CALL PRINT_STUDENTS;

25    CALL READ_A_STUDENT;
26  END;

27  RETURN;
```

```
28  PRINT_STUDENTS:   PROCEDURE;

     /********************************************************************/
     /*                                                                */
     /*                      PRINT_STUDENTS                       */
     /*                                                                */
     /* FUNCTION:   TO PRINT THE COURSE INFORMATION FOR A STUDENT      */
     /*                                                                */
     /* GLOBAL VARIABLES:     STUDENT_SSN                          */
     /*                       COURSE_ID                            */
     /*                       COURSE_CREDIT_HOURS                  */
     /*                       COURSE_GRADE                         */
     /*                                                                */
     /* SUBROUTINES CALLED: PRINT_HEADINGS                        */
     /*                                                                */
     /* INPUT:    NONE                                            */
     /*                                                                */
     /* OUTPUT:   THE STUDENT COURSE INFORMATION IS PRINTED       */
     /*                                                                */
     /********************************************************************/

29  DECLARE
       LINE_COUNT                    FIXED DEC (3) INIT (51) STATIC;

30  IF LINE_COUNT > 50 THEN
       CALL PRINT_HEADINGS;

31  PUT SKIP LIST (STUDENT_SSN, COURSE_ID, COURSE_CREDIT_HOURS,
       COURSE_GRADE);
```

```
   32  LINE_COUNT = LINE_COUNT + 1;

   33  RETURN;


PL/I OPTIMIZING COMPILER        /*   CH6EX2 - NESTED PROCEDURES */         PAGE   7
   STMT
   34  PRINT_HEADINGS:   PROCEDURE;

       /*****************************************************************/
       /*                                                             */
       /*                    PRINT_HEADINGS                           */
       /*                                                             */
       /*  FUNCTION: TO PRINT THE HEADING LINES OF THE STUDENT GRADE  */
       /*            REPORT.                                          */
       /*                                                             */
       /*  GLOBAL VARIABLES:   LINE_COUNT                            */
       /*                                                             */
       /*  SUBROUTINES CALLED: NONE                                  */
       /*                                                             */
       /*  INPUT:    NONE                                            */
       /*                                                             */
       /*  OUTPUT:   THE FIRST THREE LINES OF THE GRADE REPORT.      */
       /*                                                             */
       /*****************************************************************/

   35  DECLARE
          PAGE_COUNT                 FIXED DEC (5) INIT (1) STATIC;
   36  PUT PAGE LIST (' ', ' ', 'STUDENT GRADE REPORT');
   37  PUT SKIP LIST ('PAGE: ', PAGE_COUNT);

   38  PUT SKIP(2) LIST ('STUDENT SSN', 'COURSE ID', 'CREDIT HOURS',
          'GRADE RECEIVED');

   39  PUT SKIP LIST(' ');

   40  LINE_COUNT = 5;

   41  RETURN;
```

```
42  END PRINT_HEADINGS;
43  END PRINT_STUDENTS;


44  END PROCESS_STUDENT;
```

```
    STMT
    45  COMPUTE_HONOR_POINTS:  PROCEDURE;


    /*******************************************************************/
    /*                                                               */
    /*                    COMPUTE_HONOR_POINTS                       */
    /*                                                               */
    /* FUNCTION: TO COMPUTE THE HONOR POINTS FOR EACH STUDENT       */
    /*                                                               */
    /* GLOBAL VARIABLES:    TOTAL_STUDENTS_PROCESSED                */
    /*                      TOTAL_HONOR_POINTS                      */
    /*                      TOTAL_CREDIT_HOURS                      */
    /*                      COURSE_CREDIT_HOURS                     */
    /*                      COURSE_GRADE                            */
    /*                                                               */
    /* SUBROUTINES CALLED: NONE                                     */
    /*                                                               */
    /* INPUT:    NONE                                               */
    /*                                                               */
    /* OUTPUT:   NONE                                               */
    /*                                                               */
    /*******************************************************************/


    46  DECLARE
            HONOR_POINTS                 FIXED DEC (5) INIT (0);

    47  TOTAL_STUDENTS_PROCESSED = TOTAL_STUDENTS_PROCESSED + 1;
    48  HONOR_POINTS = COURSE_CREDIT_HOURS * COURSE_GRADE;
    49  TOTAL_HONOR_POINTS = TOTAL_HONOR_POINTS + HONOR_POINTS;
    50  TOTAL_CREDIT_HOURS = TOTAL_CREDIT_HOURS + COURSE_CREDIT_HOURS;

    51  RETURN;
    52  END COMPUTE_HONOR_POINTS;
```

FIGURE 6–3 *(continued)*

```
PL/I OPTIMIZING COMPILER          /*   CH6EX2 - NESTED PROCEDURES */          PAGE   9


   STMT
    53  PRINT_GPA:   PROCEDURE;
        /***********************************************************************/
        /*                                                                   */
        /*                       PRINT_GPA                                    */
        /*                                                                   */
        /*  FUNCTION: TO CALCULATE AND PRINT THE GPA FOR THE STUDENT.        */
        /*                                                                   */
        /*  GLOBAL VARIABLES:   TOTAL_HONOR_POINTS                           */
        /*                      TOTAL_CREDIT_HOURS                           */
        /*                                                                   */
        /*  SUBROUTINES CALLED: NONE                                         */
        /*                                                                   */
        /*  INPUT:    NONE                                                   */
        /*                                                                   */
        /*  OUTPUT:   THE GPA LINE FOR THE STUDENT.                          */
        /*                                                                   */
        /***********************************************************************/

    54  DECLARE
          STUDENT_GPA                 FIXED DEC (4,3) INIT (0);

    55  STUDENT_GPA = TOTAL_HONOR_POINTS / TOTAL_CREDIT_HOURS;

    56  PUT SKIP (2) LIST ('TOTAL CREDIT HOURS = ', TOTAL_CREDIT_HOURS);
    57  PUT SKIP LIST ('TOTAL HONOR POINTS = ', TOTAL_HONOR_POINTS);

    58  PUT SKIP LIST ('THE STUDENT''S GPA = ', STUDENT_GPA);

    59  PUT SKIP LIST (' ');

    60  RETURN;
    61  END PRINT_GPA;
    62  END CH6EX2;


PL/I OPTIMIZING COMPILER          /*   CH6EX2 - NESTED PROCEDURES */          PAGE  10
NO MESSAGES PRODUCED FOR THIS COMPILATION
```

```
                              STUDENT GRADE REPORT

PAGE:                    1

STUDENT SSN           COURSE ID            CREDIT HOURS          GRADE RECEIVED

233453234            MATH345                   3                     3
233453234            ENGL203                   4                     4
233453234            GEOG333                   3                     3
233453234            JOUR234                   4                     4

TOTAL CREDIT HOURS =           14
TOTAL HONOR POINTS =           50
THE STUDENT'S GPA =      3.571

345322344            MATH222                   4                     4
345322344            ENGL100                   4                     4
345322344            BIOG200                   4                     4

TOTAL CREDIT HOURS =           12
TOTAL HONOR POINTS =           48
THE STUDENT'S GPA =      4.000

654234545            ENGL100                   4                     0
654234545            BIOG101                   3                     0
654234545            MATH134                   4                     0

TOTAL CREDIT HOURS =           11
TOTAL HONOR POINTS =            0
THE STUDENT'S GPA =      0.000

TOTAL NUMBER OF RECORDS PROCESSED =           10
```

6.5

Debugging in PL/I

By way of background to debugging, you need to be aware of the fact that PL/I can detect errors at two distinct times in the total run process. First of all, the compiler is able to detect errors and possible errors of certain types and to generate messages describing what it found. Second, errors can be detected at execution time, as the machine code produced by the compiler is executed. Third, your program might very well contain what are errors from your point of view (i. e., producing incorrect results) but that PL/I has no means of detecting.

PL/I offers a number of aids to detecting those errors that it recognizes during execution. We introduce them before discussing debugging.

Conditions

As we noted in regard to ENDFILE in Chapter 5, during execution a number of situations can arise that will cause a program interrupt. When one of these conditions occurs, we say that it **raises** a condition. Furthermore, a condition is said to be **enabled** if raising it causes an action to occur; if raising the condition does not cause an action, the condition is **disabled**. All PL/I conditions are either enabled or disabled by default, but which are enabled and which are disabled depend partly on the compiler. The following discussion presumes the optimizing compiler. Of the conditions that are enabled by default, some can be disabled by the programmer and some cannot; conditions disabled by default can be enabled by the programmer. The ENDFILE condition, for example, is always enabled; it cannot be disabled.

A condition is explicitly enabled or disabled by a condition prefix, which can be a part of prefix list. A **condition prefix** is an appropriate PL/I keyword attached to the beginning of any PL/I statement except DECLARE and ENTRY statements (see Chapter 14), and (not discussed in this text) DEFAULT statements. The condition prefix must be enclosed in parentheses and the parentheses followed by a colon.

```
(SIZE):   RESULT = (R ** 3);
(CHECK):  CONDEX:  PROCEDURE OPTIONS (MAIN);
```

The keyword to disable a condition is a NO placed with no intervening blanks in front of the keyword to enable it (SIZE; NOSIZE).

The scope of a prefix (the portion of the program it applies to) depends on what kind of statement it is attached to.

1. Attached to a PROCEDURE or BEGIN statement, the prefix applies to all statements through the corresponding END, including any PROCEDURE or BEGIN statements nested within the block.

2. Attached to a DO (of any type) or a SELECT statement, the prefix applies to that statement only, not to the group of statements between DO/SELECT and the corresponding END.

3. Attached to a compound statement (IF, WHEN, OTHERWISE, ON), the prefix does not apply to the statement(s) contained within the compound statement.

A default (implicit) action exists for every condition, and this action occurs whenever an enabled condition is raised but for which no explicit action has been specified. To specify an alternative action, you must use an ON statement.

The conditions (and their abbreviations) that help you most in debugging are CONVERSION (CONV), FIXEDOVERFLOW (FOFL), SIZE (none), and ZERODIVIDE (ZDIV). A fifth condition, CHECK (none), is the subject of Section 6.5.5.

CONVERSION is raised when an invalid conversion is attempted on character data during an internal operation or during I/O. The condition is enabled by default but can be disabled with NOCONVERSION (NOCONV).

FIXEDOVERFLOW is raised when the length of the result of a fixed-point arithmetic operation exceeds the maximum permissible length for fixed-point numbers (15 for decimal values and 31 for binary values). The condition is enabled by default but can be disabled with NOFIXEDOVERFLOW (NOFOFL).

SIZE is raised when the high-order (leftmost) significant digit(s) is lost as the result of an assignment to a variable or intermediate result or during I/O. The condition is disabled by default.

ZERODIVIDE is raised by an attempt to divide by zero in either fixed- or floating-point division. The condition is enabled by default but can be disabled with NOZERODIVIDE (NOZDIV).

To close out our discussion of conditions, we point out that any enabled condition can be raised with the SIGNAL statement, which has the syntax of:

```
SIGNAL condition;
```

The SIGNAL statement is used primarily in testing your program to make sure that the action you have established through the ON statement actually occurs. If the condition specified on the SIGNAL statement is disabled, the statement is ignored.

6.5.2

Example Program— Debugging

Example program CH6EX3 (see Fig. 6–4) is the basis of nearly all examples given in the remainder of the chapter. As you can see, it is basically the final program from Chapter 5, now modified slightly to perform the additional task of computing the class average from the students' final averages. The added task necessitates the extra variables of CLASS_SUM, CLASS_AVG, and NUM_STUDENTS. If you look carefully at CLASS_SUM, you will discover that to create an error it has not been initialized.

```
        /* CHAPTER 6 EXAMPLE PROGRAM USING CHECK */

    1   (CHECK (TEST_AVG, PGM_AVG, FINAL_AVG, CLASS_SUM, CLASS_AVG)):
        CH6EX3:   PROCEDURE OPTIONS (MAIN);

        /*******************************************************************/
        /*                                                                 */
        /****** NOTE:   THIS PROGRAM CONTAINS AN INTENTIONAL ERROR *********/
        /*                                                                 */
        /*    FUNCTION:  THIS PROGRAM PRODUCES A REPORT ON THE STUDENTS'   */
        /*               STATUS IN A SINGLE CLASS AND CLASS AVERAGE        */
        /*    INPUT:     FOR EACH STUDENT A RECORD CONTAINING THE          */
        /*               STUDENT'S NAME, TEST AVERAGE, AND PROGRAM         */
        /*               AVERAGE                                          */
        /*                                                                 */
        /*    OUTUT:     THE INPUT AND AN APROPRIATE MESSAGE, AND THE      */
        /*               CLASS AVERAGE                                     */
        /*                                                                 */
        /*    NOTES:     1.  THE MINIMUM PASSING GRADE IS 68%             */
        /*               2.  THE STUDENT MUST HAVE A PASSING AVERAGE ON    */
        /*                   BOTH THE EXAMS AND THE PROGRAMS IN ORDER TO   */
        /*                   PASS THE COURSE                               */
        /*                                                                 */
        /*******************************************************************/

    2   DCL STUDENT_NAME    CHARACTER    (20);

    3   DCL TEST_AVG      FIXED DECIMAL (3);
    4   DCL PGM_AVG       FIXED DECIMAL (3);
    5   DCL FINAL_AVG     FIXED DECIMAL (5,2);
```

FIGURE 6–4
CH6EX3 (the use of CHECK in PL/I).

```
6        DCL CLASS_SUM         FIXED DECIMAL (5);
7        DCL CLASS_AVG         FIXED DECIMAL (5,2);

8        DCL PASS_COUNT        FIXED BINARY  (15)     INIT (0);
9        DCL FAIL_COUNT        FIXED BINARY  (15)     INIT (0);
10       DCL NUM_STUDENTS      FIXED BINARY  (15)     INIT (0);

11       DCL MORE_RECS         BIT           (1);
12       DCL  YES              BIT           (1)      INIT ('1'B);
13       DCL  NO               BIT           (1)      INIT ('0'B);

14           MORE_RECS = YES;
15           ON ENDFILE (SYSIN) MORE_RECS = NO;
16           PUT LIST (' ', 'GRADE REPORT', ' ', 'SPRING 1987');
17           PUT SKIP (2) LIST (' ', ' ', 'COMPUTER SCIENCE 101');
18           PUT SKIP (3) LIST ('STUDENT-NAME', 'TEST AVERAGE',
                 'PROGRAM AVERAGE', 'FINAL AVERAGE', 'MESSAGE');
19           PUT SKIP;

20           GET LIST (STUDENT_NAME, TEST_AVG, PGM_AVG);
```

PL/I OPTIMIZING COMPILER   /* CHAPTER 6 EXAMPLE PROGRAM USING CHECK */   PAGE   3
   STMT

```
21  LOOP:    DO WHILE (MORE_RECS);
22               NUM_STUDENTS = NUM_STUDENTS + 1;
23               FINAL_AVG = (TEST_AVG + PGM_AVG) / 2.0;
24               CLASS_SUM = CLASS_SUM + FINAL_AVG;
25               PUT SKIP LIST (STUDENT_NAME, TEST_AVG, PGM_AVG,
                               FINAL_AVG);

26               IF TEST_AVG < 68.0 | PGM_AVG < 68.0 | FINAL_AVG < 68.0
                 THEN
    COND1:       DO;
27                   PUT SKIP (0) LIST (' ', ' ', ' ', ' ',
                                               'FAILED CSCI 101');
28                   FAIL_COUNT = FAIL_COUNT + 1;
29               END COND1;

30               ELSE
    COND2:       DO;
```

FIGURE 6-4 *(continued)*

```
31                      PUT SKIP (0) LIST (' ', ' ', ' ', ' ',
                                   'PASSED CSCI 101');
32                    PASS_COUNT = PASS_COUNT + 1;
33                  END COND2;

34                  GET LIST (STUDENT_NAME, TEST_AVG, PGM_AVG);
35              END LOOP;

36              PUT SKIP (3) LIST (' ', ' ', 'NUMBER PASSING IS',
                               PASS_COUNT);
37              PUT SKIP LIST (' ', ' ', 'NUMBER FAILING IS', FAIL_COUNT);
38              CLASS_AVG = CLASS_SUM / NUM_STUDENTS;
39              PUT SKIP (2) LIST ('CLASS AVERAGE IS', CLASS_AVG);
40          END CH6EX3;
```

```
PL/I OPTIMIZING COMPILER        /* CHAPTER 6 EXAMPLE PROGRAM USING CHECK */    PAGE   4
COMPILER DIAGNOSTIC MESSAGES
ERROR ID L   STMT    MESSAGE DESCRIPTION


COMPILER INFORMATORY MESSAGES

IEL0533I I           NO 'DECLARE' STATEMENT(S) FOR 'SYSPRINT','SYSIN'.

END OF COMPILER DIAGNOSTIC MESSAGES


                    GRADE REPORT                              SPRING 1987
                                        COMPUTER SCIENCE 101


STUDENT-NAME           TEST AVERAGE        PROGRAM AVERAGE       FINAL AVERAGE         MESSAGE

TEST_AVG=    93;

PGM_AVG=    84;

FINAL_AVG=   88.50;

IBM537I  'ONCODE'=8097  DATA EXCEPTION
    IN STATEMENT 24 AT OFFSET +0003BE IN PROCEDURE WITH ENTRY CH6EX3
```

CH6EX3 was run on the PL/I Optimizing Compiler, Version 1, Release 3.1, with the compiler options GOSTMT, SMESSAGE, and NEST; all others were set to the IBM defaults. If you are a beginning programmer, this information means nothing to you, and you need not concern yourself with it. We present it here only because the exact format and content of the error messages and data produced by PL/I varies according to a wide variety of factors. If you are using the PL/C compiler, for example, the error messages (and to some extent their placement in the source code listing) will bear little resemblance to what we illustrate here.

6.5.3

Syntax and Logic Error Messages

Turning now to error messages produced by PL/I, the errors detected by the compiler, as we noted in Chapter 2, are referred to as **syntax errors** and involve violations of the established rules for the language. Placing a character other than a legitimate carriage control character in column 1 is one example of a syntax error. Other examples are misspelling keywords, failing to include the proper number of END statements, punctuation errors, and writing an assignment statement backwards (X + Y = Z instead of Z = X + Y). These errors are usually rather easy to locate and correct, although at times they can be tricky.

When the compiler detects an error, or the possibility of an error, it prints a message following the listing of the source code. The format of this output resembles the following:

```
COMPILER DIAGNOSTIC MESSAGES
ERROR ID L   STMT    MESSAGE DESCRIPTION
```

If the compiler detects one or more "unrecoverable," "severe," or "error" level errors, the next line reads

```
SEVERE AND ERROR DIAGNOSTIC MESSAGES
```

"Unrecoverable," "severe" and "error" level error messages are printed here; if none are encountered, the header is not printed. If the compiler generates any "warning" or "informatory" level messages, they are printed following this header:

```
COMPILER INFORMATORY MESSAGES
```

If the compiler generates no messages at all, the four lines above are replaced with

```
NO MESSAGES PRODUCED FOR THIS COMPILATION
```

The format of the messages themselves is suggested by the column headers, and for illustration we use the "informatory" message produced in CH6EX1. The ERROR ID is IEL0533I. IEL identifies the compiler producing the message as the optimizing compiler. You can use the entire error-id to look up the error in the appropriate manual for a more complete discussion of the problem and suggestions for correcting it. The column header L (level) indicates the severity of the error:

I: The compiler detected a possible inefficiency or is providing other information to the programmer. In CH6EX1 it is informing us that we did not declare the I/O data sets.
W: The PL/I statement referred to is syntactically correct, but it might contain an error.
E: The compiler has found an error but has "corrected" it according to built-in assumptions. Keep in mind that those assumptions might not be what you intended.
S: The compiler has detected an error that it is unable to "correct"; the program will execute, but very likely give incorrect results. (Note: At many installations default values do not allow programs with 'S' level errors to execute.)
U: The compiler has found an error that does not allow compilation to continue.

To illustrate one rather common syntax error, we deliberately coded the following statement (Statement 20) with the "G" in column 1.

```
GET LIST (STUDENT_NAME, TEST_AVG, PGM_AVG);
```

Under SEVERE AND ERROR DIAGNOSTIC MESSAGES, PL/I printed two messages:

```
IEL0234I S  20    INVALID SYNTAX.     'ET' IGNORED.
IEL0304I S  20    INVALID SYNTAX AFTER 'TEST_AVG, PGM_AVG)'.
                  'LIST (STUDENT_NAME, TEST_AVG, PGM_AVG)' IGNORED.
```

Under WARNING DIAGNOSTIC MESSAGES it printed:

```
IEL0372I I        INVALID CARRIAGE CONTROL CHARACTER 'G'.
```

This particular set of messages makes it quite clear what the problem is: The message points you to Statement 20, and by combining the "invalid" 'ET' with the warning of an invalid "G" carriage control character you can see immediately what must be done to correct the error.

An example of a common syntax error that might puzzle you the first time you encounter it is placing the opening comment delimiter (/*) in columns 1 and 2, instead of 2 and 3; in this case the compiler was unable to issue a clear error message. When we deliberately coded

```
/* NOTES:   1.   THE MINIMUM PASSING GRADE IS 68%          */
/*          2.   THE STUDENT MUST HAVE A PASSING AVERAGE   */
```

PL/I faithfully printed the source code as far as

```
/* NOTES:   1.   THE MINIMUM PASSING GRADE IS 68%          */
```

and then stopped printing source code. As an error message PL/I printed

```
IEL0241I          1 'END' STATEMENT(S) ASSUMED.
```

This message is of little help to the inexperienced programmer, but because the combination of a slash in column one and an asterisk in column 2 is a special signal meaning "end of data set," the compiler assumed it had encountered the end of the source code; therefore, it stopped processing at once. Since it had encountered a PROCEDURE statement with no matching END, it produced this message. Also, since a comment is not a PL/I statement, PL/I could not supply a statement number for the error message.

Some programmers prefer to code a single DECLARE, followed by each variable and its attributes and with each line separated from the next by a comma. One drawback of this technique is that the entire list of variables becomes a single PL/I statement. If a syntax error occurs anywhere in the DECLARE, PL/I can give you only the statement number that covers the entire declaration. This, in turn, means that you must hunt through the entire set of variables and their attributes in an attempt to find the error. If you make the declaration of each variable a separate statement by starting each line with DECLARE (or DCL) and ending it with a semicolon, each will have a separate statement number, and the compiler will be able to pinpoint the variable at fault.

To illustrate the difference between the two formats, we modified the CH6EX1 DECLARE by putting in an extraneous comma, and changing to a single DECLARE:

```
PASS_COUNT   FIXED BINARY (15),   INIT (0),
```

The 'S' level message that resulted indicates that the error occurred in Statement 2, and goes on to say:

```
LOWER BOUND 1 OF 'INIT' GREATER THAN HIGHER BOUND.
```

Given a single DECLARE statement for all the variables, STMT 2 embraces everything from DECLARE STUDENT_NAME . . . to NO . . . , which gives you 12 lines of code to search in order to find the error.

We then changed the declaration to read as it does in the printed version (i.e., each variable declared in a separate statement). This produced the same error message, with one important difference: The compiler is now able to tell us that the error is in STMT 8; since STMT 8 embraces only the single line of code declaring PASS_COUNT, we are led directly to the offending statement.

You will have to determine for yourself, of course, whether or not the advantage of being able to find syntax errors in the DECLAREs outweighs the extra typing involved in making each variable a separate declaration. As for the meaning of the error message itself, after you have studied Chapter 11 you will understand that PL/I thinks INIT is an array name.

Once you have rid your program of syntax errors, your program will execute, and you can begin the hunt for the **logic** errors detected (if they are detected at all) during execution. There are two particularly troublesome aspects of logic errors in PL/I, both of which have been mentioned previously but are worth repeating once more.

1. Your program can contain errors which are logic errors from your perspective but not from the perspective of the computer. Such errors manifest themselves either by terminating execution or as erroneous output data. The logic error in CH6EX3, of course, terminated execution. A simple example of the kind of error that produces only incorrect results is coding a subtract symbol instead of an add symbol. *You must check your output carefully.*

2. If in the executable instructions you misspell a declared variable, PL/I will implicitly declare a new variable, assigning as its name the misspelling of the declared variable; it might do so, moreover, with no warning message to you. The following example illustrates the problem.

```
DCL RECEIVER        FIXED BINARY (15) INIT (0);
RECIEVER = 1;
DO WHILE (RECIEVER < 10);

  RECIEVER = RECIEVER + 1;
END;
```

PL/I creates a second variable named RECIEVER and gives it the default attributes of FLOAT DECIMAL (6) based on the letter with which the variable begins. Its value is used to control the iterations of the DO WHILE. Meanwhile, it is RECEIVER that is being incremented each time through the loop. Because the value in RECIEVER remains constant at 1, the net result is an infinite loop that will continue until your program has used the amount of central processing unit time it has been allotted. You will learn rather quickly (probably from frustrating experience) when a misspelling might be involved in creating a bug. Remember, however, that the human eye tends to "see" what it expects to see, and it is quite possible you will stare at the two variables without noticing the spelling variation.

To illustrate the manner in which PL/I handles a logic error, we submitted CH6EX3 for execution, and it gave us the following error message the first time it encountered the CLASS_SUM = statement (Statement 24):

```
IBM5371  'ONCODE'=8097  DATA EXCEPTION
IN STATEMENT 24 AT OFFSET +0003BE IN PROCEDURE WITH ENTRY CH6EX3
```

The only information given here that is of value to the beginning programmer is the reference to Statement 24. This tells you that the error occurred when Statement 24 executed, but it does *not* necessarily mean that it is Statement 24 that must be corrected. Check Statement 24 first, of course, and if you find no error, work backward through the code in the order it executed to try to find the source of the error.

In addition to the conditions described earlier, PL/I provides two easily used tools to aid us in some situations, and we now turn to them.

6.5.4

PUT DATA

A second type of stream I/O (in addition to list-directed) is data-directed I/O; therefore, it is one of the standard output methods provided by PL/I. Data-directed I/O is so rigid in its format that it is seldom used as a standard I/O method. Data-directed input is so seldomly used, in fact, that we do not bother to introduce it. Data-directed output, on the other hand, is widely used as a debugging tool. The statement has two formats.

```
PUT DATA;
```

and

```
PUT DATA (data list);
```

where "data list," as it is in LIST I/0, is a series of one or more variables separated by commas. As you can see, the second format is identical to the list-directed output statement, except that DATA replaces LIST. PUT DATA resembles PUT LIST also in that it has the same formatting options of SKIP, LINE, and PAGE. Finally, PUT DATA uses the same tab positions (columns 25, 49, 73, and 97) as PUT LIST. There are several differences between the two, however. First, PUT LIST does not have a counterpart to the first format of PUT DATA. Second, only variables are permitted in the PUT DATA data list; no literals or expressions are allowed. Third, the output from PUT DATA has the format of variable = value.

The chief difference between the two data-directed output formats is that the first prints the name and value of all variables known to PL/I when the statement is encountered, while the second prints the name and value of only those variables included in the data list.

To see how data-directed output statements can be an aid to debugging, let us apply it to the instance of the misspelled variable. Let us assume that you have run the program, discovered the infinite loop, checked your source code, but failed to notice the misspelling. At this point you need to insert the PUT DATA statement somewhere within the range of the loop, probably immediately after the DO WHILE. Notice, however, that in this case the second format of the PUT DATA statement is less likely to help you because, being unaware of the misspelling, you would code either

```
PUT DATA (RECEIVER);
```

or

```
PUT DATA (RECIEVER);
```

but not both. When you examine the output from this run, you notice (if you coded RECEIVER) that its value was incremented properly, and you are no closer to understanding the source of the error than before. If, on the other hand, you placed RECIEVER in the data list, you discover that its value remained constant, but again you do not understand why. The solution, of course, is to use PUT DATA with no data list. The output then lists the value of both RECEIVER and RECIEVER each time through the loop, and you can detect the source of the bug almost at a glance.

PUT DATA is a valuable tool for debugging. It can be coded as many times as desired in the same program. If you are uncertain where your program is abending, you can insert a series of PUT DATA statements at strategic points in the program. Those appearing later in the program than the statement causing the abend will not execute. When we inserted a PUT

DATA in CH6EX3 following Statement 21, PUT DATA executed once, listing values for STUDENT_NAME, TEST-AVG, and PGM AVG. FINAL_AVG is shown to have no value, and CLASS_SUM is not included. This illustrates, then, how a series of PUT DATA statements can help you pinpoint the offending statement. As an aid in debugging, the PUT DATA statement can be used in many more ways than we have illustrated here; these uses will occur to you as the occasion arises. Finally, you should realize that the output from PUT DATA is printed intermixed with the regular output from your program; for this reason, as well as others, you will want to remove the PUT DATA statement from your program once you are getting the desired results.

CHECK

Like other conditions, CHECK is specified as a condition prefix. It is disabled by default, and it can be coded alone or with a data list. The syntax for CHECK with a data list is:

```
(CHECK (data list)):
```

Notice that two sets of parentheses are required. Also, because it slows execution and requires additional storage, CHECK should always be removed when the program is executing properly; therefore, it is often coded on a line by itself to facilitate its removal. The following appears in CH6EX1:

```
(CHECK (TEST_AVG, PGM_AVG, FINAL_AVG, CLASS_SUM, CLASS_AVG)):
CH6EX3:  PROCEDURE OPTIONS (MAIN);
```

The function of CHECK is to print the value of each variable in the data list. The value of a given variable is printed each time the variable is encountered in the executable code, provided that two conditions are present:

1. The value of the variable has *changed* since it was last encountered.

2. The variable contains a defined value.

A **defined value** is any value that has been assigned to a variable by any one of (1) the INITIAL attribute, (2) an assignment statement, or (3) an input statement. It is important to understand that the statement containing the variable list is executed first; then CHECK is executed. The first time a variable is encountered its value is considered to be "changed"; therefore, on the first encounter CHECK will list the value for all defined variables in the data list. On succeeding encounters PL/I will print the values of only the variables whose values have changed.

To illustrate the output of CHECK we coded the CHECK given above in CH6EX3. The output from CHECK tells us that TEST−AVG = 93; PGM_ AVG = 84; and FINAL_AVG = 88.50. A scan of the code tells us that Statement 23 has executed successfully. Because CHECK gives no value for CLASS_SUM, it seems obvious that Statement 24 is the offending statement and that CLASS_SUM is the specific cause.

Both PUT DATA and CHECK are very useful debugging tools, but as you can see, CHECK is considerably more flexible when it is appended to the PROCEDURE statement label. If you misjudge the source of your error, it is quite easy to code PUT DATA at a point where it will do little or no good. In using either PUT DATA or CHECK, however, you should give some thought to its placement in the program. With a small program such as CH6EX3, you can put CHECK on the PROCEDURE statement with no real concern. With a larger, more complex program, on the other hand, doing so might generate far too much output. Finally, let us remind you once more to remove either CHECK or PUT DATA from your program when you have finished debugging.

6.6

Summary

Having been presented (in Chapter 5) with the information necessary to implement each of the structures allowed to structured programming, you should begin to look beyond the basic problem of simply getting correct results. A truly successful program must meet the four goals of accuracy, reliability, readability, and maintainability. Accuracy is not only getting the "correct" results but also the results desired by the end user, and accuracy is achieved through good communication with the end user, a thorough analysis of the problem, and a careful check of the output. Reliability can be thought of as accuracy over the life of the program and is produced by a careful and exhaustive testing of the program. As many paths of executions as is feasible should be tested and in various combinations. Test data should be supplied which falls within, on, and outside the acceptable parameters, again in several combinations.

Readability is marked, first of all, by a program that enables a reader readily to find any variable in the DECLARE; this, in turn, is made possible by including only one variable per line, lining up the scale and base attributes in a chosen column, and also lining up the INITIAL attribute for initialized variables in a column. Any one of several orders of listing the variables is satisfactory, but you should be consistent in your order. A second factor involved in readability is that among the executable statements indentation should be used to show clearly the logic of the program and to show continuation of statements; the indentation, too, should be consistent. The code in a readable program is single-spaced, but you should mark off blocks of related statements by a double space. Page breaks should be used when necessary, generally to separate subroutine procedures. Double spacing can be specified by placing a 0 in column one, and a page break by putting a 1 in column one.

In the world of business and industry, programs have a life expectancy of as long as 20 years; over that life span, however, the program ordinarily must be modified periodically to reflect various changes. This process is called maintenance, and all programs should be written with maintainability in mind. No program is easily maintainable unless it is, in the first place, a structured program. A structured program is marked by these attributes:

1. It contains only the allowable structures of sequence, decision, and iteration.

2. All its loops are top-driven; that is, the condition controlling the loop is tested at the top of the loop, the statements within the loop are executed only if the condition is true, and it is possible that the loop will never be entered (i.e, when the condition is false on the very first test).

3. The program itself and any loops within it have one entrance and one exit, which effectively prohibits the use of the GO TO statement.

4. The program must be designed according to the principles of top-down development, a progression from the general to the specific.

Also, to be easily maintainable a program must be thoroughly documented. We recommend a box of comments at the head of the program. We further recommend the headings of FUNCTION (one sentence describing what problem the program is to solve), INPUT (a general description of the contents of the records in the input data set), OUTPUT (a general description of the output format), and, if needed, NOTES (any additional information that will help the reader understand the program). Only rarely, in our opinion, should comments be interspersed with the code. There are no universal standards for comments, however. Finally, no program is easily maintainable unless it is also easily readable.

Once a program has been designed, coded, and entered into the computer, it is processed, and the debugging process begins. Program errors are of two types: syntax errors, detected by the compiler; and logic errors, detected during execution. Syntax errors are generally quite easy to locate and correct, but logic errors can be more difficult because often they manifest themselves only in incorrect results. Debugging is not complete until all messages of severity 'W' and higher have been eliminated.

Two highly useful debugging tools are PUT DATA and CHECK. PUT DATA has two formats:

```
PUT DATA;
PUT DATA (data list);
```

PUT DATA lists the value of each variable known at the time the PUT DATA statement executes; PUT DATA (data list) lists the value of each variable in the data list if it is known when the statement executes.

CHECK has the formats of

```
(CHECK):
(CHECK (data list)):
```

Notice carefully that both CHECK and "data list" are enclosed in parentheses and that the prefix is delimited by a colon. CHECK prints the value of each defined variable in its data list or each defined variable (in the case of CHECK:) each time a statement with the variable is encountered, *provided that the value of the variable has changed since the last encounter*. The value is considered to be "changed" the first time the variable is encountered. Also, the statement containing the variable is executed *before* CHECK is executed.

The secret to the efficient use of PUT DATA and CHECK as debugging tools is determining which will best give you the needed information; this is learned primarily through experience. Both PUT DATA and CHECK should be removed from the code once the debugging process is complete.

Your debugging tools might well be put to a real test once you attempt to implement the nested IFs and DO WHILEs presented in Chapter 7; this nesting represents a quantum leap in complexity of program logic.

6.7

Exercises

1. Explain briefly the meaning of cost "efficiency" in today's programming environment.

2. Define the program goals of accuracy, reliability, readability, and maintainability. Explain why each is important.

3. Why is it wise to use meaningful variable names even if no one but you will ever read the source code?

4. List a number of aspects of coding that should be consistent. Why is consistency so important?

5. Choose one of the DECLARE formats discussed above (or another one) and defend your choice.

6. Choose a plan for inserting comments in your program and explain why you chose this particular plan.

7. Explain the concept of top-down development.

8. Assume that your employer asks you to write a payroll program. List and explain the steps you will go through prior to coding.

9. Explain how DO UNTIL and GOTO work. Defend either their inclusion or exclusion in structured programming.

10. List the two times at which PL/I can detect errors or possible errors and explain the nature of the errors in each case.

11. What is the advantage of declaring each variable in a statement by itself?

12. List the syntax rules for using CHECK.

13. Give some examples of how CHECK and PUT DATA can help you in debugging.

For the following programming assignments, use internal subroutine procedures and global variables. The main routine should do nothing but call the subroutines in the proper order.

14. Rewrite the code for Exercise 6 of Chapter 5, using four subroutines: (a) a subroutine to print the column headers, (b) a subroutine to read the records, (c) a subroutine to do the computations, and (d) a subroutine to print the detail lines.

15. Rewrite the code for Exercise 7 of Chapter 5. Provide a subroutine to (a) read the records, (b) print the detail lines, (c) keep the running totals, and (d) print the "totals" lines.

16. Rewrite the code for Exercise 8 of Chapter 5, providing subroutines as necessary. A subroutine should carry out a single function.

17. Write a billing program for a company that rents computer time. The input will have the following format.

Customer name	20 characters
Starting time	nnnn
Ending time	nnnn

The time is specified using the 24-hour clock (e.g., 1:00 P.M. is 1300). No customer will have a block of time greater than 24 hours. The company charges $450.00 per hour if the computer is used between 0800 and 1700; from 1701 to 0759 the charge is $275.00. Your program should produce a report that prints one line for each customer, listing the input and the charge. Provide appropriate subroutines.

C H A P T E R 7

MORE COMPLEX

CONDITIONAL

EXECUTION

In this chapter we present modifications of the decision and iteration structures that, while increasing the complexity of program coding considerably, also allow for the efficient solution of more complex problems. The first of these, nested IF statements, can be avoided; any problem that might suggest the use of nested IF statements can be solved without their use. Nevertheless, there are several reasons why the use of nested IF statements should not be avoided; we develop these reasons as we go along. The modification of the iteration structure, nested DO WHILE statements, on the other hand, solve commonly encountered problems that can be solved in no other manner. A second modification of the decision structure, the case structure, might also be seen as a modification of the nested IF. It is the final topic of the chapter.

**Nested IF
Statements**

You will perhaps recall that in Chapter 5, when we introduced the implementation of the decision structure, we told you that for the time being we would assume that the IF-THEN and ELSE would be followed only by an imperative or a DO and more than one imperative. It is now time to drop that restriction and to realize that the IF-THEN or the ELSE, or both, can be followed by a condition (that is, another IF-THEN and perhaps its optional ELSE) as well as by an imperative. When an IF-THEN or an ELSE is followed by another IF-THEN (and ELSE), the result is a **nested IF** statement.

To illustrate the nested IF let us suppose we are to write a program to read a file of students who have requested permission to enroll in Computer Science 400. Since the demand for the class is greater than the available space, it is necessary to rank the students, and it has been determined that the two factors used for the ranking will be whether or not the student has completed Computer Science 395 and whether his or her computer science grade point average is 3.50 or higher. The following table shows the four possible categories of students and the priorities.

Priority 1: 395 yes & GPA > 3.49
Priority 2: 395 yes & GPA < 3.50
Priority 3: 395 no & GPA > 3.49
Priority 4: 395 no & GPA < 3.50

In the following examples we present only that portion of the program that is concerned with assigning a priority to each student. Solving this problem without the use of nested IF statements requires the following code:

```
IF COURSE = 395 & GPA > 3.49 THEN
    PRIORITY = 1;
IF COURSE = 395 & GPA < 3.50 THEN
    PRIORITY = 2;
IF COURSE ^= 395 & GPA > 3.49 THEN
    PRIORITY = 3;
IF COURSE ^= 395 & GPA < 3.50 THEN
    PRIORITY = 4;
```

Implementing the solution using nested IF statements, on the other hand, produces this code:

```
(1)   IF COURSE = 395 THEN
(2)       IF GPA > 3.49 THEN
              PRIORITY = 1;
```

```
(2)     ELSE
            PRIORITY = 2;
(1) ELSE
(3)     IF GPA > 3.49 THEN
            PRIORITY = 3;
(3)     ELSE
            PRIORITY = 4;
```

Your first thought might be that nested IF statements are used to reduce the amount of coding required, but, of course, a glance at the two examples seems to rule out that motivation: The second example has two more *lines* of code, and each contains the same number of *statements*. There are, however, two other reasons for using nested IF statements. First of all, for anyone familiar with nested IF statements, the logic of the second example is a good deal clearer than that of the first example. An even more important reason to use nested IF statements, however, is the fact that the code will execute more quickly if they are used. With nested IF statements *all following statements within the nested construct are skipped as soon as a false condition is encountered*. This means that at most only two comparisons will have to be made for each student. Let us assume that the first student has completed Computer Science 395 and has a GPA of 2.80. A comparison is made with IF_1, and the result is true; a comparison is then made with IF_2, and the result is false. $ELSE_2$ is immediately executed, and $ELSE_1$ is bypassed. Let us now assume that the second student read from the file has not completed Computer Science 395 and has a GPA of 3.70. The condition of IF_1 is tested and found to be false; therefore, control jumps instantly to $ELSE_1$, and the IF_3 condition is tested and found to be true. A priority of 3 is assigned, and $ELSE_3$ is bypassed. In contrast to the situation just outlined, consider the fact that in the first example, in which the nested IF is not used, four comparisons must be made for every student, even those who have completed Computer Science 395 and have a GPA above 3.49. Given a large enough input file and a situation in which the program is run many, many times, the accumulated savings of two comparisons over four become quite significant.

In order to make use of nested IF statements you must always be aware of two rules regarding them. First, as we mentioned above, when a false condition is encountered in the IF-THEN, no more tests are made; instead control passes immediately to the corresponding ELSE. On the other hand, as you recall from Chapter 5, when the IF condition is true the corresponding ELSE statement is bypassed, and in the case of nested IF statements all higher-level (outer) ELSE statements are also bypassed. The second rule to remember is that each ELSE corresponds to the most immediately preceding IF-THEN that does not already have a corresponding ELSE. We cannot hope to illustrate all the possibilities of the nested IF, but perhaps the following will help clarify this point.

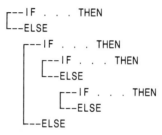

```
┌--IF . . . THEN
│  ┌--IF . . . THEN
│  │  ┌--IF . . . THEN
│  │  │  ┌--IF . . . THEN
│  │  │  └--ELSE
│  │  └--ELSE
│  └--ELSE
└--ELSE
```

One more example:

```
┌--IF . . . THEN
└--ELSE
      ┌--IF . . . THEN
      │  ┌--IF . . . THEN
      │  └--ELSE
      │     ┌--IF . . . THEN
      │     └--ELSE
      └--ELSE
```

For purposes of illustration we have shown IF-THEN statements nested four deep. In practice, however, most PL/I programmers would set three as the maximum number of levels for nesting. Beyond three the logic often becomes difficult to follow. In many instances it is better to use the case structure for handling deeper levels of nesting.

Nested IF statements are usually confusing to the beginning programmer when they are first encountered, but they are less confusing than they appear at first glance. It takes only a relatively small amount of experience with them, moreover, to feel completely at home with nested IF statements. In addition, the execution efficiency and "clean code" made possible by nested IF statements more than compensate for whatever frustration you might experience in first using them.

7.1.1

The Null ELSE in Nested IF Statements

In addition to the two rules for nested IF statements given above, there is a third that is a potential source of trouble. The third rule states that an IF-THEN *must* have a corresponding ELSE when the next higher level IF-THEN has a corresponding ELSE. This means, of course, that when the highest level IF-THEN has a corresponding ELSE each lower level must also have a corresponding ELSE. The requirement can create a problem, however, when you wish to take no action in the case that the lowest-level IF condition is false. To surmount this potential problem, PL/I offers us the **null ELSE,** which is a statement coded simply to meet the PL/I syntax rule regarding the correspondence of the ELSE with the IF-THEN. The syntax of the null ELSE is simply an ELSE immediately followed by a semicolon.

```
ELSE;
```

Going back to our example of assigning priorities to requests for Computer Science 400, let us assume that, rather than assigning priorities, we want to print the student-ID of the students that have a priority of 1, 3, or 4 in the original example, nothing is to be done in the case of students with a priority of 2. Our nested IF code now looks like this.

```
(1)    IF COURSE = 395 THEN
(2)        IF GPA > 3.49 THEN
(3)            PUT LIST (STUDENT_ID);
(2)        ELSE;
(1)    ELSE
(3)        IF GPA > 3.49 THEN
               PUT LIST (' ', STUDENT_ID);
(3)        ELSE
               PUT LIST (' ', ' ', STUDENT_ID);
```

Notice that were ELSE$_2$ not present ELSE$_1$ would correspond to IF$_2$, destroying the logic entirely. Finally, although it is not intended to reflect the example we have been discussing, the generalized flowchart shown in Figure 7–1 indicates the situation in which a null ELSE might provide the best solution. Keep in mind that the flowchart represents only what might be the two lowest levels of nested IF statements; there can be as many nested higher levels as the situation warrants.

As we noted earlier, nested IF statements are a valuable aid to both "clean code" and efficient execution. Our next topic, nested DO WHILE statements, is an even more indispensable tool.

7.1.2

Example Program—
Nested IF

The CH7EX1 program (see Fig. 7–2) solves the problem of assigning letter grades to the students in class based on their percentage average. The logic of the program is straightforward nested IF statements, and nothing more need

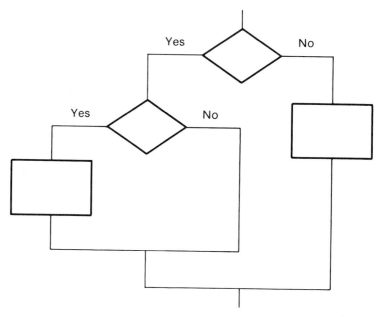

FIGURE 7–1
Nesting IF-THEN-ELSE statements.

be said about that. Notice, however, that the pattern of the IF statements and their corresponding ELSE statements differs from any we have so far presented. One other point to notice is the elementary error checking carried out by the initial IF (Statement 34) and the final ELSE (Statement 54) of PROCESS-RTN. Unless the input data has gone through a previous program that would validate it, it is always desirable to do at least some minimal error checking of this sort. The pseudocode follows.

```
CH7EX1
    Call HDR_RTN
    Call READ_RTN
    DO WHILE (more input)
        Call PROCESS_RTN
        Call READ_RTN
    ENDDO
    Call SUMMARY_RTN
STOP
HDR_RTN
    Print page, column headers
```

```
                EXIT
                READ_RTN
                   Read a record
                EXIT
                PROCESS_RTN
                   IF (percentage grade < 101 and > -1)
                      IF (percentage grade > 91)
                         Print student name, grade of 'A'
                         A-tally ← A-tally + 1
                      ELSE
                         IF (percentage grade < 92 and > 83)
                            Print student name, grade of 'B'
                            B-tally ← B-tally + 1
                         ELSE
                            IF (percentage grade < 84 and > 75)
                               Print student name, grade of 'C'
                               C-tally ← C-tally + 1
                            ELSE
                               IF (percentage grade < 76 and > 67)
                                  Print student name, grade of 'D'
                                  D-tally ← D-tally + 1
                               ELSE
                                  Print student name, grade of 'F'
                                  F-tally ← F-tally + 1
                               ENDIF
                            ENDIF
                         ENDIF
                      ENDIF
                   ELSE
                      Print student name, error message
                   ENDIF
                EXIT
                SUMMARY_RTN
                   Print number of students for each letter grade
                EXIT
```

7.2

Nested DO WHILE Statements

One of the most commonly encountered situations is one in which the iteration structure must be carried out at two separate levels. In other words, it is often necessary to enter a loop and, without leaving that loop, enter a second loop, thus creating a "loop within a loop." The generalized flowchart shown in Figure 7–3 illustrates the logic of the loop within a loop.

```
     STMT

          /* CH7EX1 USING NESTED IF */
     1  CH7EX1:   PROCEDURE OPTIONS (MAIN);
          /**********************************************************************/
          /*                                                              */
          /*   FUNCTION:  THIS PROGRAM COMPUTES A LETTER GRADE FOR EACH   */
          /*              STUDENT BASED ON THE STUDENT'S PERCENTAGE       */
          /*              AVERAGE AND TALLIES THE NUMBER OF STUDENTS       */
          /*              RECEIVING EACH GRADE                            */
          /*                                                              */
          /*   INPUT:     AN UNKNOWN NUMBER OF RECORDS, EACH CONTAINING   */
          /*              A STUDENT NAME AND A PERCENTAGE AVERAGE         */
          /*                                                              */
          /*   OUTPUT:    A REPORT LISTING THE STUDENT NAME AND          */
          /*              THE LETTER GRADE ASSIGNED                      */
          /*                                                              */
          /*   NOTES:     LETTER GRADES ARE COMPUTED ON THE FOLLOWING SCALE */
          /*                100-92 = A                                   */
          /*                 91-84 = B                                   */
          /*                 83-76 = C                                   */
          /*                 75-68 = D                                   */
          /*              BELOW 68 = F                                   */
          /*                                                              */
          /**********************************************************************/
     2      DCL STUDENT_NAME    CHARACTER     (20);
     3      DCL PERCENT_GRADE   FIXED DECIMAL (3);
     4      DCL A_TALLY         FIXED BINARY  (15)    INIT (0);
     5      DCL B_TALLY         FIXED BINARY  (15)    INIT (0);
     6      DCL C_TALLY         FIXED BINARY  (15)    INIT (0);
     7      DCL D_TALLY         FIXED BINARY  (15)    INIT (0);
     8      DCL F_TALLY         FIXED BINARY  (15)    INIT (0);
```

FIGURE 7–2
CH7EX1 (nested IF statements).

```
  9     DCL MORE_RECS         BIT            (1);
 10     DCL   YES             BIT            (1)      INIT ('1'B);
 11     DCL   NO              BIT            (1)      INIT ('0'B);
 12           MORE_RECS = YES;
 13           ON ENDFILE (SYSIN) MORE_RECS = NO;
 14           CALL HDR_RTN;
 15           CALL READ_RTN;
 16 LOOP:     DO WHILE (MORE_RECS);
 17              CALL PROCESS_RTN;
 18              CALL READ_RTN;
 19           END LOOP;
 20           CALL SUMMARY_RTN;
 21           RETURN;
```

```
 22 HDR_RTN:  PROCEDURE;
    /*****************************************************************/
    /*                                                             */
    /*  FUNCTION:  TO PRINT THE PAGE AND COLUMN HEADERS            */
    /*                                                             */
    /*  GLOBAL VARIABLES:  NONE                                   */
    /*                                                             */
    /*  SUBROUTINES CALLED:  NONE                                 */
    /*                                                             */
    /*  INPUT:     NONE                                           */
    /*                                                             */
    /*  OUTPUT:    TWO LINES OF PAGE HEADER AND ONE LINE OF COLUMN */
    /*             HEADERS, WITH TWO BLANK LINES FOLLOWING        */
    /*                                                             */
    /*****************************************************************/
 23           PUT LIST (' ', 'GRADE REPORT', ' ', 'SPRING 1987');
 24           PUT SKIP (2) LIST (' ', ' ', 'COMPUTER SCIENCE 101');
 25           PUT SKIP (2) LIST (' ', ' ', 'STUDENT-NAME', ' ',
                            'FINAL GRADE');
 26           PUT SKIP (2);
 27           RETURN;
 28        END HDR_RTN;

 29 READ_RTN:  PROCEDURE;
```

FIGURE 7–2 *(continued)*

```
/*******************************************************************/
/*                                                              */
/*  FUNCTION:  TO READ A SINGLE RECORD                          */
/*                                                              */
/*  GLOBAL VARIABLES:  STUDENT_NAME                             */
/*                     PERCENT_GRADE                            */
/*                                                              */
/*  SUBROUTINES CALLED:  NONE                                   */
/*                                                              */
/*  INPUT:  THE INPUT VARIABLES                                 */
/*                                                              */
/*  OUTPUT: NONE                                                */
/*                                                              */
/*******************************************************************/
30         GET LIST (STUDENT_NAME, PERCENT_GRADE);
31           RETURN;
32       END READ_RTN;

33  PROCESS_RTN:  PROCEDURE;
```

```
/*******************************************************************/
/*                                                              */
/*  FUNCTION:  TO CALCUALTE A LETTER GRADE FROM A PERCENTAGE    */
/*             GRADE (SEE ABOVE FOR SCALE) AND PRINT A DETAIL   */
/*             LINE                                             */
/*                                                              */
/*  GLOBAL VARIABLES:  STUDENT_NAME                             */
/*                     PERCENT_GRADE                            */
/*                     A_TALLY                                  */
/*                     B_TALLY                                  */
/*                     C_TALLY                                  */
/*                     D_TALLY                                  */
/*                     F_TALLY                                  */
/*                                                              */
/*  SUBROUTINES CALLED:  NONE                                   */
/*                                                              */
/*  INPUT:    NONE                                              */
```

```
      /*                                                           */
      /*  OUTPUT:    A DETAIL LINE CONTAINING A STUDENT NAME AND THE   */
      /*            GRADE                                          */
      /*                                                           */
      /*****************************************************************/
  34            IF PERCENT_GRADE < 101 & PERCENT_GRADE > -1 THEN
                   IF PERCENT_GRADE > 91 THEN
   IF_A:          DO;
  35                  PUT SKIP LIST (' ', ' ', STUDENT_NAME, ' ', 'A');
  36                  A_TALLY = A_TALLY + 1;
  37               END IF_A;
  38            ELSE
                   IF PERCENT_GRADE < 92 & PERCENT_GRADE > 83 THEN
   IF_B.          DO;
  39                  PUT SKIP LIST (' ', ' ', STUDENT_NAME,' ',
                                     'B');
  40                  B_TALLY = B_TALLY + 1;
  41               END IF_B;
  42            ELSE
                   IF PERCENT_GRADE < 84 & PERCENT_GRADE > 75 THEN
   IF_C:          DO;
  43                  PUT SKIP LIST (' ', ' ', STUDENT_NAME, ' ',
                                     'C');
  44                  C_TALLY = C_TALLY + 1;
  45               END IF_C;
  46            ELSE
                   IF PERCENT_GRADE < 76 & PERCENT_GRADE > 67
                   THEN
   IF_D:          DO;
  47                 PUT SKIP LIST (' ', ' ', STUDENT_NAME,
                                     ' ', 'D');
  48                  D_TALLY = D_TALLY + 1;
  49               END IF_D;
  50            ELSE
   IF_F:          DO;
  51                 PUT SKIP LIST (' ', ' ', STUDENT_NAME,
```

```
                                                ' ', 'F');
  52                  F_TALLY = F_TALLY + 1;
```

FIGURE 7–2 *(continued)*

```
53                          END IF_F;
54              ELSE

                PUT SKIP LIST (STUDENT_NAME,
                              '*** OUT OF VALID GRADE RANGE ***');
55          RETURN;
56          END PROCESS_RTN;

57  SUMMARY_RTN:  PROCEDURE;
    /********************************************************************/
    /*                                                                */
    /*  FUNCTION:  TO PRINT THE SUMMARY PAGE                         */
    /*                                                                */
    /*  GLOBAL VARIABLES:  A_TALLY                                   */
    /*                     B_TALLY                                   */
    /*                     C_TALLY                                   */
    /*                     D_TALLY                                   */
    /*                     F_TALLY                                   */
    /*                                                                */
    /*  SUBROUTINES CALLED:  NONE                                    */
    /*                                                                */
    /*  INPUT:     NONE                                              */
    /*                                                                */
    /*  OUTPUT:    FIVE LINES GIVING THE NUMBER OF STUDENTS          */
    /*             PROCESSED AND THE NUMBER RECEIVING EACH GRADE     */
    /*                                                                */
    /********************************************************************/
58          PUT SKIP (3) LIST (' ', 'NUMBER OF A STUDENTS IS', A_TALLY);
59          PUT SKIP LIST (' ', 'NUMBER OF B STUDENTS IS', B_TALLY);
60          PUT SKIP LIST (' ', 'NUMBER OF C STUDENTS IS', C_TALLY);
61          PUT SKIP LIST (' ', 'NUMBER OF D STUDENTS IS', D_TALLY);
62          PUT SKIP LIST (' ', 'NUMBER OF F STUDENTS IS', F_TALLY);
63          RETURN;
64        END SUMMARY_RTN;
65  END CH7EX1;
```

```
PL/I OPTIMIZING COMPILER        /* CH7EX1 USING NESTED IF */              PAGE   6
COMPILER DIAGNOSTIC MESSAGES
ERROR ID L   STMT    MESSAGE DESCRIPTION
```

```
COMPILER INFORMATORY MESSAGES

IEL0533I I          NO 'DECLARE' STATEMENT(S) FOR 'SYSIN','SYSPRINT'.

END OF COMPILER DIAGNOSTIC MESSAGES


             GRADE REPORT                         SPRING 1987

                            COMPUTER SCIENCE 101

                            STUDENT-NAME                  FINAL GRADE


                            CARLSON, AMY                      A
                            DAWSON, DARYL                     F
                            EVANS, DALIA                      C
                            GILBERT, GARY                     B
                            ISHAM, JOE                        D
KERRY, BOB        *** OUT OF VALID GRADE RANGE ***
MONSON, MARY      *** OUT OF VALID GRADE RANGE ***
                            NELSON, FRED                      A
                            PAUL, JUDY                        B
                            ROSE, IRIS                        C
                            STEVENS, LYLE                     F
                            THOMS, TOM                        A
                            UDELL, GERRY                      B


             NUMBER OF A STUDENTS IS        3
             NUMBER OF B STUDENTS IS        3
             NUMBER OF C STUDENTS IS        2
             NUMBER OF D STUDENTS IS        1
             NUMBER OF F STUDENTS IS        2
```

7.2.1

*A Loop Within
a Loop*

There is nothing particularly complex about the logic of nested DO WHILE
statements. In fact, they are probably less confusing to the beginner than are

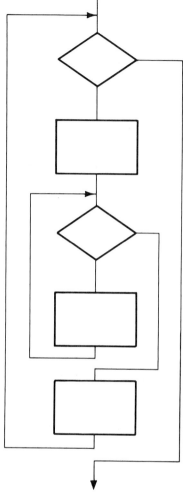

FIGURE 7–3
Nesting DO WHILE constructs.

nested IF statements. It is crucial, however, that you understand the exact order in which the statements will execute and the exact number of times any statement will execute. To explain the order of execution we assume DO WHILE statements nested two deep. DO WHILE$_1$ (the higher level or outer DO WHILE) is entered, and its statements are executed as usual. At some point within DO WHILE$_1$ the second loop (DO WHILE$_2$) is encountered. The statements within the range of DO WHILE$_2$ are then executed repeatedly until the DO WHILE$_2$ condition becomes false. When this occurs, control passes to the statement following END$_2$, and the rest of the statements in DO WHILE$_1$ are executed. (Let us assume that among the statements within the range of DO WHILE$_1$ is one that changes the value of a variable, which makes the DO WHILE$_2$ condition once again true.) When END$_1$ is encountered, control, of course, is passed to DO WHILE$_1$, and the cycle continues again, to be repeated until the DO WHILE$_1$ condition becomes false. At this time, control is transferred to the statement following END$_1$. The following block of code and table of values should render this discussion clear. The code is intended strictly for purposes of illustration, of course, and not for any actual application.

```
              COUNT_1 = 1;
LOOP1:        DO WHILE (COUNT_1 < 4);
              COUNT_2 = 1;
LOOP2:           DO WHILE (COUNT_2 < 5);
                    COUNT_2 = COUNT_2 + 1;
                 END LOOP2;
                 COUNT_1 = COUNT_1 + 1;
              END LOOP1;
```

The following listing shows the successive values of COUNT_1 and COUNT_2 during the time LOOP1 is executing.

```
COUNT_1 = 1
    COUNT_2 = 1
    COUNT_2 = 2
    COUNT_2 = 3
    COUNT_2 = 4
COUNT_1 = 2
    COUNT_2 = 1
    COUNT_2 = 2
    COUNT_2 = 3
    COUNT_2 = 4
COUNT_1 = 3
    COUNT_2 = 1
    COUNT_2 = 2
    COUNT_2 = 3
    COUNT_2 = 4
```

Suppose we added a third level DO WHILE to the example:

```
                         COUNT_3 = 1;
        LOOP3:           DO WHILE (COUNT_3 < 6);
                             COUNT_3 = COUNT_3 + 1;
                         END LOOP3;
```

Now while COUNT_1 and COUNT_2 are held at a value of 1, COUNT_3 takes on the successive values of 1 to 5. COUNT_2 is then incremented to 2, and again COUNT_3 is incremented from 1 to 5 (i.e., LOOP3 executes five times). When COUNT_2 contains the value of 4, COUNT_1 is incremented by 1, and the entire process begins again. In general, then, we can say of the nested DO WHILE that the statements within its range will execute the number of times implied by its condition for each time the DO WHILE at the next higher level executes once. To determine how many times a given DO WHILE will be *entered*, multiply together the number of times each of the higher level DO WHILE statements will be executed. To find the number of times the statements of a given loop will *execute*, multiply the product from the higher levels by the number implied by the condition at the given level. In the three level example discussed above, LOOP3 will be entered $3 \times 4 = 12$ times, and its statements will execute $3 \times 4 \times 5 = 60$ times. Finally, you should always keep two points clearly in mind.

1. The range of a lower-level loop can never exceed that of the loop at the next higher level. The fact that an END corresponds to the nearest preceding DO that does not already have a corresponding END makes it impossible.

2. When the condition of the DO WHILE at a given level is false, none of the lower-level loops can be entered.

The logic of nested DO WHILE statements, like that of the nested IF, might appear confusing at first. Any initial confusion can be cleared up, however, by simply reviewing the rules above. If necessary, you might also compute the number of times the loop will be entered and/or its statements will execute.

7.2.2

Example Program—
Nested DO WHILE

In the process of illustrating nested DO WHILE statements, CH7EX2 (see Fig. 7–4) outlines the logic of one of the most common problems in business data processing. We have been using the term "report" in a sense that makes it nearly synonymous with whatever is the output from the program. In connection with CH7EX2, however, the term takes on a far more restrictive meaning. "Report" in this context means printed output that results from considering the records in the input file as belonging to several groups; after the last record of a group has printed, one or more subtotals are usually presented. In CH7EX2, for example, we must consider all the records of Section 1 of Computer Science 101 to be a group. The records for Sections 2 and 3 constitute two more groups. The subtotals spoken of are represented here by the number of students enrolled and the section average.

The logic involved in producing a report of this type is often referred to as **report logic**, and it involves an understanding of several terms. A single line of output, such as the student-ID, test grades, and average for each student in CH7EX2, is called a **detail line**. The point at which one group of records ends and another begins is a **control break**, and the subtotals are often referred to as **footings**. Footings (or grand totals) often appear also at the very end of the report; therefore, end of file also constitutes a control break. Notice, then, that we are concerned here with three units in the input file: individual records, the records for a section (forming a group), and all records in the file (also forming a group). There are two more considerations regarding the input file for a report. Each record must contain a field, called a **control field**, which defines the group that record belongs to. The input for CH7EX2 uses the SECTION_# field for this purpose. The second consideration is that all records containing the same value in the control field must be adjacent in the input file; in the case of CH7EX2, all records for Section 1 must be together, and the same is true for Sections 2 and 3. It is usually the case also that the groups are in a particular order, such as Section 1 first, followed by Section 2 and Section 3.

The program that implements report logic begins like any other program: It does an initial read and begins a loop controlled by the ENDFILE condition. Immediately within the loop, however, we notice something new: The page headers are printed *inside* the loop. This is done for two reasons.

1. The output from each group of records (that is, between control breaks) is usually placed on a separate page; therefore, the page headers are printed not once as in past examples but several times.

2. The page header also normally contains a variable into which the value from the control field from the input file is placed in order to identify the group printed on this page.

Statement 37 of CH7EX2 carries out this operation by including SECTION_# in the output data list.

It is also possible, of course, that the number of detail lines required for one group of records will be greater than the the number of lines per page of output. If the group is carried over to a subsequent page, we still want a page header. We allow for this by referencing the ENDPAGE condition in Statement 19. As you see in Statement 19, the syntax of the ON ENDPAGE statement is

```
ON ENDPAGE (filename) imperative;
```

The ENDPAGE condition is raised when an output statement attempts to start a new line beyond the number of lines specified (explicitly or by default) for the page. The default number of lines per page is usually 60. After ENDPAGE is raised, a new page can be started by the execution of PAGE or by the execution of LINE (n), where n is less than the current line number. ENDPAGE is enabled by default and cannot be disabled. You can prevent any action from occurring when ENDPAGE is raised, however, by coding a null ON statement:

```
ON ENDPAGE (SYSPRINT);
```

Notice that ENDPAGE, like ENDFILE, requires the name of the file involved; if it is not supplied, SYSPRINT is assumed, and a warning message is produced.

The next difference you will notice is Statement 23:

```
SAVE_SEC = SECTION_#;
```

If we are to provide a control break at the proper point in processing the input file, we must have a means of knowing when that point occurs, and the method we use for determining a control break is to compare the contents of the control field on each record with the contents of the control field on the previous record. Notice that it is the contents of the control field (along with the ENDFILE condition) that controls the lower level DO WHILE (SEC_LP). We want to remain in the lower level DO WHILE until we encounter a control break, and this logic allows us to do so.

The statements within the range of SEC_LP are straightforward and require no comment except to point out that they illustrate one important generalization regarding report logic: It is the lowest level DO WHILE that reads the input file and produces the detail lines; the case remains the same no matter how many levels of control breaks (and consequently nested DO WHILE statements) are involved. The end-of-file produces a control break, of course, but there is no change in the value of the control field; therefore, the ENDFILE condition must be one of the conditions controlling all loops in the processing portion of the program.

Notice that Statement 28 is a part of EOF_LP (but not a part of

SEC_LP), which means that we have read a record having a new value in the control field. SUM_SEC_RTN calculates the values for the section footings and calls PRNT_SEC_RTN to print them. There is nothing unusual about this code. Pay special attention to Statement 57, however, since it represents a detail that is easily overlooked. Because we have finished processing a section, we must set COUNT and SECTION_TOTAL back to 0. Any field involved in producing subtotals must be reinitialized when the control break is reached. Notice, by the way, that both fields are reinitialized in the same statement. Any number of variables can be included on the left of the assignment symbol, so long as they are all to receive the same value.

Finally, when the ENDFILE condition is raised, control passes to Statement 30, FINAL_SUM_RTN is called, and control returns to the operating system from Statement 31. The pseudocode is as follows:

```
CH7EX2
    Call READ_RTN
    DO WHILE (more input)
        Call HDR_RTN
        Save the section number
        DO WHILE (same section number and more input)
            Call PROC-SEC_RTN
            Call READ_RTN
        ENDDO
        Call SUM_SEC_RTN
    ENDDO
    Call FINAL_SUM_RTN
STOP
READ_RTN
    Read a record
EXIT
HDR_RTN
    Print page and column headers
EXIT
PROC_SEC_RTN
    Compute student average = sum of test grades / 2
    Section total grade ← section total grade + student avg
    Student count ← student count + 1
    Call PRNT_DETAIL_RTN
EXIT
PRNT_DETAIL_RTN
    Print a detail line
EXIT
```

```
SUM_SEC_RTN
    Compute section average = section total grade / student
            count
    Total average ← total average + section average
    Call PRNT_SEC_RTN
    Student count, section total grade ← 0
    Number of sections ← number of sections + 1
EXIT
PRNT_SEC_RTN
    Print student count, section average
EXIT
FINAL_SUM_RTN
    Compute class average = total average / 3
    Print summary-page header
    Print class average
EXIT
```

We can now make the following generalizations about report logic.

1. Each DO WHILE level except the lowest must first do certain "house-keeping" chores to prepare for the next lower level. These chores include printing headers and saving the value of the field that controls the next lower level loop. In CH7EX2 we chose to *initialize* the variables used in producing footings as part of the DECLARE statement and to *reinitialize* them when control returned from the lower-level loop. It is also quite possible to initialize (and reinitialize) them as part of the housekeeping chores.

2. As soon as these housekeeping chores are completed, we need to enter the next-lower-level loop.

3. Individual records are read and processed within the lowest-level loop.

4. When control returns to a loop from a lower-level loop, any required footings must be produced. This seems to us the best time to reinitialize the subtotal variables, but as we noted, it could be done before we enter the lower-level loop. You must think through carefully the level at which various subtotal variables are to be reinitialized.

5. Normally some sort of grand totals are presented as the last section of the report. This often makes it necessary to have one variable to contain a given subtotal and another to contain the same item as a grand total.

6. Although we have presented report logic in a rather rigid format, there is room for more variation than we have suggested. It is these generalizations that cannot be violated if you wish to produce a successful report.

OPTIONS SPECIFIED
60STMT;

                      SOURCE LISTING
     STMT

```
         /* CH7EX2 USING NESTED DO WHILE */
   1  CH7EX2:   PROCEDURE OPTIONS (MAIN);
      /******************************************************************/
      /*                                                              */
      /*   FUNCTION:   THIS PROGRAM COMPUTES A LETTER GRADE FOR EACH   */
      /*               STUDENT BASED ON THE STUDENT'S PERCENTAGE       */
      /*               AVERAGE AND TALLIES THE NUMBER OF STUDENTS      */
      /*               RECEIVING EACH GRADE                           */
      /*                                                              */
      /*   INPUT:      AN UNKNOWN NUMBER OF RECORDS, EACH CONTAINING   */
      /*               A SECTION NUMBER, STUDENT NAME, AND A PERCENTAGE */
      /*               AVERAGE.  THE RECORDS ARE SORTED ON SECTION.     */
      /*                                                              */
      /*   OUTPUT:     A REPORT (GROUPED BY SECTION) LISTING THE       */
      /*               NAME AND THE LETTER GRADE AVERAGE              */
      /*                                                              */
      /*   NOTES:      LETTER GRADES ARE COMPUTED ON THE FOLLOWING    */
      /*               SCALE                                          */
      /*               100-92 = A                                     */
      /*                91-84 = B                                     */
      /*                83-76 = C                                     */
      /*                75-68 = D                                     */
      /*             BELOW 68 = F                                     */
      /*                                                              */
      /******************************************************************/
   2     DCL STUDENT_NAME      CHARACTER    (20);
   3     DCL SECTION_#         FIXED DECIMAL (3);
   4     DCL SAVE_SEC          FIXED DECIMAL (3);
   5     DCL TEST1             FIXED DECIMAL (3);
   6     DCL TEST2             FIXED DECIMAL (3);
```

FIGURE 7–4
CH7EX2 (nested DO WHILE statements).

FIGURE 7–4 *(continued)*

```
    7     DCL STUDENT_AVG      FIXED DECIMAL (5,2);
    8     DCL SECTION_TOTAL    FIXED DECIMAL (7,2)   INIT (0);
    9     DCL SECTION_AVG      FIXED DECIMAL (5,2);
   10     DCL AVG_TOTAL        FIXED DECIMAL (7,2)   INIT (0);
   11     DCL CLASS_AVG        FIXED DECIMAL (5,2);
   12     DCL #_SECTIONS       FIXED DECIMAL (3,1)   INIT (0);
   13     DCL STDNT_COUNT      FIXED BINARY  (15)    INIT (0);
   14     DCL MORE_RECS        BIT           (1);
   15     DCL   YES            BIT           (1)     INIT ('1'B);
   16     DCL   NO             BIT           (1)     INIT ('0'B);
   17           MORE_RECS = YES;
   18           ON ENDFILE (SYSIN) MORE_RECS = NO;
   19           ON ENDPAGE (SYSPRINT) CALL HDR_RTN;
   20           CALL READ_RTN;
```

PL/I OPTIMIZING COMPILER   /* CH7EX2 USING NESTED DO WHILE */   PAGE   3
    STMT

```
   21 EOF_LP:   DO WHILE (MORE_RECS);
   22                CALL HDR_RTN;
   23                SAVE_SEC = SECTION_#;
   24 SEC_LP:       DO WHILE (SECTION_# = SAVE_SEC & MORE_RECS);
   25                  CALL PROC_SEC_RTN;
   26                  CALL READ_RTN;
   27                END SEC_LP;
   28                CALL SUM_SEC_RTN;
   29            END EOF_LP;
   30            CALL FINAL_SUM_RTN;
   31            RETURN;

   32 READ_RTN: PROCEDURE;
      /****************************************************************/
      /*                                                              */
      /*  FUNCTION:  TO READ A SINGLE RECORD                          */
      /*                                                              */
      /*  GLOBAL VARIABLES:    SECTION_#                              */
      /*                       STUDENT_NAME                           */
      /*                       TEST1                                  */
      /*                       TEST2                                  */
      /*                                                              */
      /*  SUBROUTINES CALLED: NONE                                    */
      /*                                                              */
```

```
        /*   INPUT:     THE INPUT RECORD CONTAINING A SECTION NUMBER,      */
        /*              STUDENT NAME, AND TWO TEST SCORES                   */
        /*                                                                  */
        /*   OUTPUT:    NONE                                                */
        /*                                                                  */
        /********************************************************************/
 33             GET LIST (SECTION_#, STUDENT_NAME, TEST1, TEST2);
 34             RETURN;
 35             END READ_RTN;

 36  HDR_RTN:  PROCEDURE;
        /********************************************************************/
        /*                                                                  */
        /*  FUNCTION:  TO PRINT THE PAGE AND COLUMN HEADERS                 */
        /*                                                                  */
        /*  GLOBAL VARIABLES:   SECTION_#                                   */
        /*                                                                  */
        /*  SUBROUTINES CALLED: NONE                                        */
        /*                                                                  */
        /*  INPUT:     NONE                                                 */
        /*                                                                  */
```

```
        /*  OUTPUT:    A ONE-LINE PAGE HEADER AND A ONE-LINE COLUMN         */
        /*             HEADER                                               */
        /*                                                                  */
        /********************************************************************/
 37             PUT PAGE LIST (' ', 'CSCI 101 GRADE REPORT', 'SECTION',
                          SECTION_#);
 38             PUT SKIP (2) LIST (' ', 'NAME', 'TEST 1', 'TEST 2',
                          'AVERAGE');
 39             PUT SKIP (2);
 40             RETURN;
 41             END HDR_RTN;

 42  PROC_SEC_RTN:  PROCEDURE;
        /********************************************************************/
        /*                                                                  */
        /*  FUNCTION:  TO COMPUTE EACH STUDENT'S AVERAGE GRADE, TALLY       */
        /*             THE GRADES FOR EACH SECTION AND THE STUDENTS PER     */
        /*             SECTION                                              */
```

FIGURE 7-4 *(continued)*

```
    /*                                                         */
    /* GLOBAL VARIABLES:   STUDENT_AVG                         */
    /*                     TEST1                               */
    /*                     TEST2                               */
    /*                     SECTION_TOTAL                       */
    /*                                                         */
    /* SUBROUTINES CALLED: PRNT_DETAIL_RTN                     */
    /*                                                         */
    /* INPUT:    NONE                                          */
    /*                                                         */
    /* OUTPUT:   NONE                                          */
    /*                                                         */
    /**********************************************************/
43          STUDENT_AVG = (TEST1 + TEST2) / 2.0;
44          SECTION_TOTAL = SECTION_TOTAL + STUDENT_AVG;
45          STDNT_COUNT = STDNT_COUNT + 1;
46          CALL PRNT_DETAIL_RTN;
47          RETURN;
48          END PROC_SEC_RTN;

49  PRNT_DETAIL_RTN:  PROCEDURE;
    /**********************************************************/
    /*                                                         */
    /* FUNCTION:  TO PRINT A DETAIL LINE                       */
    /*                                                         */
    /* GLOBAL VARIABLES:   STUDENT_NAME                        */
    /*                     TEST1                               */
    /*                     TEST2                               */
```

```
    /*                     STUDENT_AVG                         */
    /*                                                         */
    /* INPUT:    NONE                                          */
    /*                                                         */
    /* OUTPUT:   A SINGLE LINE OF PRINT LISTING THE VALUE OF THE */
    /*           ABOVE VARIABLES                               */
    /*                                                         */
    /**********************************************************/
```

```
50          PUT SKIP LIST (' ', STUDENT_NAME, TEST1, TEST2, STUDENT_AVG);
51          RETURN;
52          END PRNT_DETAIL_RTN;

53  SUM_SEC_RTN:  PROCEDURE;
    /***********************************************************/
    /*                                                         */
    /* FUNCTION:  TO COMPUTE THE AVERAGE GRADE FOR ONE SECTION OF */
    /*            STUDENTS                                     */
    /*                                                         */
    /* GLOBAL VARIABLES:    SECTION_AVG                        */
    /*                      SECTION_TOTAL                      */
    /*                      AVG TOTAL                          */
    /*                      STDNT_COUNT                        */
    /*                      #_SECTIONS                         */
    /*                                                         */
    /* SUBROUTINES CALLED: PRNT_SEC_RTN                        */
    /*                                                         */
    /* INPUT:     NONE                                         */
    /*                                                         */
    /* OUTPUT:    NONE                                         */
    /*                                                         */
    /***********************************************************/
54          SECTION_AVG = SECTION_TOTAL / STDNT_COUNT;
55          AVG_TOTAL = AVG_TOTAL + SECTION_AVG;
56          CALL PRNT_SEC_RTN;
57          STDNT_COUNT, SECTION_TOTAL = 0;
58          #_SECTIONS = #_SECTIONS + 1;
59          RETURN;
60          END SUM_SEC_RTN;

61  PRNT_SEC_RTN:  PROCEDURE;
    /***********************************************************/
    /*                                                         */
    /* FUNCTION:  TO PRINT THE FOOTINGS AT THE END OF EACH SECTION */
    /*            REPORT                                       */
    /*                                                         */
    /* GLOBAL VARIABLES:    STDNT_COUNT                        */
    /*                      SECTION_AVG                        */
```

FIGURE 7–4 *(continued)*

```
     /*                                                              */
     /*  SUBROUTINES CALLED: NONE                                    */
     /*                                                              */
     /*  INPUT:     NONE                                             */
     /*                                                              */
     /*  OUTPUT:    TWO LINES, THE FIRST LISTING THE NUMBER OF       */
     /*             STUDENTS ENROLLED AND THE SECOND LISTING THE     */
     /*             SECTION AVERAGE                                  */
     /*                                                              */
     /****************************************************************/
62           PUT SKIP (4) LIST (' ', 'STUDENTS ENROLLED =', STDNT_COUNT);
63           PUT SKIP LIST (' ', 'THE SECTION AVERAGE =', SECTION_AVG);
64           RETURN;
65           END PRNT_SEC_RTN;

66  FINAL_SUM_RTN: PROCEDURE;
     /****************************************************************/
     /*                                                              */
     /*  FUNCTION:  TO PRINT A SUMMARY-PAGE HEADER AND THE CLASS     */
     /*             AVERAGE                                          */
     /*                                                              */
     /*  GLOBAL VARIABLES:   AVG_TOTAL                               */
     /*                      CLASS_AVG                               */
     /*                      #_SECTIONS                              */
     /*                                                              */
     /*  SUBROUTINES CALLED: NONE                                    */
     /*                                                              */
     /*  INPUT:     NONE                                             */
     /*                                                              */
     /*  OUTPUT:    A SUMMARY-PAGE HEADER AND A LINE LISTING THE     */
     /*             CLASS AVERAGE GRADE                              */
     /*                                                              */
     /****************************************************************/
67           CLASS_AVG = AVG_TOTAL / #_SECTIONS;
68           PUT PAGE LIST (' ', 'COMPUTER SCIENCE 101 GRADE SUMMARY');
69           PUT SKIP (5) LIST (' ', 'THE CLASS AVERAGE IS', CLASS_AVG);
70           RETURN;
71           END FINAL_SUM_RTN;
72  END CH7EX2;
```

COMPILER DIAGNOSTIC MESSAGES
ERROR ID L   STMT   MESSAGE DESCRIPTION

COMPILER INFORMATORY MESSAGES

IEL0533I I   NO 'DECLARE' STATEMENT(S) FOR 'SYSIN','SYSPRINT'.

END OF COMPILER DIAGNOSTIC MESSAGES

             CSCI 101 GRADE REPORT   SECTION   1

             NAME   TEST 1   TEST 2   AVERAGE

             DILLMAN, ERICA   88   73   80.50
             FREEMAN, JOHN   91   96   93.50
             HEINZ, KARL   77   80   78.50
             KNUDSON, GARY   67   76   71.50
             MASON, TERRI   99   100   99.50

             STUDENTS ENROLLED =   5
             THE SECTION AVERAGE =   84.69

             CSCI 101 GRADE REPORT   SECTION   2

             NAME   TEST 1   TEST 2   AVERAGE

             OBERT, ODELL   90   80   85.00
             QUIRT, LINDA   84   86   85.00
             ROBERTS, TOM   65   66   65.50

             STUDENTS ENROLLED =   3
             THE SECTION AVERAGE =   78.50

FIGURE 7–4 *(continued)*

```
         CSCI 101 GRADE REPORT    SECTION                3

         NAME                     TEST 1                 TEST 2              AVERAGE


         UBLE, JACK                 77                     66                71.50
         VERI, CARL                 90                     80                85.00
         WALTERS, JAYNE             87                     78                82.50
         YOUNG, BETTY               77                     77                77.00



         STUDENTS ENROLLED =              4
         THE SECTION AVERAGE =     79.00
```

```
         COMPUTER SCIENCE 101 GRADE SUMMARY




         THE CLASS AVERAGE IS      80.73
```

7.3

Case Structure in PL/I

A **case structure** is said to exist when a multidirectional branch is to be made, the particular branch depending on the value of a particular expression—for example, a field in an input record. Suppose, for example, that the field PART_CODE can have a value of 1–4 inclusive and that the code to be executed depends on which of the four values happens to be on each record; this is a multidirectional branch because control can branch to any one of several lines or blocks of code. In other words, the contents of a field are examined, and one (and only one) statement or group of statements is executed, which statement(s) depending on the current value in the field. The case structure is implemented in PL/I in either of two ways: through nested IF statements or through the SELECT statement.

7.3.1

Using IF to Implement the Case

It might be open to question whether or not it is accurate to speak of implementing a case structure with IF statements, but it is certainly accurate to say that any problem that can be solved by the case structure implementation (the SELECT statement) can also be implemented by using IF statements. According to the definition we have given, moreover, it is accurate to say the case structure can be implemented with IF statements. For the sake of execution efficiency, a multidirectional branch implemented with IF statements should be implemented with nested IF statements. To illustrate the relationship of nested IF statements to the SELECT statement as an implementation of the case structure, we have chosen to solve the same problem (assigning letter grades based on percentage grades) in CH7EX3 as we did in CH7EX1. The CH7EX1 program, of course, solves the problem using nested IF statements, whereas CH7EX3 uses the SELECT statement.

One problem that arises in implementing the case structure with nested IF statements is that when normal indentation is used, the space to the right margin quickly becomes so short as to make coding rather difficult. (Consider, for example, a ten-way branch in which each involves coding a lengthy literal.) To avoid this problem the following coding format can be used.

```
IF condition THEN
    Imperative(s);
ELSE IF condition THEN
    Imperative(s);
ELSE IF condition THEN
    Imperative(s);
ELSE IF condition THEN
    Imperative(s);
ELSE
    Imperative(s);
```

The basic pattern can be modified, of course, in its IF-ELSE order to meet the needs of the particular situation.

7.3.2

The SELECT
Statement

The one thing you must keep in mind in reading this and the next two sections is that *the SELECT statement is not available to PL/C.* This restriction on PL/C is, of course, one of the reasons we stressed case structure implementation with nested IF statements. Even though SELECT is unavailable to PL/C, we feel it is superior to nested IF statements as an implementation of the case structure. Also, we are writing this book under the assumption that its readers will eventually code in PL/I; therefore, we feel that a detailed discussion of the SELECT statement is warranted.

The SELECT statement has two formats; the first is

```
SELECT;
    WHEN (expression)
        Imperative(s);
    WHEN (expression)
        Imperative(s);
    WHEN (expression)
        Imperative(s);
    OTHERWISE
        Imperative(s);
END;
```

Three initial points need to be made regarding this syntax:

1. The keyword SELECT serves merely to indicate the beginning of a case structure.

2. The expression must be a comparison expression.

3. The WHEN can be followed by multiple expressions separated by commas. In this case, the expressions are treated as if they were joined by ORs.

```
WHEN (expression-1, expression-2, ..., expression-n)
    Imperative(s);
```

During execution the expression following WHEN$_1$ is evaluated. If it evaluates to true, the imperative(s) is executed, and control passes to the statement following the END; if the expression evaluates to false, the expression following WHEN$_2$ is evaluated. In general, we can say that the expressions are evaluated in the order in which they are encountered until a true expression is found, at which time its imperative(s) is executed, and control passes out of the SELECT group. The OTHERWISE statement is optional; it need not be included. If it is included, its imperative(s) is executed if none of the WHEN expressions evaluate to true. If OTHERWISE is omitted and all WHEN expressions evaluate to false, the program is in error, and processing will terminate; therefore, it would seem worthwhile always to code the few extra lines required by OTHERWISE. If more than one imperative is included following a WHEN or OTHERWISE, it is included, as always, within a DO-group. Notice that the SELECT is itself a group, using END as it does for one of its delimiters. (As you have probably inferred by now, a **group** is one or more PL/I statements delimited at the start by keywords DO or BEGIN and delimited at the end by END.) PL/I provides for a null OTHERWISE just as it does for a null ELSE; the syntax is

```
OTHERWISE;
```

See CH7EX4 for an example of the null OTHERWISE. This first SELECT format is illustrated in CH7EX3.

The syntax of the second SELECT format is

```
SELECT (expression);
    WHEN (expression)
        Imperative(s);
    WHEN (expression)
        Imperative(s);
    etc.
    OTHERWISE
        Imperative(s);
END;
```

The difference between the two formats, as you can see, is that the second has an expression following the SELECT, whereas the first does not. In this format the expression following the WHEN can be of any type supported by PL/I, but it is preferable that it be the same type as that following the SELECT. Again, the WHEN can be followed by multiple expressions, separated by commas.

Execution of this format SELECT causes, first of all, an evaluation of the SELECT expression. The expression following the first WHEN is then evaluated; if it evaluates to a value equal to the value the SELECT expression evaluated to, the imperative(s) is executed, and control passes out of the SELECT group. If the first WHEN expression is not equal to the SELECT value, successive WHEN expressions are evaluated until one does equal the SELECT value. If none of the WHEN values equals the SELECT value, the OTHERWISE imperative(s) is executed. If OTHERWISE is omitted and no WHEN values equal the SELECT value, again the ERROR condition is raised. In other words execution of the second format SELECT is the same as the first format execution except that the SELECT expression must be evaluated, and a WHEN condition must equal that value.

Case structure can be represented in a flowchart in the same manner as nested IF statements because it is, after all, simply a modification of that structure. Many professionals, however, including the two of us, prefer a different approach to flowcharting the case structure. In the diagram shown in Figure 7–5 we represent a four-way branch; case structure, of course, is not limited to four. The procedure boxes, moreover, can represent any number of statements (including I/O statements) to be executed. The new

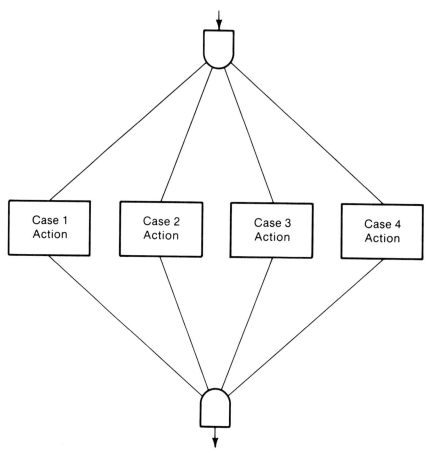

FIGURE 7–5
The CASE construct.

symbol at the top and bottom of the case structure shows that there is a single entrance into the structure and a single exit from it.

Our suggestion for flowcharting is this. If you implement nested IF statements using normal indentation, use nested IF flowchart structure. If, on the other hand, you implement the structure using the ELSE IF condition format, use the case structure shown in Figure 7–5. If you are coding in PL/I and use SELECT to implement case structure, again use the Figure 7–5 structure. In pseudocode, again there is no fully standard format; the one we have chosen is

```
CASE
    Case 1: (condition)
        Imperative(s)
    Case 2: (condition)
        Imperative(s)

                .

    Case n: (condition)
ENDCASE
```

7.3.3

*Using SELECT to
Implement the Case*

The second SELECT format is considerably more restrictive in its application than is the first. The second SELECT is used primarily when a field in each input record contains an absolute value. To illustrate let us assume that the input record contains a student identification number and a final grade (GRADE) represented by a number: 4 = A; 3 = B; 2 = C; 1 = D; and 0 = F. We want to compute the honor points (grade multiplied by the credit hours of the course); we further assume that the class is worth four credit hours. Although we could use either, we would use the second format SELECT to solve the problem because it involves less coding.

```
SELECT (GRADE);
    WHEN (4)
        HONOR_POINTS = 16;
    WHEN (3)
        HONOR_POINTS = 12;
    WHEN (2)
        HONOR_POINTS = 8;
    WHEN (1)
        HONOR_POINTS = 4;
    WHEN (0)
        HONOR_POINTS = 0;
    OTHERWISE
        PUT SKIP LIST ('*** CREDIT_HOURS OUT OF RANGE');
END;
```

The second-format SELECT would never have to be used; all problems to which SELECT is applicable can be solved with the first format. As we noted above, the first-format SELECT is illustrated in CH7EX3, where we want to know if the value in a field lies within a certain *range* of values. To solve this problem we must use the first-format SELECT because the SELECT expression cannot represent a range of values. A compound expression is often required in order to establish a range (e.g., $> 0 \ \& < 10$), and this suggests a second application of the first-format SELECT. Suppose we are again assigning priorities to students requesting admission to a class with limited enrollment. The criteria for determining priority is senior or junior standing and a grade of A, B, or C in a prior course. The two highest priorities go to seniors with grades of A and B, then juniors with A and B grades, followed by seniors and juniors with grades of C. Both the grades (4–2) and class standing (4, 3) are represented by numbers.

```
SELECT;
    WHEN (GRADE = 4 & CLASS = 4)
        PRIORITY = 1;
    WHEN (GRADE = 3 & CLASS = 4)
        PRIORITY = 2;
    WHEN (GRADE = 4 & CLASS = 3)
        PRIORITY = 3;
    WHEN (GRADE = 3 & CLASS = 3)
        PRIORITY = 4;
    WHEN (GRADE = 2 & CLASS = 4)
        PRIORITY = 5;
    WHEN (GRADE = 2 & CLASS = 3)
        PRIORITY = 6;
    OTHERWISE
        PUT SKIP LIST ('*** INVALID INPUT DATA');
    END;
```

Note that the action (PRIORITY = 6;) of the last WHEN could be the action following OTHERWISE. We have reserved it, however, for unusual conditions.

When you consider the fact that a WHEN expression must equal the second-format SELECT value, you can begin to see why this format is restrictive in its application. Obviously, the SELECT expression cannot evaluate to a range of values as is required in CH7EX3. We cannot say, of course, that one SELECT format is "better" than the other; each has its proper application, as our examples above illustrate.

7.3.4

Example Program—
SELECT Statement

As we noted earlier, CH7EX3 (see Fig. 7–6) solves the same problem as does CH7EX1, now using a select-group instead of nested IF statements. The logic of the program is quite straightforward, and the flowchart in Appendix A (see Fig. A–9) is included simply to illustrate the case structure integrated into a total program flowchart. We include here the pseudocode for the case structure to show you one possibility of how it is constructed.

```
CASE
     Case 1:  (percentage grade < 101 and > 91)
              Print student name, grade of 'A'
              A-tally ← A-tally + 1
     Case 2:  (percentage grade < 92 and > 83)
              Print student name, grade of 'B'
              B-tally ← B-tally + 1
     Case 3:  (percentage grade < 84 and > 75)
              Print student name, grade of 'C'
              C-tally ← C-tally + 1
     Case 4:  (percentage grade < 76 and > 67)
              Print student name, grade of 'D'
              D-tally ← D-tally + 1
     Case 5:  (percentage grade < 68 and > -1)
              Print student name, grade of 'F'
              F-tally ← F-tally + 1
     OTHERWISE
              Print student name, error message
ENDCASE
```

```
PL/I OPTIMIZING COMPILER  VERSION 1 RELEASE 5.1  TIME: 17.32.76  DATE: 15 FEB 87  PAGE    1
OPTIONS SPECIFIED
GOSTMT;


PL/I OPTIMIZING COMPILER          /* CH7EX3 USING SELECT */                    PAGE    2
               SOURCE LISTING
    STMT


        /* CH7EX3 USING SELECT */
    1  CH7EX3:   PROCEDURE OPTIONS (MAIN);
        /***************************************************************/
        /*                                                          */
        /*   FUNCTION:  THIS PROGRAM COMPUTES A LETTER GRADE FOR EACH  */
```

FIGURE 7–6
CH7EX3 (the SELECT statement).

FIGURE 7–6 *(continued)*

```
      /*            STUDENT BASED ON THE STUDENT'S PERCENTAGE         */
      /*            AVERAGE AND TALLIES THE NUMBER OF STUDENTS         */
      /*            RECEIVING EACH GRADE                               */
      /*                                                              */
      /*  INPUT:    AN UNKNOWN NUMBER OF RECORDS, EACH CONTAINING      */
      /*            A STUDENT NAME AND A PERCENTAGE AVERAGE            */
      /*                                                              */
      /*  OUTPUT:   A REPORT LISTING THE STUDENT NAME AND             */
      /*            THE LETTER GRADE ASSIGNED                          */
      /*                                                              */
      /*  NOTES:    LETTER GRADES ARE COMPUTED ON THE FOLLOWING SCALE */
      /*            100-92 = A                                         */
      /*             91-84 = B                                         */
      /*             83-76 = C                                         */
      /*             75-68 = D                                         */
      /*          BELOW 68 = F                                         */
      /*                                                              */
      /*****************************************************************/
   2      DCL STUDENT_NAME      CHARACTER     (20);
   3      DCL PERCENT_GRADE     FIXED DECIMAL (3);
   4      DCL A_TALLY           FIXED BINARY  (15)    INIT (0);
   5      DCL B_TALLY           FIXED BINARY  (15)    INIT (0);
   6      DCL C_TALLY           FIXED BINARY  (15)    INIT (0);
   7      DCL D_TALLY           FIXED BINARY  (15)    INIT (0);
   8      DCL F_TALLY           FIXED BINARY  (15)    INIT (0);
   9      DCL MORE_RECS         BIT           (1);
  10      DCL   YES             BIT           (1)     INIT ('1'B);
  11      DCL   NO              BIT           (1)     INIT ('0'B);
  12          MORE_RECS = YES;
  13          ON ENDFILE (SYSIN) MORE_RECS = NO;
  14          CALL HDR_RTN;
  15          CALL READ_RTN;
  16  LOOP:   DO WHILE (MORE_RECS);
  17             CALL PROCESS_RTN;
  18             CALL READ_RTN;
  19          END LOOP;
  20          CALL SUMMARY_RTN;
  21          RETURN;
```

```
 22  HDR_RTN:  PROCEDURE;
     /*********************************************************************/
     /*                                                                */
     /*  FUNCTION:   TO PRINT THE PAGE AND COLUMN HEADERS              */
     /*                                                                */
     /*  GLOBAL VARIABLES:  NONE                                      */
     /*                                                                */
     /*  SUBROUTINES CALLED:  NONE                                    */
     /*                                                                */
     /*  INPUT:     NONE                                              */
     /*                                                                */
     /*  OUTPUT:    TWO LINES OF PAGE HEADER AND ONE LINE OF COLUMN   */
     /*             HEADERS, WITH TWO BLANK LINES FOLLOWING           */
     /*                                                                */
     /*********************************************************************/
 23          PUT LIST (' ', 'GRADE REPORT', ' ', 'SPRING 1987');
 24          PUT SKIP (2) LIST (' ', ' ', 'COMPUTER SCIENCE 101');
 25          PUT SKIP (2) LIST (' ', ' ', 'STUDENT-NAME', ' ',
                         'FINAL GRADE');
 26          PUT SKIP (2);
 27          RETURN;
 28      END HDR_RTN;

 29  READ_RTN:  PROCEDURE;
     /*********************************************************************/
     /*                                                                */
     /*  FUNCTION:   TO READ A SINGLE RECORD                          *
     /*                                                                */
     /*  GLOBAL VARIABLES:   STUDENT_NAME                            */
     /*                      PERCENT_GRADE                           */
     /*                                                                */
     /*  SUBROUTINES CALLED:  NONE                                    */
     /*                                                                */
     /*  INPUT:  THE INPUT VARIABLES                                 */
     /*                                                                */
     /*  OUTPUT:  NONE                                               */
     /*                                                                */
     /*********************************************************************/
 30          GET LIST (STUDENT_NAME, PERCENT_GRADE);
```

FIGURE 7–6 *(continued)*

```
31              RETURN;
32          END READ_RTN;

33 PROCESS_RTN:  PROCEDURE;
```

PL/I OPTIMIZING COMPILER   /* CH7EX3 USING SELECT */   PAGE   4
  STMT

```
     /**********************************************************************/
     /*                                                                  */
     /*  FUNCTION:  TO CALCULATE A LETTER GRADE FROM A PERCENTAGE         */
     /*             GRADE (SEE ABOVE FOR SCALE) AND PRINT A DETAIL        */
     /*             LINE                                                  */
     /*                                                                  */
     /*  GLOBAL VARIABLES:  STUDENT_NAME                                  */
     /*                     PERCENT_GRADE                                 */
     /*                     A_TALLY                                       */
     /*                     B_TALLY                                       */
     /*                     C_TALLY                                       */
     /*                     D_TALLY                                       */
     /*                     F_TALLY                                       */
     /*                                                                  */
     /*  SUBROUTINES CALLED:  NONE                                        */
     /*                                                                  */
     /*  INPUT:    NONE                                                   */
     /*                                                                  */
     /*  OUTPUT:   A DETAIL LINE CONTAINING A STUDENT NAME AND THE        */
     /*            GRADE                                                  */
     /*                                                                  */
     /**********************************************************************/
  34 SELEKT:    SELECT;
  35              WHEN (PERCENT_GRADE < 101 & PERCENT_GRADE > 91)
     IF_A:        DO;
  36                  PUT SKIP LIST (' ', ' ', STUDENT_NAME, ' ', 'A');
  37                  A_TALLY = A_TALLY + 1;
  38              END IF_A;
  39              WHEN (PERCENT_GRADE < 92 & PERCENT_GRADE > 83)
     IF_B:        DO;
  40                  PUT SKIP LIST (' ', ' ', STUDENT_NAME,' ', 'B');
  41                  B_TALLY = B_TALLY + 1;
```

```
42                    END IF_B;
43                WHEN (PERCENT_GRADE < 84 & PERCENT_GRADE > 75)
   IF_C:          DO;
44                    PUT SKIP LIST (' ', ' ', STUDENT_NAME, ' ', 'C');
45                    C TALLY = C_TALLY + 1;
46                    END IF_C;
47                WHEN (PERCENT_GRADE < 76 & PERCENT_GRADE > 67)
   IF_D:          DO;
48                    PUT SKIP LIST (' ', ' ', STUDENT_NAME, ' ', 'D');
49                    D_TALLY = D_TALLY + 1;
50                    END IF_D;
51                WHEN (PERCENT_GRADE < 68 & PERCENT_GRADE > -1)
   IF_F:          DO;
52                    PUT SKIP LIST (' ', ' ', STUDENT_NAME, ' ', 'F');
53                    F_TALLY = F_TALLY + 1;
54                    END IF_F;
55                OTHERWISE
                  PUT SKIP LIST (STUDENT_NAME,
                                 '*** OUT OF VALID GRADE RANGE ***');
56                END SELEKT;
57            RETURN;
```

```
   STMT
    58            END PROCESS_RTN;

    59  SUMMARY_RTN: PROCEDURE;
       /******************************************************************/
       /*                                                              */
       /*  FUNCTION:  TO PRINT THE SUMMARY PAGE                        */
       /*                                                              */
       /*  GLOBAL VARIABLES:  A_TALLY                                  */
       /*                     B_TALLY                                  */
       /*                     C_TALLY                                  */
       /*                     D_TALLY                                  */
       /*                     F_TALLY                                  */
       /*                                                              */
       /*  SUBROUTINES CALLED:  NONE                                   */
       /*                                                              */
       /*  INPUT:    NONE                                              */
       /*                                                              */
```

FIGURE 7–6 *(continued)*

```
      /*  OUTPUT:    FIVE LINES GIVING THE NUMBER OF STUDENTS          */
      /*             PROCESSED AND THE NUMBER RECEIVING EACH GRADE      */
      /*                                                                */
      /******************************************************************/
60              PUT SKIP (3) LIST (' ', 'NUMBER OF A STUDENTS IS', A_TALLY);
61              PUT SKIP LIST (' ', 'NUMBER OF B STUDENTS IS', B_TALLY);
62              PUT SKIP LIST (' ', 'NUMBER OF C STUDENTS IS', C_TALLY);
63              PUT SKIP LIST (' ', 'NUMBER OF D STUDENTS IS', D_TALLY);
64              PUT SKIP LIST (' ', 'NUMBER OF F STUDENTS IS', F_TALLY);
65              RETURN;
66          END SUMMARY_RTN;
67   END CH7EX3;


PL/I OPTIMIZING COMPILER          /* CH7EX3 USING SELECT */                    PAGE   6
COMPILER DIAGNOSTIC MESSAGES
ERROR ID L   STMT    MESSAGE DESCRIPTION


COMPILER INFORMATORY MESSAGES


IEL0533I I           NO 'DECLARE' STATEMENT(S) FOR 'SYSIN','SYSPRINT'.


END OF COMPILER DIAGNOSTIC MESSAGES
```

The function of the program, of course, is to assign a letter grade based on a percentage grade and to keep a tally of the students receiving each letter grade. The program cannot be run under PL/C because PL/C does not support the SELECT statement.

7.4

Example Program—Combining Nested IF, DO WHILE, and SELECT

Program CH7EX4, shown in Figure 7–7, illustrates the use of nested IF statements, nested DO WHILE statements, and SELECT groups in a single program. Again, because it uses the SELECT statement, it cannot be run under PL/C; SELECT is available only to PL/I.

The problem we are dealing with in the program is to process student applications for internships. Internships are offered in three areas—systems analysis and design, database, and applications programming—and the student supplies a first and second choice of areas. They are given an internship in their second choice area when their first choice area has no openings; if their second choice area also has no openings, they are denied an internship altogether. The process is complicated by the fact that each internship area has a single faculty advisor, and the advisor is assigned no more than two interns.

Notice, first of all, the keyword COPY in the GET LIST statement (Statement 41). It causes the input to be "echoed" to the printer and is used here as a means of verifying the input data.

The main routine contains a DO WHILE (PROC_LP, Statement 23), within which a call is made to CHOICE_RTN (Statement 26). CHOICE_RTN itself contains a loop; therefore, we have our familiar loop within a loop.

The logic of the lower level DO WHILE (in CHOICE_RTN) is, first of all, to attempt to give the student her or his first choice of areas in which to serve the internship. Thus, each WHEN under the SELECT checks for the student's preference among the three areas and also checks to see whether the maximum of two students have been assigned yet (i.e., checks the compound condition of each WHEN). If both conditions on a given WHEN are found to be true, the student is assigned an internship in that area; the appropriate "prof_count" is incremented by 1, and MORE-TO-SCHED is set to NAY. This, in turn, causes the SCHED_LP DO WHILE condition to evaluate to false, and control passes to the IF statement (Statement 64). This IF statement evaluates to true, and so PRNT_CHOICE_RTN is called. Notice that the SLCT_1 SELECT is a format-one SELECT, while the SLCT_2 SELECT (in PRNT_CHOICE_RTN) is of format two. The area the student has selected is printed when the proper WHEN condition is encountered. Because MORE_TO _SCHED now contains NAY, the SCHED_LP DO WHILE condition is now false. Control therefore drops through SCHED_LP to Statement 74, which resets MORE_TO _SCHED. Control then returns to Statement 27 (CALL READ_RTN), and the process begins with another student.

If the student cannot be assigned her or his first choice, however, the null OTHERWISE (Statement 62) is executed (i.e., nothing is done except to pass control to the following statement). Now the IF condition of Statement 64 is false, and control passes to the ELSE of Statement 65. We could have provided a third SELECT following the ELSE, but we chose instead to include another (that is, nested) IF-THEN-ELSE. We are now attempting to give the student a second choice of internship area. The first time (for each student) this section of code is entered, CHOICE_NUMBER has a value of 1. CHOICE_CODE is assigned the value of SECOND_CHOICE_CODE (Statement 66), and CHOICE_NUMBER is given the value of 2 (Statement 67). If the student can be given her or his second choice, the SLCT_1 and SLCT_2 blocks of code handle the request just as they do when the student's first choice is granted. If the student cannot be granted the second choice, control passes to the ELSE of Statement 69, PRNT_FULL_MSG_RTN is called, and a message to that effect is written. The pseudocode is as follows:

```
CH7EX4
    Call PG_HDR_RTN
    Call READ_RTN
    DO WHILE (more input)
        Choice code ← first choice
        Choice number ← 1
        Call CHOICE_RTN
        Call READ_RTN
    ENDDO
STOP
PG_HDR_RTN
    Print page, column headers
EXIT
READ_RTN
    Read a record
EXIT
CHOICE_RTN
    DO WHILE (schedule-done flag not 'Yes')
        CASE
            Case 1:  (prof-1 count < prof max and choice is
                            Systems Analysis & Design)
                    Print detail line
                    Prof-1 count ← prof-1 count + 1
                    Schedule-done flag ← 'Yes'
            Case 2:  (prof-2 count < prof max and choice is
                            Database)
                    Print detail line
                    Prof-2 count ← prof-2 count + 1
                    Schedule-done flag ← 'Yes'
            Case 3:  (prof-3 count < prof max and choice is
                            Applications Programming)
                    Print detail line
                    Prof-3 count ← prof-3 count + 1
                    Schedule-done flag ← 'Yes'
        ENDCASE
        IF (schedule-done flag = 'Yes')
            Call PRNT_CHOICE_RTN
        ELSE
            IF (choice number = 1)
```

```
                         Choice code ← second choice code
                         Choice number ← 2
                    ELSE
                       IF (choice number = 2)
                          Call PRNT_FULL_MSG_RTN
                          Schedule-done flag ← 'Yes'
                       ENDIF
                    ENDIF
                 ENDDO
                 Schedule-done flag ← 'Yes'
           EXIT
           PRNT_CHOICE_RTN
              CASE
                 Case 1:  (choice code = database)
                          Print choice
                 Case 2:  (choice code = systems analysis & design)
                          Print choice
                 Case 3:  (choice code = applications programming)
                          Print choice
              ENDCASE
           EXIT
           PRNT_FULL_MSG_RTN
              Print student name, advisory messages
           EXIT
```

```
PL/I OPTIMIZING COMPILER  VERSION 1 RELEASE 5.1  TIME: 17.33.23  DATE: 15 FEB 87  PAGE   1
OPTIONS SPECIFIED
GOSTMT;


PL/I OPTIMIZING COMPILER          /* CH7EX4 USING NESTED IF, DO WHILE & SELECT */ PAGE   2
                 SOURCE LISTING
     STMT


     /* CH7EX4 USING NESTED IF, DO WHILE & SELECT */
   1 CH7EX4: PROC OPTIONS (MAIN);
     /*****************************************************************/
     /* FUNCTION:  TO PROCESS STUDENT INTERNSHIP APPLICATIONS.       */
     /*                                                              */
```

FIGURE 7–7
**CH7EX4 (nested IFs, nested DO WHILES, and the SELECT
statement).**

FIGURE 7-7 *(continued)*

```
    /* INPUT:      FOR EACH STUDENT A RECORD CONTAINING              */
    /*                 (1) STUDENT'S NAME                           */
    /*                 (2) CREDIT HOURS FOR INTERNSHIP              */
    /*                 (3) STUDENT'S FIRST INTERNSHIP CHOICE        */
    /*                 (4) STUDENT'S SECOND INTERNSHIP CHOICE       */
    /*                                                              */
    /* OUTPUT:     FOR EACH STUDENT A RECORD CONTAINING             */
    /*                 (1) STUDENT'S NAME                           */
    /*                 (2) FACULTY ADVISOR                          */
    /*                 (3) COMPANY OFFERING INTERNSHIP              */
    /*                 (4) HOURS CREDIT                             */
    /*                 (5) SUBJECT AREA                             */
    /*                                                              */
    /* NOTES:                                                       */
    /*           STUDENTS HAVE A CHOICE OF THREE SUBJECT AREAS FOR  */
    /*           THEIR INTERNSHIPS:  SYSTEMS ANALYSIS & DESIGN,     */
    /*           DATABASE, AND APPLICATIONS PROGRAMMING.  THEY MAY  */
    /*           CHOOSE ONE SUBJECT AREA AND AN ALTERNATE.  ONE     */
    /*           FACULTY ADVISOR IS ASSIGNED TO EACH SUBJECT AREA.  */
    /*           HE/SHE MAY SPONSOR A MAXIMUM OF TWO INTERNS.       */
    /*           STUDENTS ARE GIVEN THEIR ALTERNATE CHOICE WHEN     */
    /*           THERE ARE NO MORE OPENINGS FOR THEIR FIRST CHOICE. */
    /*           IF THEIR ALTERNATE CHOICE IS ALSO FULL, THEY ARE   */
    /*           NOT GIVEN AN INTERNSHIP.  INPUT CODES FOR THE THREE */
    /*           SUBJECT CODES ARE                                  */
    /*               SYSTEMS ANALYSIS & DESIGN  = 01                */
    /*               DATABASE                   = 10                */
    /*               APPLICATION PROGRAMMING    = 11                */
    /****************************************************************/
2   DCL STU_NAME                CHARACTER (20);
3   DCL HOURS_CREDIT            FIXED DEC (1);
4   DCL FIRST_CHOICE_CODE       FIXED DEC (3);
5   DCL SECOND_CHOICE_CODE      FIXED DEC (3);
6   DCL CHOICE_CODE             FIXED DEC (3);
7   DCL CHOICE_NUMBER           FIXED BIN (15) INIT (0);
8   DCL PROF_1_COUNT            FIXED BIN (1)  INIT (0);
9   DCL PROF_2_COUNT            FIXED BIN (15) INIT (0);
10  DCL PROF_3_COUNT            FIXED BIN (15) INIT (0);
11  DCL PROF_MAX                FIXED BIN (15) INIT (2);
12  DCL MORE_TO_SCHED           BIT      (1);
```

```
   13    DCL    YEA                       BIT        (1)  INIT ('1'B);
```
```
   STMT
   14    DCL    NAY                       BIT        (1)  INIT ('0'B);
   15    DCL MORE_RECS                    BIT        (1);
   16    DCL    YES                       BIT        (1)  INIT ('1'B);
   17    DCL    NO                        BIT        (1)  INIT ('0'B);
   18            MORE_RECS = YES;
   19            MORE_TO_SCHED = YEA;
   20            ON ENDFILE (SYSIN) MORE_RECS = NO;
   21            CALL PG_HDR_RTN;
   22            CALL READ_RTN;
   23 PROC_LP:  DO WHILE (MORE_RECS);
   24                CHOICE_CODE = FIRST_CHOICE_CODE;
   25                CHOICE_NUMBER = 1;
   26                CALL CHOICE_RTN;
   27                CALL READ_RTN;
   28            END PROC_LP;
   29            PUT SKIP (2) LIST (' ', '===> SCHEDULING COMPLETED <===');


   30 PG_HDR_RTN:  PROCEDURE;
      /******************************************************************/
      /*                                                              */
      /*  FUNCTION:  TO PRINT PAGE AND COLUMN HEADERS                  */
      /*                                                              */
      /*  GLOBAL VARIABLES:    NONE                                   */
      /*                                                              */
      /*  SUBROUTINES CALLED: NONE                                    */
      /*                                                              */
      /*  INPUT:     NONE                                             */
      /*                                                              */
      /*  OUTPUT:    A TWO-LINE PAGE HEADER AND A TWO-LINE COLUMN      */
      /*             HEADER                                           */
      /*                                                              */
      /******************************************************************/
   31        PUT LIST ('','','   NORTHERN ILLINOIS UNIVERSITY');
   32        PUT SKIP LIST ('','','COMPUTER SCIENCE INTERNSHIP PROGRAM');
   33        PUT SKIP LIST ('','','      FALL SEMESTER');
   34        PUT SKIP (2) LIST ('STUDENT','FACULTY','','HOURS','SUBJECT');
   35        PUT SKIP    LIST ('NUMBER', 'ADVISOR', 'COMPANY',
                              'CREDIT', 'AREA');
   36        PUT SKIP    LIST ((120)'-');
```

FIGURE 7–7 *(continued)*

```
37          PUT SKIP;
38          RETURN;
39          END PG_HDR_RTN;


40  READ_RTN:  PROCEDURE;
```

PL/I OPTIMIZING COMPILER   /* CH7EX4 USING NESTED IF, DO WHILE & SELECT */ PAGE   4
   STMT

```
     /*******************************************************************/
     /*                                                                 */
     /*  FUNCTION:  TO READ A SINGLE RECORD                            */
     /*                                                                 */
     /*  GLOBAL VARIABLES:    STU_NAME                                 */
     /*                       HOURS_CREDIT                             */
     /*                       FIRST_CHOICE_CODE                        */
     /*                       SECOND_CHOICE_CODE                       */
     /*                                                                 */
     /*  SUBROUTINES CALLED: NONE                                      */
     /*                                                                 */
     /*  INPUT:     THE INPUT RECORDS CONTAINING A VALUE FOR EACH      */
     /*             VARIABLE ABOVE                                     */
     /*                                                                 */
     /*  OUTPUT:    NONE                                               */
     /*                                                                 */
     /*******************************************************************/
41          GET LIST (STU_NAME,HOURS_CREDIT,FIRST_CHOICE_CODE,
                        SECOND_CHOICE_CODE) COPY;
42          RETURN;
43          END READ_RTN;


44  CHOICE_RTN:  PROCEDURE;
     /*******************************************************************/
     /*                                                                 */
     /*  FUNCTION:  TO ASSIGN STUDENTS THEIR CHOICE OF INTERNSHIPS OR */
     /*             DETERMINE THEY CANNOT BE ASSIGNED ONE              */
     /*                                                                 */
     /*  GLOBAL VARIABLES:    PROF_1_COUNT                            */
     /*                       CHOICE_CODE                             */
     /*                       STU_NAME                                */
```

```
        /*                     HOURS_CREDIT                      */
        /*                     PROF_2_COUNT                      */
        /*                     PROF_3_COUNT                      */
        /*                     CHOICE_NUMBER                     */
        /*                     SECOND_CHOICE_CODE                */
        /*                                                       */
        /*  SUBROUTINES CALLED: PRNT_CHOICE_RTN                  */
        /*                      PRNT_FULL_MSG_RTN                */
        /*                                                       */
        /*  INPUT:    NONE                                       */
        /*                                                       */
        /*  OUTPUT:   NONE                                       */
        /*                                                       */
        /*************************************************************/
   45  SCHED_LP:    DO WHILE (MORE_TO_SCHED);
   46  SLCT_1:          SELECT;
   47                       WHEN (PROF_1_COUNT < PROF_MAX & CHOICE_CODE = 01)
        IF_1_OK:             DO;
   48                           PUT SKIP LIST (STU_NAME, 'B. J. HARING',
```

```
                                'J. C. PENNEY', HOURS_CREDIT);
   49                           PROF_1_COUNT = PROF_1_COUNT + 1;
   50                           MORE_TO_SCHED = NAY;
   51                       END IF_1_OK;
   52                       WHEN (PROF_2_COUNT < PROF_MAX & CHOICE_CODE = 10)
        IF_2_OK:             DO;
   53                           PUT SKIP LIST (STU_NAME, 'B. R. R. NUYL',
                                    'SEARS', HOURS_CREDIT);
   54                           PROF_2_COUNT = PROF_2_COUNT + 1;
   55                           MORE_TO_SCHED = NAY;
   56                       END IF_2_OK;
   57                       WHEN (PROF_3_COUNT < PROF_MAX & CHOICE_CODE = 11)
        IF_3_OK:             DO;
   58                           PUT SKIP LIST (STU_NAME, 'S. A. WRITE',
                                    'SOFTECH, INC', HOURS_CREDIT);
   59                           PROF_3_COUNT = PROF_3_COUNT + 1;
   60                           MORE_TO_SCHED = NAY;
   61                       END IF_3_OK;
   62                   OTHERWISE;
```

FIGURE 7-7 *(continued)*

```
63                    END SLCT_1;
64                    IF MORE_TO_SCHED = NAY THEN
                          CALL PRNT_CHOICE_RTN;
65                    ELSE
                          IF CHOICE_NUMBER = 1 THEN
     IF_1:                  DO;
66                            CHOICE_CODE = SECOND_CHOICE_CODE;
67                            CHOICE_NUMBER = 2;
68                          END IF_1;
69                    ELSE
                          IF CHOICE_NUMBER = 2 THEN
     IF_2:                  DO;
70                            CALL PRNT_FULL_MSG_RTN;
71                            MORE_TO_SCHED = NAY;
72                          END IF_2;
73              END SCHED_LP;
74              MORE_TO_SCHED = YEA;
75          RETURN;
76          END CHOICE_RTN;


77  PRNT_CHOICE_RTN:  PROCEDURE;
    /*****************************************************************/
    /*                                                             */
    /*  FUNCTION:  TO PRINT THE SUBJECT AREA TO WHICH THE STUDENT   */
    /*             IS ASSIGNED                                      */
    /*                                                             */
    /*  GLOBAL_VARIALBES:   CHOICE_CODE                            */
    /*                                                             */
```

PL/I OPTIMIZING COMPILER   /* CH7EX4 USING NESTED IF, DO WHILE & SELECT */ PAGE   6
    STMT
```
    /*  SUBROUTINES CALLED: NONE                                   */
    /*                                                             */
    /*  INPUT:    NONE                                             */
    /*                                                             */
    /*  OUTPUT:   A PRINT FIELD CONTAINING THE STUDENT'S SUBJECT    */
    /*            AREA                                             */
    /*                                                             */
    /*****************************************************************/
```

```
78  SLCT_2:   SELECT (CHOICE_CODE);
79            WHEN (10) PUT LIST ('DATABASE');
80            WHEN (01) PUT LIST ('SYS ANALYSIS & DESIGN');
81            WHEN (11) PUT LIST ('APPLICATIONS PROGRAMMING');
82            END SLCT_2;
83            RETURN;
84            END PRNT_CHOICE_RTN;

85  PRNT_FULL_MSG_RTN:  PROCEDURE;
    /*******************************************************************/
    /*                                                              */
    /*  FUNCTION:  TO PRINT A MESSAGE ADVISING A STUDENT THAT HER/   */
    /*             HIS INTERNSHIP IS NOT AVAILABLE                   */
    /*                                                              */
    /*  GLOBAL VARIABLES:   STU_NAME                                */
    /*                                                              */
    /*  SUBROUTINES CALLED: NONE                                    */
    /*                                                              */
    /*  INPUT:     NONE                                             */
    /*                                                              */
    /*  OUTPUT:    TWO PRINT LINES CONSISTING OF A STUDENT NAME      */
    /*             AND TWO ADVISORY MESSAGES                        */
    /*                                                              */
    /*******************************************************************/
86            PUT SKIP LIST (STU_NAME,
                             '==> ALL FALL INTERNSHIPS FULL <==');
87            PUT SKIP LIST (' ', 'SEE YOUR ACADEMIC ADVISOR');
88            RETURN;
89            END PRNT_FULL_MSG_RTN;
90  END CH7EX4;
```

PL/I OPTIMIZING COMPILER   /* CH7EX4 USING NESTED IF, DO WHILE & SELECT */ PAGE   7
COMPILER DIAGNOSTIC MESSAGES
ERROR ID L   STMT   MESSAGE DESCRIPTION

COMPILER INFORMATORY MESSAGES

IEL0533I I   NO 'DECLARE' STATEMENT(S) FOR 'SYSIN','SYSPRINT'.

END OF COMPILER DIAGNOSTIC MESSAGES

FIGURE 7–7 *(continued)*

```
                        NORTHERN ILLINOIS UNIVERSITY
                      COMPUTER SCIENCE INTERNSHIP PROGRAM
                               FALL SEMESTER

STUDENT              FACULTY                             HOURS        SUBJECT
NUMBER               ADVISOR         COMPANY             CREDIT       AREA
-----------------------------------------------------------------------------------------

'LARRY WOJICKY' 3 10 11

LARRY WOJICKY        B. R. R. NUYL   SEARS                 3          DATABASE
'KARLA ANDEERS' 3 11 10

KARLA ANDEERS        S. A. WRITE     SOFTECH, INC          3          APPLICATIONS PROGRAMMING

'JIM LOCKARD' 3 11 10

JIM LOCKARD          S. A. WRITE     SOFTECH, INC          3          APPLICATIONS PROGRAMMING

'GARY DOBSON' 3 10 11

GARY DOBSON          B. R. R. NUYL   SEARS                 3          DATABASE
'ARLENE NEHR' 3 11 10

ARLENE NEHR          ==> ALL FALL INTERNSHIPS FULL <==
                     SEE YOUR ACADEMIC ADVISOR
'VERA BENSON' 3 11 10

VERA BENSON          ==> ALL FALL INTERNSHIPS FULL <==
                     SEE YOUR ACADEMIC ADVISOR
'CURTIS ROJO' 3 01 10

CURTIS ROJO          B. J. HARING    J. C. PENNEY          3          SYS ANALYSIS & DESIGN
'TANYA IVANOVA' 3 11 01

TANYA IVANOVA        B. J. HARING    J. C. PENNEY          3          SYS ANALYSIS & DESIGN

                     ===> SCHEDULING COMPLETED <===
```

(Note: This program has been reduced for design purposes)

Summary

Nested IF statements provide an efficient means of implementing the solution to a problem requiring a series of decisions. The same solution can be implemented with a series of discrete IF-THEN statements, but this approach requires the computer to make more comparisons. In nested IF statements, as soon as a false condition is found, all code between that IF and its corresponding ELSE (including any lower level IF-THEN-ELSE statements) are bypassed. When a true condition is found, all statements following the corresponding ELSE (again including any lower-level IF-THEN-ELSE statements) are bypassed. Two rules to keep in mind when using nested IF statements are:

1. An ELSE corresponds to the nearest preceding IF statement that has no corresponding ELSE.

2. When an IF at any level has a corresponding ELSE, the IF statements at all lower levels must also have an ELSE.

The null ELSE, which has no function but to meet the demands of PL/I syntax rules, is used to overcome problems raised by these two rules.

Nested DO WHILE statements result from the common situation in which it is necessary to enter a second loop while in the range of a first loop. There are two keys to using nested DO WHILE statements. The first is to make a careful consideration of the logic so that each statement is within the proper loop; subtotals, for example, must be initialized and/or reinitialized at the proper point. The second key is keeping track of the number of times each loop will be entered and how many times its statements will be executed. In general, a loop will be entered once for each successive value of the first term of the DO WHILE condition at the next higher level. More specifically, to find the number of times a loop will be entered, multiply together the number of times the statements at each higher level will be executed. This product multiplied by the number of times the given-level statements will execute each time the loop is entered gives you the total number of times the code at the given level will execute. Remember, too, that when the DO WHILE condition is false, neither it nor any lower level nested under it will be entered.

A case structure (which is a variation of the IF-THEN-ELSE statement) should ordinarily be used to code a multidirectional branch. If you are coding in PL/I, you should use the true case structure format, headed by the keyword SELECT. The SELECT group has two formats, and the first is

```
SELECT;
    WHEN (expression)
        Imperative(s);
    WHEN (expression)
        Imperatives);

        .
        .
        .

    OTHERWISE
        Imperative(s);
END;
```

The expression following each WHEN is evaluated in turn until one is found
to be true. At this point its imperative(s) is executed, and control passes out
of the SELECT group. The OTHERWISE statement need not be coded, but if
it isn't and all expressions are false the ERROR condition is raised. The
OTHERWISE can also be a null OTHERWISE, existing for the sole purpose
of satisfying the demands of PL/I syntax.

The second SELECT format is

```
SELECT (expression);
    WHEN (expression)
        Imperative(s);
    WHEN (expression)
        Imperative(s)

        .
        .
        .

    ELSE
        Imperative(s);
END;
```

The second format SELECT works exactly the same as the first except that
the expression following SELECT is evaluated first. The imperative(s) executed is the one(s) following the WHEN whose expression equals in value
that of the SELECT. If none agree, the OTHERWISE imperative(s), if any,
are executed.

The SELECT statement is not available to PL/C, and a multi-directional
branch, if coded with indented nested IF statements, can quickly move the
left (indented) margin too far to the right. Therefore, the "case structure" in
PL/C can be coded as

```
IF condition THEN
    Imperative(s);
ELSE IF condition THEN
    Imperative(s);
ELSE IF condition THEN
    Imperative(s);

    .
    .
    .

ELSE
    Imperative(s);
```

Nested IF and DO WHILE statements will appear regularly in the solutions to your problems. It is important that you understand both clearly before proceeding to further topics.

7.6

Exercises

1. Write the segment of code necessary to accomplish the following:

   R = 1 when X = Y and A = Z

   R = 2 when X = Y and A ^= Z

   R = 3 when X ^= Y

2. Write the section of code necessary to assign the following consequences to insurance applicants, based on the applicant's age and driving record.

Under 26 and No Accidents	Under 26 and Accident(s)	Over 25 and No Accidents	Over 25 and Accident(s)
Insure: Rate 1	Assign to High-Risk Pool	Insure: Rate 2	Insure: Rate 3

3. Show the results for each of the following insurance applicants based on the code you wrote for Exercise 1. NA = no accidents; AC = Accident(s)
   (a) ADLER   26   NA
   (b) BAKER   18   AC
   (c) CARSON   25   NA

(d) DIEHL	42	AC
(e) EIKES	24	AC
(f) FARGO	63	AC
(g) GRAY	21	NA
(h) HAWKS	25	AC

4. Show what will be printed by these PL/I statements when the variables have the values given below.

```
DCL (A1, A2, A3, A4, A5) CHAR (2);
IF A1 < A2 | A3 > A4 THEN
   IF A2 > A3 & A3 > A4 THEN
      PUT DATA (A1);
   ELSE
      IF A4 < A5 THEN
         PUT DATA (A2);
      ELSE
         PUT DATA (A3;
ELSE
   PUT DATA (A4);
PUT DATA (A5);
```

(a) A1 = 'AB'; A2 = 'BC'; A3 = 'CD'; A4 = 'DE'; A5 = 'EF';
(b) A1 = 'EF'; A2 = 'DE'; A3 = 'CD'; A4 = 'BC'; A5 = 'AB';
(c) A1 = 'AB'; A2 = 'AB'; A3 = 'AB'; A4 = 'AB'; A5 = 'AB';
(d) A1 = 'AB'; A2 = 'EF'; A3 = 'BC'; A4 = 'DE'; A5 = 'CD';

5. Show the output from each of the following segments of code. Use the same declaration for each.

```
DCL (N1, N2)   FIXED BINARY (15);
```

a.
```
        N = 1;
   LP_1: DO WHILE (N1 <= 3);
            PUT DATA (N1);
            N1 = N1 + 1;
            N2 = 4;
   LP_2:    DO WHILE (N2 <= 7);
               PUT DATA (N2);
               N2 = N2 + 1;
            END LP_2;
```

```
                N1 = N1 + 1;
            END LP_1;
b.          N1 = 5;
    LP_1: DO WHILE (N1 <= 5);
            PUT DATA (N1);
            N2 = 4;
    LP_2:   DO WHILE (N2 < 7);
                N2 = N2 + 1:
            END LP_2;
            PUT DATA (N2);
            N1 = N1 + 1;
        END LP_1;
c.          N1 = 1;
            N2 = 1;
    LP_1: DO WHILE (N1 < 100);
            PUT DATA (N1);
    LP_2:   DO WHILE (N2 < 50);
                PUT DATA (N2);
                N2 = N2 + 10;
            END LP_2;
            N1 = N1 + 25;
        END LP_1;
d.          N1 = 10;
    LP_1: DO WHILE (N1 < 20);
            PUT DATA (N1);
            N2 = 50;
    LP_2:   DO WHILE (N2 < 50);
                PUT DATA (N2);
            END LP_2;
            N1 = N1 + 10;
        END KP_1;
```

6. Write a program to calculate workers' gross pay. Each input record will contain a worker's shift code, social security number, name, and number of hours worked. Your program must be prepared to handle overtime at shift-rate * 1.5. The shift codes and their rates are 'D' day shift 10.75; 'S' swing shift 11.50; 'G' graveyard shift 12.25.

7. Write a program to produce a credit report on prospective house buyers. The input records consist of the customer's name, monthly payment on loan sought, monthly income, and credit rating. The categories and actions to take are

Payment	Credit Rating	Action
<= 25% of income	Good	Grant loan
<= 25% of income	Bad	Investigate more
> 25% of income	Bad	Deny loan
> 25% of income	Good	Investigate more

8. Write a program to prepare an accounts receivable statement for a group of physicians in a clinic. The input records contain the doctor's name, patient's name, and the amount due. Print the input as detail lines, the total due each doctor as subtotals, and total due to all doctors in the clinic.

9. Write a program to determine the extent to which reference books in the library are being used in a single day. The references are categorized as follows:

Dictionaries	Indexes	Handbooks
English	Humanities	Language
Foreign	Science	Science

The input consists of records giving a category and subcategory code (e.g., 'D' 'E' = Dictionary English), the name of the reference, and the number of times it was used that day. Your program should report the input as detail lines and the number for each subcategory, category, and all reference books. The input records will be grouped according to category and subcategory.

10. Write a program to prepare a billing report for a computer consulting company. The input records contain a service code indicating the type of service performed, the number of hours invested, and the client's name and address. The codes, services, and charges are as follows:

Code	Service	Charge per Hour
MAIN	Maintenance	47.95
EVAL	Evaluation of needs	52.50
APLP	Applications programming	57.57
SD&L	Systems analysis & design	63.00
SPGM	Systems programming	98.00

Assume that any one client used only one service. Use the SELECT statement to implement your solution.

11. Write a program to assign a rating to major league batters. The input consists of records containing the player name, number of "at bats," and number of hits. The categories are based on batting average as follows: Over .300 = Superstar; .250−.299 = All star; .225−.249 = Outstanding; .200−.224 = Good; Under .200 = Trade bait. Use SELECT to implement your solution.

12. Write a program to make change. The input consists of a series of numbers <= 99, representing a purchase. In each case the customer tendered a one-dollar bill. For each purchase, your program is to indicate the number of quarters, dimes, nickels, and pennies required to make the correct change. If a coin is not required, leave it blank. Be sure to compute the change so that it requires the fewest number of coins.

C H A P T E R 8

MORE ON

CHARACTER

VARIABLES

In this chapter we explore the power of PL/I to manipulate character strings. In Chapter 3 we explained the declaration of character variables and the INITIAL attribute. Another attribute is presented here. Also, this chapter deals with the built-in functions available for use with character strings. Before starting the discussion, we suggest that you review the example program below and return to it whenever necessary during the course of the chapter.

8.1

Example Program—Character Strings

In example program CH8EX1 (see Fig. 8–1) the student names in the student information system of a university are divided into two fields: first name and last name. A list of students needs to be printed in order to find the student's identification number when the name is known; for example,

Last Name, First Name	Student ID Number
JONES, BOB	2345
SMITH, JIM	5563
THOMSON, KAREN	2158

The input data is already in alphabetical order as follows:

'first name' 'last name' 'student ID number'

'BOB' 'JONES' '2345'

'JIM' 'SMITH' '5563'

'KAREN' 'THOMSON' '2158'

CH8EX1 reads in the names and ID numbers of the students, places a comma and a blank between the last name and first name of the students, and prints the resulting name followed by the student ID number. The pseudocode is shown next.

```
CH8EX1
      PRINT the page heading
      READ the first student
      DO WHILE (end-of-file has not been reached)
            Rearrange the student's name
            PRINT the name and identification number
            READ the next student
      ENDDO
      PRINT the trailer message
END CH8EX1.
```

```
PL/I OPTIMIZING COMPILER   VERSION 1 RELEASE 5.1  TIME: 15.17.14  DATE: 13 JAN 87  PAGE   1
OPTIONS USED


PL/I OPTIMIZING COMPILER       /* CH8EX1 USING CHARACTER VARIABLES */        PAGE   2
                 SOURCE LISTING
     STMT
```

FIGURE 8-1
CH8EX1 (character variables).

FIGURE 8-1 *(continued)*

```
    /* CH8EX1 USING CHARACTER VARIABLES */
1   CH8EX1: PROCEDURE OPTIONS (MAIN);
    /****************************************************************/
    /*                                                            */
    /*   FUNCTION:  TO PRINT A LIST OF STUDENT NAMES AND          */
    /*              THE CORRESPONDING STUDENT ID NUMBERS          */
    /*                                                            */
    /*   INPUT:     AN ALPHABETIC LIST OF STUDENTS AS:            */
    /*              FIRST_NAME LAST_NAME STUDENT_ID               */
    /*                                                            */
    /*   OUTPUT:    AN ALPHABETIC LIST OF STUDENTS AS:            */
    /*              LAST_NAME, FIRST_NAME STUDENT_ID              */
    /*                                                            */
    /****************************************************************/

2   DECLARE
        FIRST_NAME              CHARACTER    (10) VARYING,
        LAST_NAME               CHARACTER    (20) VARYING,
        FINAL_NAME              CHARACTER    (30) VARYING,
        STUDENT_ID              CHARACTER    (4),

        EOF                     BIT          (1) INIT ('0'B),
          YES_EOF               BIT          (1) INIT ('1'B),
          NOT_EOF               BIT          (1) INIT ('0'B);

3       ON ENDFILE(SYSIN) EOF = YES_EOF;

4   PUT PAGE LIST (' ','STUDENT LIST');
5   PUT SKIP(2) LIST ('STUDENT NAME','STUDENT ID NUMBER');
6   PUT SKIP(0) LIST ('_____','_____');
7   PUT SKIP LIST(' ');

8     GET LIST (FIRST_NAME, LAST_NAME, STUDENT_ID);

9   READ_LOOP:  DO WHILE (EOF = NOT_EOF);
10                FINAL_NAME = LAST_NAME || ', ' || FIRST_NAME;

11                PUT SKIP LIST (FINAL_NAME, STUDENT_ID);
12                GET LIST (FIRST_NAME, LAST_NAME, STUDENT_ID);
13              END READ_LOOP;
```

```
 14   PUT SKIP(2) LIST ('**** END OF STUDENTS ****');
 15   END CH8EX1;
```

PL/I OPTIMIZING COMPILER   /* CH8EX1 USING CHARACTER VARIABLES */   PAGE   3
COMPILER DIAGNOSTIC MESSAGES
ERROR ID L   STMT   MESSAGE DESCRIPTION

COMPILER INFORMATORY MESSAGES

IEL0533I I   NO 'DECLARE' STATEMENT(S) FOR 'SYSIN','SYSPRINT'.

END OF COMPILER DIAGNOSTIC MESSAGES

                    STUDENT LIST

STUDENT NAME   STUDENT ID NUMBER

ANDERSON, SHANE   9976
BUTTS, EDWARD   4436
DONNELLY, JOHN   8659
HARVEY, SUSAN   0642
JONES, BOB   5678
JONES, KARA   2345
O'HARA, RORY   8740
SMITH, JIM   5563
THOMSON, KAREN   2158

**** END OF STUDENTS ****

Before we present the other attribute for character strings, let us discuss the $||$ symbol (two bars with no blank between them) used in CH8EX1 (Statement 10). This symbol denotes the **concatenation** of two character strings. Concatenation can be thought of as an operator that "glues" two strings together in the order they are listed within the statement. It can be used in the assignment statement; for example,

```
DECLARE NAME CHARACTER (15) INITIAL (' ');
NAME = 'JONES' || ', ' || 'TOM';
```

produces

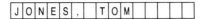

Note the blank following the comma in the literal. Concatenation does not automatically leave a blank between the items being combined. Therefore, if the blank following the comma were omitted, the result would be JONES,TOM. Concatenation can also be used in the output statement, and we illustrate that in the following section.

The DECLARE Statement Revisited for Character VARYING

During the introduction to character variable declaration in Chapter 3, we made the following points: (1) If a value being assigned to a character variable is shorter than the declared length, padding on the right with blanks occurs; (2) if a value assigned to a character variable is longer than the declared length, the value is truncated on the right; and (3) an apostrophe within a character string is represented by two single quotation marks (or apostrophes). Using these facts as a starting point, let us continue the discussion of character strings.

As you have probably discovered, a problem can occur with the blanks padded on the right of a character variable, especially when you are trying to print several character variables side by side. The blanks are considered part of the character string in the variable and therefore take up space in the output. For example,

```
DECLARE LAST_NAME   CHARACTER (15) INITIAL (' ');
DECLARE FIRST_NAME  CHARACTER (10) INITIAL (' ');
DECLARE FINAL_NAME  CHARACTER (25) INITIAL (' ');
LAST_NAME = 'SMITH';
FIRST_NAME = 'BOB';
FINAL_NAME = LAST_NAME || ', ' || FIRST_NAME;
PUT SKIP LIST (FINAL_NAME);
```

causes the following output to print

```
SMITH          , BOB
```

SMITH is followed by ten blank characters, and BOB is followed by seven. We think you will agree that this result is less than desirable.

To remedy the situation the **VARYING** attribute can be used when declaring character variables. A character variable declared as VARYING is not padded on the right with blanks when it is assigned a value shorter than the variable's declared length. The length of a VARYING character variable is in fact zero until a value is placed in it; when it receives a value, its length is precisely the length of the actual assigned value. Thus, if the variables in the above example had the VARYING attribute, the comma would be printed immediately following the last name:

```
JONES, BOB
```

The VARYING attribute follows the length attribute and is prior to the INITIAL attribute in the DECLARE statement. Some examples:

```
DECLARE LAST_NAME   CHARACTER (15) VARYING INITIAL (' ');
DECLARE ADDRESS     CHARACTER (50) VARYING INITIAL (' ');
DECLARE NAME        CHARACTER (15)         INITIAL ('ANDREWS');
LAST_NAME = 'ANDERSON';            /* LENGTH = 8  */
LAST_NAME = 'JONES';               /* LENGTH = 5  */
LAST_NAME = 'SMYTHE-ANDERSON';     /* LENGTH = 15 */
LAST_NAME = 'O''CONNELL';          /* LENGTH = 9  */
ADDRESS = '3475 N. FIRST ST., APT #6., ROCKFORD, IL 60445';
                                   /* LENGTH = 46 */
```

The above examples all involve assigning a literal to a character-varying variable. The length of the character-varying variable is the exact length of the literal. Examine the following code

```
LAST_NAME = 'JONES    ';           /* LENGTH = 9  */
```

In this case the LAST_NAME field is nine characters in length and does include the four blanks at the end. The reason is that this is the exact value being assigned to the LAST_NAME variable. The varying attribute *does not* remove blanks in a variable during the assignment operation. All the varying attribute allows us to do is to keep character values of different lengths in the same variable. When we actually move blanks into the character-varying variable, they will be stored. In order to avoid the padded blanks at the end of a character field, we must remove these blanks prior to assigning the data to the character-varying variable. The character built-in functions can assist us in the removal of the blanks.

The following example shows the character varying with two variables in the assignment statement.

```
LAST_NAME = NAME;                  /* LENGTH = 15 */
```

Here, NAME is declared without the VARYING attribute with a length of 15. The value in NAME after the initialization is

A	N	D	R	E	W	S								

Even though LAST_NAME is declared as character-varying, the value in LAST_NAME after the assignment statement is the same as the value in NAME above. Remember, the varying attribute does not remove blanks; it just allows us to store different-length character strings in the same variable.

Looking back at CH8EX1, we can see the usefulness of the varying attribute. It allows us to prevent any padding in the LAST_NAME field so that the comma is placed in the correct location. In CH8EX1 and the example above, LAST_NAME is concatenated to a comma and a blank, which in turn is concatenated to FIRST_NAME to produce FINAL_NAME. The difference is that the names are not padded with blanks in CH8EX1.

8.1.2

A Technical Aspect of Character VARYING

We mention a technical point here as to exactly how PL/I manages the character variable fields because a problem can occur (especially in character parameters—see Chapter 10). The memory allocated to character strings, both regular and varying, is one byte for each character up to the maximum number of characters specified in the declare statement. In character variables with the varying attribute, however, a halfword (two bytes) is added to the beginning of the variable (invisible to everyone except PL/I). This halfword specifies the number of bytes that is currently being occupied by data. Many times this extra halfword has caused programmers headaches because it is easily forgotten. We come back to this discussion at the time when knowing that the halfword exists is very important.

8.1.3

Rules of Character VARYING

To continue our discussion of character-varying fields, we remind you of these points.

1. The length used in character-varying declarations is the *maximum* number of characters allowed in the field.

2. No padding is done to extend a value to the maximum field length. The length of the field is the length of the value it contains.

3. Truncation occurs in character-varying variables as in regularly declared variables (on the right) when the declared length of the field is exceeded.

4. A character-varying field can contain no characters at all. This is called a **NULL** string, and the current length of the field is zero. The NULL string is denoted by two consecutive single quotation marks with no blank between.

```
DCL NAME CHAR (20) INIT ('');     /* NULL STRING */
NAME = '';                        /* NULL STRING */
```

We explain other aspects of the NULL string in the next sections.

8.2

Using Character Variables

Character variables can be used much the same as numeric variables except that they are not allowed in arithmetic operations. They can be read as input, compared, and printed. The following sections describe the uses of character variables that have not been previously discussed. First, GET LIST is revisited for use with character variables. The comparison of character variables is then presented, and concluding this section is a description of data conversions that involve character variables.

8.2.1

Character Strings as Input

In presenting GET LIST, we gave many examples of using numeric data as input and a few using character input. In this section, character string input is discussed in detail.

When the GET LIST statement is used, the input data must be enclosed in single quotation marks and follow the rules of character literals as described in Chapter 3. Below is a valid GET LIST data stream.

```
GET LIST (ADDRESSES);
Data:    'JOHN C. ANDERSON;2345 FIRST STREET;ANYTOWN, NY 99999'
         'MARY O''CONNELL;7764 N. MAIN ST.;NEW YORK, NY 09876'
         'JEFF MANNY;123 E. STATE APT. #5;MAINTOWN, IL 63076'
```

8.2.2

Comparison of Character Strings

Character strings are compared in terms of what is called the **collating sequence** of symbols. An ordering is assigned to all characters in the PL/I character set for the collating sequence; in order from lowest to highest, the sequence is:

1. Special characters:

blank ¢ . < (+ | & ! $ *) ; ^ / , % _ > ? : # @ ' = "

2. Lowercase letters:

a b c d e f g h i j k l m n o p q r s t u v w x y z

3. Uppercase letters:

A B C D E F G H I J K L M N O P Q R S T U V W X Y Z

4. Character digits:

0 1 2 3 4 5 6 7 8 9

Below are some examples of comparing character strings and the results.

'(' < '&'	True
')' < '+'	False
'M' > 'm'	True
' ' < any other symbol	True
'O' < 'M'	False

An important fact about character strings is that the NULL string is *not* the same as a string containing all blanks. The NULL string contains no value whatsoever, whereas a string of blanks does contain a value, that being blanks. When you use a string in the condition of an IF statement, you must take care to specify the correct literal.

```
DECLARE NAME CHARACTER (20) VARYING;
IF NAME = ' ' THEN . . .        /* IS NAME ALL BLANKS? */
IF NAME = '' THEN . . .         /* IS NAME NULL?       */
```

Naturally, we can compare strings longer than one character in length. In the following PL/I code, NAME_ONE is compared to NAME_TWO. The characters in the first positions of both strings, the 'J's, are compared first, followed by the 'O's, 'H's, and 'N's. At this point the character strings are equivalent. However, you remember character strings are padded on the right with blanks; therefore, the next two characters compared are the 'S' from NAME_TWO and a blank from NAME_ONE. Since 'S' is greater (in collating sequence) than a blank, the result of the IF statement is FALSE.

```
DECLARE NAME_ONE CHARACTER (20) INIT ('JOHN');
DECLARE NAME_TWO CHARACTER (20) INIT ('JOHNSON');
IF NAME_ONE > NAME_TWO THEN . . .
```

When character strings are different lengths, the shorter string is padded on the right with blanks to make it equal in length to the longer string. For example, in the IF statement below, PL/I looks at the corresponding characters of 'P', 'R', 'E', 'S', and 'S' in STRING1 and STRING2. At this point the STRING1 character is a blank, while the STRING2 character is an 'M', and so the result of the IF is FALSE.

```
DECLARE STRING1 CHARACTER (5) INIT ('PRESS');
DECLARE STRING2 CHARACTER (8) INIT ('PRESSMAN');
IF STRING1 = STRING2 THEN . . .
```

8.2.3

Data Conversion to and from Character Strings

In this section we discuss data-type conversions with respect to character variables. The discussion is included to assist you in the further understanding of character variables. However, please heed the following warning: *We do not recommend mixing data types.* Complex problems can arise from the mixing of data types, problems that are usually not easily located and solved. There are three main reasons for this section: (1) Not all languages allow these conversions to take place, and to be a good programmer you should know all the aspects of a language; (2) in our programming experience we have come across at least one application where the mixing of data types and the subsequent data conversion could not be avoided; and (3) if you become involved in maintaining programs written by others, you will almost inevitably come across mixed data types.

To summarize what we have learned about numbers and characters: A character variable containing a number is not the same as a numeric variable containing the same number. The first is a character string on which no arithmetic operations should be performed. The latter can be an operand in an arithmetic expression. However, it may be necessary to convert a character variable containing numbers into a numeric variable, and PL/I allows just such a conversion. The reverse—numeric to character—is also allowed. The following are the rules that apply to the data conversions involving character strings.

1. *Character to arithmetic*: A character string that contains only valid numeric values, with surrounding blanks (but none embedded), can be converted successfully to the corresponding number. (A numeric value contains digits, an optional decimal point, and an optional sign.)

2. *Arithmetic to character*: An arithmetic data item is converted to the corresponding character string. A decimal point and the negative sign, if in the data item, appears in the character string. The length of the

resulting character string is based upon the precision of the item, one byte per digit declared for arithmetic data item.

3. *Character to bit*: Only the characters 0 and 1 are converted. As a result of the conversion the character 0 becomes the bit 0, and the character 1 becomes the bit 1.

4. *Bit to character*: The bit string is converted by the bit 0 becoming the character 0 and the bit 1 becoming the character 1.

The following are examples of the conversions discussed above.

```
DECLARE
        DEC_NUM1            FIXED DECIMAL (6),
        DEC_NUM2            FIXED DECIMAL (6),
        BIN_NUM1            FIXED BINARY  (15),
        BIN_NUM2            FIXED BINARY  (15),
        CNUM               CHARACTER      (6);
DEC_NUM1 = 4:
DEC_NUM2 = DEC_NUM1 + 5;           /* DEC_NUM2 = 000009 */
CNUM = '6';
DEC_NUM2 = DEC_NUM1 + CNUM;        /* DEC_NUM2 = 000010 */
BIN_NUM1 = 12;
BIN_NUM2 = BIN_NUM1 + CNUM;        /* BIN_NUM2 = 18      */
CNUM = DEC_NUM2;                   /* CNUM = 'bbbb10'    */
```

8.2.4

Converting FIXED DECIMAL Numbers to Character Strings for Printing

As we noted in Section 4.1.2 of Chapter 4, numbers that are declared as FIXED DECIMAL are converted to character strings for either list-directed or data-directed output. A length is assigned to the character string based upon the precision of the FIXED DECIMAL variable as follows:

length of character string = precision of fixed dec + 3

The three extra characters are needed for the following.

1. A decimal point.

2. A minus sign if the number is negative. (If the number is positive, then this position is blank).

3. A leading blank.

    This will explain why your columns of FIXED DECIMAL numbers have not aligned with the column headers (which are character strings) when you have used PUT LIST. Study the example below.

```
DECLARE STUDENT_GPA FIXED DECIMAL (5,3) INIT (3.32);
PUT SKIP LIST ('GPA FOR STUDENT');
PUT SKIP LIST (STUDENT_GPA);
```

The output looks like the following.

```
GPA FOR STUDENT
↑↑↑ 3.320
  │ │ └_____ Leading zeros are printed as blanks.
  │ └_____ The blank for a positive number.
  └_____ The leading blank.
```

8.3

Built-in Functions for Character Strings

As we noted in Chapter 3, a function is a block of code that performs a single process, returns a single result, and can be invoked from anywhere within a program. Built-in functions are those that have been supplied with the language. PL/I has several built-in functions that use character strings as arguments and return values based upon the process performed by the function. The functions are most commonly invoked as:

```
Variable = function name (arguments);
```

The built-in functions we discuss in this section are LENGTH, REPEAT, INDEX, VERIFY, and SUBSTR, respectively. The following DECLARE statements are used to illustrate each function.

```
DECLARE   NAME      CHARACTER (20) VARYING,
          ADDRESS   CHARACTER (22) VARYING,
          TITLE     CHARACTER (15) VARYING,
          PHONE     CHARACTER (10),
          AREA      CHARACTER (3),
          V         FIXED BINARY (15);
```

Two notes before continuing: (1) Some of the following functions return a number in FIXED BINARY (15) format. Therefore, a variable should be declared with that scale, base, and precision to avoid data conversion of the result of the function. (2) LENGTH, INDEX, VERIFY, TRANSLATE, and CHAR are able to accept a bit string as input. The process performed by these functions on bit strings is identical to that performed on character strings as described below.

8.3.1

LENGTH

The LENGTH function has the form of LENGTH (char-string) and returns the length of the character string as a FIXED BINARY (15) number. The character string can be either a character variable or a literal.

```
V = LENGTH ('KARA JOHNSON');                /* V = 12  */
NAME = 'SHANE EDWARDS O''HARA';
V = LENGTH (NAME);                           /* V = 20  */
ADDRESS = '2457 N. FIRST ST., APT. #6';
V = LENGTH (ADDRESS);               /* V = 22 (truncated) */
NAME = '';
V = LENGTH (NAME);                           /* V = 0  */
```

The LENGTH function is very useful with character-varying fields because the length of the field is the actual length of the data within the field. The argument of LENGTH does not need to be a character string because conversion will take place prior to the invocation of the LENGTH function for any data item not in character string format (see the conversion rules above).

If the LENGTH function is used with character variables that are declared without the varying attribute, the length returned is the length specified in the declare statement. For example, since PHONE doesn't have the varying attribute, LENGTH would return the following:

```
PHONE = '756-3567';
V = LENGTH (PHONE);                          /* V = 10  */
```

8.3.2

REPEAT

The form of the REPEAT function is REPEAT (char-string, x). Char-string represents a character string to be concatenated with itself x number of times to produce a new string. (Note that the final string contains x + 1 number of the original character string.)

```
TITLE = REPEAT ('BANG', 1);   /* TITLE = BANGBANG        */
TITLE = REPEAT ('BANG ', 2)   /* TITLE = BANG BANG BANG */
TITLE = REPEAT ('A', 8);      /* TITLE = AAAAAAAAA       */
```

The REPEAT function is very useful when you need to underline a list of words.

```
PUT SKIP LIST ('CLASS LIST');
PUT SKIP (0) LIST (REPEAT ('_', 9);
```

prints <u>CLASS LIST</u>

INDEX

The form of the INDEX function is INDEX (c, s), where c is a character string to be searched, and s is a character string to use in the searching of c. INDEX looks for an exact match between s and some part of c. INDEX returns the following information as a FIXED BINARY (15) value.

1. If a match is found, s is somewhere in c, and INDEX returns the starting location (from 1) of the leftmost character matched in c.

2. If a match is not found, the value zero is returned.

The following examples illustrate the power of the INDEX function.

```
NAME = 'JONES, BOB';
V = INDEX (NAME, ' ');              /* V = 7  */
V = INDEX (NAME, 'B');              /* V = 8  */
V = INDEX (NAME, 'JON');            /* V = 1  */
V = INDEX (NAME, 'Z');              /* V = 0  */
ADDRESS = '24 EAST ST., NEW YORK';
V = INDEX (ADDRESS, ',');           /* V = 12 */
```

VERIFY

The VERIFY function has a format similar to the INDEX function, but a very different purpose. The format is VERIFY (c, s), where c is a character string that will be used one character at a time in the verification and s is the string that is checked to determine if it contains only valid characters as determined by c. VERIFY returns one of the following FIXED BINARY (15) values.

1. If every character in string c appears in string s, a value of zero is returned.

2. Otherwise, the location (starting from 1) of the first character in c that does not appear in s is returned.

• EXAMPLES.   In one example we are checking to see whether NAME contains only alphabetic characters or a blank. If you notice, there is a blank in the first position of the string s.

```
NAME = 'JOHN DOE';
V = VERIFY (NAME, ' ABCDEFGHIJKLMNOPQRSTUVWXYZ');
                              /* IS NAME ALL ALPHABETIC? */
                                    /* V = 0 */
```

In the following example we determine whether a phone number is comprised of numerical digits.

```
PHONE = '8005551326';
V = VERIFY (PHONE, '0123456789');
                              /* IS PHONE ALL DIGITS?   */
                                    /* V = 0 */
PHONE = '555-1326';
V = VERIFY (PHONE, '0123456789');   /* V = 4 */
```

A major use of the VERIFY function, of course, is testing a field to see whether it contains a valid set of characters. Suppose, for example, that a product number is coded to match with the department that sells it; that is, all products in department 21 must be composed of some combination of the letters A, L, X, and the numbers 3, 6, and 9. We can then check the product number by using the VERIFY.

```
IF DEPT_NUMBER = '21' THEN
    LEGAL_CODE = VERIFY (PROD_CODE, 'ALX369');
```

If LEGAL_CODE is 0, the product code is valid; otherwise, it is an invalid code.

The process of verifying field c is done by taking one character at a time from c (starting with the leftmost) and trying to find a match in s. Do not confuse this with the INDEX function, which takes the entire string of s and tries to find a match in c.

8.3.5

SUBSTR

The SUBSTR function has three arguments. Its format is SUBSTR (c, sp, l), where c is the character string from which a substring will be extracted, sp is the starting point of the substring , and l is the length of the substring. You may omit l from the SUBSTR function call; you will then get the default length, which is the end of the string c. SUBSTR returns a new character string as a result.

```
NAME = 'BOB JONES';
LAST_NAME = SUBSTR (NAME, 5);        /* LAST_NAME = JONES */
FIRST_NAME = SUBSTR (NAME, 1, 3);   /*  FIRST_NAME = BOB */
PHONE = '8005551212';
PUT LIST (SUBSTR (PHONE, 1, 3) || '-' ||
          SUBSTR (PHONE, 4, 3) || '-' ||
          SUBSTR (PHONE, 7));        /* 800-555-1212      */
```

As the last example shows, SUBSTR is a very powerful tool for editing fields for output.

We can also use the SUBSTR built-in function on the left-hand side of an assignment statement. Look at the example below.

```
DECLARE STRING1 CHARACTER (10) INIT ('ABCDEFGHIJ');
SUBSTR(STRING1,3,2) = '**';
PUT LIST (STRING1);              /* STRING1 = 'AB**EFGHIJ' */
```

If we omitted the length of the substring, the result would blank out the rest of STRING1.

```
SUBSTR(STRING1,3) = '**';       /* STRING1 = 'AB**        ' */
```

8.3.6.

TRANSLATE

The format of this function is TRANSLATE (ss, rs, ps), where ss is the string in which replacement takes place, and rs is the string that replaces the string ps. The TRANSLATE performs a translation process on a single character; however, more than one single character can be translated during one invocation of the TRANSLATE function. Here are some examples.

```
DECLARE
      CHAR_FLD1        CHARACTER (4),
      CHAR_FLD2        CHARACTER (5),
      CHAR_FLD3        CHARACTER (2);
CHAR_FLD1 = '1234';
CHAR_FLD2 = TRANSLATE (CHAR_FLD1, '*', '2');
                        /* CHAR_FLD2 = 1*34   */
```

```
CHAR_FLD3 = TRANSLATE (CHAR_FLD1, '*-', '24');
                    /* CHAR_FLD3 = 1*3-  */
```

The TRANSLATE function might be useful in translating leading zeros in a number to asterisks for printing. For example,

```
DECLARE
        CHECK_AMOUNT        FIXED DECIMAL (6),
        CHECK_CHAR          CHARACTER    (6),
        PRINT_CHECK         CHARACTER    (6);
CHECK_AMOUNT = 3458;            /* CHECK_AMOUNT = 003458  */
CHECK_CHAR = CHECK_AMOUNT;      /* CHECK_CHAR = '003458'  */
PRINT_CHECK = TRANSLATE (CHECK-CHAR, '*', '0');
                               /* PRINT_CHECK = '**3458'  */
```

(One note: Any embedded or following zeros will also be translated into '*'.)

8.3.7

DATE and TIME

The DATE and TIME functions are not used to manipulate character strings and do not accept any arguments. However, when they are invoked, they return a character string that can then be used like any other character string (e.g., as an argument to the SUBSTR function). The computer has two special internal devices that allow programmers to obtain the current date and current time. These devices are accessed via various methods within different programming languages. PL/I uses built-in functions to access them.

The DATE function returns a six-character field as yymmdd, where yy is the last two digits of the current year, mm is the current month expressed as two digits, and dd is the day of the month. An example of using the DATE function is

```
OUT_DATE = SUBSTR (DATE, 3, 2) || '/' ||
           SUBSTR (DATE, 5, 2) || '/' ||
           SUBSTR (DATE, 1, 2);
PUT SKIP LIST (OUT_DATE);                 /* mm/dd/yy */
```

Assuming this code executed on July 4, 1976, the output would be 07/04/76.

The TIME function returns a nine-character string in the form of hhmmssttt, where hh is hours, mm is minutes, ss is seconds, and ttt is milliseconds (thousandths of a second). An example of the use of the TIME function is

```
OUT_TIME = TIME;
PRINT_TIME = SUBSTR (OUT_TIME, 1, 2) || ':' ||
             SUBSTR (OUT_TIME, 3, 2) || ':' ||
             SUBSTR (OUT_TIME, 5, 2) || ':' ||
             SUBSTR (OUT_TIME, 7);
PUT SKIP LIST (PRINT_TIME);          /* hh:mm:ss:ttt */
```

8.3.8

CHAR and
BIT

The CHAR and BIT functions can be used to perform data conversions when you are mixing data types rather than having PL/I automatically perform them. These functions have two formats: the first has one argument, and the second has two arguments. The first format is CHAR (x) or BIT (x). The x is a value, in some other data type, that is to be converted into either character or bit data. Here the length of the result is the length of the argument. The second format is CHAR (x, n) or BIT (x, n). The x is the same as in the first format, and the n is a decimal integer constant specifying the length of the result. The result of either function is the conversion into the data type denoted by the name of the function, CHAR or BIT. Some examples:

```
DECLARE
    NUM_ONE        FIXED DECIMAL (5)   INIT (145),
    NUM_TWO        FIXED BINARY  (15)  INIT (4500),
    CHAR_NUM       CHARACTER     (5);
CHAR_NUM = CHAR (NUM_ONE);          /* CHAR_NUM = '  145'  */
CHAR_NUM = CHAR (NUM_TWO, 4);       /* CHAR_NUM = '4500 '  */
NUM_TWO = BIT (NUM_ONE);            /* NUM_TWO LENGTH = 15 */
```

8.3.9

Example Program—
Combining the
Built-in Functions

The combining of the built-in functions within PL/I creates a capability for string manipulation that is not matched in most languages. CH8EX2 (see Fig. 8–2) illustrates just a small portion of these capabilities.

A junk mailing company just received a list of addresses from a source (to remain anonymous). The addresses are not in the correct format for producing the mailing labels needed. The input addresses are as follows:

```
'last name, first name initial;address;city, state zip'
```

```
'JONES, BOB C.;236 N. FIRST ST.;ANYTOWN, IL 61111'
'ZUNY, TOM F.;777 MAIN STREET;NEW YORK, NY 08888'
```

The mailing labels must be in the following format.

```
first name initial last name
street address
city, state zip

BOB C. JONES
236 N. FIRST ST.
ANYTOWN, IL 61111

TOM F. ZUNY
777 MAIN STREET
NEW YORK, NY 08888
```

The zip codes within the data received are not necessarily accurate; they must be checked to be sure they are five characters long and that they are composed entirely of numbers.

CH8EX2 uses several of the built-in functions to solve the problem described above. An important point to remember when looking at CH8EX2 is that the INDEX function starts its count of positions from the beginning of the string it is searching. For example,

```
TITLE = 'OLD MACDONALD''S FARM';
I = INDEX (TITLE, ' ');                    /* I = 4 */
J = INDEX (SUBSTR (TITLE, I + 1), ' ');    /* J = 12 */
```

The second INDEX statement above starts its count at position 1 of the substring 'MACDONALD''S FARM', not at position 1 of TITLE, as does the first INDEX statement. Therefore, in CH8EX2 the starting position of the subsequent INDEX statements must include the results of the previous INDEX statements.

If you study the example program, you will find the following substrings of INPUT_ADDR being searched.

1. In Statement 8 the entire INPUT_ADDR (starting position = 1) is searched: 'last name, first name initial;street address;city, state zip'.

2. In Statement 11 the INDEX starts after LAST_NAME (starting position = I + 2 to miss the comma and blank): 'first name initial;street address;city, state zip'.

3. In Statement 14 the INDEX starts after FIRST_NAME (starting position = I + 2 + J to miss the semicolon: 'street address;city, state zip'.

4. In Statement 17 the starting position is after street address (starting position − I + 2 + J + K to miss the semicolon); 'city, state zip'.

5. Starting after CITY_STATE (starting position = I + 2 + J + K + 4 to pass over city, state and blank), we obtain 'zip' in Statement 20.

Notice also that the zip code has been declared as length 10 instead of length 5. In order to be able to check the length of the zip code, we need to obtain all the rest of the characters in the address line for zip code. The LENGTH function is then applied in Statement 21 to the zip code to see whether it is too long.

The CH8EX2 pseudocode is as follows:

```
CH8EX2:
      PRINT the heading
      READ the first address
      DO WHILE (end-of-file has not been reached)
            Increment the label counter
            Obtain the information to print
            IF (the zip code is not valid)
                  set the zip code to '*****'
            ENDIF
            PRINT the label
            READ the next address
      ENDDO
      PRINT the label counter
      PRINT the number of invalid zip codes
END CH8EX2.
```

PL/I OPTIMIZING COMPILER VERSION 1 RELEASE 5.1 TIME: 14.54.11 DATE: 1 MAR 87 PAGE   1
OPTIONS USED

PL/I OPTIMIZING COMPILER /* CH8EX2 - BUILTIN FUNCTIONS FOR CHAR VARIABLES */   PAGE   2
               SOURCE LISTING
    STMT

FIGURE 8–2
CH8EX2 (character variable built-in functions).

FIGURE 8–2 *(continued)*

```
     /* CH8EX2 - BUILTIN FUNCTIONS FOR CHAR VARIABLES */
  1  CH8EX2:  PROCEDURE OPTIONS (MAIN);
     /*********************************************************************/
     /*                                                                 */
     /*  FUNCTION:   TO PRINT MAILING LABELS FROM A LIST OF ADDRESSES   */
     /*                                                                 */
     /*  INPUT:      A LIST OF NAMES AND ADDRESSES OF THE FORM:         */
     /*                 LAST NAME, FIRST NAME INITIAL;ADDRESS;CITY, ST  */
     /*                    ZIP                                          */
     /*                                                                 */
     /*  OUTPUT:     MAILING LABELS OF THE FORM:                        */
     /*                 FIRST NAME INITIAL LAST NAME                    */
     /*                 STREET ADDRESS                                  */
     /*                 CITY, ST ZIP                                    */
     /*                                                                 */
     /*********************************************************************/


  2  DECLARE
        INPUT_ADDR                 CHARACTER (80) VARYING,
        FIRST_NAME                 CHARACTER (30) VARYING,
        LAST_NAME                  CHARACTER (30) VARYING,
        NAME_LINE                  CHARACTER (60) VARYING,
        STREET_ADDR                CHARACTER (60) VARYING,
        CITY_STATE_ZIP             CHARACTER (60) VARYING,
        CITY_STATE                 CHARACTER (60) VARYING,
        ZIP                        CHARACTER (10) VARYING,

        INVALID_ZIP_COUNT          FIXED DECIMAL (7) INITIAL (0),
        LABELS_COUNT               FIXED DECIMAL (7) INITIAL (0),

        I                          FIXED BINARY (15) INITIAL (0),
        J                          FIXED BINARY (15) INITIAL (0),
        K                          FIXED BINARY (15) INITIAL (0),
        L                          FIXED BINARY (15) INITIAL (0),
        START_POS                  FIXED BINARY (15) INITIAL (0),

        INDEX                      BUILTIN,
        SUBSTR                     BUILTIN,
        LENGTH                     BUILTIN,
        VERIFY                     BUILTIN,
```

```
        EOF                           BIT (1) INITIAL ('0'B),
          NOT_EOF                     BIT (1) INITIAL ('0'B),
          YES_EOF                     BIT (1) INITIAL ('1'B);

    3    ON ENDFILE (SYSIN) EOF - YES_EOF;

    4    PUT LIST ('**** START OF MAILING LABELS ****');

    5    GET LIST (INPUT_ADDR);
```

```
    STMT
    6  R_LOOP: DO WHILE (EOF = NOT_EOF);
    7            LABELS_COUNT = LABELS_COUNT + 1;

    8            I = INDEX (INPUT_ADDR, ',');
    9            LAST_NAME = SUBSTR (INPUT_ADDR, 1, I-1);

   10            START_POS = I + 2;
   11            J = INDEX (SUBSTR (INPUT_ADDR, START_POS), ';');
   12            FIRST_NAME = SUBSTR (INPUT_ADDR, START_POS, J-1);

   13            START_POS = START_POS + J;
   14            K = INDEX (SUBSTR (INPUT_ADDR, START_POS), ';');
   15            STREET_ADDR = SUBSTR (INPUT_ADDR, START_POS, K-1);

   16            START_POS = START_POS + K;
   17            L = INDEX (SUBSTR (INPUT_ADDR, START_POS), ', ');
   18            CITY_STATE = SUBSTR (INPUT_ADDR, START_POS, L+4);

   19            START_POS = START_POS + L + 4;
   20            ZIP = SUBSTR (INPUT_ADDR, START_POS);

   21            IF LENGTH (ZIP) > 5 |
                    VERIFY (ZIP, '0123456789') ^= 0 THEN
                      DO;
   22                    ZIP = '*****';
   23                    INVALID_ZIP_COUNT = INVALID_ZIP_COUNT + 1;
   24                  END;

   25            NAME_LINE = FIRST_NAME || ' ' || LAST_NAME;
   26            PUT SKIP(3) LIST (NAME_LINE);
```

FIGURE 8–2 *(continued)*

```
27          PUT SKIP LIST (STREET_ADDR);

28          CITY_STATE_ZIP = CITY_STATE || ' ' || ZIP;
29          PUT SKIP LIST (CITY_STATE_ZIP);

30          GET LIST (INPUT_ADDR);

31       END R_LOOP;

32    PUT PAGE LIST ('**** NUMBER OF LABELS PRINTED = ', LABELS_COUNT);
33    PUT SKIP LIST ('**** NUMBER OF INVALID ZIP CODES = ',
                      INVALID_ZIP_COUNT);

34    END CH8EX2;
```

```
PL/I OPTIMIZING COMPILER  /* CH8EX2 - BUILTIN FUNCTIONS FOR CHAR VARIABLES */    PAGE   4
COMPILER DIAGNOSTIC MESSAGES
ERROR ID L   STMT    MESSAGE DESCRIPTION

WARNING DIAGNOSTIC MESSAGES

IEL0239I W   1       COMMENTS IN OR FOLLOWING STATEMENT CONTAIN ONE OR MORE SEMICOLONS.
IEL0238I W   11, 14   CHARACTER STRING CONSTANT CONTAINS SEMICOLON.

COMPILER INFORMATORY MESSAGES

IEL0533I I           NO 'DECLARE' STATEMENT(S) FOR 'SYSIN','SYSPRINT'.

END OF COMPILER DIAGNOSTIC MESSAGES


**** START OF MAILING LABELS ****


MARY A. ANDERSON
3674 N. B STREET
RALEIGH, NC  98383
```

```
JEFF B. BOUEY
848 SOUTH STREET
NEW YORK, NY  *****


JOHN X. CHANNEL
777 MAIN AVE.
INDIANAPOLIS, IN  *****


KATHERYN L. MAXWELL
6743 HOME STREET
CHICAGO, IL  87655


MAX NOTTINGHAM
8976 WEST AVE
SEATTLE, WA  87654


SHANE E. THOMSON
#54 TERRACE AVE.
SAN DIEGO, CA  98765


LANCE J. WESSLEY
8764 JONES AVE.
ST. PETERSBURG, FL  54328

----------------------------------------------------------------------
**** NUMBER OF LABELS PRINTED =                    7
**** NUMBER OF INVALID ZIP CODES =                 2
```

8.4

Example Program—Character Variables

A university wishes to send a letter to students who have been accepted by the admissions office. This letter will notify the students of an orientation date. The acceptance letters go out to all students accepted, regardless of the semester for which they applied; the letter must therefore be flexible enough to allow for any semester and for two orientation dates per semester. The orientation dates are assigned according to the first character of the student's last name. A–M receives the first date and N–Z the second.

CH8EX3 uses SUBSTR, DATE, concatenation, and the comparison of characters (see Fig. 8–3). The I/O is list-directed.

The CH8EX3 pseudocode follows.

```
CH8EX3
    READ the dates
    READ the first student's record
    DO WHILE (end-of-file has not been reached)
        READ the department record
        PRINT the letter
        Format the semester date line
        PRINT the rest of the letter
        READ the next student's record
    ENDDO
END CH8EX3.
```

```
PL/I OPTIMIZING COMPILER  VERSION 1 RELEASE 5.1  TIME: 14.54.31  DATE: 1 MAR 87  PAGE   1
OPTIONS USED


PL/I OPTIMIZING COMPILER         /* CH8EX3 - USING CHARACTER VARIABLES */        PAGE   2
                SOURCE LISTING
    STMT


    /* CH8EX3 - USING CHARACTER VARIABLES */
  1 CH8EX3:  PROCEDURE OPTIONS (MAIN);
    /*******************************************************/
    /*                                                     */
    /*  FUNCTION:  TO PRINT A LETTER INFORMING STUDENTS THAT*/
    /*             THEY HAVE BEEN ACCEPTED AT THE UNIVERSITY*/
    /*                                                     */
    /*  INPUT:     THE FIRST THREE LINES:                  */
    /*                 TWO DATES TO REPORT TO THE UNIVERSITY */
    /*                     FALL FOR A-M, FALL FOR N-Z       */
    /*                     SPRING FOR A-M, SPRING FOR N-Z   */
    /*                     SUMMER FOR A-M, SUMMER FOR N-Z   */
    /*                                                     */
    /*             THE REST OF THE INPUT LINES:            */
    /*                 THE STUDENTS TO RECEIVE THE LETTER  */
```

FIGURE 8–3
CH8EX3 (printing a letter using character variables).

```
      /*              TWO INPUT LINES/STUDENT AS FOLLOWS: */
      /*              STUDENT_NAME, STUDENT_ADDR,         */
      /*                  STUDENT_CITY_STATE_ZIP          */
      /*              MAJORING_DEPARTMENT,                */
      /*                  APPLIED_SEMESTER, APPLIED_YEAR  */
      /*                                                  */
      /*  OUTPUT:    THE FORM LETTER TO THE STUDENT       */
      /*                                                  */
      /****************************************************/

  2  DECLARE
       FALL_DATE_AM            CHARACTER (20) VARYING,
       FALL_DATE_NZ            CHARACTER (20) VARYING,
       SPRING_DATE_AM          CHARACTER (20) VARYING,
       SPRING_DATE_NZ          CHARACTER (20) VARYING,
       SUMMER_DATE_AM          CHARACTER (20) VARYING,
       SUMMER_DATE_NZ          CHARACTER (20) VARYING,

       STUDENT_NAME            CHARACTER (30) VARYING,
       STUDENT_ADDR            CHARACTER (25) VARYING,
       STUDENT_CSZ             CHARACTER (20) VARYING,
       DEPT_NAME               CHARACTER (30) VARYING,
       APPL_SEM                CHARACTER (1),
       APPL_YEAR               CHARACTER (4),

       FIRST_CHAR_LAST         CHARACTER (1),
       SALUTATION              CHARACTER (30) VARYING,
       DATE_OUT                CHARACTER (8),
       LINE1                   CHARACTER (80) VARYING,
       LINE2                   CHARACTER (80) VARYING,
       LINE3                   CHARACTER (80) VARYING,
       LINE4                   CHARACTER (80) VARYING,
       LINE5                   CHARACTER (80) VARYING,

       EOF                     BIT     (1) INIT ('0'B),
         NOT_EOF               BIT     (1) INIT ('0'B),
```

PL/I OPTIMIZING COMPILER   /* CH8EX3 - USING CHARACTER VARIABLES */   PAGE   3
  STMT

```
         YES_EOF               BIT     (1) INIT ('1'B),
```

FIGURE 8–3 *(continued)*

```
          SUBSTR                 BUILTIN,
          INDEX                  BUILTIN,
          DATE                   BUILTIN;

  3   ON ENDFILE (SYSIN) EOF = YES_EOF;

  4  GET LIST (FALL_DATE_AM, FALL_DATE_NZ);
  5  GET LIST (SPRING_DATE_AM, SPRING_DATE_NZ);
  6  GET LIST (SUMMER_DATE_AM, SUMMER_DATE_NZ);

  7  GET LIST (STUDENT_NAME, STUDENT_ADDR, STUDENT_CSZ);

  8  R_LOOP: DO WHILE (EOF = NOT_EOF);
  9          GET LIST (DEPT_NAME, APPL_SEM, APPL_YEAR);

 10          DATE_OUT = SUBSTR (DATE, 3, 2) || '/' ||
                       SUBSTR (DATE, 5, 2) || '/' ||
                       SUBSTR (DATE, 1, 2);

 11          PUT PAGE  LIST (DATE_OUT);
 12          PUT SKIP (2) LIST ('EXCELL UNIVERSITY');
 13          PUT SKIP LIST ('ADMISSIONS OFFICE');
 14          PUT SKIP LIST ('ST.MARY, KT 22222');

 15          PUT SKIP(2) LIST (STUDENT_NAME);
 16          PUT SKIP LIST (STUDENT_ADDR);
 17          PUT SKIP LIST (STUDENT_CSZ);

 18          SALUTATION = 'DEAR ' || STUDENT_NAME || ',';
 19          PUT SKIP (2) LIST (SALUTATION);

 20          LINE1 = '    CONGRATULATIONS ' ||
                    SUBSTR (STUDENT_NAME, 1,
                         INDEX (STUDENT_NAME, ' ') - 1) ||
                    '!';
 21          PUT SKIP (2) LIST (LINE1);

 22          LINE2 = 'YOU HAVE BEEN ACCEPTED INTO OUR ' ||
                    DEPT_NAME;
 23          PUT SKIP (2) LIST (LINE2);
```

```
 24              FIRST_CHAR_LAST = SUBSTR (STUDENT_NAME,
                            INDEX (STUDENT_NAME, ' ') + 1, 1);


 25 SEM_SEL: SELECT (APPL_SEM);
 26          WHEN ('1')
    SEL1:       DO;
 27              LINE3 = 'DEPARTMENT FOR ' || 'FALL ' || APPL_YEAR;

 28 DATE_SEL1:    SELECT;
```

```
 29                  WHEN (FIRST_CHAR_LAST >= 'A' &
                          FIRST_CHAR_LAST <= 'M')
                       LINE4 = 'PLEASE REPORT TO CAMPUS ON ' ||
                              FALL_DATE_AM;

 30                  WHEN (FIRST_CHAR_LAST >= 'N')
                       LINE4 = 'PLEASE REPORT TO CAMPUS ON ' ||
                              FALL_DATE_NZ;

 31              END DATE_SEL1;
 32            END SEL1;

 33          WHEN ('2')
    SEL2:       DO;
 34              LINE3 = 'DEPARTMENT FOR ' || 'SPRING ' || APPL_YEAR;

 35 DATE_SEL2:    SELECT;

 36                  WHEN (FIRST_CHAR_LAST >= 'A' &
                          FIRST_CHAR_LAST <= 'M')
                       LINE4 = 'PLEASE REPORT TO CAMPUS ON ' ||
                              SPRING_DATE_AM;

 37                  WHEN (FIRST_CHAR_LAST >= 'N')
                       LINE4 = 'PLEASE REPORT TO CAMPUS ON ' ||
                              SPRING_DATE_NZ;

 38              END DATE_SEL2;
 39            END SEL2;
```

FIGURE 8-3 *(continued)*

```
40              WHEN ('3')
   SEL3:        DO;
41                 LINE3 = 'DEPARTMENT FOR ' !! 'SUMMER ' !! APPL_YEAR;

42  DATE_SEL3:    SELECT;

43                   WHEN (FIRST_CHAR_LAST >= 'A' &
                           FIRST_CHAR_LAST <= 'M')
                       LINE4 = 'PLEASE REPORT TO CAMPUS ON ' !!
                                 SUMMER_DATE_AM;

44                   WHEN (FIRST_CHAR_LAST >= 'N')
                       LINE4 = 'PLEASE REPORT TO CAMPUS ON ' !!
                                 SUMMER_DATE_NZ;
45                 END DATE_SEL3;
46              END SEL3;
47          END SEM_SEL;

48          PUT SKIP LIST (LINE3);
49          PUT SKIP LIST (LINE4);

50          PUT SKIP LIST ('AT 10:00 A.M. IN THE BALLROOM OF THE');
51          PUT SKIP LIST ('STUDENT CENTER FOR ORIENTATION.');
```

PL/I OPTIMIZING COMPILER   /* CH8EX3 - USING CHARACTER VARIABLES */   PAGE   5
```
 STMT
52          PUT SKIP (2) LIST ('IF YOU HAVE ANY QUESTIONS, CALL');
53          PUT SKIP LIST ('800-555-4444 ANYTIME BETWEEN 8:00 AM.');
54          PUT SKIP LIST ('AND 4:30 PM.  MONDAY THROUGH FRIDAY.');
55          PUT SKIP (2) LIST ('LOOKING FORWARD TO HAVING YOU AS');
56          PUT SKIP LIST ('ONE OF OUR STUDENTS.');

57          PUT SKIP(2) LIST ('SINCERELY,');
58          PUT SKIP(3) LIST ('MS. SUSAN ANDERSON');
59          PUT SKIP LIST ('ADMISSIONS OFFICER');

60          GET SKIP LIST (STUDENT_NAME, STUDENT_ADDR, STUDENT_CSZ);
61       END R_LOOP;

62  END CH8EX3;
```

COMPILER INFORMATORY MESSAGES

IEL0533I I   NO 'DECLARE' STATEMENT(S) FOR 'SYSIN','SYSPRINT'.

END OF COMPILER DIAGNOSTIC MESSAGES

03/01/87

EXCELL UNIVERSITY
ADMISSIONS OFFICE
ST.MARY, KT 22222

JEREMIAH JOHNSON
2367 N. FIRST ST.
CHICAGO, IL 67584

DEAR JEREMIAH JOHNSON,

     CONGRATULATIONS JEREMIAH!

YOU HAVE BEEN ACCEPTED INTO OUR COMPUTER SCIENCE
DEPARTMENT FOR FALL 1985
PLEASE REPORT TO CAMPUS ON AUGUST 12, 1985
AT 10:00 A.M. IN THE BALLROOM OF THE
STUDENT CENTER FOR ORIENTATION.

IF YOU HAVE ANY QUESTIONS, CALL
800-555-4444 ANYTIME BETWEEN 8:00 AM.
AND 4:30 PM. MONDAY THROUGH FRIDAY.

LOOKING FORWARD TO HAVING YOU AS
ONE OF OUR STUDENTS.

SINCERELY,

MS. SUSAN ANDERSON
ADMISSIONS OFFICER

Summary

In this chapter we explored character strings and character variables. We presented the advantage of using VARYING—namely, to eliminate the padded blanks when necessary. The rules that apply to a variable with the VARYING attribute are:

1. The declared length is the maximum number of characters allowed in the variable.

2. No padding with blanks is performed to extend the length of the character string to the declared maximum.

3. Truncation occurs on the right.

4. A character variable can contain the NULL string—no value (and a length of zero).

GET LIST requires that the variable be enclosed in single quotation marks following the rules for character literals found in Chapter 3: 'MARY SMITH'.

Comparison of character strings takes advantage of the collating sequence: a blank is first in the sequence, followed by special characters, lowercase letters, uppercase letters, and the digits. You must take care when using the NULL value in a comparison because a character string containing a NULL is not equal to one containing one or more blanks.

The data conversions allowed by PL/I involving character strings are character to arithmetic, arithmetic to character, character to bit, and bit to character. Again, we stress that data conversion should be avoided whenever possible, as it can lead to some problems that are difficult to solve.

The built-in functions discussed for character strings are

1. LENGTH—to obtain the length of a character VARYING string.

2. REPEAT—to create a repetition of a character string a specified number of times.

3. INDEX—to search a character string for a given value.

4. VERIFY—to compare one character string with another, looking for places of mismatch.

5. SUBSTR—to subdivide a character string.

6. TRANSLATE—to translate the specified characters of a given character string into another specified set of characters.

7. DATE and TIME— to obtain the current date and time.

8. CHAR and BIT—to convert a value to a character or bit string.

The example programs we presented in this chapter reflect the power that PL/I has in manipulating character strings.

8.6

Exercises

1. Show the result of the following DECLARE statements.

```
a.  DCL   PRODUCT_NAME   CHAR (10) INIT (' ');
b.  DCL   NAME           CHAR (20) VAR INIT ('HALLEY');
c.  DCL   STREET_ADDR    CHAR (15) VAR
                         INIT ('3456 SOUTH STATE STREET, APT # 5');
d.  DCL TITLE            CHAR (24) VAR
                         INIT ('KATRINA''S MOONTRIP');
```

2. Using the given declarations, show the results of the following assignment statements:

```
DCL   ACCOUNT_NUMBER    CHAR (10) INIT (' '),
      CUSTOMER_NAME      CHAR (20) VAR INIT (''),
      ITEM_CODE          CHAR (6) VAR;
```

```
a.  ACCOUNT_NUMBER = '2387-929';
b.  CUSTOMER_NAME = 'MS. MARY SMITH';
c.  ITEM_CODE = '928383920';
d.  CUSTOMER_NAME = ' ';
e.  CUSTOMER_NAME = '';
```

3. What does TITLE contain after the following GET statements execute?

```
DCL   TITLE     CHAR (20) INIT ('');
```

```
a.  GET DATA (TITLE);        TITLE = 'WITCH WORLD';
b.  GET LIST (TITLE);        'RETURN TO WITCH WORLD'
c.  GET LIST (TITLE);        ' '
d.  GET LIST (TITLE);        'ANSWER TO JANE''S PROBLEM'
```

4. State whether each of the comparisons below evaluates to TRUE or FALSE.

```
a.  IF 'X' = 'x' THEN            e.  IF '*' < '%' THEN
b.  IF '3' > '1' THEN            f.  IF ' ' < '0' THEN
c.  IF 'MAY' < 'JUNE' THEN       g.  IF 'a' > 'A' THEN
d.  IF 'JUNE' < 'JULY' THEN      h.  IF '&' > ';' THEN
```

5. Using the given declaration, show the results after all the statements within each part of this question have executed.

```
DCL   SSN CHAR        (11) INIT ('456-334-4444'),
      ZIP CHAR        (9)  INIT ('092838484'),
      I   FIXED BIN (15) INIT (0);
```

    a.   I = INDEX (SSN, '-');
    b.   I = INDEX (SSN, '-');
           J = INDEX (SUBSTR (SSN, I + 1), '-');
    c.   I = INDEX (SSN, '-');
           J = INDEX (SUBSTR (SSN, I), '-');
    d.   I = VERIFY (ZIP, '0123456789ABCDEF');
    e.   I = VERIFY (SSN, '0123456789');
    f.   I = LENGTH (SUBSTR (ZIP, 4));
    g.   I = LENGTH (SSN);
    h.   ZIP = TRANSLATE (ZIP, '8', '*');

6. Using the given DECLARE statements, code the statements necessary to perform the functions listed. Include any further DECLARE statements you need.

```
DCL   SSN              CHAR (9)       INIT ('0987654567'),
      NAME             CHAR (20) VAR  INIT (''),
      ACCOUNT_NUMBER CHAR (10)        INIT ('  ');
```

    a.   Print the SSN above with hyphens in the appropriate places.
    b.   Read in names that contain a title ('MS.' or 'MR.') as the first three characters. The title is followed by a blank, the first name, another blank, and the last name. Print a name field containing just the title and the last name.
    c.   Check the account number above to be sure that it contains only numbers and a '/'.
    d.   Obtain the date and time of a financial transaction and print these appropriately edited for easy reading.

7. Write a program that will produce class lists. The list for each class shows the name of the course, and the number of students enrolled. This information is followed by the ID numbers and names of the students enrolled, along with the name of their major department. The input is a header record for each class followed by a record for each student enrolled in that class. The format of the header is

'C' 'course name' number of students enrolled [nn]

The format of the student records is

'student ID' 'student name' 'major name'

and continues until another header is encountered. Print the class lists, one class per page, with appropriate headings.

8. Add the following modifications to the program in Problem 7.
   a. Read in the student name as first name followed by last name, and print it as 'last name, initial' (of first name).
   b. Print the date at the top of every page.
   c. Make sure that the student ID number is all digits.
   d. Add a hyphen ('-') after the first two digits of the student ID number before printing.

9. Write a program to produce a customer billing statement for the accounts receivable system for THE LITTLE TOYSHOP. The input is on two lines, as follows:

Line 1:
Customer number	CHAR (6)
Customer last name	CHAR (20)
Customer first name	CHAR (10)
Customer beginning balance	99999.99
Payment made	99999.99
Up to five charges for each customer:	
Charge date	CHAR (8)
Charge amount	999.99

Line 2:
Customer street address	CHAR (20)
Customer city	CHAR (15)
Customer state	CHAR (2)
Customer zip	CHAR (5)

The output should resemble the following:

```
01/23/87
THE LITTLE TOYSHOP
STATEMENT FOR JANUARY 1987
MARY JOHNSON
3245 HOME STREET
CHICAGO, IL 70994
CUSTOMER NUMBER =        23-43-67
```

```
OLD BALANCE =                  36.00
CHARGES:
01/02/87                        2.98
01/10/87                       56.67
01/15/87                      109.00
PAYMENT  =                     50.00
NEW BALANCE =                 154.65
```

10. Write a program similar to the one in Exercise 9, adding the following modification: THE LITTLE TOYSHOP now adds 1.5% interest to the unpaid balance each month. Include this in the printed statement.

11. The library of a university needs a list of all the books checked out during any given day. Therefore, a daily report is to be produced that shows the books checked out, the patron's ID number, the status of the patron (student, faculty, or staff), and the book's due date. The information printed for each checkout is the title and ISBN. The due date for books depends upon the status of the person checking out the book. Students have two weeks, staff members have six months, and faculty have one year in which to return their books. The input data must include enough information to print the daily report needed. Be sure to include the date of the report on the output.

C H A P T E R 9

EDIT

I/O

In Chapter 4 we pointed out that PL/I allows for both stream and record I/O. Stream I/O, in turn, can be any one of three types: data-directed, list-directed, or edit-directed. Data-directed output for debugging purposes was presented in Chapter 6, and list-directed I/O was presented in Chapter 4. It is now time to consider edit-directed I/O.

9.1
Introduction to EDIT I/O

Although we develop specific advantages and disadvantages in later sections of the chapter, we can say at this point that the advantage EDIT input has over LIST input is that much more data can be stored on a record; moreover, EDIT output has the advantage of providing us with much more formatting flexibility than does LIST output. The chief disadvantage of both input and output in edit-directed I/O is that there is much more detail to be coded; consequently, EDIT I/O opens up a much greater potential for coding errors.

The PL/I statements for edit-directed I/O have the following syntax.

```
GET EDIT (data list) (format list);
PUT EDIT (data list) (format list);
```

The data list in EDIT I/O statements functions exactly the same as you are used to in LIST I/O; that is, it names the variables into which values are placed on input and the variables from which values are taken on output. The format list, on the other hand, is new and somewhat complex; certainly it requires a good deal of attention to detail. As you will see in the following sections, the values for certain items in the format list are enclosed in parentheses. When the format list ends with one of these values, it is a common mistake to forget to provide two closing parentheses, one for the value and the other for the entire format list.

9.2

The Format List

The function of the format list is to provide detailed information regarding each field in the input or output record. A format list is made up of items that are separated by commas and optionally by one or more blanks. The items in the list detail either the type of data and the width of the field (**data item**) or describe such operations as skipping to a new page and horizontal or vertical spacing of the output (**control item**). Collectively, they are referred to as **format items**.

9.2.1

Format List Data Items

Data items within the format list describe the data fields in an input or output record in terms of the width of the field and the type of data contained by the field. The format and meaning of each code is presented in the following table. In each case the letters in parentheses represent integers greater than zero.

Format	Meaning
A	Character data
A(w)	Character data (field width)
B	Bit data
B(w)	Bit data (field width)
E(w,d)	Floating-point data (field width, decimal places)
E(w,d,s)	Floating-point data (field width, decimal places, significant digits)
F(w)	Fixed-point data (field width)
F(w,d)	Fixed-point data (field width, decimal places)

The field width specified in the format list need not be the same as that specified in the DECLARE for the variable that contains the data, nor need the data type necessarily be the same. The data item in the format list describes the *external representation* of the value; the DECLARE attributes describe its *internal representation,* with the exception of decimal to binary. We very strongly recommend, however, that the data type of the format list and the variable always be the same, and we observe that restriction in all of our examples. For a detailed discussion of each format list item, see the appropriate subsections in Section 9.4 or 9.5.

9.2.2

Format List
Control Items

With two exceptions you are already familiar with the keywords used as control items in the format list; nevertheless, to collect them all in one place, we present the syntax of each control item in the following table.

Format	Meaning
COLUMN (n)	Absolute starting column of data
LINE (n)	Absolute line to skip to before printing
PAGE	Skip to new page before printing
SKIP (n)	Relative line to skip to before printing
X (n)	Skip n columns in the record (horizontal spacing); on output, the effect is to create a blank field of n columns

By comparing the table of format items with the table of control items, you will see that the opening parenthesis for the value can be separated from the keyword or code by blanks, or it can be coded with no intervening blanks.

The two control items entirely new to you are, of course, COLUMN and X. COLUMN (n) tells PL/I to skip to column n before starting to read or print. The absolute column specified by n is normally the column in which the first field begins. Subsequent fields in the record are most often specified as relative to this first field; however, COLUMN (which can be abbreviated COL) can be coded as many times as necessary in a single format list. X tells PL/I to skip n columns before reading or printing, and in printing the skipped columns will contain blanks. There can be as many X control items in the format list as necessary. Note, then, that to begin reading or printing in column ten you can specify either COLUMN (10) or X(9) before the first data item in the format list. If current position has passed the column specified when COLUMN is encountered, there is a skip to the designated column in the following line.

The other control items, LINE, PAGE, and SKIP, function in edit-directed I/O exactly the same as they do in list-directed I/O. SKIP, you will recall, is valid for both input and output records, while LINE and PAGE are valid only for records sent to the printer.

9.3

Some Rules for Edit-Directed I/O

Before turning to the specifics of the edit-directed I/O statements, we need to call your attention to a number of rules for edit-directed statements that apply equally to both input and output operations.

1. Each item in the data list has a corresponding data item in the format list. If the coding does not explicitly allow for this correspondence, certain rules will come into play to make this rule true. (See rule 5.)

2. The relationship between items in the data list and data items in the format list is *positional*; that is, the first data item describes the first item in the data list.

3. I/O continues until the data list is exhausted. This rule, of course, is nothing new; it is just our reminder to you that EDIT I/O is still stream I/O and naturally conforms to its qualities. It is also closely related to rules 4 and 5.

4. If the format list contains more items than the data list, the extra items in the format list are ignored. When the data list is exhausted, the computer makes no attempt to do anything with any remaining items in the format list.

5. If the format list contains fewer items than the data list, the format list is used over again until the data list is exhausted. Suppose the data list contains six items and the format list contains two data items. Items 1, 3, and 5 of the data list then correspond to data item 1 of the format list, and items 2, 4, and 6 of the data list correspond to data item 2 of the format list.

6. An **iteration factor**, which indicates the number of times the item in the format list is repeated, can be specified for any item in the format list. The iteration factor can be coded in either of two ways: in front of the item in the format list and separated from it by a space—for example, 5 F(3)—or again in front of the item in the format list and enclosed in parentheses—for example, (5)F(3). An iteration factor can also be used with a series of items in the format list by enclosing them in a set of parentheses:

```
PUT EDIT ('A', 'B', 'C') ((3) (A(1), X(5)));
```

The execution of this statement produces

```
AbbbbbBbbbbbCbbbbb
```

(The print position is in column 19 after execution.)

7. The field size specified in the DECLARE statement for a variable need not be the same as that specified in the format list. For example, CHAR_DATA might be declared to have the attribute of CHARACTER and a length of 20; in the format list, however, the field width might be specified as being, for example, 15 or 25.

9.4

The GET EDIT Statement

The syntax of the edit-directed input statement is

```
GET EDIT (data list) (format list);
```

One of the big advantages of edit-directed input is that the input data fields need not be separated by a blank or comma as is the case with list-directed input fields; this in turn means that it is possible to place much more data in an input record. If we assume an 80-column record containing the maximum number of two-digit integers, the maximum number of fields in a list-directed record is 27. An 80-column record for edit-directed input, on the other hand, could contain 40 such fields. The reason no field delimiters are required in edit-directed input records is that the format list tells the computer exactly how many columns of input data to place in each variable and exactly where these columns are relative to column 1 of the record. Just as this feature has the advantage of saving space, it also has the disadvantage of requiring that the data be placed in the *exact* columns in the input record that the format list specifies. In other words, when a GET EDIT statement executes, the computer goes to the columns in the input record specified in the format list and reads in as many columns of data as is specified in the field width portion of the format specification. If, for example, one column of character data is accidentally placed in a field designated as numeric in the format list, the CONVERSION condition will be raised. Therefore, edit-directed input records must be constructed with great care.

One final note of caution regarding edit-directed input: You must be very conscious of when you want to read from a new record, and put a SKIP or COLUMN on that GET statement. The following example illustrates why. Suppose we have two input records with the data starting in column 2

```
10 12 14

20 22 24
```

If we code

```
GET LIST (A, B, C);

GET LIST (A, B, C);
```

we will have no problem because with list input PL/I scans blank columns until it comes to the next nonblank field. (Remember, the record is 80 columns wide.) Thus, the first GET puts 10 in A, 12 in B, and 14 in C; and the next GET puts 20 in A, 22 in B, and 24 in C. However, if we code

```
GET EDIT (A, B, C) (F(3), F(3), F(3));

GET EDIT (A, B, C) (F(3), F(3), F(3));
```

the first GET produces the same results as the GET LIST. After this GET EDIT executes, PL/I marks the column following 14 (column 10), and when the next GET EDIT is encountered it moves columns 10–12 into A, 13–15 into B, and 16–18 into C. You, in turn, sit there wondering why you have zeros in your variables. In other words, with edit-directed input, PL/I does not scan to the next nonblank column; instead it places in the variable the data from the columns you have specified in the format list, even if those columns contain all blanks.

9.4.1

Fixed-Decimal Input

Fixed-decimal data can be specified in the format list in either of two ways. The first, F(w), affects w digits in the input record. PL/I assumes the digits comprise an integer, with an implied decimal point to the right of the rightmost digit. If the receiving field is larger than w, the remaining columns to the left of the leftmost digit are filled with zeros; if the receiving field is smaller than w, the SIZE condition (if enabled) is raised. A fixed-point decimal can have an explicit plus or minus sign in front of it; when this is the case, w must be at least one larger than the number of digits in the input record field. The following example helps to illustrate these concepts. Suppose the input record contains (starting in column 1)

```
123456789-123
```

and we code

```
DCL FLD1 FIXED DECIMAL (3);
DCL FLD2 FIXED DECIMAL (4);
DCL FLD3 FIXED DECIMAL (5);
DCL FLD4 FIXED DECIMAL (4);
GET EDIT (FLD1, FLD2, FLD3, FLD4) (COL(1), F(3), F(3), F(3),
         F(4));
```

The following table shows the contents of each variable after the GET EDIT statement has executed.

Variable	Contents
FLD1	123
FLD2	0456
FLD3	00789
FLD4	−123

In the second fixed-point decimal format, F(w,d), w is the width of the field, and d is the number of digits to the right of the decimal point. There is no need for a decimal point to be placed in the input data, and when there is none, we need not account for the *implied* decimal point. (Remember, no decimal is stored with the data, but PL/I "remembers" where the decimal is supposed to be.) If there is a decimal point physically present in the input record, w must be large enough to account for it. For example, for an input field containing 123.4, w has to be at least 5. Also, if the decimal point is physically present and its location does not agree with d, the physical decimal point overrides the format specification. Thus, if an input field contains 123.4 and the format specification is F(5,3), the resulting value placed in the variable is 123.4, not 1.234, as the format specification suggests. Keep in mind, too, that the two values you specify in the data item are total field width and digits to the right of the decimal point. This means that as many digits as necessary from the input field will be used to fill the d specification, with any remaining used for the whole number (implied) specification. In the following table, notice that in each case the data in the input record is located somewhere within a five-column field. Remember, too, that no decimal point is physically present in the variable even if one is physically present in the input record data. We use a vertical line (|) to indicate the location of the implied decimal point. Finally, we assume that the DECLARE attributes agree exactly with the data item specification.

Input Field	Format Specification	Result
2	F(5,5)	|00002
22	F(5,4)	0|0022
222	F(5,3)	00|222
2222	F(5,2)	022|22
22222	F(5,1)	2222|2
22.22	F(5,5)	22.22

The final entry in the table demonstrates that a decimal point physically present in the input field overrides the format specification.

Actual GET EDIT statements for fixed-point decimal input, as they might appear in your program, follow. They all refer to the same following input, in which b indicates a blank column.

```
bbb123b4567bb890bbbbb
GET EDIT (FLD1, FLD2, FLD3, FLD4) (X(3), F(3), X(1), F(3,1),
         F(1), X(2), F(3,1));
```

This statement produces the following results:

```
X(3)      causes a skip to column 4
F(3)      places 123 in FLD1
X(1)      causes one column to be skipped
F(3,1)    places 45|6 in FLD2
F(1)      places 7 in FLD3
X(2)      causes two columns to be skipped
F(3,1)    places 89|0 in FLD4
```

```
GET EDIT (FLD1, FLD2, FLD3, FLD4) (COL(4), F(3), X(1),
         F(4,4), X(2), F(3), F(5));
```

This statement produces these results:

```
COL(4)    causes a skip to column 4
F(3)      places 123 in FLD1
X(1)      causes one column to be skipped
F(4,4)    places |4567 in FLD2
X(2)      causes two columns to be skipped
F(3)      places 890 in FLD3
F(5)      places 00000 in FLD4
```

Normally, of course, an edit-directed input record containing nothing but fixed-length numeric fields will not include any blank columns. We have inserted them here primarily to illustrate the X control item in GET EDIT statements.

Our final example illustrates the use of the iteration factor. Suppose we have a record containing a 20-column name field and five 3-column fields, each holding a percentage grade. The GET EDIT statement could be coded as

```
GET EDIT (NAME, G1, G2, G3, G4, G5) (COL(1), A(20), F(3),
        F(3), F(3), F(3), F(3));
```

Using the iteration factor, on the other hand, simplifies the statement somewhat.

```
GET EDIT (NAME, G1, G2, G3, G4, G5) (COL(1), A(20), (5)F(3));
```

Both examples of the GET EDIT statement achieve the same results.

9.4.2

Float Decimal Input

The float decimal data item is coded E(w,d); E(w,d,s) can also be coded, but when it appears on the GET EDIT statement the s is ignored. This data item tells the computer that the data in the input record is in float decimal form. Remember: It does not indicate anything about the form in which the data is to be stored internally; that is the function of the DECLARE attributes. The w in the data item specifies the total width of the field in the input record. It must allow for a decimal point (if physically present in the input field), any sign physically present in the input field, and all digits. If the exponent is signed, the letter E can be omitted; therefore, the *minimum* size of w must be d + 2. Keep in mind that both the mantissa and the exponent can have signs; in their absence a positive value is assumed in both cases. The d again specifies the number of digits to the right of the implied decimal point. This designation is overridden by a physical decimal point in the input field. A floating-point decimal data item that corresponds to a blank field in the input record will cause the CONVERSION condition to be raised; this is different from the situation with fixed-point decimal input, where blanks are read in as zeros.

The following example illustrates the points we have made regarding floating-point decimal input.

```
Input: 17575E217575+3bb-17575E-41.7575E-2
GET EDIT (FLD1,  FLD2,  FLD3, FLD4) (COL(1), E(7,3), E(7,4),
              X(2), E(9,1), E(5,2));
```

The results of execution of the GET EDIT are

```
COL(1)      specifies the data starts in column 1
E(7,3)      places 17|575+02 in FLD1
E(7,4)      places 1|7575+03 in FLD2
X(2)        causes two columns to be skipped
E(9,1)      places 1757|5-04 in FLD3
E(9,5)      places 1.7575-02 in FLD4
```

9.4.3

Character
Input

Earlier we demonstrated the savings in input record space when edit-directed is used over list-directed input. The space savings can be even more dramatic in the case of character data. In an input record formatted for list-directed input, you recall, each character data field must be delimited by single quotation marks and separated from the following field by a comma or a blank. This means, of course, that each field is effectively three columns wider than the field would have to be if edit-directed input were employed. Consider, for example, an 80-column record containing the maximum number of four-character fields: such a record formatted for list-directed input contains 11 fields (with five remaining, unusable columns), while the same record formatted for edit-directed input contains 20 fields.

Earlier we pointed out that a data item describing character data can be coded as either A or A(w). The A form is not valid for input; therefore, we discuss only A(w) here. You no doubt recall from Chapter 8 that a variable given the CHARACTER attribute in the declaration can contain any character the computer recognizes (256 in all) but that if the field contains numbers they should not be used for arithmetic operations; the same is true for the contents of a field described by A(w) because it is intended for a CHARACTER variable.

When A(w) is encountered in the format list, w characters of data are read into the corresponding variable. If the declared length of the variable is greater than w, blanks are placed to the right of the rightmost character (which, of course, might itself be a blank) read in; in other words, character data is left-justified in the variable. If the declared length is less than w, characters are truncated on the right.

In the following example assume the data starts in column one and the declared length of each variable is the same as its corresponding w in the format list; we continue to use b to represent a blank.

```
Input:ABC123bbb4DEFGbb56@#HIJKL890
GET EDIT (FLD1, FLD2, FLD3) (COL(6), A(5), X(2), A(9),
         A(7));
COL(6)    causes a skip to column 6 (even though data is
          being skipped)
A(5)      places 3bbb4 in FLD1
X(2)      causes the two columns containing DE to be skipped
A(9)      places FGbb56@#H in FLD2
A(7)      places IJKL890 in FLD3
```

The iteration factor can also be used with character data, of course.

9.4.4

Example Program—
GET EDIT

The CH9EX1 program (see Fig. 9–1) solves the problem of computing an itemized statement of semester fees for students. The college in question allows for single or double occupancy of dormitory rooms, with a different fee for each. Students enrolled for six or fewer semester credit hours are considered part-time students, and their tuition is computed at a rate per credit hour. Those enrolled for seven or more hours are considered full-time students, and their tuition is a set amount. The college has two different laboratory fees, and the input record indicates how many classes the student is enrolled in that assess one or the other of the fees. At the time of this writing, the cost of attending college is changing regularly (always upward, never downward); therefore, it is unwise to hard-code the various rates in the program, and we avoid doing so by placing that data in a header record. The header record and a sample student record are shown below, in order to correlate them with their respective GET EDIT statements.

```
Column
        1         2         3         4         5         6
1234567890123456789012345678901234567890123456789012345
220000170000500051000160022200
ROBERT JONES      414 PARK AVE   DE KALB   IL6011511521
```

When the header record is correlated with the data list and format list of its GET EDIT (Statement 28), we can see that execution produces these results (again we use ⎮ to represent the implied decimal point).

```
COL(1)      data begins in column 1
F(6,2)      places 2200|00 in SINGLE_RATE
F(6,2)      places 1700|00 in DOUBLE_RATE
F(4,2)      places 50|00 in HOURS_RATE
F(5,2)      places 510|00 in FULL_RATE
F(4,2)      places 16|00 in LAB1_RATE
F(4,2)      places 22|00 in LAB2_RATE
```

In examining the code of READ_RTN (Statement 42) of CH9EX1, notice that the last data item—(2)F(1)—illustrates the iteration factor. When this statement executes, it has this result.

```
COL(1)      data begins in column 1
A(20)       places ROBERT JONES in NAME
A(15)       places 414 PARK AVE in STREET
A(10)       places DE KALB in CITY
A(2)        places IL in STATE
A(5)        places 60115 in ZIP
A(1)        places '1' in OCCUPANCY
F(2)        places 15 in HOURS .
(2)F(1)     places 2 in #_LAB1, and
            places 1 in #_LAB2
```

The pseudocode for CH9EX1 is as follows:

```
CH9EX1
    Call PG_HDR_RTN
    Read the header record
    Call READ_RTN
    DO WHILE (more input)
        Call PROCESS_RTN
        Call PRNT_DETAIL_RTN
        Call READ_RTN
    ENDDO
STOP
PG_HDR_RTN
    Print an underlined page header
EXIT
READ_RTN
    Read a record
EXIT
PROCESS_RTN
```

```
                    Laboratory-1, laboratory-2 fee ← 0
                    IF (single occupancy)
                        Dorm cost ← single rate
                    ELSE
                        Dorm cost ← double rate
                    ENDIF
                    Amount due ← amount due + dorm cost
                    IF (hours enrolled < 7)
                        Tuition ← rate per hour * hours enrolled
                    ELSE
                        Tuition ← full-time rate
                    ENDIF
                    Amount due ← amount due + tuition
                    IF (number lab-1 classes > 0)
                        Lab-1 fee ← lab-1 fee + (lab-1 rate * number lab-1
                            classes)
                    ENDIF
                    IF (number lab-2 classes > than 0)
                        Lab-2 fee ← lab-2 fee + (lab-2 rate * number lab-1
                            classes)
                    ENDIF
                    Amount due ← amount due + lab-1 fee + lab_2 fee
                EXIT
                PRNT_DETAIL_RTN
                    Print student name, address
                    Print label, dorm fee
                    Print label, tuition
                    IF (number lab-1 classes > 0)
                        Print label, lab-1 fee
                    ENDIF
                    IF (number lab-2 classes > 0)
                        Print label, lab-2 fee
                    ENDIF
                    Print label, total amount due
                EXIT
```

PL/I OPTIMIZING COMPILER VERSION 1 RELEASE 5.1 TIME: 18.08.54 DATE: 15 FEB 87 PAGE 1
OPTIONS SPECIFIED
GOSTMT;

FIGURE 9–1
CH9EX1 (the GET EDIT statement).

```
PL/I OPTIMIZING COMPILER            /* CH9EX1 USING GET EDIT */                    PAGE   2
                  SOURCE LISTING
    STMT


        /* CH9EX1 USING GET EDIT */
    1  CH9EX1:   PROCEDURE OPTIONS (MAIN);
        /******************************************************************/
        /*                                                              */
        /*   FUNCTION:  TO COMPUTE SEMESTER CHARGES FOR STUDENTS AND    */
        /*             GENERATE A REPORT                                */
        /*                                                              */
        /*   INPUT:       (1) A HEADER RECORD CONTAINING               */
        /*                    SINGLE OCCUPANCY DORMITORY RATE           */
        /*                    DOUBLE OCCUPANCY DORMITORY RATE           */
        /*                    COST PER CREDIT HOUR FOR PART TIME        */
        /*                       STUDENTS                               */
        /*                    TUITION FOR FULL TIME STUDENTS            */
        /*                    LABORATORY I FEE RATE                     */
        /*                    LABORATORY II FEE RATE                    */
        /*                (2) FOR EACH STUDENT A RECORD CONTAINING      */
        /*                    NAME                                      */
        /*                    HOME ADDRESS (STREET, CITY, STATE, ZIP)   */
        /*                    SINGLE OR DOUBLE DORMITORY OCCUPANCY      */
        /*                    NUMBER OF CREDIT HOURS ENROLLED FOR       */
        /*                    NUMBER OF LAB-1 FEE COURSES ENROLLED IN   */
        /*                    NUMBER OF LAB-2 FEE COURSES ENROLLED IN   */
        /*                                                              */
        /*   OUTPUT:    FOR EACH STUDENT:  NAME, ADDRESS, AND AN        */
        /*                  ITEMIZED STATEMENT                          */
        /*                                                              */
        /******************************************************************/
    2      DCL NAME          CHARACTER    (20);
    3      DCL STREET        CHARACTER    (15);
    4      DCL CITY          CHARACTER    (10);
    5      DCL STATE         CHARACTER    (2);
    6      DCL ZIP           CHARACTER    (5);
    7      DCL OCCUPANCY     CHARACTER    (1);
    8      DCL HOURS         FIXED DECIMAL (3);
    9      DCL SINGLE_RATE   FIXED DECIMAL (7,2);
   10      DCL DOUBLE_RATE   FIXED DECIMAL (7,2);
```

```
11    DCL DORM_COST    FIXED DECIMAL (7,2);
12    DCL HOURS_RATE   FIXED DECIMAL (5,2);
13    DCL FULL_RATE    FIXED DECIMAL (5,2);
14    DCL TUITION      FIXED DECIMAL (5,2);
15    DCL LAB1_RATE    FIXED DECIMAL (5,2);
16    DCL LAB2_RATE    FIXED DECIMAL (5,2);
17    DCL #_LAB1       FIXED DECIMAL (1);
18    DCL #_LAB2       FIXED DECIMAL (1);
19    DCL LAB1_FEE     FIXED DECIMAL (5,2);
20    DCL LAB2_FEE     FIXED DECIMAL (5,2);
21    DCL AMNT_DUE     FIXED DECIMAL (7,2);
```

```
STMT
22    DCL MORE_RECS    BIT          (1);
23    DCL   YES        BIT          (1)      INIT ('1'B);
24    DCL   NO         BIT          (1)      INIT ('0'B);
25          MORE_RECS = YES;
26          ON ENDFILE (SYSIN) MORE_RECS = NO;
27          CALL PG_HDR_RTN;
28          GET EDIT (SINGLE_RATE, DOUBLE_RATE, HOURS_RATE, FULL_RATE,
                      LAB1_RATE, LAB2_RATE) (COL(1), F(6,2), F(6,2),
                      F(4,2), F(5,2), F(4,2), F(4,2));
29          CALL READ_RTN;
30 LOOP:    DO WHILE (MORE_RECS);
31             CALL PROCESS_RTN;
32             CALL PRNT_DETAIL_RTN;
33             CALL READ_RTN;
34          END LOOP;
35          RETURN;

36 PG_HDR_RTN: PROCEDURE;
   /*******************************************************************/
   /*                                                                 */
   /* FUNCTION:  TO PRINT THE PAGE HEADER                             */
   /*                                                                 */
   /* GLOBAL VARIABLES:   NONE                                        */
   /*                                                                 */
   /* SUBROUTINES CALLED: NONE                                        */
   /*                                                                 */
   /* INPUT:    NONE                                                  */
   /*                                                                 */
   /* OUTPUT:   AN UNDERLINED PAGE HEADER                             */
```

FIGURE 9–1 *(continued)*

```
        /*                                                          */
        /******************************************************************/
37          PUT SKIP LIST (' ', ' ', 'STUDENT BILLING REPORT');
38          PUT SKIP (0) LIST (' ', ' ', '_____ _____ _____');
39          RETURN;
40          END PG_HDR_RTN;

41  READ_RTN:  PROCEDURE;
        /******************************************************************/
        /*                                                          */
        /*  FUNCTION:  TO READ A SINGLE RECORD                      */
        /*                                                          */
        /*  GLOBAL VARIABLES:    NAME                               */
        /*                       STREET                             */
        /*                       CITY                               */
```

PL/I OPTIMIZING COMPILER   /* CH9EX1 USING GET EDIT */   PAGE   4
  STMT

```
        /*                       STATE                              */
        /*                       ZIP                                */
        /*                       OCCUPANCY                          */
        /*                       HOURS                              */
        /*                       #_LAB1                             */
        /*                       #_LAB2                             */
        /*                                                          */
        /*  SUBROUTINES CALLED: NONE                                */
        /*                                                          */
        /*  INPUT:    AN INPUT RECORD CONTAINING A VALUE FOR EACH   */
        /*            VARIABLE ABOVE                                */
        /*                                                          */
        /*  OUTPUT:   NONE                                          */
        /*                                                          */
        /******************************************************************/
42          GET EDIT (NAME, STREET, CITY, STATE, ZIP, OCCUPANCY, HOURS,
                    #_LAB1, #_LAB2) (COL(1), A(20), A(15), A(10),
                    A(2), A(5), A(1), F(2), (2)F(1));
43          RETURN;
44          END READ_RTN;
```

```
45  PROCESS_RTN:  PROCEDURE;
    /*******************************************************************/
    /*                                                               */
    /*  FUNCTION:   TO CALCULATE THE STUDENT'S ITEMIZED AND TOTAL    */
    /*              FEES                                             */
    /*                                                               */
    /*  GLOBAL VARIABLES:    LAB1_FEE                                */
    /*                       LAB2_FEE                                */
    /*                       OCCUPANCY                               */
    /*                       DORM_COST                               */
    /*                       SINGLE_RATE                             */
    /*                       DOUBLE_RATE                             */
    /*                       AMNT_DUE                                */
    /*                       HOURS                                  */
    /*                       TUITION                                */
    /*                       HOURS_RATE                             */
    /*                       FULL_RATE                              */
    /*                       #_LAB1                                 */
    /*                       LAB1_RATE                              */
    /*                       #_LAB2                                 */
    /*                       LAB2_FEE                               */
    /*                       LAB2_RATE                              */
    /*                                                             */
    /*  SUBROUTINES CALLED: NONE                                    */
    /*                                                             */
    /*  INPUT:     NONE                                            */
    /*                                                             */
    /*  OUTPUT:    LABELS, AN INVOICE NUMBER, AND STUDENT ADDRESS   */
    /*                                                             */
```

PL/I OPTIMIZING COMPILER   /* CH9EX1 USING GET EDIT */   PAGE   5

```
   STMT
        /*******************************************************************/
    46          LAB1_FEE, LAB2_FEE,  AMNT_DUE = 0;
    47          IF OCCUPANCY = '1' THEN
                    DORM_COST = SINGLE_RATE;
    48          ELSE
                    DORM_COST = DOUBLE_RATE;
    49          AMNT_DUE = AMNT_DUE + DORM_COST;
    50          IF HOURS < 7 THEN
                    TUITION = HOURS_RATE * HOURS;
```

THE GET EDIT STATEMENT   295

FIGURE 9–1 (continued)

```
51              ELSE
                   TUITION = FULL_RATE;
52              AMNT_DUE = AMNT_DUE + TUITION;
53              IF #_LAB1 > 0 THEN
                   LAB1_FEE = LAB1_FEE + (LAB1_RATE * #_LAB1);
54              IF #_LAB2 > 0 THEN
                   LAB2_FEE = LAB2_FEE + (LAB2_RATE * #_LAB2);
55              AMNT_DUE = AMNT_DUE + LAB1_FEE + LAB2_FEE;
56              RETURN;
57              END PROCESS_RTN;

58  PRNT_DETAIL_RTN:  PROCEDURE;
    /********************************************************************/
    /*                                                                */
    /*  FUNCTION:  TO PRINT THE DETAIL LINES OF THE INVOICE           */
    /*                                                                */
    /*  GLOBAL VARIABLES:    NAME                                     *
    /*                       STREET                                   */
    /*                       CITY                                     */
    /*                       STATE                                    */
    /*                       ZIP                                      */
    /*                       DORM_COST                                */
    /*                       TUITION                                  */
    /*                       #_LAB1                                   */
    /*                       LAB1_FEE                                 */
    /*                       #_LAB2                                   */
    /*                       LAB2_FEE                                 */
    /*                       AMNT_DUE                                 */
    /*                                                                */
    /********************************************************************/
59          PUT SKIP (3) LIST (NAME, STREET, CITY, STATE, ZIP);
60          PUT SKIP LIST (' ', 'DORMITORY FEE', DORM_COST);
61          PUT SKIP LIST (' ', 'TUITION FEE', TUITION);
62          IF #_LAB1 > 0 THEN
               PUT SKIP LIST (' ', 'LABORATORY I FEE', LAB1_FEE);
63          IF #_LAB2 > 0 THEN
```

```
                    PUT SKIP LIST (' ', 'LABORATORY II FEE', LAB2_FEE);
    64              PUT SKIP (0) LIST (' ', ' ', '_____');
    65              PUT SKIP (2) LIST (' ', 'BALANCE DUE', AMNT_DUE);
    66              RETURN;
    67              END PRNT_DETAIL_RTN;
    68   END CH9EX1;
```

PL/I OPTIMIZING COMPILER   /* CH9EX1 USING GET EDIT */   PAGE   7
COMPILER DIAGNOSTIC MESSAGES
ERROR ID L   STMT   MESSAGE DESCRIPTION

COMPILER INFORMATORY MESSAGES

IEL05331 I   NO 'DECLARE' STATEMENT(S) FOR 'SYSIN','SYSPRINT'.

END OF COMPILER DIAGNOSTIC MESSAGES

                                        STUDENT BILLING REPORT

ROBERT JONES   414 PARK AVE   DE KALB   IL   60115
                      DORMITORY FEE   2200.00
                      TUITION FEE   510.00
                      LABORATORY I FEE   32.00
                      LABORATORY II FEE   22.00
                                             _____

                      BALANCE DUE   2764.00

BETTY SMITH   86 STONE DR   MALCOLM   NE   68123
                      DORMITORY FEE   1700.00
                      TUITION FEE   300.00
                                             _____

                      BALANCE DUE   2000.00

MONICA MOODY   1066 ROYAL ST   ACKROYD   RI   12345
                      DORMITORY FEE   1700.00
                      TUITION FEE   510.00
                      LABORATORY I FEE   32.00
                      LABORATORY II FEE   44.00
                                             _____

                      BALANCE DUE   2286.00
```
```

A question that might be occurring to you at this point is why input fields containing numeric data are placed in variables with the CHARACTER attribute and described with A(w) data items. The answer is that it is a common practice to define numeric data *not used in arithmetic operations* as character data. Notice that in this case we then compare the contents of the variable with a character literal in Statement 47; this is not required, but it does avoid PL/I having to make a data conversion before the comparison is made. Certainly, however, it is also possible to give them FIXED DECIMAL attributes and describe them with F(w).

9.5

The PUT EDIT Statement

The syntax of the edit-directed output statement is the same as for the edit-directed input statement; that is,

```
PUT EDIT (data list) (format list);
```

The major advantage of edit-directed over list-directed output is that it gives the programmer complete control over the formatting of the output. The major disadvantage is that once again it is necessary to write considerable extra code, which is very detailed and hence subject to error.

In the GET EDIT statement the format list describes the data as it appears in the input record. In the PUT EDIT statement, on the other hand, the format list describes the data in the output record; for our purposes this means the way the data will appear in the printed output, since our output is invariably sent to the printer. To aid you in formatting the output, PL/I, as we mentioned earlier, supplies you with two keywords (LINE and PAGE) that do not apply to the input statement. These keywords, along with SKIP, can be coded outside the format list as you are used to in coding list-directed output statements; in this case they are called **control options**. They can also be coded within the format list, and are then called **control items**. The following examples illustrate some of the combinations.

```
PUT PAGE EDIT (NUM, ALPHA) (LINE(5), F(5), A);
PUT EDIT (NUM, ALPHA) (LINE(5), F(5), LINE(10), A);
PUT SKIP (2) EDIT (NUM, ALPHA) (F(5), SKIP(2), A);
PUT PAGE SKIP (3) EDIT (NUM, ALPHA) (F(5), LINE(5), A);
```

Keep in mind, first of all, that control options and control items are executed in the order in which they are encountered. Thus,

```
PUT LINE (5) EDIT (NUM, ALPHA) (PAGE, (F(5), A);
```

causes the printer first to advance to line 5, then advance to the top of the next page, then print the contents of NUM and ALPHA. If the printer is on line 6 or farther down the page, LINE (5) will cause a skip to a new page; PAGE, of course, will cause a skip to the top of the next page, resulting in one entirely blank page. The control items X(w) and COLUMN, as we noted above, are available to both input and output.

9.5.1

Numeric Output

Numeric output is essentially the same in reverse as numeric input. The number is stored internally with the attributes in the DECLARE statement for that variable. When the PUT EDIT statement executes, the contents of the variable are translated from the declared attributes to those specified in the data item. Keep in mind, however, that if the translation cannot occur (e.g., character to numeric) the CONVERSION condition is raised. Except for binary to decimal, the data item should differ from the declared attribute only in (at most) field size; you should not hesitate to describe a different size (particularly larger) for the field, however.

Fixed-point decimal data items, you recall, can be expressed as either F(w) or F(w,d). When F(w) is specified, the decimal point is assumed to be to the right of the rightmost digit, and it is not printed. Values are placed right-justified in the field. If w is larger than the number of digits to be printed, the leading zeros (all zeros to the left of the leftmost significant digit) are automatically suppressed; if w is smaller than the number of digits to be printed, the SIZE condition (if enabled) is raised. If a value is less than zero, it is preceded by a minus sign (−), and this space must be accounted for in determining w; in other words, if there is any chance that a negative value will appear in this field, w must be at least one greater than the total possible number of digits. Because the values are right justified in the field created by w, you will need to think in terms of the right end of the field when you are formatting the printed page. Consider the following examples:

Internal Value	Format Item	Output
987	F(3)	987
987	F(6)	bbb987
−987	F(3)	error
−987	F(4)	−987

One instance in which it might make sense to code w as larger than the largest number of digits that will appear in the field is when several numbers will be printed on the same line. Suppose, for example, that 12345 678 90 is to be printed on one line and that you want four blanks between each number. One way of accomplishing this is to code

```
PUT SKIP EDIT (NUM1,  NUM2,  NUM3) (F(5),  X(4), F(3), X(4),
          F(2));
```

Another method of attaining the same result is to code

```
PUT SKIP EDIT (NUM1, NUM2, NUM3) (F(5), F(7), F(6));
```

The NUM1 value completely fills F(5) of course, but NUM2, being right-justified in its field, produces bbbb678, thereby effectively creating the needed four blanks. F(6) also leaves four blank spaces before the 90 prints. As you can see, the second method can reduce significantly the amount of code required, but it does so at some expense to program readability. It might be that as a beginner you will want to use the X(w) to avoid confusion.

When F(w,d) is specified as the data item, d specifies the number of digits to the right of the decimal point. In this case the decimal point prints, and the values are aligned on the decimal point. Since a physical decimal point is now involved, w must always be at least d + 1. Also, if there is any chance that the field will contain the maximum number of declared digits, w must be the declared precision + 1. If a negative number can appear in the field, the fact that the sign also prints must be taken into account in computing w. Again, zero suppression is automatic except that when a value is less than 1 a single zero is placed to the left of the decimal point. If d is larger than the number of fractional digits as specified in the declaration, the field is padded on the right with zeros. If w is not large enough to contain the whole number portion of the value, the field is filled with asterisks and the SIZE condition (if enabled) is raised. On the other hand, if d is less than the number of fractional digits specified in the declaration, rounding occurs; thus if the leftmost digit to be truncated is 5 or greater, 1 is added to the digit on its left (in other words, the printed value is rounded up at $=> 5$, and rounded down at $=< 4$). The following examples should help to clarify these rules. If a series of fractional decimal numbers is to be placed on single line of output, the necessity of coding the X(w) control item can again be avoided by making w large enough to produce the desired number of blank columns.

As we noted earlier, the data item for floating-point decimals can be

Internal Value	Format Item	Output
123|45	F(6,2)	123.45
123|45	F(8,2)	bb123.45
-123|45	F(6,2)	******
-123|45	F(7,2)	−123.45
|12345	F(7,5)	0.12345
123|45	F(6,1)	b123.5
123|45	F(8,3)	b123.450
123|45	F(5,2)	*****

either E(w,d) or E(w,d,s). When s is coded, it specifies the *total* number of significant digits to be printed. When s is not coded, the total number of significant digits defaults to d + 1; in other words, unless s is coded, there is always one digit (which can be 0) to the left of the decimal point. Also, when a floating-point decimal is printed, the exponent is always four characters (E±nn) (nn can equal 00), and they must be taken into account in computing w. Therefore, in all cases w must be, as a minimum, four greater than s. In addition, if the mantissa is negative, it is printed with the minus sign; then, of course, w must be => s + 5. The following table illustrates the rules for computing w in floating-point decimal output.

Internal Value	Format Item	Output	Comment	Algorithm
123|45+5	F(9,0,5)	12345e+03	d=0	w=s+4
−123|45+5	F(10,0,5)	−12345E+03	neg.; d=0	w=s+5
|12345+5	F(10,5)	1.2345E+04	s=1 (default)	w=s+5
−|12345+5	F(11,5)	−1.2345E+04	neg; s=1	w=s+6
123|45	F(10,2,5)	123.45E+05	d<s	w=s+5
123|45	F(10,5,5)	0.12345E+05	d=s	w=s+6
−123|45	F(11,5,5)	−0.12345E+05	neg; d=s	w=s+7
000|00E+05	F(9,0,5)	0E+05	0; d=0	w=s+4
000|00+5	F(11,2,5)	0.00000E+05	0; d<s	w=s+6

Assuming, then, that the input values are unknown and that you want a minimum of one blank column between each of a series of floating-point decimals, a "safe" figure for w is s + 8. Finally, if w is such that any significant digit(s) or the sign is lost, the SIZE condition (if enabled) is raised.

Character
Output

The data item for output character data also has two forms: A(w) and A. The first, A(w), is used when the field width as specified by the data item differs from the field width as specified by the declaration of the corresponding variable. The data is taken from the variable and placed left-justified in the field specified by the data item. If w is greater than the number of characters (including blanks) moved to it, the remaining space on the right is filled with blanks. If w is less than the number of characters moved in, characters are truncated on the right, but the SIZE condition is *not* raised. Output fields for character data also are often made larger than the declaration in order to provide for blank columns between two character strings that will be printed on the same line. In this case, A(w) must be used as the data item. PL/I knows, of course, the width specified in the declaration; therefore, when you want the output field to be the same width, you need to specify only A. PL/I in effect supplies the (w) from its knowledge of the declared width. Suppose your program contains the following declaration.

```
DCL ALPHA CHARACTER (10);
DCL BETA CHARACTER (12);
```

Each of the following PUT EDIT statements produces the same results.

```
PUT EDIT (ALPHA, BETA) (COL(10), A(15), A(12));
PUT EDIT (ALPHA, BETA) (COL(10), A(15), A);
PUT EDIT (ALPHA, BETA) (X(9), A(15), A);
PUT EDIT (ALPHA, BETA) (COL(10), A, COL(26), A);
PUT EDIT (ALPHA, BETA) (X(9),A,X(5),A);
```

In each case the value in ALPHA begins in column 10, and the value in BETA begins in column 26.

Suppose now that the variable in question contains the value PRO-GRAMMING LANGUAGE I; the following table illustrates the result of PUT EDIT statements with several specific data items.

Format Item	Result
A	PROGRAMMING LANGUAGE I
A(22)	PROGRAMMING LANGUAGE I
A(25)	PROGRAMMING LANGUAGE Ibbb
A(10)	PROGRAMMIN
A(13)	PROGRAMMING L

Another possibility for the items in the table is that instead of the character string being the value in a variable it might be a literal in the PUT EDIT statement itself:

```
PUT EDIT ('PROGRAMMING LANGUAGE I') (COL(10), A);
```

or

```
PUT EDIT ('PROGRAMMING LANGUAGE I') (COL(10), A(22));
```

The other format specifications in the table produces the same result with the literal as it does with the value in a variable.

9.5.3

Example Program—
PUT EDIT

With a few exceptions the logic of the PUT EDIT example program CH9EX2 is identical to that of the GET EDIT program (CH9EX1). There are some other significant differences in the problem, however: first, we designed CH9EX1 to produce a report of fees due from the students (perhaps intended for the bursar's office). We want CH9EX2, on the other hand, to produce the billing statements themselves to be sent to the students. Second, the output from CH9EX2 is to be printed on preprinted forms; therefore, both the horizontal and vertical spacing must be exact. Since preprinted forms are not available, however, the program must in fact build the "preprinted" forms as it proceeds to fill them in. In other words, you are to imagine the following preprinted form:

```
                                        INVOICE # _____

                    NAME
                    STREET
                    ADDRESS
```

with space below for the itemized statement.

The purpose of supposing a preprinted form is to emphasize the detailed and exact formatting of the output that is required in some situations and to demonstrate that these requirements can be met using edit-directed output. Notice, also, that the student's name, street, and address are each placed on a separate line. We are supposing these statements will be mailed in "window" envelopes; therefore, the placement of these items must be exact in order to show through the window. The labels (NAME, STREET, and ADDRESS) will not show in the window.

In order for you to see more clearly the results produced by the PUT

EDIT statement, we present in Figure 9–2 the output for one student. The CH9EX2 program is shown in Figure 9–3.

To solidify your understanding of the PUT EDIT statement we first examine Statement 43 (in PRNT_ADDR_RTN) in detail. The statement is

```
PUT PAGE EDIT ('INVOICE # ', INVOICE_#) (SKIP (20),
                COL(60), A, F(5));
```

PAGE   causes a skip to the top of a new page; thus each statement is printed on a separate page

SKIP(20)   causes a skip of the first 19 lines on the page (in other words, a skip to line 20) and provides approximate vertical centering

COL(60)   causes a skip to column 60 of line 20

A   causes INVOICE # to be printed starting in column 60; this is a 10-column literal (note the blank between # and the closing quotation mark)

F(5)   causes the value in the variable INVOICE_# (currently 101) to be printed starting in column 70. Because we have specified a five character field and we have a three digit value, it actually begins in column 73. Due to the automatic zero suppression of PL/I, the first two columns contain blanks.

```
                              INVOICE #   101


        NAME        ROBERT JONES
        STREET      414 PARK AVE
        ADDRESS     DE KALB    IL 60115


                    DORMITORY FEE      2200.00
                    TUITION FEE         510.00
                    LABORATORY I FEE     32.00
                    LABORATORY II FEE    22.00


                    BALANCE DUE        2764.00
```

FIGURE 9–2
Report logic.

The pseudocode for CH9EX2 is as follows:

```
CH9EX2
    Read the header record
    Call READ_RTN
    DO WHILE (more input)
        Call PRNT_ADDR_RTN
        Call CALC_FEE_RTN
        Call PRNT_FEE_RTN
        Call READ_RTN
    ENDDO
STOP
READ_RTN
    Read a record
EXIT
PRNT_ADDR_RTN
    Invoice number ← invoice number + 1
    Print invoice label, number
    Print labels and student name and address
EXIT
CALC_FEE_RTN
    Laboratory-1, laboratory-2 fee ← 0
    IF (single occupancy)
        Dorm cost ← single rate
    ELSE
        Dorm cost ← double rate
    ENDIF
    Amount due ← amount due + dorm cost
    IF (hours enrolled < 7)
        Tuition ← rate per hour   hours enrolled
    ELSE
        Tuition ← full-time rate
    ENDIF
    Amount due ← amount due + tuition
    IF (number lab-1 classes > 0)
        Lab-1 fee <- lab-1 fee + (lab-1 rate * number lab-1
            classes)
    ENDIF
    IF (number lab-2 classes > than 0)
        Lab-2 fee ← lab-2 fee + (lab-2 rate * number lab-1
            classes)
    ENDIF
    Amount due ← amount due + lab-1 fee + lab-2 fee
EXIT
PRNT_FEE_RTN
```

```
                          Print labels, dorm cost, tuition
                          IF (number lab-1 classes > 0)
                             Print label, lab-1 fee
                          ENDIF
                          IF (number lab-2 classes > 0)
                             Print label, lab-2 fee
                          ENDIF
                          Print label, total amount due
                       EXIT
```

PL/I OPTIMIZING COMPILER VERSION 1 RELEASE 5.1 TIME: 18.10.13 DATE: 15 FEB 87 PAGE   1
OPTIONS SPECIFIED
GOSTMT:

PL/I OPTIMIZING COMPILER   /* CH9EX2 USING PUT EDIT */   PAGE   2
                  SOURCE LISTING

     STMT

         /* CHAPTER 9 EXAMPLE PROGRAM 2 USING PUT EDIT */
     1 CH9EX2:   PROCEDURE OPTIONS (MAIN):
        /**/
        /*   */
        /*   FUNCTION: TO COMPUTE SEMESTER CHARGES FOR STUDENTS AND */
        /*   GENERATE INVOICES ON (SUPPOSEDLY) PREPRINTED */
        /*   FORMS (ACTUALLY, IT CONSTRUCTS THE "PREPRINTED */
        /*   FORM ALSO)   */
        /*   */
        /*   INPUT: (1) A HEADER RECORD CONTAINING   */
        /*   SINGLE OCCUPANCY DORMITORY RATE   */
        /*   DOUBLE OCCUPANCY DORMITORY RATE   */
        /*   */
        /*   COST PER CREDIT HOUR FOR PART TIME   */
        /*   STUDENTS   */
        /*   TUITION FOR FULL TIME STUDENTS   */
        /*   LABORATORY I FEE RATE   */
        /*   LABORATORY II FEE RATE   */
        /*   (2) FOR EACH STUDENT A RECORD CONTAINING   */
        /*   NAME   */
```

**FIGURE 9–3**
**CH9EX2 (the PUT EDIT statement).**
```

```
/*                      HOME ADDRESS                          */
/*                      SINGLE OR DOUBLE DOMITORY OCCUPANCY    */
/*                      NUMBER OF CREDIT HOURS ENROLLED FOR    */
/*                      OPTIONALLY A 1 FOR EITHER OR BOTH      */
/*                        LAB-1 AND LAB-2 FEES                 */
/*                                                            */
/*  OUTPUT:   FOR EACH STUDENT:  NAME, ADDRESS, AND AN        */
/*                      ITEMIZED STATEMENT                    */
/*                                                            */
/************************************************************/
2      DCL NAME          CHARACTER    (20);
3      DCL STREET        CHARACTER    (15);
4      DCL CITY          CHARACTER    (10);
5      DCL STATE         CHARACTER    (2);
6      DCL ZIP           CHARACTER    (5);
7      DCL OCCUPANCY     CHARACTER    (1);
8      DCL HOURS         FIXED DECIMAL (3);
9      DCL SINGLE_RATE   FIXED DECIMAL (7,2);
10     DCL DOUBLE_RATE   FIXED DECIMAL (7,2);
11     DCL DORM_COST     FIXED DECIMAL (7,2);
12     DCL HOURS_RATE    FIXED DECIMAL (5,2);
13     DCL FULL_RATE     FIXED DECIMAL (5,2);
14     DCL TUITION       FIXED DECIMAL (5,2);
15     DCL LAB1_RATE     FIXED DECIMAL (5,2);
16     DCL LAB2_RATE     FIXED DECIMAL (5,2);
17     DCL #_LAB1        FIXED DECIMAL (1);
18     DCL #_LAB2        FIXED DECIMAL (1);
19     DCL LAB1_FEE      FIXED DECIMAL (5,2);
```

```
STMT
20     DCL LAB2_FEE      FIXED DECIMAL (5,2);
21     DCL AMNT_DUE      FIXED DECIMAL (7,2);
22     DCL INVOICE_#     FIXED BINARY  (15)    INIT (00100);
23     DCL MORE_RECS     BIT          (1);
24     DCL   YES         BIT          (1)    INIT ('1'B);
25     DCL   NO          BIT          (1)    INIT ('0'B);
26         MORE_RECS = YES;
27         ON ENDFILE (SYSIN) MORE_RECS = NO;
28         GET LIST (SINGLE_RATE, DOUBLE_RATE, HOURS_RATE, FULL_RATE,
                   LAB1_RATE, LAB2_RATE);
```

FIGURE 9–3 *(continued)*

```
 29            CALL READ_RTN;
 30  LOOP:     DO WHILE (MORE_RECS);
 31               CALL PRNT_ADDR_RTN;
 32               CALL CALC_FEE_RTN;
 33               CALL PRNT_FEE_RTN;
 34               CALL READ_RTN;
 35            END LOOP;
 36            RETURN;

 37  READ_RTN:  PROCEDURE;
     /********************************************************************/
     /*                                                                  */
     /*   FUNCTION:  TO READ A SINGLE RECORD                             */
     /*                                                                  */
     /*   GLOBAL VARIABLES:    NAME                                      */
     /*                        STREET                                    */
     /*                        CITY                                      */
     /*                        STATE                                     */
     /*                        ZIP                                       */
     /*                        OCCUPANCY                                 */
     /*                        HOURS                                     */
     /*                        #_LAB1                                    */
     /*                        #_LAB2                                    */
     /*                                                                  */
     /*   SUBROUTINES CALLED: NONE                                       */
     /*                                                                  */
     /*   INPUT:     A RECORD CONTAINING A VALUE FOR EACH OF THE         */
     /*              VARIABLES ABOVE                                     */
     /*                                                                  */
     /*   OUTPUT:    NONE                                                *
     /*                                                                  */
     /********************************************************************/
 38            GET LIST (NAME, STREET, CITY, STATE, ZIP, OCCUPANCY, HOURS,
                    #_LAB1, #_LAB2);
 39            RETURN;
```

PL/I OPTIMIZING COMPILER   /* CH9EX2 USING PUT EDIT */   PAGE   4
   STMT

```
 40            END READ_RTN;
```

```
41  PRNT_ADDR_RTN: PROCEDURE;
    /******************************************************************/
    /*                                                              */
    /* FUNCTION:  TO PRINT THE INVOICE NUMBER AND THE  STUDENT'S    */
    /*            ADDRESS                                           */
    /*                                                              */
    /* GLOBAL VARIABLES:    INVOICE_#                               */
    /*                      NAME                                    */
    /*                      STREET                                  */
    /*                      CITY                                    */
    /*                      STATE                                   */
    /*                      ZIP                                     */
    /* INPUT:    NONE                                               */
    /*                                                              */
    /* OUTPUT:   AN INVOICE NUMBER AND THE STUDENT'S ADDRESS IN     */
    /*           STANDARD ADDRESS FORMAT (ALL WITH LABELS)          */
    /*                                                              */
    /******************************************************************/
42          INVOICE_# = INVOICE_# + 1;
43          PUT PAGE EDIT ('INVOICE # ', INVOICE_#) (SKIP(20),
                        COL(60), A, F(5));
44          PUT SKIP (0) EDIT ('_____') (COL(70), A);
45          PUT SKIP (3) EDIT ('NAME', NAME) (COL(35), A, X(7), A);
46          PUT SKIP EDIT ('STREET', STREET) (COL(35), A, X(5), A);
47          PUT SKIP EDIT ('ADDRESS', CITY, STATE, ZIP) (COL(35),
                        A, X(4), A, X(1), A, X(1), A);
48          RETURN;
49          END PRNT_ADDR_RTN;


50  CALC_FEE_RTN: PROCEDURE;
    /******************************************************************/
    /*                                                              */
    /* FUNCTION:  TO COMPUT AN ITEMIZED LIST OF STUDENT CHARGES     */
    /*            AND THE TOTAL CHARGES                             */
    /*                                                              */
    /* GLOBAL VARIABLES:    OCCUPANCY                               */
    /*                      DORM_COST                               */
    /*                      SINGLE_RATE                             */
    /*                      DOUBLE_RATE                             */
    /*                      HOURS                                   */
    /*                      TUITION                                 */
    /*                      HOURS_RATE                              */
```

FIGURE 9–3 (continued)

```
      /*                      FULL_RATE                            */
      /*                      #_LAB1                               */
      /*                      LAB1_FEE                             */


PL/I OPTIMIZING COMPILER        /* CH9EX2 USING PUT EDIT */                PAGE   5
   STMT
      /*                      LAB1_RATE                            */
      /*                      #_LAB2                               */
      /*                      LAB2_FEE                             */
      /*                      LAB2_RATE                            */
      /*                                                           */
      /*****************************************************************/
   51         LAB1_FEE, LAB2_FEE,  AMNT_DUE = 0;
   52          IF OCCUPANCY = '1' THEN
                  DORM_COST = SINGLE_RATE;
   53          ELSE
                  DORM_COST = DOUBLE_RATE;
   54         AMNT_DUE = AMNT_DUE + DORM_COST;
   55         IF HOURS < 7 THEN
                  TUITION = HOURS_RATE * HOURS;
   56          ELSE
                  TUITION = FULL_RATE;
   57         AMNT_DUE = AMNT_DUE + TUITION;
   58       IF #_LAB1 > 0 THEN
              LAB1_FEE = LAB1_FEE + (LAB1_RATE * #_LAB1);
   59       IF #_LAB2 > 0 THEN
              LAB2_FEE = LAB2_FEE + (LAB2_RATE * #_LAB2);
   60       AMNT_DUE = AMNT_DUE + LAB1_FEE + LAB2_FEE;
   61       RETURN;
   62       END CALC_FEE_RTN;


   63 PRNT_FEE_RTN:  PROCEDURE;
      /*****************************************************************/
      /*                                                           */
      /* FUNCTION:  TO PRINT THE ITEMIZED LIST OF CHARGES AND THE  */
      /*            TOTAL AMOUNT DUE                               */
      /*                                                           */
      /* GLOBAL VARIABLES:   DORM_COST                             */
      /*                     TUITION                               */
```

```
      /*                   LAB1_FEE                      */
      /*                   LAB2_FEE                      */
      /*                   AMNT_DUE                      */
      /*                                                 */
      /* SUBROUTINES CALLED: NONE                        */
      /*                                                 */
      /* INPUT:    NONE                                  */
      /*                                                 */
      /* OUTPUT:   AN ITEMIZED LIST OF CHARGES AND THE TOTAL   */
      /*                                                 */
      /**********************************************************************/
  64          PUT SKIP (5) EDIT ('DORMITORY FEE', DORM_COST) (COL(50), A,
                        COL(68), F(7,2));
```

```
    STMT
  65          PUT SKIP EDIT ('TUITION FEE', TUITION) (COL(50), A,
                        COL(68), F(7,2));
  66      IF #_LAB1 > 0 THEN
              PUT SKIP EDIT ('LABORATORY I FEE  ', LAB1_FEE) (COL(50),
                        A, F(7,2));
  67      IF #_LAB2 > 0 THEN
              PUT SKIP EDIT ('LABORATORY II FEE ', LAB2_FEE) (COL(50),
                        A, F(7,2));
  68          PUT SKIP (0) EDIT ('_____') (COL(68), A);
  69          PUT SKIP (2) EDIT ('BALANCE DUE', AMNT_DUE) (COL(50),
                        A, COL(68), F(7,2));
  70          RETURN;
  71          END PRNT_FEE_RTN;
  72  END CH9EX2;
```

```
COMPILER DIAGNOSTIC MESSAGES
ERROR ID L   STMT    MESSAGE DESCRIPTION


COMPILER INFORMATORY MESSAGES

IEL0533I I        NO 'DECLARE' STATEMENT(S) FOR 'SYSIN','SYSPRINT'.

END OF COMPILER DIAGNOSTIC MESSAGES
```

As the example programs CH9EX1 and CH9EX2 illustrate, it is quite possible to specify list-directed input with edit-directed output and vice versa in the same program. Although we do not illustrate it in our example programs, it is also possible to use both list-directed and edit-directed output to the same file. Statement 43 of CH9EX2, for example, could be coded as:

```
PUT PAGE LIST ('INVOICE # ', INVOICE_#);
```

while the other output statements remain as PUT EDIT.

Edit-directed I/O provides a number of special features, and we turn now to a consideration of some of them.

9.6

Special Features

Edit-directed I/O in PL/I offers a variety of special features to facilitate the coding of I/O operations. Among these special features are expressions and BUILTIN functions in output statements, multiple data and format lists in a single input or output statement, and the remote format. The use of these special features is never required; expressions and BUILTIN functions, nevertheless, do provide an alternative and sometimes handier method of achieving the same results.

9.6.1

Expressions in Output

Rather than assigning the result of the evaluation of an expression to an intermediate variable, then printing the contents of the variable, you can have the result placed directly in the output stream by including the expression in the PUT EDIT statement. For example, to compute and print the area of a triangle, you might code

```
AREA = BASE * .5 * ALTITUDE;
PUT EDIT ('AREA =', AREA) (A(7), F(5,1));
```

You could save yourself one line of code and one declaration, on the other hand, by coding

```
PUT EDIT ('AREA =', BASE * .5 * ALTITUDE) (A(7), F(5,1));
```

In this case no intermediate variable is required because the value resulting from the evaluation of the expression immediately becomes a part of the output stream. You must provide a format item to describe the value resulting from the evaluation of the expression, however, just as you must for any other output item.

BUILTIN Functions in Output

When a BUILTIN function is included as a data item in an output statement, the function is invoked and the value the function returns is a part of the output data, and again a format item must describe that value. To illustrate the BUILTIN function in a PUT EDIT statement, let us suppose we want simply to read in a name and print it. Suppose, too, that the input looks like this:

```
BABBAGECHARLES
```

and that we have read it into the variable NAME. We want the printed output to appear as CHARLES BABBAGE. As you saw in Chapter 8, one way of doing this would be as follows:

```
FIRST_NAME = SUBSTR (NAME, 8, 7);
LAST_NAME = SUBSTR (NAME, 1, 7);
PUT EDIT (FIRST_NAME, LAST_NAME) (A(8), A);
```

It is also possible, however, to include the SUBSTR function in the PUT EDIT statement itself:

```
PUT EDIT (SUBSTR (NAME,8,7), SUBSTR (NAME,1,7)) (A(8), A);
```

In this example we have eliminated two statements and the declaration of two variables.

An example of a numeric BUILTIN function in the PUT EDIT statement is

```
PUT EDIT ('SQUARE ROOT =', SQRT (NUMBER)) (A(14), F(8,3));
```

BUILTIN functions can also be combined with expressions in the PUT EDIT statement. In this case, the expression is evaluated, the function is invoked, and the value it returns is a part of the output stream. Suppose the variable DIAMETER contains a value representing the diameter of a circle, and we want to compute and print out, rounded to two places, the area of the circle. The problem can be solved by coding the following:

```
PUT EDIT (ROUND ((DIAMETER / 2) ** 2 * 3.1416, 2)) (F(6,2));
```

If you use these special features, be sure to keep your parentheses balanced; the PL/I compiler will not accept unbalanced (that is, a different number of opening than closing) parentheses.

Many programmers (including the two of us) discourage the use of expressions and references to built-in functions in PUT EDIT statements on the grounds that it increases considerably the difficulty of reading the program.

9.6.3

Multiple Data and Format Lists

All I/O statements thus far have consisted of a single data list followed by a single format list. However, an I/O statement can contain any number of data and format list pairs, so long as the two alternate:

```
GET EDIT (data list) (format list) (data list) (format list);
```

The items in the first format list correspond to the items in the first data list, and so forth. Be aware, moreover, that the rules concerning too few format items in the format list apply to each individual format list in the multiple data and format lists. Multiple data and format lists are used primarily in those cases in which there are a large number of data items. A single data list followed by a single format list then requires a very careful count to determine which format item describes a given data item. By way of illustration, compare the two following statements.

```
PUT EDIT (A1, N1, A2, A3, N2, N3, N4) (COL(10), A(5),
          F(4,2), X(5), A(6), X(2), A(4), X(1), F(7),
          F(8,3), F(6));
PUT EDIT (A1, N1)    (COL(10), A(5), F(4,2))
         (A2, A3)    (X(5), A(6), X(2), A(4))
         (N2, N3, N4) (X(1), F(7), F(8,3), F(6));
```

You could go so far as to put each data item in a list by itself, followed by its format list, but most PL/I programmers would say that doing so is going too far in the other direction. The alternating data and format lists could also be strung out on one or two lines, but maintaining the alignment we have illustrated probably makes matching the format items to the correct data items somewhat more convenient.

The Remote
Format Item

In addition to our having repeatedly told you so, you have probably discovered through your experience that the more code you have to write, the greater the chance for error. This problem can be especially troublesome when identical code must be repeated at various points in the program. To reduce the problem for data format items and control format items, PL/I offers the **remote format item** and the corresponding FORMAT statement. The FORMAT statement consists of a label, the keyword FORMAT, and a format list.

```
label:  FORMAT (list);
```

The FORMAT statement is not coded as a part of any I/O statement, but it can be referenced from almost any such statement by coding a remote format item—R—followed by the label from the FORMAT statement in parentheses. Suppose the following FORMAT statement:

```
RMT1:  FORMAT (F(2), X(3), A);
```

If we then code:

```
PUT EDIT (NUM, ALPHA) (R (RMT1));
```

it will have the same effect as if we had coded:

```
PUT EDIT (NUM, ALPHA) (F(2), X(3), A);
```

A remote format item can also be combined with other format items:

```
GET EDIT (D, NUM, ALPHA, K)
         (SKIP, A(4), R(RMT1), F(7));
```

The remote format item has two restrictions on its use, however, of which you must be aware.

1. The remote format item cannot appear in a FORMAT statement.

```
INVLD:  FORMAT (F(8,4), R(RMT1));
        /* INVALID USE OF REMOTE FORMAT ITEM */
```

2. The remote format item cannot appear in an I/O statement which is the *single* statement of an on-condition.

```
ON ENFILE (SYSIN)
        PUT EDIT (NUM, ALPHA) (R(RMT1));
                /* INVALID USE OF REMOTE FORMAT ITEM */
```

9.7

Example Program—
EDIT I/O

We want a program to compute grades and "quality points" for students enrolled in a computer science internship program. The input is to be a series of records, each containing a student number, name, street address, city-state-zip address, and student codes expressed as a string of seven binary digits. The input records are shared by several application programs, and therefore they must contain all the information needed by each of the programs that use them, but some programs do not require all the input data items. As a matter of fact, the CH9EX3 program (see Fig. 9–4) has no use for the student address, and consequently these items are ignored when the records are read. Note, however, that they are included in the documented description of the input record. The series of bit string variables (STU_CODES to G_MASK) in the declaration are used in CH9EX3 as a method of compacting a great deal of data into very little space. In the space of seven bits, for example, we have included the internship-area code, the hours-credit code, and the grade-received code. If we were to use fixed-decimal integers to represent the codes, we would use three times as much space as we presently do. In other words, each single-digit fixed decimal requires as much space as the entire set of codes does in bit strings.

```
PL/I OPTIMIZING COMPILER  VERSION 1 RELEASE 5.1  TIME: 18.11.11  DATE: 15 FEB 87 PAGE   1
OPTIONS SPECIFIED
GOSTMT;



PL/I OPTIMIZING COMPILER        /* CH9EX3 USING EDIT I/O */                    PAGE   2
                SOURCE LISTING
     STMT

        /* CHAPTER 9 EXAMPLE PROGRAM 3 USING EDIT I/O */
      1 CH9EX3: PROC OPTIONS (MAIN);
```

FIGURE 9–4
CH9EX3 (the use of EDIT I/O).

```
/*****************************************************************/
/*                                                               */
/* FUNCTION:   TO COMPUTE GRADES FOR STUDENTS IN AN INTERNSHIP    */
/*             PROGRAM IN COMPUTER SCIENCE.                       */
/*                                                               */
/* INPUT:      FOR EACH STUDENT A RECORD CONTAINING              */
/*                1) STUDENT'S IDENTIFICATION NUMBER             */
/*                2) STUDENT'S NAME (LAST, FIRST, MIDDLE)        */
/*                3) STUDENT'S STREET ADDRESS                    */
/*                4) STUDENT'S CITY, STATE, ZIP CODE             */
/*                5) STU_CODES:  BIT STRING OF CODES TO BE       */
/*                               PROCESSED                      */
/*                                                               */
/* OUTPUT:     FOR EACH STUDENT A RECORD CONTAINING             */
/*                1) STUDENT'S IDENTIFICATION NUMBER             */
/*                2) STUDENT'S NAME                             */
/*                3) NAME OF ADVISING PROFESSOR                  */
/*                4) COMPANY NAME WHERE INTERNSHIP OCCURRED      */
/*                5) SUBJECT AREA OF INTERNSHIP                  */
/*                6) LETTER GRADE STUDENT EARNED                 */
/*                7) QUALITY POINTS STUDENT EARNED               */
/*                                                               */
/* NOTES:      STU_CODES IS A SEVEN-BIT STRING COMPOSED AS       */
/*             FOLLOWS:                                          */
/*             BITS 1 & 2:    INTERNSHIP SUBJECT AREA CODE       */
/*             (L TO R)          10 = DATABASE                   */
/*                               01 = SYSTEMS ANALYSIS & DESIGN  */
/*                               11 = APPLICATIONS PROGRAMMING   */
/*             BITS 3 & 4:    HOURS CREDIT CODE(VARIABLE FROM     */
/*                            1 TO 3)                           */
/*                               01 = ONE HOUR CREDIT            */
/*                               10 = TWO HOURS CREDIT           */
/*                               11 = THREE HOURS CREDIT         */
/*             BITS 5,6 & 7:  GRADE RECEIVED CODES               */
/*                111 = A    110 = B    101 = C    100 = D       */
/*                010 = F    001 = I    000 = W                  */
/*                                                               */
/*             SAMPLE BIT STRING: 1011110  = DATABASE, 3HRS.     */
/*                                GRADE OF B                     */
/*             BIT POSITIONS----> 1234567                        */
/*                                                               */
/*        *** BECAUSE THE STUDENT ADDRESS IS NOT USED IN THIS    */
```

FIGURE 9–4 (*continued*)

```
      /*            APPLICATION. IT IS IGNORED WHEN INPUT IS READ    */
      /*                                                             */
      /***************************************************************/
  2      DCL STU_NUMBER   FIXED DEC (9);
  3      DCL QUALITY_PTS  FIXED DEC (3);
```

PL/I OPTIMIZING COMPILER   /* CH9EX3 USING EDIT I/O */   PAGE   3
  STMT
```
  4      DCL HOURS        FIXED DEC (1);
  5      DCL GPTS         FIXED DEC (1);
  6      DCL STU_NAME     CHAR      (20);
  7      DCL PROF_NAME    CHAR      (20);
  8      DCL COMPANY_NAME CHAR      (20);
  9      DCL WORK_AREA    CHAR      (23);
 10      DCL GRADE        CHAR      (1);
 11      DCL STU_CODES    BIT       (7);
 12      DCL I_CODE       BIT       (7);
 13      DCL H_CODE       BIT       (7);
 14      DCL G_CODE       BIT       (7);
 15      DCL I_MASK       BIT       (7) INIT ('1100000'B);
 16      DCL H_MASK       BIT       (7) INIT ('0011000'B);
 17      DCL G_MASK       BIT       (7) INIT ('0000111'B);
 18      DCL MORE_RECS    BIT       (1);
 19      DCL  YES         BIT       (1) INIT ('1'B);
 20      DCL  NO          BIT       (1)  INIT ('0'B);
 21           MORE_RECS = YES;
 22           ON ENDFILE (SYSIN) MORE_RECS = NO;
 23           CALL PG_HDR_RTN;
 24           CALL READ_RTN;
 25  PROC_LP: DO WHILE (MORE_RECS);
 26              CALL GET_WRK_AREA_RTN;
 27              CALL CALC_GRADE_RTN;
 28              CALL PRNT_DETAIL_RTN;
 29              CALL READ_RTN;
 30           END PROC_LP;
 31           RETURN;

 32  PG_HDR_RTN: PROCEDURE;
```

```
/*********************************************************************/
/*                                                                   */
/*  FUNCTION:  TO PRINT PAGE AND COLUMN HEADERS                      */
/*                                                                   */
/*  GLOBAL VARIABLES:   NONE                                         */
/*                                                                   */
/*  SUBROUTINES CALLED: NONE                                         */
/*                                                                   */
/*  INPUT:     NONE                                                  */
/*                                                                   */
/*  OUTPUT:    A THREE-LINE PAGE HEADER AND A TWO-LINE COLUMN        */
/*             HEADER                                                */
/*                                                                   */
/*********************************************************************/
 33          PUT EDIT ('NORTHERN ILLINOIS UNIVERSITY')
```

PL/I OPTIMIZING COMPILER   /* CH9EX3 USING EDIT I/O */   PAGE   4
   STMT

```
                          (COL(46),A(28));
 34          PUT SKIP EDIT ('COMPUTER SCIENCE INTERNSHIP PROGRAM')
                     (COL(43),A(35));
 35          PUT SKIP EDIT ('FALL SEMESTER')
                     (COL(54),A(13));
 36          PUT SKIP(2) EDIT ('STUDENT','STUDENT','FACULTY','COMPANY')
                      (A(7),X(5),A(7),X(17),A(7),X(15),A(7))
                      ('SUBJECT','GRADE','QUALITY')
                      (X(16),A(7),X(17),A(5),X(1),A(7));
 37          PUT SKIP EDIT ('NUMBER','NAME','NAME')
                      (A(6),X(6),A(4),X(20),A(4))
                      ('NAME','AREA','POINTS')
                      (X(18),A(4),X(19),A(4),X(26),A(6));
 38          PUT SKIP (2);
 39          RETURN;
 40          END PG_HDR_RTN;

 41  READ_RTN: PROCEDURE;
     /*********************************************************************/
     /*                                                                   */
     /*  FUNCTION:  TO READ A SINGLE RECORD                              */
     /*                                                                   */
     /*  GLOBAL VARIABLES:   STU_NUMBER                                   */
```

FIGURE 9–4 *(continued)*

```
      /*                    STU_NAME                          */
      /*                    STU_CODES                         */
      /*                                                      */
      /*  SUBROUTINES CALLED: NONE                            */
      /*                                                      */
      /*  INPUT:    AN INPUT RECORD WITH A VALUE FOR EACH OF THE  */
      /*            ABOVE VARIABLES                           */
      /*                                                      */
      /*  OUTPUT:   NONE                                      */
      /*                                                      */
      /*******************************************************/
  42          GET EDIT (STU_NUMBER,STU_NAME)   (COL(1),F(9),A(20))
                       (STU_CODES)             (COL(70), B(7));
  43          RETURN;
  44          END READ_RTN;

  45  GET_WRK_AREA_RTN:  PROCEDURE;
      /*******************************************************/
      /*                                                      */
      /*  FUNCTION:  TO GET AREA OF INTERNSHIP FROM CODES     */
      /*                                                      */
      /*  GLOBAL VARIABLES:   I_CODE                          */
      /*                      I_MASK                          */
      /*                      STU_CODES                       */
```

```
PL/I OPTIMIZING COMPILER       /* CH9EX3 USING EDIT I/O */              PAGE   5
   STMT
      /*                    WORK_AREA                         */
      /*                    PROF_NAME                         */
      /*                    COMAPNY_NAME                      */
      /*                                                      */
      /*  SUBROUTINES CALLED: NONE                            */
      /*                                                      */
      /*  INPUT:    NONE                                      */
      /*                                                      */
      /*  OUTPUT:   NONE                                      */
      /*                                                      */
      /*******************************************************/
```

```
46              I_CODE = I_MASK & STU_CODES;
47   WRK_AREA:  SELECT (I_CODE);
48                WHEN ('1000000'B)
     IF_DB:         DO;
49                   WORK_AREA = 'DATABASE';
50                   PROF_NAME = 'B. R. R. NUYL';
51                   COMPANY_NAME = 'SEARS';
52                 END IF_DB;
53                WHEN ('0100000'B)
     IF_SAD:        DO;
54                   WORK_AREA    = 'SYSTEMS ANALYSIS/DESIGN';
55                   PROF_NAME    = 'B. J. HARING';
56                   COMPANY_NAME = 'J. C. PENNEY';
57                 END IF_SAD;
58                WHEN ('1100000'B)
     IF_AP:         DO;
59                   WORK_AREA    = 'APPLICATION PROGRAMMING';
60                   PROF_NAME    = 'S. A. WRITE';
61                   COMPANY_NAME = 'SOFTECH, INC.';
62                 END IF_AP;
63              END WRK_AREA;
64              RETURN;
65              END GET_WRK_AREA_RTN;


66   CALC_GRADE_RTN:  PROCEDURE;
     /*********************************************************************/
     /*                                                              */
     /*  FUNCTION:  TO CALCULATE LETTER GRADES, GRADE POINTS, AND    */
     /*             QUALITY POINTS FROM GRADE CODES                  */
     /*                                                              */
     /*  GLOBAL VARIABLES:    G-CODE                                 */
     /*                       G_MASK                                 */
     /*                       STU_CODES                              */
     /*                       GRADE                                  */
     /*                       GPTS                                   */
```

```
     /*                       H_CODE                                 */
     /*                       H_MASK                                 */
     /*                       HOURS                                  */
```

FIGURE 9–4 *(continued)*

```
       /*                                                             */
       /*  SUBROUTINES CALLED: NONE                                   */
       /*                                                             */
       /*  INPUT:     NONE                                            */
       /*                                                             */
       /*  OUTPUT:    NONE                                            */
       /*                                                             */
       /**************************************************************/
 67              G_CODE = G_MASK & STU_CODES;
 68  GET_GRD:    SELECT (G_CODE);
 69                  WHEN ('0000111'B)
                         GRADE = 'A';
 70                  WHEN ('0000110'B)
                         GRADE = 'B';
 71                  WHEN ('0000101'B)
                         GRADE = 'C';
 72                  WHEN ('0000100'B)
                         GRADE = 'D';
 73                  WHEN ('0000010'B)
                         GRADE = 'F';
 74                  WHEN ('0000001'B)
                         GRADE = 'I';
 75                  WHEN ('0000000'B)
                         GRADE = 'W';
 76              END GET_GRD;
 77  GET_GPTS:   SELECT (GRADE);
 78                  WHEN ('A')
                         GPTS = 4;
 79                  WHEN ('B')
                         GPTS = 3;
 80                  WHEN ('C')
                         GPTS = 2;
 81                  WHEN ('D')
                         GPTS = 1;
 82                  OTHERWISE
                         GPTS = 0;
 83              END GET_GPTS;
 84              H_CODE = H_MASK & STU_CODES;
 85  GET_HRS:    SELECT (H_CODE);
 86                  WHEN ('0001000'B)
```

```
                        HOURS = 1;
    87            WHEN ('0010000'B)
                        HOURS = 2;
    88            WHEN ('0011000'B)
                        HOURS = 3;
    89         END GET_HRS;
```

```
    90            QUALITY_PTS = HOURS * GPTS;
    91            RETURN;
    92            END CALC_GRADE_RTN;

    93  PRNT_DETAIL_RTN:  PROCEDURE;
        /*******************************************************************/
        /*                                                               */
        /*  FUNCTION:  TO PRINT A DETAIL LINE                            */
        /*                                                               */
        /*  GLOBAL VARIABLES:    STU_NUMBER                             */
        /*                       STU_NAME                               */
        /*                       PROF_NAME                              */
        /*                       COMPANY_NAME                           */
        /*                       WORK_AREA                              */
        /*                       GRADE                                 */
        /*                       QUALITY_PTS                           */
        /*                                                               */
        /*  SUBROUTINES CALLED: NONE                                    */
        /*                                                               */
        /*  INPUT:     NONE                                             */
        /*                                                               */
        /*  OUTPUT:    A DETAIL LINE FOR ONE STUDENT                    */
        /*                                                               */
        /*******************************************************************/
    94            PUT EDIT (STU_NUMBER,STU_NAME,PROF_NAME)
                       (SKIP(2),F(9),X(3),A(20),X(3),A(20),X(3))
                       (COMPANY_NAME,WORK_AREA,GRADE,QUALITY_PTS)
                       (A(20),X(3),A(23),X(5),A(1),X(5),F(2));
    95            RETURN;
    96            END PRNT_DETAIL_RTN;
    97  END CH9EX3;
```

FIGURE 9–4 (continued)

```
PL/I OPTIMIZING COMPILER        /* CH9EX3 USING EDIT I/O */                    PAGE   8
COMPILER DIAGNOSTIC MESSAGES
ERROR ID L  STMT   MESSAGE DESCRIPTION


COMPILER INFORMATORY MESSAGES


IEL0533I I          NO 'DECLARE' STATEMENT(S) FOR 'SYSIN','SYSPRINT'.


END OF COMPILER DIAGNOSTIC MESSAGES
```

```
                         NORTHERN ILLINOIS UNIVERSITY
                       COMPUTER SCIENCE INTERNSHIP PROGRAM
                                FALL SEMESTER
```

STUDENT NUMBER	STUDENT NAME	FACULTY NAME	COMPANY NAME	SUBJECT AREA	GRADE	QUALITY POINTS
111111111	SMITH, DAVID B.	B. R. R. NUYL	SEARS	DATABASE	A	12
222222222	JONES, DON R.	S. A. WRITE	SOFTECH, INC.	APPLICATION PROGRAMMING	B	9
333333333	BOBBY, BETTY B.	S. A. WRITE	SOFTECH, INC.	APPLICATION PROGRAMMING	A	12
444444444	LONG, BOB. T.	B. J. HARING	J. C. PENNEY	SYSTEMS ANALYSIS/DESIGN	I	0
555555555	SHORT, I. M.	B. R. R. NUYL	SEARS	DATABASE	B	3
666666666	APPLE, CLEO M.	B. J. HARING	J. C. PENNEY	SYSTEMS ANALYSIS/DESIGN	W	0

In the executable code, the first reference to a bit string is in READ_RTN (Statement 42), which reads STU_CODES from the input record. The next reference is in Statement 46, where we find

```
I_CODE = I_MASK & STU_CODES;
```

Looking in the declaration statement we find that I_MASK has been initialized to '1100000'B; from the input data we discover that STU-CODES for the first student has the value of 1011111. The NOTES section of the documentation tells us that this pattern of bits means that the student was in the database internship area (10) for three hours of credit (11) and received a grade of A (111). The ampersand (&) tells the computer to carry out a **boolean operation** (named for the English mathematician George Boole [1815–1864]) called ANDing. In an **AND**ing operation a pair of bit strings are compared bit by bit. The bit strings themselves are not modified, but a third bit string is created according to these rules: If both bit strings have a 1 in the same position, a 1 is placed in that position in the third bit string; otherwise (i.e., both 0 or 0 and 1) a 0 is placed at that position in the third bit string. Let us illustrate with the data from CH9EX3.

```
I_MASK:        1100000
STU_CODES:     1011111
I_CODE:        1000000
```

The value in I_CODE then becomes the basis for determining which of the nested IF statements in GET_WRK_AREA_RTN will execute for a given student. The first statement of CALC_GRADE_RTN is

```
G_CODE = G_MASK & STU_CODES;
```

which gives us (again for the first student):

```
G_MASK:        0000111
STU_CODES:     1011111
G_CODE:        0000111
```

When we look at the case structure (Statements 68–75), we learn that this student will be assigned a grade of 'A'.

As for the edit-directed I/O itself, there are only two features of CH9EX3 that are new as compared to CH9EX1 and CH9EX2. The CH9EX3 program incorporates both GET EDIT and PUT EDIT in the same program, and it illustrates multiple data and format lists (Statements 36, 37, 42, and 94) in the same GET or PUT EDIT statement. The flowchart for CH9EX3 is shown in Figure A–15.

9.8

Summary

Edit-directed is the third type (after data- and list-directed) of stream I/O. The primary advantage of edit-directed input is that, since no delimiters need separate the data items on the input record, much more data can be stored on a record. The primary advantage of edit-directed output is that the programmer has complete control over the format of the printed output. The primary disadvantage of edit-directed input, on the other hand, is that it requires that the data be placed in the *exact* columns described by the format list of the GET EDIT statement. Therefore, it is more likely that errors will occur in constructing the input record. The second disadvantage, which applies equally to input and output, is that it involves a considerable amount of very detailed coding, again increasing the possibility of programmer error.

The syntax of the edit-directed I/O statement is

```
GET [or PUT] EDIT (data list) (format list);
```

The data list is made up of variables and/or literals, while the format list is composed of data items and control items. Data items describe the size of the fields in the input or output record and the type of data the field can contain; control items describe the relative location of the fields (both horizontally and vertically on print output). The following table summarizes the data item codes.

Code	Data Type	Comments
A	CHARACTER	Output only; the field width is taken from declaration
A(w)	CHARACTER	w specifies the field width
E(w,d)	FLOAT	w specifies the total field width; d specifies number of digits to the right of the implied decimal point
E(w,d,s)	FLOAT	s is ignored on input; on output specifies total number of significant digits
F(w)	FIXED DECIMAL	w specifies the field width
F(w,d)	FIXED DECIMAL	d specifies number of digits to right of implied decimal point

The control items and their meanings are as follows:

Code	Comments
COLUMN(n)	Abbreviated COL; n specifies absolute starting column of data
LINE(n)	Absolute line on which data is to appear
PAGE	Force a new page before printing
SKIP(n)	Skip to nth line from current line
X(n)	Skip n columns; on output these columns contain blanks

LINE and PAGE are restricted to print output; the rest are valid for both input and output. Any of them except X can be coded as control options preceding the keyword EDIT or as control items within the format list; in either case they are executed as they are encountered.

There are several rules concerning edit-directed I/O of which you need to be aware.

1. Each variable or literal in the data list has a corresponding data item in the format list.

2. If there are more data items in the format list than items in the data list, the extra data items are ignored.

3. If the format list contains fewer data items than items in the data list, the data items are used over again until the data list is exhausted.

4. The field size in the declaration need not be the same as that specified in the format list.

5. I/O continues until the data list is exhausted.

6. An iteration factor can be specified in conjunction with one or more items in the format list; it creates as many fields as the iteration factor specifies. The iteration factor appears to the left of the item(s) it controls, either separated by a blank or enclosed within parentheses.

7. A physical decimal point takes up a column, but an implied decimal point does not.

PL/I provides a number of special features to facilitate I/O operations; three of these are

1. expressions in output statements,

2. BUILTIN functions in output statements, and

3. multiple data and format lists in either input or output statements.

In no case is the use of any of the special features required by PL/I.

Both expressions and BUILTIN functions save on the amount of coding required and reduce the number of variables that must be declared; at the same time, their use reduces readability somewhat. When an expression is encountered in the output data list, it is evaluated and the result becomes an item in the output stream. When a BUILTIN function is encountered, the function is invoked and the value returned is placed in the data stream. In either case, a data item in the format list must describe the resulting value.

Expressions and functions can be combined in the same PUT EDIT statement. For example,

```
PUT EDIT (MAX (G_1, G_2, G_3, G_4, G_5)) (COL(45), F(7,2));
```

Multiple data and format lists are used primarily when the data list contains several items. Keeping the data item closer to the variable or literal it describes makes it easier to determine the proper relationship. The data lists and format lists must alternate, with data and control items in the format list describing the variables and/or literals in the immediately preceding data list.

```
PUT EDIT (NAME, PHONE, SSN)        (A(25), A(13), X(5), A(11))
             (STREET, ADDRESS, INCOME) (A(30), A(30), F(9,2));
```

The remote format item can be a useful tool if format and control items repeat several times in the program. Its use requires a FORMAT statement, which has the following syntax:

```
label:  FORMAT (list);
```

The remote format item consists of R (label), and it brings into the I/O statement containing the remote format item the list from the FORMAT statement.

9.9

Exercises

1. Using the following declarations, input statements, and input data, show the contents of the variables after each GET EDIT statement executes. Use a stroke (|) to show the implied decimal point and a b to show blanks. The input data always starts in column 1.

```
DCL VAR_1 FIXED DEC (5),
    VAR_2 FIXED DEC (6,2),
    VAR_3 CHAR       (10),
    VAR_4 CHAR       (5);
```

Input: 1234567890ABCDEFGHIJKL

```
a.  GET EDIT (VAR_1, VAR_3) (COL(5), F(3), A(8));
b.  GET EDIT (VAR_2, VAR_3) (COL(1), F(5,3), X(5), A(10));
c.  GET EDIT (VAR_1, VAR_4) (COL(3), F(5), X(10), A(2));
```

2. Using the given declarations, input data, and results, code the GET EDIT statement required to produce these results. The input and output both start in column 1.

```
DCL NAME        CHAR        (10),
    CLASS_NM    CHAR        (4),
    CLASS_NUM   FIXED DEC   (3),
    SEEC        FIXED DEC   (2);
```

Input: ANDERSENbbCSCI20013

a. NAME=ANDERSEN   CLASS_NM=CSCI   CLASS_NUM=200   SEC=13
b. NAME=ANDERbbb   CLASS_NM=CSbb   CLASS_NUM=020   SEC=00
c. NAME=SENbbbbb   CLASS_NM=CSCb   CLASS_NUM=000   SEC=01
d. NAME=ANDbbbbb   CLASS_NM=Cbbb   CLASS_NUM=002   SEC=13

3. Show the results of executing each of the following statements.
a. PUT EDIT ('COMPUTERS ARE FAST') (COL(8), A);
b. PUT EDIT ('COMPUTERS ARE FAST') (COL(8), A(18));
c. PUT EDIT ('COMPUTERS ARE FAST') (COL(2), A(10));
d. PUT EDIT ('COMPUTERS' 'ARE' 'FAST') (COL(10), A, SKIP,
         COL(19), A, SKIP, COL(22), A);
e. PUT EDIT ('COMPUTERS', 'ARE FAST') (LINE(5), A, A);

4. For each format list control item below state what the control item would cause the printer to do and what and where (i.e., page, line, column) each format item would cause to print. If a format item causes an error, state the reason and do no more on that format list.

```
DCL A1   CHAR        (8)         INIT ('HOLY COW'),
    N1   FIXED DEC (4,2)         INIT (69.13),
    N2   FIXED DEC (4,1)         INIT (136.9);
PUT EDIT (A1, N1, N2) (format list)
```

a. (COL(20), A(10), F(4,2), X(3), F(4,1));
b. (PAGE, A(8), F(5,2), F(5,1));
c. (LINE(7), A(4), SKIP(3), F(7,2), LINE(15), F(8,4));
d. (COL(10), F(5,2), X(5), F(5,1));
e. (COL(10), A(8), COL(10), F(5,2), COL(10), F(5,1));
f. (LINE(5), A(10), SKIP, F(5,2), LINE(5), F(7,3));

5. Each item below shows the output for one PUT EDIT statement. Write the proper PUT EDIT to produce this output. The starting column for the first output character is column 1. Use the declarations

```
DCL  EIN   FIXED DEC (6,3)    INIT (106.661),
     ZWEI  FIXED DEC (4)      INIT (1789),
     ALPHA CHAR      (5)      INIT ('HIJKL'),
     BETA  CHAR      (12)     INIT ('COMPUTERS SI');
```

a. 178900106.661COMP
b. bbCOMPUTERS SI
    bbbHIJ
c. COMPUTERSbbb1789HIJKL
d. COMPUTERSbb106bb1789bbbH
e. COMPUTERS SI
    106.661000
    bbbHIJKL
    1789.0000

6. Write the PUT EDIT statement for each of the following, using the appropriate BUILTIN function.

    a.   Put the last four characters of the CHAR (10) variable ALPHA in a four-character print field.

    b.   The value in FIX1 is 72.9834; print it rounded as nn.nn.

    c.   Print in the form of nnnn the smallest of the values in GR_1, GR_2, GR_3.

    d.   Print in the form of nn the number of characters in ZETA.

7. Write the PUT EDIT statement for each of the following, including in each the appropriate expression and function (if required).

    a.   Print in the form of nn.nn the result of adding B_2, B_3, B_4 and dividing the result by Q_2.

    b.   Print in the form of nnnnn the result of raising M_1 to the power of K_9.

    c.   Print as nnn the square root of the sum of V_1 and V_2.

8. Change the following single data and format lists to a multiple list in each case. Put two items in each variable list (except, possibly, the last). Assume that all variables have been properly declared.

    a.   PUT EDIT (NAME, SSN, EMP_#, GROSS, TAX_RATE, TAX_$,
            NET_$) (PAGE, LINE(3), A(25), A(15), A,
            F(9,2), F(2,2), F(6,2), F(9,2));

    b.   PUT EDIT (PROD_#, PROD_NAME, PURCH_COST, SALE_PRICE,
            WHSE_#, SHIP_BY) (COL(10), F(5), X(2),
            A(20), X(3), F(5,2), X(4), SKIP(1), F(6,2),
            X(5), F(2));

    c.   PUT EDIT (ALPHA, BETA, F_A, F_B, ZETA, X_1, X_2, X_3)
            (COL(6), A(10), X(2), A(3), COL(20), F(5),
            F(7,2), X(7), A(9), X(3), F(2), F(3), F(4));

9. Write a program to produce a unified inventory report on the inventory in the three warehouses owned by the company. The input has the following format.

Columns 1–5   product number
         6–8   number on hand in warehouse A
       9–11   number on hand in warehouse B
    12–14   number on hand in warehouse C

Clearly, you will have to use GET EDIT to read the data. Use PUT LIST to create a report format resembling the following:

```
                         GIZMO MFG. CO.
                      UNIFIED INVENTORY REPORT
    PRODUCT-#      WHSE-A        WHSE-B          WHSE-C
    ---------      ------        ------          ------
    12345              10           260               7
    12354             180            72              61
```

10. Make the following modifications to the program in Exercise 9.
    a. Use PUT EDIT as well as GET EDIT.
    b. In the column headers put the city in which each warehouse is located (New York, Chicago, and San Francisco).
    c. Add a column to each detail line presenting the total in all warehouses for that item.
    d. On a separate page print a summary giving the total number of all products in each warehouse and a grand total giving the total number of all products in all the warehouses.
    e. For the output,
        (1) center the page header over the detail lines
        (2) start the column headers in columns 10, 20, 28, 36, 55
        (3) place the rightmost digit in each column of the detail lines directly under the rightmost character of its column header
        (4) on the summary page
            (a) centered on line 10, print UNIFIED INVENTORY REPORT SUM-MARY
            (b) leave three blank lines
            (c) starting in column 25 print WAREHOUSE TOTALS
            (d) leave one blank line
            (e) for each warehouse print city, total
            (f) leave 5 blank lines
            (g) starting in column 25 print THE GRAND TOTAL IS grand total

11. Using both GET EDIT and PUT EDIT, write a program to create a billing report. The input consists of

Columns   1–5   customer account number
            6–25   customer name

          26–31 balance
          32–37 charges
          38–43 payment

The last three items are in dollars-and-cents format. The output consists of the input plus the interest charge and the payment due. If balance + charges − payment is greater than zero, assess an interest charge of 1.5% (monthly). Thus, the (new) balance = (old) balance + new charges − payment + interest charge. If a payment reduces the balance to less than zero, append CREDIT to the balance, and show it as a positive number. The minimum payment schedule, computed always on the (new) balance is

Balance = $1.00–10.00   payment in full
          10.01–100.00 $10.00
          over $100.00   10% of the balance

The page header (centered) should be BILLING REPORT FOR T. H. E. COMPANY. Provide adequate column headers and produce an attractive and easily readable format.

12. Many retail stores now use a microcomputer and printer to print invoices for their customers. Typically the unit cost is stored in the computer; in our case, however, it will be part of the input. The input, then, will have the format of

Columns   product number; if the first digit is a 0, the item is nontaxable;
1–4   otherwise it is taxed at 6%.

4–8   not used

9–25   product name

26–27   number of units

28–29   not used

30–35   cost per unit

Your program should print out the input data, listing in order on a single line the (1) number of units, (2) name of product, (3) cost subtotal, and (4) tax. The end of one customer transaction is signaled by all zeros in the product number field and blanks in all other fields. At this point, skip a line; then print (with labels) the total tax and the total amount owed.

C H A P T E R   1 0

MORE ON

SUBROUTINE

PROCEDURES

Previously we discussed internal subroutine procedures, which are nested within a MAIN procedure and must be compiled along with it. Also, the internal subroutine procedures can be called only by the MAIN procedure and by other subroutine procedures that are internal to the same MAIN program. In this chapter we present external subroutine procedures, sometimes called **subprograms** because they are compiled separately from any other procedure. As you will see, these external subprograms can also be called from any other program within the system.

First, however, we introduce the topic of arguments and parameters. They can (and many programmers feel should) be used with internal subroutine procedures. They are a necessity, however, when using subprograms because they constitute the only means by which the MAIN program and the subprogram can share data.

The next topic we discuss is the differences between internal subroutine procedures and external subprograms. Naturally, more differences exist than just the naming conventions we chose to use. Actually, we chose these names for the two types of subroutines just to assist you in distinguishing between them. We show you how to declare and call external subprograms and how to use arguments and parameters with subprograms.

Finally, we present a short description of "recursive" procedures within PL/I. This topic is not meant to be a complete discussion of recursion but only of recursion within PL/I. You might need to refer to other sources for a more thorough understanding of recursion itself.

10.1

Arguments and Parameters

In order for some procedures to function properly, they have to be able to access variables within the calling procedure. The subroutine procedure's purpose is to use the values in the variables and to perform some operations upon these values to obtain a certain result. We know that internal subroutines can access variables within other procedures by using global variables (any variable declared in the MAIN procedure, you recall, is global). However, global variables are not always the best method (and might not always work) to provide a subroutine with the data it needs. Consider the following example.

```
MAIN:  PROCEDURE OPTIONS(MAIN);
         .
         .

READSTUDENT:  PROCEDURE;
       DCL  STUDENT_NAME        CHAR (30);
         .
         .

END READSTUDENT;
PRINTSTUDENT:  PROCEDURE;
         .
         .

END PRINTSTUDENT;
END MAIN;
```

In this example, PRINTSTUDENT cannot access the STUDENT_NAME variable because it is declared in the READSTUDENT subroutine. One way to allow PRINTSTUDENT access to STUDENT_NAME is to move the declaration of STUDENT_NAME to the MAIN procedure instead of within READSTUDENT. This way STUDENT_NAME becomes a global variable accessible by any other subroutine.

Another way to use STUDENT_NAME is to pass it to READSTUDENT via the CALL statement, thereby allowing the PRINTSTUDENT sub-routine to access the value in the STUDENT_NAME variable residing in the READSTUDENT subroutine.

In the following section we use the term **procedure** to denote a PL/I procedure, whether it be an internal subroutine or an external subprogram.

10.1.1

CALL Statement with Arguments

The CALL statement is used to transfer control from one procedure to another procedure. Examine the CALL statement below.

```
CALL SUB1 (ARG1, ARG2, ARG3, ..., ARGn);
```

As you can see, it differs from the previous CALL statements that we have used. The items within the parentheses are called **arguments**. The entire list of arguments is called an **argument list**. Arguments are variables declared within the calling procedure to which the procedure is given access. The procedure can obtain the values located within the arguments at the time the CALL statement is executed and use them in its processing.

Arguments can be declared using any data type that is available within PL/I, as shown in the list below.

Arguments can be

Constants, Variables, Expressions, Arrays, Structures,
Built-In Function Names, and File Names

In the following example CALL statements you will see examples of valid arguments:

```
CALL SUB1 (LAST_NAME, 2, 15 * 6);
CALL SUB2 (SQRT(X), INPUT_FILE, ANSWER);
CALL SUB3 ('TODAY''S DATE IS ', '11/05/87');
```

In the first CALL statement, LAST_NAME, the value 2, and the value 90 (expressions are evaluated first and the answer is available to the procedure) are the arguments passed to SUB1. In the second CALL statement above, you will notice the use of the built-in function SQRT. In this case, the built-in function is invoked using the value of X; the result of the SQRT function is the actual argument value available to SUB2. In the third CALL statement, we show you that literals can be used as arguments.

Each argument corresponds to a parameter in the called procedure, as you will learn in the following section.

Parameters are the variables through which a called procedure accesses the arguments in the calling procedure. Parameters are designated on the PROCEDURE statement as follows:

```
SUB1:   PROCEDURE (PARM1, PARM2, PARM3, ..., PARMn);
```

The correspondence between arguments and parameters is one-to-one, from left to right (i.e., PARM1 refers to the value in ARG1, PARM2 refers to the value in ARG2, etc.). In reality, arguments and parameters refer to the same storage area, that being the area declared for the arguments within the invoking procedure. Therefore, when you use a parameter name within the invoked procedure, you are indirectly referring to the storage area assigned to the corresponding argument, whereas the argument directly references the storage area assigned to it. Even with this correspondence, arguments and parameters can have different names. The correspondence is made via position of the argument within the argument list of the CALL statement and the position of the parameter within the **parameter list** of PROCEDURE statement, not via the names.

You might want to go back and examine the example CALL statements shown above, for we now give you the corresponding PROCEDURE statements for receiving the parameters.

```
SUB1:   PROCEDURE (LAST_NAME, NUM_OF_CHILDREN, AMT_SOLD);
SUB2:   PROCEDURE (SQ_ROOT, READ_FILE, RESULT);
SUB3:   PROCEDURE (HEADING, TODAY_DATE);
```

As you have probably gathered, since arguments can be of any data type, so can parameters. The data type of an argument and its corresponding parameter *must* match. Parameters must be defined in the called procedure using a DECLARE statement just as is the case with any other variable. The name of the parameter need not be the same as the argument. As a matter of fact, it is probably best to give the parameter a slightly different name than the argument name (especially in internal subroutines) so that the names can aid in debugging a program.

It is important for you to remember that when a parameter is declared, new storage is not allocated. The declaration of a parameter simply tells the compiler to recognize the variable name. When the program is executed, the use of the parameter name refers to the storage allocated to the corresponding argument in the invoking procedure. Therefore, you *cannot* use the INITIAL attribute on the declaration of a parameter. It can be used when the argument is declared in the invoking procedure, but *never* for the parameter.

Parameters are declared in the same manner as any other type of variable. However, we highly recommend that you declare all parameters immediately following the PROCEDURE statement. Then you can declare any other variables (i.e., local variables) needed by the invoked procedure. For example,

```
SUB1:  PROCEDURE (LAST_NAME, NUM_OF_CHILDREN, AMT_SOLD);
       DCL  LAST_NAME          CHAR (25),
            NUM_OF_CHILDREN     FIXED DEC (2),
            AMT_SOLD            FLOAT DEC (3);


SUB2:  PROCEDURE (SQ_ROOT, READ_FILE, RESULT);
       DCL  SQ_ROOT            FLOAT (6),
            READ_FILE          FILE,
            RESULT             FIXED DEC (7);
```

Notice in this declaration that the data type of SQ_ROOT must agree with the data type of the value returned by the SQRT built-in function.

```
SUB3:  PROCEDURE (HEADING, TODAY_DATE);
       DCL  HEADING            CHAR(•),
            TODAY_DATE         CHAR(8);
```

In this example, we introduced a new length value, the '*' (asterisk), which can be used for character parameters when the actual length of the character argument is not known. In this case PL/I uses the actual length of the argument as the length assigned to the corresponding parameter.

The *most important* thing for you to remember when using arguments and parameters is that the data type *must* match between the argument and its corresponding parameter. If it does not, an error can occur, and the invoked procedure might not execute properly.

10.1.3

Example Program—
Arguments and
Parameters in Internal
Subroutines

The CH10EX1 program (see Fig. 10–1) is a modification of CH7EX4. In CH10EX1 we show the use of arguments and parameters with internal subroutine procedures. You will notice that none of the subroutines use any global variables; all variable values are passed in via the parameters. Examine subroutine CHOICE_RTN. Notice that it passes the parameters it received as arguments to the subroutines it calls. CHC_CHOICE_CODE is received as a parameter and then passed to the print subroutine via the CALL statement as follows:

```
CALL PRNT_CHOICE_RTN (CHC_CHOICE_CODE);
```

The other thing to realize about this program is that the use of arguments and parameters is not really necessary with the use of global variables (as in CH7EX4). However, when parameters are used with a slightly different name, they can make locating a bug easier because each subroutine references the variables by a different name. Remember, too, that the parameter is a *local* variable; other procedures (including MAIN) will not recognize this parameter name.

```
PL/I OPTIMIZING COMPILER  VERSION 1 RELEASE 5.1  TIME: 14.10.53  DATE: 13 JAN 87  PAGE   1
OPTIONS SPECIFIED
GOSTMT;



PL/I OPTIMIZING COMPILER        /* CH10EX1 USING ARGUMENTS AND PARAMETERS */   PAGE   2
                SOURCE LISTING
    STMT


     /* CH10EX1 USING ARGUMENTS AND PARAMETERS */
  1  CH10EX1: PROC OPTIONS (MAIN);
     /*****************************************************************/
     /* FUNCTION:  TO PROCESS STUDENT INTERNSHIP APPLICATIONS.        */
     /*                                                               */
     /* INPUT:     FOR EACH STUDENT A RECORD CONTAINING               */
     /*                (1) STUDENT'S NAME                             */
     /*                (2) CREDIT HOURS FOR INTERNSHIP                */
     /*                (3) STUDENT'S FIRST INTERNSHIP CHOICE          */
     /*                (4) STUDENT'S SECOND INTERNSHIP CHOICE         */
     /*                                                               */
     /* OUTPUT:    FOR EACH STUDENT A RECORD CONTAINING               */
     /*                (1) STUDENT'S NAME                             */
     /*                (2) FACULTY ADVISOR                            */
     /*                (3) COMPANY OFFERING INTERNSHIP                */
     /*                (4) HOURS CREDIT                               */
     /*                (5) SUBJECT AREA                               */
```

FIGURE 10-1
CH10EX1 (arguments and parameters in internal subroutines).

```
       /*                                                        */
       /* NOTES:                                                 */
       /*           STUDENTS HAVE A CHOICE OF THREE SUBJECT AREAS FOR   */
       /*           THEIR INTERNSHIPS:  SYSTEMS ANALYSIS & DESIGN,      */
       /*           DATABASE, AND APPLICATIONS PROGRAMMING.  THEY MAY   */
       /*           CHOOSE ONE SUBJECT AREA AND AN ALTERNATE.  ONE      */
       /*           FACULTY ADVISOR IS ASSIGNED TO EACH SUBJECT AREA.   */
       /*           HE/SHE MAY SPONSOR A MAXIMUM OF TWO INTERNS.        */
       /*           STUDENTS ARE GIVEN THEIR ALTERNATE CHOICE WHEN      */
       /*           THERE ARE NO MORE OPENINGS FOR THEIR FIRST CHOICE.  */
       /*           IF THEIR ALTERNATE CHOICE IS ALSO FULL, THEY ARE    */
       /*           NOT GIVEN AN INTERNSHIP.  INPUT CODES FOR THE THREE */
       /*           SUBJECT CODES ARE                                   */
       /*              SYSTEMS ANALYSIS & DESIGN  = 01                  */
       /*              DATABASE                   = 10                  */
       /*              APPLICATION PROGRAMMING    = 11                  */
       /***********************************************************************/
   2    DCL STU_NAME                  CHARACTER (20);
   3    DCL HOURS_CREDIT              FIXED DEC (1);
   4    DCL FIRST_CHOICE_CODE         FIXED DEC (3);
   5    DCL SECOND_CHOICE_CODE        FIXED DEC (3);
   6    DCL CHOICE_CODE               FIXED DEC (3);
   7    DCL CHOICE_NUMBER             FIXED BIN (15) INIT (0);
   8    DCL PROF_1_COUNT              FIXED BIN (1)  INIT (0);
   9    DCL PROF_2_COUNT              FIXED BIN (15) INIT (0);
  10    DCL PROF_3_COUNT              FIXED BIN (15) INIT (0);
  11    DCL PROF_MAX                  FIXED BIN (15) INIT (2);
  12    DCL MORE_TO_SCHED             BIT     (1);
  13    DCL    YEA                    BIT     (1)  INIT ('1'B);
```

```
  14    DCL    NAY                    BIT     (1)  INIT ('0'B);
  15    DCL  .:ORE_RECS               BIT     (1);
  16    DCL    YES                    BIT     (1)  INIT ('1'B);
  17    DCL    NO                     BIT     (1)  INIT ('0'B);
  18         MORE_RECS = YES;
  19         MORE_TO_SCHED = YEA;
  20         ON ENDFILE (SYSIN) MORE_RECS = NO;
  21         CALL P6_HDR_RTN;
  22         CALL READ_RTN (STU_NAME, HOURS_CREDIT,
                       FIRST_CHOICE_CODE, SECOND_CHOICE_CODE);
```

FIGURE 10-1 (continued)

```
23  PROC_LP:  DO WHILE (MORE_RECS);
24            CALL CHOICE_RTN(PROF_1_COUNT, FIRST_CHOICE_CODE, STU_NAME,
                             HOURS_CREDIT, PROF_2_COUNT,
                             PROF_3_COUNT, 1,
                             SECOND_CHOICE_CODE);
25            CALL READ_RTN(STU_NAME, HOURS_CREDIT,
                             FIRST_CHOICE_CODE, SECOND_CHOICE_CODE);
26          END PROC_LP;
27          PUT SKIP (2) LIST (' ', '===> SCHEDULING COMPLETED <===');

28  PG_HDR_RTN:  PROCEDURE;
    /*****************************************************************/
    /*                                                             */
    /*  FUNCTION:  TO PRINT PAGE AND COLUMN HEADERS                */
    /*                                                             */
    /*  PARAMETERS:  NONE                                          */
    /*                                                             */
    /*  GLOBAL VARIABLES:   NONE                                   */
    /*                                                             */
    /*  SUBROUTINES CALLED: NONE                                   */
    /*                                                             */
    /*  INPUT:     NONE                                            */
    /*                                                             */
    /*  OUTPUT:    A TWO-LINE PAGE HEADER AND A TWO-LINE COLUMN    */
    /*             HEADER                                          */
    /*                                                             */
    /*****************************************************************/
29          PUT LIST ('','','    NORTHERN ILLINOIS UNIVERSITY');
30          PUT SKIP LIST ('','','COMPUTER SCIENCE INTERNSHIP PROGRAM');
31          PUT SKIP LIST ('','','        FALL SEMESTER');
32          PUT SKIP (2) LIST ('STUDENT','FACULTY','','HOURS','SUBJECT');
33          PUT SKIP    LIST ('NUMBER', 'ADVISOR', 'COMPANY',
                             'CREDIT', 'AREA');
34          PUT SKIP    LIST (((120)'-');
35          PUT SKIP;
36          RETURN;
```

```
37          END PG_HDR_RTN;

38  READ_RTN:  PROCEDURE (READ_STU_NAME, READ_HOURS_CREDIT,
                          READ_FIRST CHOICE_CODE, READ_SECOND_CHOICE_CODE);
    /*******************************************************************/
    /*                                                                 */
    /*  FUNCTION:  TO READ A SINGLE RECORD                             */
    /*                                                                 */
    /*  PARAMETERS:     READ_STU_NAME                                  */
    /*                  READ_HOURS_CREDIT                              */
    /*                  READ_FIRST_CHOICE_CODE                         */
    /*                  READ_SECOND_CHOICE_CODE                        */
    /*                                                                 */
    /*  GLOBAL VARIABLES:  NONE                                        */
    /*                                                                 */
    /*  SUBROUTINES CALLED: NONE                                       */
    /*                                                                 */
    /*  INPUT:    THE INPUT RECORDS CONTAINING A VALUE FOR EACH        */
    /*            VARIABLE ABOVE                                       */
    /*                                                                 */
    /*  OUTPUT:   NONE                                                 */
    /*                                                                 */
    /*******************************************************************/
39    DCL READ_STU_NAME              CHARACTER (20);
40    DCL READ_HOURS_CREDIT          FIXED DEC (1);
41    DCL READ_FIRST_CHOICE_CODE     FIXED DEC (3);
42    DCL READ_SECOND_CHOICE_CODE    FIXED DEC (3);

43          GET LIST (READ_STU_NAME,READ_HOURS_CREDIT,
                      READ_FIRST_CHOICE_CODE,
                      READ_SECOND_CHOICE_CODE) COPY;
44          RETURN;
45          END READ_RTN;

46  CHOICE_RTN:  PROCEDURE(CH_PROF_1_COUNT, CH_CHOICE_CODE,
                           CH_STU_NAME, CH_HOURS_CREDIT,
                           CH_PROF_2_COUNT, CH_PROF_3_COUNT,
                           CH_CHOICE_NUMBER, CH_SECOND_CHOICE_CODE);
    /*******************************************************************/
    /*                                                                 */
    /*  FUNCTION:  TO ASSIGN STUDENTS THEIR CHOICE OF INTERNSHIPS OR */
    /*             DETERMINE THEY CANNOT BE ASSIGNED ONE               */
```

FIGURE 10–1 *(continued)*

```
/*                                                             */
/*   PARAMETERS:      CH_PROF_1_COUNT                          */
/*                    CH_CHOICE_CODE                           */
/*                    CH_STU_NAME                              */
/*                    CH_HOURS_CREDIT                          */
```

PL/I OPTIMIZING COMPILER   /* CH10EX1 USING ARGUMENTS AND PARAMETERS */   PAGE   5
   STMT

```
/*                      CH_PROF_2_COUNT                        */
/*                      CH_PROF_3_COUNT                        */
/*                      CH_CHOICE_NUMBER                       */
/*                      CH_SECOND_CHOICE_CODE                  */
/*                                                             */
/*   GLOBAL VARIABLES:  MORE_TO_SCHED                          */
/*                      PROF_MAX                               */
/*                                                             */
/*   SUBROUTINES CALLED: PRNT_CHOICE_RTN                       */
/*                       PRNT_FULL_MSG_RTN                     */
/*                                                             */
/*   INPUT:     NONE                                           */
/*                                                             */
/*   OUTPUT:    NONE                                           */
/*                                                             */
/*****************************************************************/
47     DCL CH_PROF_1_COUNT              FIXED BIN (1);
48     DCL CH_CHOICE_CODE              FIXED DEC (3);
49     DCL CH_STU_NAME                 CHARACTER (20);
50     DCL CH_HOURS_CREDIT             FIXED DEC (1);
51     DCL CH_PROF_2_COUNT             FIXED BIN (15);
52     DCL CH_PROF_3_COUNT             FIXED BIN (15);
53     DCL CH_CHOICE_NUMBER            FIXED BIN (15);
54     DCL CH_SECOND_CHOICE_CODE       FIXED DEC (3);

55 SCHED_LP:     DO WHILE (MORE_TO_SCHED);
56 SLCT_1:          SELECT;
57                  WHEN (CH_PROF_1_COUNT < PROF_MAX
                        & CH_CHOICE_CODE = 01)
   IF_1_OK:            DO;
```

```
58                              PUT SKIP LIST (CH_STU_NAME, 'B. J. HARING',
                                         'J. C. PENNEY', CH_HOURS_CREDIT);
59                          CH_PROF_1_COUNT = CH_PROF_1_COUNT + 1;
60                          MORE_TO_SCHED = NAY;
61                      END IF_1_OK;
62                  WHEN (CH_PROF_2_COUNT < PROF_MAX
                        & CH_CHOICE_CODE = 10)
    IF_2_OK:            DO;
63                          PUT SKIP LIST (CH_STU_NAME, 'B. R. R. NUYL',
                                     'SEARS', CH_HOURS_CREDIT);
64                          CH_PROF_2_COUNT = CH_PROF_2_COUNT + 1;
65                          MORE_TO_SCHED = NAY;
66                      END IF_2_OK;
67                  WHEN (CH_PROF_3_COUNT < PROF_MAX
                        & CH_CHOICE_CODE = 11)
    IF_3_OK:            DO;
68                          PUT SKIP LIST (CH_STU_NAME, 'S. A. WRITE',
                                     'SOFTECH, INC', CH_HOURS_CREDIT);
69                          CH_PROF_3_COUNT = CH_PROF_3_COUNT + 1;
70                          MORE_TO_SCHED = NAY;
```

```
71                      END IF_3_OK;
72                  OTHERWISE;
73              END SLCT_1;
74          IF MORE_TO_SCHED = NAY THEN
                CALL PRNT_CHOICE_RTN(CH_CHOICE_CODE);
75          ELSE
                IF CH_CHOICE_NUMBER = 1 THEN
    IF_1:           DO;
76                      CH_CHOICE_CODE = CH_SECOND_CHOICE_CODE;
77                      CH_CHOICE_NUMBER = 2;
78                  END IF_1;
79              ELSE
                    IF CH_CHOICE_NUMBER = 2 THEN
    IF_2:               DO;
80                          CALL PRNT_FULL_MSG_RTN(CH_STU_NAME);
81                          MORE_TO_SCHED = NAY;
82                      END IF_2;
```

FIGURE 10–1 *(continued)*

```
83                END SCHED_LP;
84                MORE_TO_SCHED = YEA;
85             RETURN;
86             END CHOICE_RTN;

87   PRNT_CHOICE_RTN:  PROCEDURE(PRNT_CHOICE_CODE);
     /**********************************************************************/
     /*                                                                  */
     /*  FUNCTION:  TO PRINT THE SUBJECT AREA TO WHICH THE STUDENT       */
     /*             IS ASSIGNED                                          */
     /*                                                                  */
     /*  PARAMETERS:   PRNT_CHOICE_CODE                                  */
     /*                                                                  */
     /*  GLOBAL VARIABLES:  NONE                                         */
     /*                                                                  */
     /*  SUBROUTINES CALLED: NONE                                        */
     /*                                                                  */
     /*  INPUT:    NONE                                                  */
     /*                                                                  */
     /*  OUTPUT:   A PRINT FIELD CONTAINING THE STUDENT'S SUBJECT        */
     /*            AREA                                                  */
     /*                                                                  */
     /**********************************************************************/
88      DCL PRNT_CHOICE_CODE              FIXED DEC (3);
89   SLCT_2:   SELECT (PRNT_CHOICE_CODE);
90             WHEN (10) PUT LIST ('DATABASE');
91             WHEN (01) PUT LIST ('SYSTEMS ANALYSIS & DESIGN');
92             WHEN (11) PUT LIST ('APPLICATIONS PROGRAMMING');
93             END SLCT_2;
94             RETURN;
```

PL/I OPTIMIZING COMPILER   /* CH10EX1 USING ARGUMENTS AND PARAMETERS */   PAGE   7
   STMT

```
95             END PRNT_CHOICE_RTN;

96   PRNT_FULL_MSG_RTN:  PROCEDURE(PRNT_STU_NAME);
     /****************************************************************/
     /*                                                            */
```

```
          /* FUNCTION:  TO PRINT A MESSAGE ADVISING A STUDENT THAT HER/    */
          /*            HIS INTERNSHIP IS NOT AVAILABLE                     */
          /*                                                               */
          /* PARAMETERS:    PRNT_STU_NAME                                  */
          /*                                                               */
          /* GLOBAL VARIABLES: NONE                                        */
          /*                                                               */
          /* SUBROUTINES CALLED: NONE                                      */
          /*                                                               */
          /* INPUT:    NONE                                                */
          /*                                                               */
          /* OUTPUT:   TWO PRINT LINES CONSISTING OF A STUDENT NAME        */
          /*           AND TWO ADVISORY MESSAGES                           */
          /*                                                               */
          /*****************************************************************/
   97     DCL PRNT_STU_NAME                    CHARACTER (20);
   98         PUT SKIP LIST (PRNT_STU_NAME,
                             '==> ALL FALL INTERNSHIPS FULL <==');
   99         PUT SKIP LIST (' ', 'SEE YOUR ACADEMIC ADVISOR');
  100         RETURN;
  101         END PRNT_FULL_MSG_RTN;
  102  END CH10EX1;


PL/I OPTIMIZING COMPILER      /* CH10EX1 USING ARGUMENTS AND PARAMETERS */    PAGE   8

COMPILER DIAGNOSTIC MESSAGES
ERROR ID L   STMT    MESSAGE DESCRIPTION

COMPILER INFORMATORY MESSAGES

IEL0533I I         NO 'DECLARE' STATEMENT(S) FOR 'SYSIN','SYSPRINT'.

END OF COMPILER DIAGNOSTIC MESSAGES
```

10.2

External Subprograms

Modular programming is advantageous to top-down development. We showed you how to achieve modular programming via internal subroutines. Now we show you modular programming via external subprograms. Based upon the four goals of structured programming discussed earlier, there are several advantages to using subprograms rather than subroutines.

1. *Accuracy:* A subprogram that performs one and only one function—for example, to sort data—and is known to be accurate can be called by any program, rather than have a programmer write a new module to perform the same function, which must then be tested for accuracy.

2. *Reliability:* Once the sort program is written and tested, we can count upon its reliability to perform the proper sort over and over again without ever having to be debugged again.

3. *Readability:* A system made up of many subprograms is much easier to read, providing the subprograms perform one and only one function and are of reasonable length. The logic of the system is easy to follow because the logic of each subprogram is easy to understand. And every program needing the same function performed—for example, sorting data—calls the same subprogram (which is already known).

4. *Maintainability:* There are two reasons why maintainability is the area in which subprograms contribute the biggest benefit to structured programming. One, any time a function residing within a subprogram needs modification that does not change the input or output, only the subprogram need be changed. For example, a subprogram might have as its function to sort data, and the programmer has implemented thc function with a bubble sort. Later, it is determined that another sort algorithm would be more efficient. So long as it does not change the input or output, the new algorithm can replace the old with no effect on any other procedure, thus saving maintenance time. And, two, the results of any modifications can be tested by simply testing the subprogram that has been changed.

10.2.1

*External vs.
Internal Procedures*

As we mentioned above, one way in which external subprograms differ from internal subroutines is that subprograms are compiled separately. Another difference is that the only internal subroutines an external subprogram can call are those internal subroutines that are nested within it; it can also call other external subprograms. A third difference, therefore, is that the scope of the variables that an external subprogram can access is limited to within itself. (The scope of the subroutines within a subprogram and scope of the variables within a subprogram follow the same rules as for MAIN procedures.)

Previously, we described two stages that are involved in the processing of a PL/I program, those being compilation and execution. What you have done so far is to compile a PL/I program and then immediately submit it for execution. There can be, however, an intermediate step. A PL/I program can be compiled and saved as an **object module** on a disk until needed for execution. An object module is the output of the compiler. The compiler takes the PL/I source code as input and changes it into the machine code that the computer can understand and execute.

When a subprogram is saved on a disk as an object module, it can be invoked from any other program. The program wishing to execute the subprogram must know the name of the subprogram, and there are a few restrictions that are placed upon the subprogram. These are described within this chapter as we continue with the topic of subprograms.

External subprograms are "programs" in their own right and behave as main procedures, except there is no OPTIONS (MAIN) on the PROCEDURE statement. External subprograms can have internal subroutines, as long as they follow all the rules stated in Chapter 6. External subprograms **cannot** access any variables within the MAIN procedure as global variables, nor can the MAIN procedure access any variables declared in the subprogram. However, global variables can be declared within the external subprogram and used by the internal subroutines within the subprogram. Local variables within the subroutines can be declared and used by the subroutines (again following the scope rules introduced in Chapter 6).

10.2.2

Creating External Subprograms

The PROCEDURE statement in an external subprogram does not contain the OPTIONS (MAIN) clause of a MAIN procedure. However, because the subprogram is known as a separate entity to the system, the name of the subprogram must follow the rules for naming main procedures. The subprogram name can be only seven characters in length and must start with an alphabetic character. For example,

```
EXTSUB:    PROCEDURE;
               .
               .
               .

END EXTSUB;
```

A procedure that calls an external subprogram must inform the compiler that the name in the CALL statement is not located within the source being compiled; it is located externally, within another compiled entity in the system (maybe on disk). The compiler is informed of this fact in a DECLARE statement by using the ENTRY attribute on the name of the subprogram to be called.

```
         MAIN:   PROCEDURES OPTIONS(MAIN);
                     .
                     .
                 DECLARE EXTSUB ENTRY;
                     .
                     .
                     .
                 CALL EXTSUB;
                     .
                     .
         END MAIN;
```

```
         EXTSUB:   PROCEDURE;
                     .
                     .
                     .
             END EXTSUB;
```

The compiler must recognize all identifiers that are used within a procedure. Because EXTSUB is actually not located within the same procedure as the MAIN, we must tell the compiler not to try to recognize the name EXTSUB when it appears in the MAIN procedure code. The ENTRY attribute does just that. Thus, the recognition of external procedure names is left until a later stage in the compilation and execution process of a program.

10.2.3

The *PROCESS
Statement

There are times when you will want to compile an external subprogram in the same run as a MAIN program. You might do this during the testing stage of the subprogram until the subprogram is free from bugs. At that time the subprogram can be compiled and placed on a disk as an object module for use by other programs within the system. However, whenever any external subprograms are compiled within that same run as a MAIN procedure, PL/I requires an *PROCESS statement (starting in column 1) to be placed prior to the subprogram PROCEDURE statement.

```
         MAIN:   PROCEDURE OPTIONS(MAIN);
```

```
                    END MAIN;
                  *PROCESS
                    EXTSUB1:   PROCEDURE;
                        .
                        .
                        .
                    END EXTSUB1;
```

The *PROCESS statement begins in column 1 (the * is in column 1).
There are other attributes that can be used on this statement, but we do not
cover them in this book. Please refer to your PL/I manual for the details of
these attributes.

10.2.4

Arguments and
Parameters
in Subprograms

Arguments and parameters in subprograms work the same as they do in
internal subroutines. All the rules stated in the previous sections still apply.
The important thing to realize when using external subprograms is that
arguments and parameters are the *only* means a subprogram has of accessing
values in a calling procedure. Global variables *only* work within the program
in which they are declared and in its internal subroutine procedures, not
from one program (MAIN or subprogram) to another.

10.2.5

Example Program—
External Subprograms

The CH10EX2 program (see Fig. 10–2) is a modification of CH6EX2.
Whereas CH6EX2 uses nested internal subroutine procedures, CH10EX2
uses external subprograms to solve the same problem. If you remember, the
program reads in student information and produces a report showing the
student's social security number, a course ID, the credit hours of the course,
and the grade the student received in the course. The grade point average
(GPA) of the student is printed after all the course information for the
student. The program also reports the number of students processed.
    You might want to look over the pseudocode for CH6EX2 before study-
ing CH10EX2.

10.3

Special Require-
ments when
Using Arguments
and Parameters

As you will probably discover, using arguments and parameters can be very
tricky if not carefully thought out. In this section, we present some problem
areas that we have discovered when using arguments and parameters. And
to help you avoid the same pitfalls, we have included the necessary code.

FIGURE 10-2
CH10EX2 (external subprograms).

```
PL/I OPTIMIZING COMPILER  VERSION 1 RELEASE 5.1  TIME: 15.38.49  DATE: 13 JAN 87  PAGE   1


OPTIONS SPECIFIED
GOSTMT;



PL/I OPTIMIZING COMPILER    /*   CH10EX2:  EXAMPLE OF EXTERNAL PROCEDURES */   PAGE   2
               SOURCE LISTING
     STMT


       /*   CH10EX2:  EXAMPLE OF EXTERNAL PROCEDURES */
    1  CH10EX2:  PROCEDURE OPTIONS (MAIN);
       /***********************************************************************/
       /*                                                                   */
       /*  FUNCTION: TO PRINT A REPORT OF STUDENTS COURSE GRADES AND GPA    */
       /*                                                                   */
       /*  INPUT:     ONE CARD FOR EACH STUDENT CONTAINING                  */
       /*                   STUDENT'S SOCIAL SECURITY NUMBER                */
       /*                   THE COURSE ID                                   */
       /*                   CREDIT HOURS OF THE COURSE                      */
       /*                   COURSE GRADE AS FOLLOWS                         */
       /*                       4 FOR AN A                                  */
       /*                       3 FOR A  B                                  */
       /*                       2 FOR A  C                                  */
       /*                       1 FOR A  D                                  */
       /*                       0 FOR AN F                                  */
       /*                                                                   */
       /*  OUTPUT:  A REPORT LISTING                                        */
       /*                   EACH STUDENTS SOCIAL SECURITY NUMBER            */
       /*                   THE COURSE ID FOR EACH COURSE                   */
       /*                   THE CREDIT HOURS FOR EACH COURSE               */
       /*                   THE GRADE AS LISTED ABOVE                       */
       /*               AFTER ALL THE COURSES FOR A STUDENT ARE PRINTED    */
       /*                   THE STUDENTS GPA IS PRINTED                     */
       /*                                                                   */
       /***********************************************************************/
    2  DECLARE
           TOTAL_STUDENTS_PROCESSED         FIXED DEC (9) INIT (0);
    3  DECLARE
           PROCSTU                    ENTRY,
```

```
           READSTU                    ENTRY:

    4   CALL PROCSTU(TOTAL_STUDENTS_PROCESSED);

    5   RETURN;
    6   END CH10EX2;
```

PL/I OPTIMIZING COMPILER   /*   CH10EX1: EXAMPLE OF EXTERNAL PROCEDURES */

NO MESSAGES PRODUCED FOR THIS COMPILATION
COMPILE TIME   0.00 MINS   SPILL FILE: 0 RECORDS, SIZE 4051

PL/I OPTIMIZING COMPILER VERSION 1 RELEASE 5.1 TIME: 15.38.51 DATE: 13 JAN 87 PAGE   4
OPTIONS SPECIFIED
GOSTMT;
*PROCESS;

PL/I OPTIMIZING COMPILER   /* CH10EX2: READSTU SUBROUTINE */   PAGE   5
                  SOURCE LISTING
    STMT

        /* CH10EX2: READSTU SUBROUTINE */
    1   READSTU: PROCEDURE (STUDENT_SSN,COURSE_ID,COURSE_CREDIT_HOURS,
                   COURSE_GRADE,END_FLAG);
        /***/
        /*   */
        /* FUNCTION: TO READ THE INFORMATION FOR A SINGLE STUDENT.   */
        /*   */
        /* PARAMETERS: STUDENT_SSN   */
        /*   COURSE_ID   */
        /*   COURSE_CREDIT_HOURS   */
        /*   COURSE_GRADE   */
        /*   END_FLAG   */
        /*   */
        /* SUBROUTINES CALLED: NONE   */
        /*   */
        /* INPUT: STUDENT_SSN, COURSE_ID, COURSE_CREDIT_HOURS, AND */
        /*   COURSE_GRADE (IN NUMBER AS 0 : F, 1 : D, 2 : C, */
        /*   3 : B, AND 4 : A)   */
```
```

FIGURE 10-2 *(continued)*

```
      /*                                                              */
      /*  OUTPUT:   NONE                                              */
      /*                                                              */
      /****************************************************************/
   2  DECLARE
          STUDENT_SSN                 CHAR (9),
          COURSE_ID                   CHAR (8),
          COURSE_CREDIT_HOURS         FIXED DEC (1),
          COURSE_GRADE                FIXED DEC (1);

   3  DECLARE
          END_FLAG                    BIT (1),
              MORE_RECS               BIT (1) INIT ('1'B),
              FINISHED                BIT (1) INIT ('0'B);

   4  ON ENDFILE (SYSIN) END_FLAG = FINISHED;

   5  GET EDIT (STUDENT_SSN, COURSE_ID, COURSE_CREDIT_HOURS, COURSE_GRADE)
            (COL(1),A(9),A(8),F(1),F(1));

   6  RETURN;
   7  END READSTU;


PL/I OPTIMIZING COMPILER        /* CH10EX2:  READSTU SUBROUTINE */          PAGE   6

COMPILER DIAGNOSTIC MESSAGES
ERROR ID L   STMT    MESSAGE DESCRIPTION

COMPILER INFORMATORY MESSAGES

IEL0533I I          NO 'DECLARE' STATEMENT(S) FOR 'SYSIN'.
IEL0430I I   1      NO 'MAIN' OPTION ON EXTERNAL PROCEDURE.

END OF COMPILER DIAGNOSTIC MESSAGES


PL/I OPTIMIZING COMPILER  VERSION 1 RELEASE 5.1  TIME: 15.38.58  DATE: 13 JAN 87  PAGE   7
OPTIONS SPECIFIED
GOSTMT;
*PROCESS;
```

    STMT

```
    /*  CH10EX2:  PROCSTU SUBROUTINE */
  1 PROCSTU:   PROCEDURE (TOTAL_STUDENTS_PROCESSED);
    /**********************************************************************/
    /*                                                                  */
    /*                      PROCESS_STUDENT                             */
    /*                                                                  */
    /*  FUNCTION: TO READ AND PROCESS ALL THE STUDENTS ON THE INPUT     */
    /*            FILE.  A REPORT IS PRODUCED SHOWING THE COURSES IN     */
    /*            WHICH THE STUDENT WAS ENROLLED ALONG WITH THE GRADE    */
    /*            AND CREDIT HOURS GIVEN.  THE PURPOSE IS TO CALCULATE   */
    /*            THE GPA FOR THE STUDENT AFTER PRINTING ALL THE         */
    /*            COURSES.                                               */
    /*                                                                  */
    /*  PARAMETERS:    TOTAL_STUDENTS_PROCESSED                          */
    /*                                                                  */
    /*  SUBROUTINES CALLED: COMPUTE_HONOR_POINTS                         */
    /*                      PRINT_STUDENT                                */
    /*                      READ_A_STUDENT                               */
    /*                                                                  */
    /*  INPUT:    EACH STUDENTS INFORMATION IS READ IN ONE COURSE AT     */
    /*            A TIME BY CALLING READ_A_STUDENT                       */
    /*                                                                  */
    /*  OUTPUT:   EACH STUDENT'S COURSE INFORMATION IS PRINTED           */
    /*                                                                  */
    /**********************************************************************/
  2 DECLARE
        TOTAL_STUDENTS_PROCESSED        FIXED DEC (9);

  3 DECLARE
        STUDENT_SSN                     CHAR (9),
        COURSE_ID                       CHAR (8),
        COURSE_CREDIT_HOURS             FIXED DEC (1),
        COURSE_GRADE                    FIXED DEC (1);

  4 DECLARE
        END_FLAG                        BIT (1) INIT('1'B),
            MORE_RECS                   BIT (1) INIT ('1'B),
            FINISHED                    BIT (1) INIT ('0'B);
```

FIGURE 10–2 *(continued)*

```
 5  DECLARE
        OLD_STUDENT_SSN                 CHAR (9),
        TOTAL_HONOR_POINTS              FIXED DEC (7),
        TOTAL_CREDIT_HOURS              FIXED DEC (3);
 6  DECLARE
        LINE_COUNT                      FIXED DEC (3) INIT (51);

 7  DECLARE
        READSTU                         ENTRY;
```

PL/I OPTIMIZING COMPILER   /* CH10EX2: PROCSTU SUBROUTINE */   PAGE   9
    STMT

```
 8  CALL READSTU(STUDENT_SSN,COURSE_ID,COURSE_CREDIT_HOURS,COURSE_GRADE,
            END_FLAG);

 9  DO WHILE (END_FLAG = MORE_RECS);

10      OLD_STUDENT_SSN = STUDENT_SSN;
11      TOTAL_HONOR_POINTS = 0;
12      TOTAL_CREDIT_HOURS = 0;

13      DO WHILE (OLD_STUDENT_SSN = STUDENT_SSN &
                END_FLAG = MORE_RECS);

14          CALL COMPUTE_HONOR_POINTS;
15          CALL PRINT_STUDENTS;
16          CALL READSTU(STUDENT_SSN,COURSE_ID,
                COURSE_CREDIT_HOURS,COURSE_GRADE,END_FLAG);
17      END;

18      CALL PRINT_GPA;
19  END;
20  RETURN;
```

PL/I OPTIMIZING COMPILER   /* CH10EX2: PROCSTU SUBROUTINE */   PAGE 10
    STMT

```
 21  PRINT_STUDENTS:   PROCEDURE;
     /*******************************************************************/
     /*                                                               */
     /*                    PRINT_STUDENTS                             */
     /*                                                               */
     /* FUNCTION: TO PRINT THE COURSE INFORMATION FOR A STUDENT       */
     /*                                                               */
     /* GLOBAL VARIABLES:    STUDENT_SSN                              */
     /*                      COURSE_ID                                */
     /*                      COURSE_CREDIT_HOURS                      */
     /*                      COURSE_GRADE                             */
     /*                                                               */
     /* SUBORUTINES CALLED: PRINT_HEADINGS                           */
     /*                                                               */
     /* INPUT:    NONE                                               */
     /*                                                               */
     /* OUTPUT:   THE STUDENT COURSE INFORMATION IS PRINTED          */
     /*                                                               */
     /*******************************************************************/
 22  IF LINE_COUNT > 50 THEN
        CALL PRINT_HEADINGS;

 23  PUT SKIP EDIT (STUDENT_SSN, COURSE_ID, COURSE_CREDIT_HOURS,
        COURSE_GRADE)
                (COL(5),A,COL(20),A,COL(32),F(1),COL(43),F(1));

 24  LINE_COUNT = LINE_COUNT + 1;
 25  RETURN;
```

```
 26  PRINT_HEADINGS:   PROCEDURE;
     /*******************************************************************/
     /*                                                               */
     /*                    PRINT_HEADINGS                             */
     /*                                                               */
     /* FUNCTION: TO PRINT THE HEADING LINES OF THE STUDENT GRADE     */
     /*           REPORT.                                             */
     /*                                                               */
     /* GLOBAL VARIABLES:    LINE_COUNT                              */
```

FIGURE 10-2 *(continued)*

```
     /*                                                            */
     /*  SUBROUTINES CALLED: NONE                                  */
     /*                                                            */
     /*  INPUT:     NONE                                           */
     /*                                                            */
     /*  OUTPUT:   THE FIRST THREE LINES OF THE GRADE REPORT.      */
     /*                                                            */
     /****************************************************************/
27   DECLARE
        PAGE_COUNT                    FIXED DEC (9) INIT (0) STATIC;

28   PAGE_COUNT = PAGE_COUNT + 1;

29   PUT PAGE EDIT ('STUDENT GRADE REPORT')
                 (COL(40),A);
30   PUT SKIP EDIT ('PAGE: ', PAGE_COUNT)
                 (COL(5),A,F(8));
31   PUT SKIP(2) EDIT ('STUDENT SSN', 'COURSE ID', 'CREDIT', 'GRADE')
                   (COL(5),A,COL(20),A,COL(30),A,COL(40),A);
32   PUT SKIP(0) EDIT ('_____', '_____')
                   (COL(5),A,COL(20),A);
33   PUT SKIP EDIT ('HOURS', 'RECEIVED')
                 (COL(30),A,COL(40),A);
34   PUT SKIP(0) EDIT ('_____', '_____')
                   (COL(30),A,COL(40),A);

35   LINE_COUNT = 6;

36   RETURN;
37   END PRINT_HEADINGS;
38   END PRINT_STUDENTS;


PL/I OPTIMIZING COMPILER        /* CH10EX2: PROCSTU SUBROUTINE */        PAGE  12
   STMT


39   PRINT_GPA:    PROCEDURE;
     /****************************************************************/
     /*                                                            */
     /*                      PRINT_GPA                             */
```

```
      /*                                                          */
      /*  FUNCTION: TO CALCULATE AND PRINT THE GPA FOR THE STUDENT.   */
      /*                                                          */
      /*  GLOBAL VARIABLES:    TOTAL_HONOR_POINTS                 */
      /*                       TOTAL_CREDIT_HOURS                 */
      /*                                                          */
      /*  SUBROUTINES CALLED: NONE                                */
      /*                                                          */
      /*  INPUT:    NONE                                          */
      /*                                                          */
      /*  OUTPUT:    THE GPA LINE FOR THE STUDENT.                */
      /*                                                          */
      /**********************************************************/
   40  DECLARE
          STUDENT_GPA                  FIXED DEC (4,3) INIT (0);

   41  STUDENT_GPA = TOTAL_HONOR_POINTS / TOTAL_CREDIT_HOURS;

   42  PUT SKIP (2) EDIT ('TOTAL CREDIT HOURS = ', TOTAL_CREDIT_HOURS)
                      (COL(5),A,COL(42),F(3));
   43  PUT SKIP EDIT ('TOTAL HONOR POINTS = ', TOTAL_HONOR_POINTS)
                  (COL(5),A,COL(40),F(5));
   44  PUT SKIP EDIT ('THE STUDENT''S GPA = ', STUDENT_GPA)
                  (COL(5),A,COL(40),F(5,3));

   45  LINE_COUNT = LINE_COUNT + 4;

   46  RETURN;
   47  END PRINT_GPA;
```

```
   48  COMPUTE_HONOR_POINTS:  PROCEDURE;
       /**********************************************************/
       /*                                                        */
       /*                    COMPUTE_HONOR_POINTS                 */
       /*                                                        */
       /*  FUNCTION: TO COMPUTE THE HONOR POINTS FOR EACH STUDENT */
```

FIGURE 10-2 *(continued)*

```
    /*                                                            */
    /*  GLOBAL VARIABLES:   TOTAL_STUDENTS_PROCESSED              */
    /*                      TOTAL_HONOR_POINTS                    */
    /*                      TOTAL_CREDIT_HOURS                    */
    /*                      COURSE_CREDIT_HOURS                   */
    /*                      COURSE_GRADE                          */
    /*                                                            */
    /*  SUBROUTINES CALLED: NONE                                  */
    /*                                                            */
    /*  INPUT:    NONE                                            */
    /*                                                            */
    /*  OUTPUT:   NONE                                            */
    /*                                                            */
    /****************************************************************/
49   DECLARE
         HONOR_POINTS                    FIXED DEC (5) INIT (0);

50   TOTAL_STUDENTS_PROCESSED = TOTAL_STUDENTS_PROCESSED + 1;
51   HONOR_POINTS = COURSE_CREDIT_HOURS * COURSE_GRADE;
52   TOTAL_HONOR_POINTS = TOTAL_HONOR_POINTS + HONOR_POINTS;
53   TOTAL_CREDIT_HOURS = TOTAL_CREDIT_HOURS + COURSE_CREDIT_HOURS;

54   RETURN;
55   END COMPUTE_HONOR_POINTS;
56  END PROCSTU;


PL/I OPTIMIZING COMPILER        /* CH10EX2:  PROCSTU SUBROUTINE */         PAGE  14


COMPILER DIAGNOSTIC MESSAGES
ERROR ID L  STMT    MESSAGE DESCRIPTION

COMPILER INFORMATORY MESSAGES

IEL0533I I          NO 'DECLARE' STATEMENT(S) FOR 'SYSPRINT'.
IEL0430I I   1      NO 'MAIN' OPTION ON EXTERNAL PROCEDURE.

END OF COMPILER DIAGNOSTIC MESSAGES
```

CHARACTER VARYING
Parameters

One problem that occurred during an actual implementation of a program was the mismatch of data types within character variables. As we discovered, if an argument is declared as CHARACTER VARYING, the corresponding parameter must also be declared CHARACTER VARYING. Now this might seem rather obvious if the rule that we stated above is followed. However, the problem occurred with the VARYING option. The CHARACTER VARYING adds a two-byte field to the beginning of the variable; these two bytes contain the length of the value in the variable. It was this two-byte length field that caused us problems.

The argument was declared as CHARACTER VARYING, and the parameter did not include the VARYING attribute. Thus, the extra two-byte field existed in the storage area. When the parameter tried to refer to the character portion of the field, the first two bytes were not character, but the length. The following example is the proper way to use a CHARACTER VARYING argument/parameter.

```
MAIN:  PROCEDURE OPTIONS (MAIN);
       DCL  NAME            CHAR (25) VAR;
                 .
                 .
                 .

       CALL PRNTNME (NAME);
                 .
                 .

END MAIN;

PRNTNME:  PROC (LAST_NAME);
          DCL  LAST_NAME     CHAR (*) VAR;
                 .
                 .
                 .

          PUT LIST (LAST_NAME);
                 .
                 .

END PRNTNME;
```

More on the
ENTRY Attribute

There might be times when the attributes and data types of the parameters are different than those of the arguments. In this case, you can use the declaration of the external subprogram name to specify to PL/I that it must convert the attributes of the arguments to match the attributes of the parameters before invoking the external subprogram.

```
DCL  EXTSUB1 ENTRY (FIXED DEC (5,3), FLOAT(4));
```

The general format of this type of declaration statement is

```
DCL subprogram ENTRY (parm1 attribute, parm2 attribute, ...,
                           parmn attribute);
```

When the subprogram is called, PL/I will convert the argument attributes into the proper attributes for the parameters as specified in the declaration of the subprogram. This is done at execution time, not at compile time.

10.3.3

Using BUILTIN Functions as Arguments

As we described above, you can use a built-in function name as an argument in a CALL statement. There are two things that can happen during the execution of such a CALL, depending upon whether or not the built-in function has arguments of its own. The rules are as follows:

1. If the built-in function has arguments, PL/I will call the built-in function before invoking the subprogram. The result of the built-in function will be passed as the argument value. For example,

```
CALL EXTSUB1 (SQRT(25));
```

will pass as the value in the argument the result of 5.

2. If the built-in function does not have any arguments, its name is passed as the argument value. Thus, if

```
CALL EXTSUB1 (DATE);
```

were executed, the argument value to EXTSUB1 would be 'DATE'. If you want the built-in function invoked prior to the call to the subprogram and the results passed as the argument, include an extra set of parentheses in the CALL statement.

```
CALL EXTSUB1 (DATE());
```

When this call is executed, the built-in function DATE will be invoked first and the result passed as an argument value to EXTSUB1.

10.4

Documenting External Subprograms

We recommend that you document the external subprograms exactly as you would any other program, with the exception that now you should include information about the parameters being received. The category of PARAMETERS should be added to the documentation as follows:

```
PROCSTUD: PROCEDURE (TOTAL_STUDENTS_PROCESSED);

/**************************************************************/
/*                                                          */
/*  FUNCTION: TO READ AND PROCESS ALL THE STUDENTS ON THE INPUT  */
/*            FILE. A REPORT IS PRODUCED SHOWING THE COURSES IN   */
/*            WHICH THE STUDENT WAS ENROLLED ALONG WITH THE GRADE */
/*            AND CREDIT HOURS GIVEN.  THE PURPOSE IS TO CALCULATE*/
/*            THE GPA FOR THE STUDENT AFTER PRINTING ALL THE      */
/*            COURSES.                                      */
/*                                                          */
/*  PARAMETERS:    TOTAL_STUDENTS_PROCESSED                 */
/*                                                          */
/*  SUBROUTINES CALLED: COMPUTE_HONOR_POINTS               */
/*                      PRINT_STUDENT                       */
/*                      READ_A_STUDENT                      */
/*                                                          */
/*  INPUT:    EACH STUDENT'S INFORMATION IS READ IN, ONE COURSE  */
/*            AT A TIME BY CALLING READ_A_STUDENT.          */
/*                                                          */
/*  OUTPUT:   EACH STUDENT'S COURSE INFORMATION IS PRINTED  */
/*                                                          */
/**************************************************************/
```

10.5

Recursive Procedures

In this section we introduce the topic of recursive procedures. We briefly explain recursion and how PL/I handles it. However, if you are not familiar with the technique of recursion, please refer to other sources to assist you in understanding the ideas behind recursion.

A **recursive procedure** is a procedure that invokes itself. An excellent example of the use of recursion is computing n factorial denoted in mathematics circles as n!. If you remember, n factorial can be written as follows:

$$n! = 1 \times 2 \times 3 \times \ldots \times n$$

The computing of n! could be done by a nonrecursive procedure as (let N be the n and FACN the answer n!):

```
COMPFAC:   PROCEDURE(N,FACN);
       DCL   N            FIXED DEC(3),
             FACN         FIXED DEC(8);
       DCL   I            FIXED DEC(3) INIT (1),
             J            FIXED DEC(3) INIT (1);
       DO WHILE (I <= N);
             J = J * I;
             I = I + 1;
       END;
       FACN = J;
       RETURN;
END COMPFAC;
```

We could also code the same function, computing n!, as a recursive procedure. Compare the following procedure with the nonrecursive one above.

```
COMPFAC:   PROC (N,FACN) RECURSIVE;
       DCL       N    FIXED DEC(3),
                 FACN FIXED DEC(8);
       DCL       J    FIXED DEC(3);
       J = N - 1;
       IF J = 1 THEN
         DO;
             FACN = N;
         END;
       ELSE
         DO;
             CALL COMPFAC (J,FACN);
             FACN = N * FACN;
         END;
    RETURN;
    END COMPFAC;
```

To understand how the recursive procedure works, we want you to remember that each time a procedure is invoked, storage is assigned for the variables declared. Thus the following picture illustrates the recursive procedure when n = 4. The first time COMPFAC is invoked:

N = 4, therefore J = 3 so COMPFAC (3,) is issued.

The second time COMPFAC is invoked:

N = 3, therefore J = 2 so COMPFAC (2,) is issued.

The third time COMPFAC is invoked:

N = 2, therefore J = 1 so FACN = N; RETURN is performed.

Notice that until this point none of the calls has returned to the invoker because J has been greater than 1. We have simply gone down levels into the program by invoking COMPFAC several times. No RETURN statement has been executed as of yet. Now, the RETURN statements will be executed starting at the lowest-level invocation of COMPFAC.

The first return is issued (from third invocation of COMPFAC), where J = 1 and N = 2;

FACN = N; N = 2; : FACN = 2

is returned. After the first return,

FACN = 2, FACN = N * FACN and N = 3 : FACN = 3 * 2

After the second return (one level up),

FACN = 6, FACN = N * FACN and N = 4 : FACN = 4 * 6

After the third return (to the original invoker), the answer of

4! = 24

is returned.

Notice that during all the calls FACN did not contain any value until N became 2 and J became 1. It was at this time that a value was first assigned to FACN, which was subsequently returned to the caller and multiplied.

10.6

Summary

Procedures invoking other procedures can make data available via arguments in an argument list in the CALL statement. The invoked procedure accesses the arguments by using parameters that are in a parameter list in the PROCEDURE statement. Below is an example of a CALL statement with arguments and a PROCEDURE statement using parameters.

```
CALL PROC1 (NAME, SOC_SEC_NUM, LOAN_AMOUNT);
PROC1:  PROCEDURE (CUST_NAME, SSN, LOAN_AMOUNT);
```

There is a one-to-one correspondence between the arguments in the argument list and the parameters in the parameter list. Above, CUST_NAME accesses the storage area for NAME in the calling procedure, SSN accesses SOC_SEC_NUM, and LOAN_AMOUNT accesses LOAN_AMOUNT. Notice that parameters are just different names assigned to the same storage as the corresponding argument. Parameters and the corresponding arguments can have different names, as shown above, because the relationship is established via the order in the lists, not the names.

Arguments and parameters can be of any data type allowed in PL/I. Even built-in functions can be used as arguments with the following rules observed:

1. If the built-in function has arguments, PL/I will call the built-in function before invoking the procedure. The result of the built-in function will be the parameter value received.

2. If the built-in function does not have any arguments, its name is passed as the argument value.

Parameters must be declared within a DECLARE statement in the called procedure. However, parameters *cannot* be initialized because they are just redefining the same storage as the corresponding argument, not setting up new storage. The data type of the parameters and arguments *must* be identical; otherwise, unpredictable results might occur when using parameters.

Subprograms are separate units of PL/I code that can be compiled separately from main procedures and stored as object modules on disks. The subprograms are then invoked by using the CALL statement from other programs within the system. Subprograms fit in naturally with the ideas of structured programming in that they are small independent modules that perform one function. (If coded properly, at least, the subprograms meet these criteria.)

Subprograms can contain internal subroutines. The name of a subprogram must follow the rules of MAIN program names: only seven characters in length and the first a letter. In order to compile a subprogram in the same source listing as a MAIN procedure, an *PROCESS card must immediately precede the subprogram source code.

The ENTRY attribute is used to show the compiler that the name of a procedure used in a CALL statement is an external subprogram. The compiler will then make no effort to recognize the subprogram name (the recognition comes later in the processing of the program). Also, the ENTRY attribute can be used to show the attributes of the parameters (i.e., FIXED

DEC(6), CHAR(20)). If used, PL/I will convert the attributes of the arguments to the attributes stated on the ENTRY.

A recursive procedure is a procedure that calls itself. In PL/I recursive procedures have the word RECURSIVE on the PROCEDURE statement following the parameter list. Recursive procedures are very valuable when processing repeats the same function over and over again. One example is computing the n factorial. Other examples will be shown in Chapter 15.

10.7

Exercises

1. What are the differences between internal subroutine procedures and external subprograms?

2. How do external subprograms fit into the structured programming methodology?

3. Describe the use of the *PROCESS statement.

4. Given the following PL/I statements, show the corresponding PROCEDURE and DECLARE statements for the subprograms:

```
DECLARE
        NAME                    CHAR (20) INIT (' '),
        BALANCE_DUE             FIXED DEC (7,2),
        STREET_ADDRESS          CHAR (25) VARYING,
        EXTSUB1                 ENTRY,
        EXTSUB2                 ENTRY;
```
   a. CALL EXTSUB1 (NAME, BALANCE_DUE);
   b. CALL EXTSUB2 (NAME, STREET_ADDRESS, BALANCE_DUE);

5. What value is the corresponding parameter given access to in the following CALL statements?
   a. CALL SUB1 (SQRT, DATE());
   b. CALL PROC1 (MAX(5,3), TIME);
   c. CALL EXTSUB1 (MIN, DATE, ABS);

6. What is the most important thing you should remember when using arguments and parameters?

7. Redo Exercise 19 in Chapter 6 using external subprograms.

8. Redo Exercise 11 in Chapter 7 using external subprograms.

9. Using external subprograms, write a PL/I program that will write a sales report showing the sales figures for all the salespeople at City Auto Sales. The input data is

1 – 5	Salesperson's ID
6 – 25	Salesperson's name
26 – 28	The number of units sold by the salesperson
29 – 37	Total amount of sales in dollars

The sales report should contain the following information: each salesperson's figures, along with a message beside the top and bottom salespersons, and a grand totals page showing the total number of units sold and the total sales in dollars. Also, the amount of the average sale should be printed.

C H A P T E R   1 1

ARRAYS

AND THE

ITERATIVE DO

One of the more common tasks in data processing is that of sorting data in some particular order. Suppose, for example, that as students enroll for a particular class, their names are entered into a data set. When the registration period has closed, the administration wants to provide the instructor with an alphabetized listing of the students enrolled. If we write a program in which we read each name consecutively into the same variable (e.g., NAME), only one name is available at any one time because as each new input record is read, it overwrites the previous contents of the variable; yet obviously we must have all the names in memory at the same time in order to make the comparisons and move the data around to produce an alphabetized list. At first glance, it might seem that the answer to the problem is to declare a separate variable for each student. Second thoughts make us aware, however, that to do so involves us in a very tedious process: Suppose the enrollment in the class is 40 students. We must declare 40 separate variables. Note also that we are not able to put our read statement in a loop because the name of the variable changes for each read statement. Clearly, this is not a practical solution. To solve this kind of problem, PL/I provides us with the array. An **array** is a series of logically consecutive memory locations, whose contents all have the same attributes; additionally, the entire set of locations can be referenced with a single symbolic name. Data items grouped together in such a manner that they can be referred to either individually or collectively are called **aggregates**, and the array is an example of an aggregate. The first part of this chapter is devoted to explaining how arrays are created and referenced; the mechanics of sorting are left to the last part of the chapter. Along the way we introduce the iterative DO, a convenient way of manipulating arrays, discuss some array built-in functions, and present two methods of searching an array for a given value.

11.1

One-Dimensional Arrays

One-dimensional arrays (which are often referred to as **tables**) can be thought of as having a single row of logically consecutive memory locations (each of which is called an **element**). Suppose we have an array of ten elements that have the attributes of FIXED DEC (2); assuming some arbitrary data in the array, conceptually it would look like this.

| 25 | 33 | 45 | 21 | 13 | 99 | 62 | 50 | 11 | 88 |

11.1.1

Declaring and Initializing Arrays

Before we present the syntax for declaring a one-dimensional array, it is necessary to introduce some terminology associated with arrays. Each element in an array is referred to by its position within the array, and its **bounds** are the integers representing the first element (the **lower bound**) and the last element (the **upper bound**). Bounds can be expressed as literals, variables, expressions, or asterisks (asterisks are covered in Section 11.2.4). The **extent** of the array is the total number of elements in the array.
The simplest syntax for declaring an array is

```
array name (upper bound) attributes;
```

and a specific example is

```
DCL MONTH_ARRAY (31) FIXED DECIMAL (2);
```

In this declaration we are telling the compiler to set aside memory for 31 two-digit, fixed-decimal numbers and that we will refer to the array with the array name MONTH_ARRAY. Because we do not specify a lower bound, the compiler assumes it to be 1. As you can see, in this declaration the upper bound and the extent have the same value (31). Finally, by coding a single bounds specification, we are saying that this is a one-dimensional array because there must be a bounds specification for each dimension.

In some instances it is more convenient for the programmer if the lower bound has a value other than 1, and we can accomplish this by declaring the bounds as (lower bound:upper bound). The colon makes it clear to the compiler that we are explicitly declaring the lower bound. The two integers separated by a colon still constitute a single bounds specification. Some examples of explicit lower-bound declarations are

```
DCL ARRAY_1 (1:20)    FIXED DEC (4)      /* EXTENT = 20 */
DCL ARRAY_2 (60:100) CHAR       (3);     /* EXTENT = 41 */
DCL ARRAY_3 (-5:+5)  FIXED BIN (15);     /* EXTENT = 11 */
DCL ARRAY_4 (-9:0)   FIXED DEC (6,2);    /* EXTENT = 10 */
DCL ARRAY_5 (-16:-8) CHAR       (5);     /* EXTENT = 9  */
```

The declaration of the lower bound in the first example is simply an explicit declaration of the default value; it serves only the purpose of documentation. The value of the extent is once again the same as that of the upper bound (20). The second example illustrates the lower bound having a positive value greater than 1. The extent of ARRAY_2 is 41. Notice that the formula

upper bound − lower bound + 1

will always give you the extent of the array.

The remaining examples illustrate a lower bound having a negative value in conjunction with the upper bounds having a positive value, zero value, and a negative value, respectively. The following examples illustrate the use of variables and expressions to specify bounds.

```
DCL ARRAY_6 (YEAR_1:YEAR_X) FIXED DEC (2);
DCL ARRAY_7 (N+K:Y-N)       FIXED BIN (15);
```

The following rules apply to the declaration of arrays:

1. The dimension specification, in parentheses, must immediately follow the array name.

2. The number of bounds specifications indicate the number of dimensions in the array. The maximum number of dimensions is 15; on some compilers it is less.

3. The upper bound must be greater than or equal to the lower bound.

4. If only the upper bound is specified, the lower bound defaults to 1.

5. The extent of the array is equal to the number of elements in the array; both upper- and lower-bound integers represent elements.

6. If both lower and upper bounds are specified, the upper must be separated from the lower by a colon (and optional blanks).

7. The minimum value for the lower bound is −32768, and the maximum for the value for the upper bound is +32767.

Arrays can also be given the INITIAL attribute. The initialization of arrays through the INITIAL attribute most often takes one of two forms: In some cases we want to initialize every element to the same value, whereas in other cases we want to initialize each element to a different value. When we want to initialize every element to the same value, we must distinguish between arithmetic and character arrays. Suppose, for example, we have an arithmetic array of 50 elements and we want to initialize each element to zero. If we code

```
DCL ARRAY_50 (50)   FIXED DECIMAL (3) INIT (0);
```

only the first element is given the value of 0. Instead, we must provide an initial value for each element we want to initialize, and we can do this by using an iteration factor:

```
DCL ARRAY_50 (50)   FIXED DECIMAL (3) INIT ((50) 0);
```

With a character array, on the other hand, both a repetition factor and an iteration factor can be involved. If only one factor is present, PL/I assumes it to be the repetition factor. Thus,

```
DCL ARRAY_50 (50)   CHARACTER   (3) INIT ((50) 'XXX');
```

would generate a single three-character string, repeat it 50 times, and assign the entire 150 X's to the first element (obviously only three X's would be stored, however). In order to place XXX in each of the 50 elements, we would have to code:

```
DCL ARRAY_50 (50)    CHARACTER      (3) INIT ((50) (1) 'XXX');
```

or

```
DCL ARRAY_50 (50)    CHARACTER (3) INIT ((50) ('XXX'));
```

This declaration will generate XXX (from the repetition factor of 1 and the initial value); it will then iterate the character string 50 times, placing one iteration in each element in the array.

In the second case, in which we want to initialize each element of the array to a different value, we must supply those values in *positional* order. Thus,

```
DCL NUMS (10) FIXED DECIMAL (3) INIT (00,11,22,33,44,55,66,
                                      77,88,99);
```

places 00 in the first element, 11 in the second, and so on through the tenth element, which contains 99. To initialize each element of a character array, code each initial value as a separate character literal.

```
DCL SIX_MO (6) CHAR (3)
              INIT ('JAN','FEB','MAR','APR','MAY','JUN');
```

If the number of initial values supplied is greater than the extent of the array, the remaining values are ignored. If the number of initial values supplied is fewer than the extent, the remaining elements are not initialized.

Many variations of these two basic patterns are possible, and we illustrate one more. If you wish to initialize some of the elements to individual values but leave others uninitialized, you can include an asterisk in the INITIAL value list in the position corresponding to the element you want uninitialized.

```
DCL NAMES (3) CHAR (6) INIT ('JONES',*,'SMITH');
```

results in the following:

JONES		SMITH

Referencing
Arrays

You will recall that a part of the definition of an array is that the entire set of memory locations is given a single name by which to reference the array. If we want to reference every element in the array in order from lower bound to upper bound, we need use only the array name. Much more commonly, however, it is necessary to reference a single element. To reference a single element, we must use a **subscript**, which is an integer that identifies a particular element in the array. The set of valid subscript values is the set of integers starting with the lower bound integer and proceeding through the upper bound integer. The subscript is coded in parentheses with nothing intervening between it and the array name except blanks. (Array names, in fact, are often called subscripted variables.) If an array is declared, for example, as

```
DCL WRK_DAYS (1:5) CHAR (3);
```

the valid references to the elements are WRK_DAYS (1), WRK_DAYS (2), WRK_DAYS (3), WRK-DAYS (4), WRK_DAYS (5). For further clarification, we show several other bounds specifications and the valid references to their elements.

A (5:10)	B (−5:0)	C (−3:+2)
A (5)	B (−5)	C (−3)
A (6)	B (−4)	C (−2)
A (7)	B (−3)	C (−1)
A (8)	B (−2)	C (0)
A (9)	B (−1)	C (+1)
A (10)	B (0)	C (+2)

You must remember, however, that if the subscript has a value that is not within the bounds of the array, execution will end, usually with an error message that gives no clue as to the real source of the problem. In most situations you can make the error easier to locate by using the SUBSCRIPTRANGE (abbreviated SUBRG) condition while you are testing your program. SUBSCRIPTRANGE is disabled by default; therefore, you must use a condition prefix to enable it. It should be removed from the program once testing has shown the program to be reliable.

If the subscript value always had to be expressed as a literal, most of the flexibility promised by the array would be lost, but such is not the case. A subscript, in fact, can be expressed as a variable or an expression with arithmetic operators, in addition to a literal. When a variable is used, the value in the variable (which must be an integer) identifies the element to be referenced. An expression with arithmetic operators must evaluate to an integer; the expression is evaluated, and the resulting integer identifies the array element. It is when you are using a variable or arithmetic expression that you must be particularly careful that its value does not fall out of bounds. The following examples illustrate the use of variables and arithmetic expressions as subscripts:

```
ARRAY (SUB) = 1900;
IF ARRAY (SUB) < ARRAY (SUB + 1) THEN
    DO;
```

11.2
Using
Arrays

We have suggested the solution to two of the problems associated with sorting data that we raised in the introduction to this chapter: We have provided a means of storing all the data in memory at once, and, at the same time, we have avoided the problem of having to code large numbers of declarations. The third problem we raised was that of how to reference the data from within a loop. Because all the elements in the array are referenced with the same symbolic name, it is now possible to reference each from within a loop. Suppose we have several salespeople, and the input consists of the sales of each person for each day of the week. At the end of the week, we want to compute the total sales of each salesperson for a week. We must further assume that the input lists the sales figures for the salespeople in the same order for each day of the week. Given this situation, then, we can code

```
SUB = 1;
DO WHILE (MORE_INPUT);
    GET LIST (SALES_FIG);
    S_ARRAY (SUB) = S_ARRAY (SUB) + SALES_FIG;
    SUB = SUB + 1;
END;
```

This block of code works to accumulate the weekly sales total, but there is a considerably more convenient way to accomplish the same task, and we now turn our attention to it.

11.2.1
The Iterative
DO

We have been creating loops using the DO WHILE statement, and it is sufficient in many cases. However, another form of the DO, the iterative DO, is preferable in many situations, particularly in manipulating arrays. The syntax of the iterative DO is

```
DO control variable = initial value TO limit value BY
modification value;
```

To facilitate the discussion we present a specific example:

```
DO CNTRL = 1 TO 10 BY 1;
```

Before the loop is entered, the control variable (here CNTRL) is set to the initial value (here 1) and a comparison is made with the limit value (here 10). If CNTRL were to be larger than 10, the loop would not be entered at all. Since in this case CNTRL is less than or equal to the limit value, the statements in the iterative DO group execute. At the bottom of the loop, the control variable (CNTRL) is modified by adding the modification value (here 1) to it. In our specific example, CNTRL now has a value of 2. At this point control returns to the DO statement, and again the control variable is compared to the limit value. This process continues until the control variable is greater than the limit value (here 11); then when the comparison is made between the two, the DO statement evaluates to false, and control passes to the statement following the END statement which delimits the iterative DO group. Note that at this point the control variable has a value that is one greater than the limit value.

When we spoke a moment ago of the increased efficiency of the iterative DO over the DO WHILE, we were thinking primarily of coding efficiency. When using the DO WHILE, the programmer is responsible for initializing the control variable before the loop is entered and for modifying it somewhere within the loop. In the iterative DO, on the other hand, both of these actions are automatic; that is, both the initializing and the modification are built into the statement itself. Execution efficiency can be enhanced by always giving the control variable FIXED BINARY attributes.

There are a large number of variations to the basic iterative DO we have presented:

1. The initial value, limit value, and modification value can each be a numeric literal, as we have illustrated, a variable, or an arithmetic expression.

```
DO CNTRL = LOW TO HIGH BY MOD_V;
DO CNTRL = LOW - START TO HIGH + STOP BY MOD_V / #_TS;
```

2. The modification value defaults to +1; therefore, if you want the control variable modified by adding 1 to it, you can code either

```
DO CNTRL = 1 TO HIGH BY 1;
```

or

```
DO CNTRL = 1 TO HIGH;
```

Either form produces the same results.

3. The initial and/or modification values need not be a +1, and the limit value need not be a positive integer. Other possibilities include

Positive numbers:

```
DO CNTRL = 1 TO 22 BY 5;
```

In this case CNTRL takes on the successive values of 1, 6 (i.e., 1 + 5), 11, 16, 21, 26. When it reaches 26, of course, the DO group is not executed, but CNTRL has that as its final value.

Negative numbers:

```
DO CNTRL = 10 TO 1 BY -1;
```

Here CNTRL is initialized to 10 and modified by adding −1 each time through the loop. The statements in the loop execute if the control variable is *greater than* or equal to the limit value. The limit value can also be a negative number.

```
DO CNTRL = -1 TO -10 BY -2;
```

CNTRL will take on the successive values of −1, −3, −5, −7, −9, and finally −11. When CNTRL has the value of −11, the loop, however, will be bypassed.

Fractions:

```
DO CNTRL = .50 TO 2.50 BY .25;
    DO CNTRL = -.25 TO -5 BY .5;
```

In this case CNTRL will assume the values of −.25, −.75, −1.25, . . . , −5.

Multiple specifications:

```
DO CNTRL = 1 TO 5, 20 TO 40
         BY 10, 55 TO 65;
```

This loop executes 19 times: 5 times with the values of 1 through 5; 3 times with values of 20, 30, and 40; and 11 times with the values of 55 through 65. CNTRL's final value is 66.

Random values:

```
DO CNTRL = 12, 7, 36, -8, 21;
```

The loop executes five times, and CNTRL contains in turn each value listed. Its final value is 21.

In addition to the modifications just presented, you need to be aware of three other iterative DO possibilities. The first of these is adding a WHILE (condition). Thus

```
DO control variable = initial value TO limit value BY
     modification value WHILE (condition);
```

or, for example,

```
DO CNTRL = 1 TO 10 WHILE (VAR_1 <= VAR_2);
```

The test of CNTRL is first made as usual; if it evaluates to true, the WHILE condition is tested. If it, too, evaluates to true, the loop is executed. In other words, both tests must evaluate to true for the loop to execute; as soon as either becomes false, control drops through the loop. In the example above the loop executes ten times *providing VAR 1 continues to be less than or equal to VAR 2*.

The second possibility is to attach an UNTIL (condition) to the iterative DO.

```
DO CNTRL = J TO K BY Q UNTIL (LOW = HIGH);
```

In this case CNTRL is tested as usual; if it evaluates to true, the statements in the loop are executed. At the bottom of the loop the UNTIL condition is tested; if it evaluates to false, control passes again to the DO statement, where once again the control variable is tested. If the control variable evaluates to true in its first comparison with the limit value, the loop is executed once regardless of whether the UNTIL condition is true or false. You might recall from our earlier presentation of DO UNTIL that we consider the UNTIL, with its test at the bottom of the loop, to be unstructured.

Finally, the iterative DO can be nested. If you have a good understanding of nested DO WHILE statements, nested iterative DO statements should pose no problems for you. There are two aspects of the execution of nested iterative DO statements you must keep in mind. First, the lower level DO runs through its entire range (i.e., from initial to final value) for *each value* in the control variable of the higher level DO. A generalized example will help make this clear.

```
HLVL_LP:   DO OUTER = 1 TO 3;
LLVL_LP:      DO INNER= 1 TO 5;
                .
                .
                .
           END LLVL_LP;
                .
                .
                .
           END HLVL_LP;
```

When the HLVL_LP DO is first encountered, OUTER, of course, has a value of 1. It retains that value while INNER takes on the successive values of 1–5. When INNER becomes a 6, control passes to the HLVL_LP, OUTER takes on the value of 2, and again INNER acquires the values of 1 to 6. This process continues until OUTER has the value of 4, at which time control passes to the statement following END HLVL_LP;.

The second aspect of nested iterative DO statements to remember is that the range of a lower level DO cannot extend past the range of the next higher level DO. The following pairs of brackets show the permissible patterns.

The PL/I syntactical rule that an END statement corresponds to the nearest previous DO that has no corresponding END effectively prevents the range of the lower-level loop from extending past the range of the higher-level loop, but you will be in for an unpleasant surprise if you code a program in which you expect that to occur.

11.2.2

I/O of Arrays

You have seen in the discussion above that the iterative DO provides for coding efficiency by eliminating the need for the programmer to initialize and increment the subscript. It is in the I/O operation, however, that its efficiency becomes most apparent. As we noted earlier, if you want to reference every element in the array, you need code only the array name (i.e., unsubscripted). Thus, to fill an array with input values, you can code simply

```
GET LIST (array name);
```

The same can be done with any stream input. Assuming an array of 50 elements declared as CHAR (3), you code (for edit-directed input)

```
GET EDIT (array name) (A(3));
```

(Remember: The single format item is used repeatedly until the data list is exhausted.) Considerable care must be taken in using this statement for placing values in an array because if there are not enough input values to fill every element in the array, an error condition exists. By the same token, referencing the unsubscripted array name in an output statement means that the contents of every element are printed, even if no values have been placed in some elements. Referencing the unsubscripted array name is fine if every element in the array is to be used. It is more commonly the case, however, that the array is given an extent equal to the greatest number of values possible or expected to be placed in it. For example, a teacher declaring an array with an element for each week of the semester knows that 16 is the maximum possible number of weeks. He or she might want to run his or her program every four weeks, however, and for the first three runs no data is available for some array elements. The same teacher might declare an array to hold a value for each student. Since the teacher knows that the policy is to close the class at 50 students, he or she declares 50 elements in the array. For any given semester, however, the actual enrollment might range from 40 to 50 students; once again there is insufficient data to fill every element of the array.

To solve the problem of fewer data items than elements, it is a common practice to place the actual number of input data items in an input header record; that value is read into a variable, and the variable, in turn, becomes the limit value. An alternative is to perform the iterative DO WHILE (not end-of-file) and count the records. In addition, PL/I allows us to place an iterative DO in an I/O statement:

```
GET LIST ((variable (subscript) DO subscript = initial value
        TO limit value BY modification value));
PUT EDIT ((variable (subscript) DO subscript = initial value
        TO limit value)) (format list);
```

This inclusion of the iterative DO in an I/O statement is called **repetitive specification**. Perhaps the first thing to notice about the repetitive specification is that the DO has no corresponding END statement. PL/I knows that the loop is to end when the I/O operation is complete. In other words, no other PL/I statement can be a part of the iterative DO when it is used for repetitive specification. The second thing to notice is the parentheses. Each repetitive specification must be enclosed in its own set of parentheses. In the example above, we have only a single repetitive specification, but it is considered to be a part of the data list, and, of course, the data list is enclosed in parentheses, hence the two sets. Finally, the two examples above illustrate again that repetitive specification can be used with either list-directed or edit-directed I/O. In fact, it can be used with any stream I/O. Some specific examples of repetitive specification are

```
GET LIST ((WEEK_VAL (SUB) DO SUB = 1 TO #_WKS));
PUT LIST ((STUDENT_VAL (SUB) DO SUB = 1 TO #_STUDENTS));
```

Some PL/I programmers, however, feel that the use of the repetitive specification results in code that is difficult to read; they would prefer that the two examples above be

```
DO SUB = 1 TO #_WKS;
    GET LIST (WEEK_VAL (SUB));
END;
DO SUB = 1 TO #_STUDENTS;
    PUT LIST (STUDENT_VAL (SUB));
END;
```

There are occasions, on the other hand, in which the format of the output demands the repetitive specification. Supposing you want to print five numbers per line from an array and you want to start in column 46. If you code:

```
DO SUB = 1 TO 5;
    PUT EDIT (ARRAY (SUB)) (COL (46), (5)F(8));
END;
```

the COL (46) will be executed each time. Since the printer is past column 46 the second time through, it will force a new line and print the second number in column 46 of that line; thus, the desired format cannot be achieved. The only solution is to code instead

```
PUT EDIT ((ARRAY (SUB)) DO SUB = 1 TO 5)) (COL (46), (5)F(8));
```

With this format, the COL (46) will execute only once, and the five numbers will print on the same line. You should consider, then, reserving the repetitive specification for those situations that demand it for the desired formatting.

It is also possible to fill or print more than one array with a single input or output statement. Continuing the example above, we might code:

```
#_WK = 12;
#_STUDENTS = 45;
GET LIST ((WEEK_VAL (SUB) DO SUB = 1 TO #_WKS) (STUDENT_VAL
          (SUB) DO SUB = 1 TO #_STUDENTS));
```

When this statement executes, it first puts 12 input values in the first 12 elements of WEEK_VAL, then puts values in the first 45 elements of STUDENT_VAL. It is the programmer's responsibility, of course, to make sure the items are read in an order consistent with the order in which the input data is arranged.

11.2.3

Assignment Statement Revisited for Arrays

Two types of assignment operations can be specified for arrays: scalar-to-array and array-to-array. (A **scalar** is a single data item.) The scalar-to-array assignment operation is used when you want to assign the same value to each element in the array, and it has the syntax of

```
array name = value;
```

where value can be a variable, a literal, or an arithmetic expression. One common use of scalar-to-array assignment is to set all of the elements to zero so that, for example, each element can act as an accumulator. For example,

```
STUDENT_VAL = 0;
```

sets each element in STUDENT_VAL to zero. If you wish to set only one element in the array, you must specify the desired element with a subscript:

```
STUDENT_VAL (SUB) = 0;
```

In the array-to-array assignment statement, all arrays involved must have identical bounds (which means, of course, that they also have identical extents). The effect of the array-to-array assignment is to assign to each element in the array named on the left side of the assignment symbol the value of the corresponding element of the array named on the right side. Thus, if ARRAY_1 contains

3	42	−9	−12	−16	−20	22	−26

then

```
ARRAY_2 = ARRAY_1;
```

causes ARRAY_2 to contain the same values as ARRAY_1. If an array expression appears on the right side of the assignment symbol, the expression is evaluated for each element, and the result is placed in the corresponding element of the array named on the left. The array on the right remains unchanged. Thus,

```
ARRAY_1 = −ARRAY_1;
```

causes the array to change to

−3	−42	9	12	16	20	−22	26

If ARRAY_1 contains

then when

```
ARRAY_2 = ARRAY_1 / 2;
```

executes, ARRAY_2 appears as

1	2	3	4	5	6	7	8

11.2.4

Arrays as Parameters

Before discussing arrays as parameters, it might be well to review the relationship between arguments and parameters. You will recall from Chapter 10 that the relationship between the two lists is positional and that an argument and its corresponding parameter refer to exactly the same area in memory. This second relationship is what (generally) requires us to declare the parameter with the same attributes as its corresponding argument. The argument list can contain a reference to an array element (e.g., GRADES (7)) or a reference to the entire array; in either case, the corresponding parameter for the array must be *unsubscripted,* and this creates a potential problem. The array must be declared with the same attributes in the sub-routine as was declared in the invoking procedure. If the array extent is known to the subroutine and if the array is always full, there is no problem. In the invoking procedure we code, for example,

```
DCL GRADES (50) FIXED DEC (3);
```

and (if we are using PL/I in the subroutine)

```
DCL SR_GRADES (50) FIXED DEC (3);
```

As long as we are dealing with internal subroutines, this creates only one minor problem: If it becomes necessary to change the extent of the array, we have to modify the code in two places. Thus, maintainability is reduced, however slightly. We are dealing here, of course, with internal subroutines, but in the "real world" of computing such procedures are often external subroutines. External subroutines are not a part of the procedure that calls them, and they can be called by any number of main procedures. The various main procedures, moreover, can each require an array of a different extent; therefore, the subroutine has no way of knowing what extent its declared array should be. PL/I provides such an easy means of handling the extent declaration in the subroutine that it has become standard practice to use it in internal subroutines also. The extent is declared by using an asterisk:

```
DCL SR_ARRAY (*) FIXED DEC (3);
```

The statement tells PL/I that the actual extent of the array has been declared elsewhere, and, of course, PL/I can locate that information. If you are using PL/C, you will discover that you *must* declare the array with an asterisk in the subroutine. If you are using PL/I, you can choose between the asterisk and the same integer used in the main procedure.

Typically, the argument list contains a second value in addition to the array name, and it contains the number of elements in the array which contain significant values. Suppose, for example, the main procedure contains

```
DCL GRADES (50) FIXED DEC (3);
DCL #_GRDS       FIXED BIN (15);
```

Still in the main procedure, the grade for each student is read into an element of the array, and each time this occurs #_GRDS is incremented by one. The main procedure is then prepared to issue the following call to the subroutine:

```
CALL SUB_RTN (GRADES, #_GRDS);
```

If it is the function of the subroutine to determine the average of all the grades, the following statements give an accurate result. (As noted above, SR_GRADES is the parameter corresponding to the argument GRADES.)

```
SUB_RTN: PROCEDURE (SR_GRADES, #_GRADES);
    DCL   SR_GRADES(*) FIXED DEC (3);
    DCL   #_GRADES        FIXED BIN (15);
FND_AVG:    DO NDX = 1 TO #_GRDS;
                GRADE_TOTAL = GRADE_TOTAL + SR_GRADES (NDX);
            END FND_AVG;
            GRADE_AVG = GRADE_TOTAL / #_GRDS;
```

11.2.5

Example Program—
One-Dimensional Arrays

The function of the CH11EX1 program (see Fig. 11–1) is to prepare a status report on the students in any one of several classes at the end of the semester. The classes in question give grades for programming projects and for tests, and for each student we want to compute the average programming grade, the average test grade, and the final average computed in such a manner that the programming grades will count 40 percent and the test grades 60 percent, regardless of the maximum number of points for each category. This seems like a straightforward and rather simple program until we consider some complicating factors.

1. Because we will use the program in several different classes during the same semester and also over a number of semesters, we cannot hard-code the class number and semester for the page header. Also, we cannot predetermine and hard-code the number of programs and the number of tests for a given run. In addition, we want to be able to run the program in any class at any point during the semester to determine the students' standing at that point. These are common enough problems, and we solve them in commonplace fashion. The first input header record gives the class number and semester, and the second gives the number of program grades and the number of test grades. This information allows us to process the program grades separate from the test grades.

2. We have no way of knowing in advance what the maximum number of points for each programming assignment and test will be; again, we solve this problem by using a third input header record that lists the maximum possible points for each grade. This information allows us to compute the various averages because it allows us to know the total number of points possible for programs, for tests, and for the two combined.

3. The third complicating factor is that we must continue to carry on the roll those students who have officially dropped from the class. We could solve this problem by assigning those students an arbitrary grade that no enrolled student can earn (e.g., −99); instead, we decided to give them no grade. Therefore, following the header records, there are two records for those students still enrolled in the class. The first record contains a status code for the student, the student's social security number, and name. The second record contains the program scores followed by the test scores. For those students with a status code of 1 (withdrew passing) or 2 (withdrew failing), there is only the first of the two records described above. The logic of the program, therefore, must be such that there is no attempt to read a grade record for those students who have withdrawn.

List input is used to make data entry an easier process; we don't have to worry about getting the grades in the wrong columns.

Turning now to the program itself, notice, first of all, that MAIN_RTN calls all the subprograms. The first subprogram (INHDR) processes the header records, reading the maximum possible points into the array PTS_ PSBL. Statements 20–24 then compute the total maximum points each for programs and tests. This breakdown is required because the formula for determining the final average for each student is (program average * 4 + test average * 6) / 10.0. This formula, in turn, frees the user from the necessity of assuring a 40:60 ratio of total program points to test points.

On return from INHDR the main program calls the subprogram named PGHDR. This procedure simply prints the class, section, and semester; it then prints the maximum possible points for each program and test, and the maximum total points for each. Because all of this information is printed at the top of each page of output, the procedure sets a line counter to 1 before returning to the main procedure. The main procedure then reads the first status-SSN-name student record, and goes into a loop and calls AVGSRTN and PRNTRTN, which process student records. It then reads another status-SSN-name student record.

In AVGSRTN we process the student record based on the status code. If the student has withdrawn from the course, we immediately return to the main routine, which calls a subroutine, PRNTRTN, to print the student's name, social security number, and a message indicating his or her status. If the student is still enrolled, however, we read a grade record and compute the total program points earned and the total test points earned. We then compute the average for the student, using the formula given above, and again call the print subroutine.

Since we want to print the student's name and SSN in any case, that is the first thing we do in PRNTRTN. The subroutine then makes another check of the status code; if it is not 0, the subroutine prints, as we noted above, only the name, social security number, and status message. Otherwise, the final average, program average, and test average are printed, followed by the total earned program points, total earned test points, and the score on each program and test.

To accomplish all of this, two one-dimensional arrays are used (PTS_ PSBL and GRADE). We have allowed for a total of 25 grades. The pseudo-code for the MAIN and AVGSRTN routines are as follows:

```
CH11EX1
    Call INHDR
    Call PGHDR
    Read a status-SSN-name record
    DO WHILE (more input)
        Call AVGSRTN
        IF (line counter > 40
        Call PGHDR
```

```
                              ENDIF
                              Call PRNTRTN
                      STOP
                      AVGSRTN
                          Total program points earned, total test points earned
                                  ← 0
                          IF (student still enrolled)
                              FOR (subscript = 1 to number of grades by 1)
                                  Read grades array (subscript)
                              ENDFOR
                              FOR (subscript = 1 to number of programs by 1)
                                  Total program points earned ← total program
                                          points earned + grades array (subscript)
                              ENDFOR
                              FOR (subscript = number of programs + 1 to number
                                          of grades by 1)
                                  Total test points earned ← total test points
                                          earned + grades array (subscript)
                              ENDFOR
                              Program average ← total program points earned /
                                      total program points
                              Test average ← total test points earned / total
                                      test points
                              Final average ← (program average * 4 + test
                                      average * 6) / 10
                          ENDIF
                      EXIT
```

PL/I OPTIMIZING COMPILER VERSION 1 RELEASE 5.1 TIME: 13.12.43 DATE: 10 MAR 87 PAGE   1
OPTIONS SPECIFIED
GOSTMT;

PL/I OPTIMIZING COMPILER   /* CH11EX1 USING ONE-DIMENSIONAL ARRAYS */   PAGE   2
                  SOURCE LISTING
    STMT

        /* CH11EX1 USING ONE-DIMENSIONAL ARRAYS */
    1 CH11EX1:   PROCEDURE OPTIONS (MAIN);

FIGURE 11–1
CH11EX1 (one-dimensional arrays).

```
/**********************************************************************/
/*                                                                  */
/*    FUNCTION:   TO COMPUTE STUDENTS' FINAL AVERAGES                */
/*                                                                  */
/*    INPUT:    1.  THREE HEADER CARDS                               */
/*                  A.  'CLASS # (E.G. '465'), 'SECTION #' ('X'),   */
/*                      'SEMESTER' ('SPRING 19XX')                  */
/*                  B.  (1) NUMBER OF PROGRAM GRADES AND (2)        */
/*                      NUMBER OF TEST GRADES                       */
/*                  C.  MAXIMUM NUMBER OF POINTS FOR EACH PROGRAM   */
/*                      AND FOR EACH TEST                           */
/*              3.  TWO CARDS FOR EACH STUDENT                      */
/*                  A.  THE FIRST CONTAINING A STATUS,             */
/*                      0 = COMPLETED THE COURSE                    */
/*                      1 = WITHDREW PASSING                        */
/*                      2 = WITHDREW FAILING                        */
/*                  STUDENT SSN                                     */
/*                  STUDENT NAME                                    */
/*                  B.  THE SECOND CONTAINING THE STUDENT GRADES    */
/*                                                                  */
/*    OUTPUT:   A REPORT LISTING                                     */
/*                  A.  THE MAXIMUM POSSIBLE POINTS FOR EACH        */
/*                      PROGRAM AND TEST                            */
/*                  B.  THE MAXIMUM TOTAL POINTS FOR PROGRAMS AND   */
/*                      FOR TESTS                                   */
/*                  C.  FOR EACH STUDENT                            */
/*                      1.  NAME, SSN, AND                          */
/*                          IF (COURSE NOT COMPLETED) STATUS        */
/*                          ELSE                                    */
/*                          FINAL AVERAGE, PROGRAM AVERAGE, AND     */
/*                           TEST AVERAGE                           */
/*                      2.  THE TOTAL POINTS EARNED FOR PROGRAMS    */
/*                          AND THE POINTS EARNED FOR EACH PROGRAM  */
/*                      3.  THE TOTAL POINTS EARNED FOR TESTS AND   */
/*                          THE TOTAL TEST POINTS                   */
/*                                                                  */
/*** NOTES:     THE FINAL AVERAGE IS DETERMINED BY (PGM_AVG * 4     */
/*              TST_AVG * 6) / 10.0 (PROGRAMS = 40%, TESTS = 60%)   */
/*                                                                  */
/**********************************************************************/
2   DCL CLASS          CHARACTER    (11);
3   DCL SMSTR          CHARACTER    (11);
```

FIGURE 11–1 *(continued)*

```
        4       DCL SECTION          CHARACTER    (2);
        5       DCL NAME             CHARACTER    (25);
        6       DCL SSN              CHARACTER    (9);
        7       DCL STATUS           FIXED DECIMAL (1);


PL/I OPTIMIZING COMPILER        /* CH11EX1 USING ONE-DIMENSIONAL ARRAYS */      PAGE   3
    STMT
        8       DCL #GRADES          FIXED DECIMAL (3);
        9       DCL #PGMS            FIXED DECIMAL (3);
       10       DCL #TSTS            FIXED DECIMAL (3);

       11       DCL GRADE (25)       FIXED DECIMAL (3);
       12       DCL PTS_PSBL (25)    FIXED DECIMAL (5)   INIT (0);

       13       DCL TTL_PGM_PTS      FIXED DECIMAL (5);
       14       DCL TTL_TST_PTS      FIXED DECIMAL (5);
       15       DCL PGM_PT_TTL       FIXED DECIMAL (5);
       16       DCL TST_PT_TTL       FIXED DECIMAL (5);
       17       DCL PGM_AVG          FIXED DECIMAL (7,2);
       18       DCL TST_AVG          FIXED DECIMAL (7,2);
       19       DCL FINAL_AVG        FIXED DECIMAL (7,2);

       20       DCL SUB              FIXED BINARY  (15);
       21       DCL NDX              FIXED BINARY  (15);
       22       DCL LINE_CNT         FIXED BINARY  (15);

       23       DCL MORE_RECS        BIT           (1);
       24       DCL    YES           BIT           (1)  INIT ('1'B);
       25       DCL    NO            BIT           (1)  INIT ('0'B);
       26       DCL    INHDR         ENTRY;
       27       DCL    PGHDR         ENTRY;
       28       DCL    AVGSRTN       ENTRY;
       29       DCL    PRNTRTN       ENTRY;

       30  MAIN_RTN: MORE_RECS = YES;
       31           ON ENDFILE (SYSIN) MORE_RECS = NO;
       32           CALL INHDR (CLASS, SMSTR, SECTION, #GRADES, #PGMS,
                                #TSTS, PTS_PSBL, TTL_PGM_PTS,
                                TTL_TST_PTS);
```

```
33              CALL PGHDR (CLASS, SMSTR, SECTION, #GRADES, PTS_PSBL,
                        TTL_PGM_PTS, TTL_TST_PTS, LINE_CNT);


34              GET LIST (STATUS, SSN, NAME);

35  PROC_LP:  DO WHILE (MORE_RECS);
36                  CALL AVGSRTN (STATUS, GRADE, #GRADES, #PGMS,
                            TTL_PGM_PTS, TTL_TST_PTS,
                            FINAL_AVG, PGM_AVG, TST_AVG,
                            PGM_PT_TTL, TST_PT_TTL);

37                  IF LINE_CNT > 40 THEN
                        CALL PGHDR (CLASS, SMSTR, SECTION, #GRADES,
                                PTS_PSBL, TTL_PGM_PTS, TTL_TST_PTS);
38                  CALL PRNTRTN (NAME, GRADE, SSN, STATUS,FINAL_AVG,
                            PGM_AVG, TST_PT_TTL, TST_AVG,
                            PGM_PT_TTL, #PGMS, #GRADES, LINE_CNT);
39                  GET LIST (STATUS, SSN, NAME);
40              END PROC_LP;
41      RETURN;
```

```
PL/I OPTIMIZING COMPILER        /* CH11EX1 USING ONE-DIMENSIONAL ARRAYS */    PAGE   4
   STMT
    42  END CH11EX1;


PL/I OPTIMIZING COMPILER        /* CH11EX1 USING ONE-DIMENSIONAL ARRAYS */    PAGE   5
COMPILER DIAGNOSTIC MESSAGES
ERROR ID L   STMT     MESSAGE DESCRIPTION

COMPILER INFORMATORY MESSAGES

IEL0533I I         NO 'DECLARE' STATEMENT(S) FOR 'SYSIN'.

END OF COMPILER DIAGNOSTIC MESSAGES


PL/I OPTIMIZING COMPILER  VERSION 1 RELEASE 5.1  TIME: 13.12.47  DATE: 10 MAR 87 PAGE   6
OPTIONS SPECIFIED
GOSTMT;
*PROCESS;
```

FIGURE 11-1 (continued)

```
PL/I OPTIMIZING COMPILER          INHDR: PROC (IH_CLASS, IH_SMSTR, IH_SECTION, PAGE   7
                  SOURCE LISTING

    STMT


    1     INHDR: PROC (IH_CLASS, IH_SMSTR, IH_SECTION, IH_#GRADES,
                       IH_#PGMS, IH_#TSTS, IH_PTS_PSBL,
                       IH_TTL_PGM_PTS, IH_TTL_TST_PTS);
        /**********************************************************************/
        /*                                                                  */
        /*                      INHDR SUBROUTINE                            */
        /*                                                                  */
        /*  FUNCTION:  TO READ AND PROCESS THE INPUT HEADER RECORDS         */
        /*                                                                  */
        /*  PARAMETERS:  IH_CLASS                                           */
        /*               IH_SMSTR                                           */
        /*               IH_SECTION                                         */
        /*               IH_#GRADES                                         */
        /*               IH_#PGMS                                           */
        /*               IH_#TSTS                                           */
        /*               IH_PTS_PSBL                                        */
        /*               IH_TTL_PGM_PTS                                     */
        /*               IH_TTL_TST_PTS                                     */
        /*                                                                  */
        /*  GLOBAL VARIABLES:   NONE                                        */
        /*                                                                  */
        /*  SUBROUTINES CALLED:  NONE                                       */
        /*                                                                  */
        /*  INPUT:    THE HEADER RECORDS                                    */
        /*                                                                  */
        /*  OUTPUT:  NONE                                                   */
        /*                                                                  */
        /**********************************************************************/


    2     DCL IH_CLASS          CHARACTER     (11);
    3     DCL IH_SMSTR          CHARACTER     (11);
    4     DCL IH_SECTION        CHARACTER     (2);

    5     DCL IH_#GRADES        FIXED DECIMAL (3);
    6     DCL IH_#PGMS          FIXED DECIMAL (3);
    7     DCL IH_#TSTS          FIXED DECIMAL (3);
```

```
 8     DCL IH_PTS_PSBL (*)      FIXED DECIMAL (5);
 9     DCL IH_TTL_PGM_PTS       FIXED DECIMAL (5);
10     DCL IH_TTL_TST_PTS       FIXED DECIMAL (5);

11   DCL NDX                    FIXED BINARY  (15);
12          IH_TTL_PGM_PTS = 0;
13          IH_TTL_TST_PTS = 0;

14          GET LIST (IH_CLASS, IH_SECTION, IH_SMSTR);
15          GET LIST (IH_#PGMS, IH_#TSTS);

16          IH_#GRADES = IH_#PGMS + IH_#TSTS;

17 RD_LP1:  DO NDX = 1 TO IH_#GRADES;
18              GET LIST (IH_PTS_PSBL (NDX));
```

PL/I OPTIMIZING COMPILER   INHDR: PROC (IH_CLASS, IH_SMSTR, IH_SECTION, PAGE   8
   STMT

```
19              END RD_LP1;

20 P_TTL:   DO NDX = 1 TO IH_#PGMS;
21              IH_TTL_PGM_PTS = IH_TTL_PGM_PTS + IH_PTS_PSBL (NDX);
22              END P_TTL;

23 T_TTL:   DO NDX = IH_#PGMS + 1 TO IH_#GRADES;
24              IH_TTL_TST_PTS = IH_TTL_TST_PTS + IH_PTS_PSBL (NDX);
25              END T_TTL;
26              RETURN;
27      END INHDR;
```

PL/I OPTIMIZING COMPILER   INHDR: PROC (IH_CLASS, IH_SMSTR, IH_SECTION, PAGE   9
COMPILER DIAGNOSTIC MESSAGES
ERROR ID L   STMT   MESSAGE DESCRIPTION

COMPILER INFORMATORY MESSAGES

IEL0533I I   NO 'DECLARE' STATEMENT(S) FOR 'SYSIN'.
IEL0430I I   1   NO 'MAIN' OPTION ON EXTERNAL PROCEDURE.

END OF COMPILER DIAGNOSTIC MESSAGES

FIGURE 11-1 (continued)

```
PL/I OPTIMIZING COMPILER  VERSION 1 RELEASE 5.1  TIME: 13.12.52  DATE: 10 MAR 87 PAGE  10
OPTIONS SPECIFIED
GOSTMT;
*PROCESS;



PL/I OPTIMIZING COMPILER              PGHDR:  PROC (PH_CLASS, PH_SMSTR,          PAGE  11
                    SOURCE LISTING
   STMT


    1     PGHDR:  PROC (PH_CLASS, PH_SMSTR, PH_SECTION, PH_#GRADES,
                        PH_PTS_PSBL, PH_TTL_PGM_PTS, PH_TTL_TST_PTS,
                        PH_LINE_CNT);
         /*******************************************************************/
         /*                                                               */
         /*                    PGHDR SUBROUTINE                           */
         /*                                                               */
         /*  FUNCTION:  TO PRINT THE PAGE HEADERS                         */
         /*                                                               */
         /*  PARAMETERS:  PH_CLASS                                        */
         /*               PH_SMSTR                                        */
         /*               PH_SECTION                                      */
         /*               PH_#GRADES                                      */
         /*               PH_PTS_PSBL                                     */
         /*               PH_TTL_PGM_PTS                                  */
         /*               PH_TTL_TST_PTS                                  */
         /*                                                               */
         /*  GLOBAL VARIABLES:   NONE                                     */
         /*                                                               */
         /*  SUBROUTINES CALLED:  NONE                                    */
         /*                                                               */
         /*  INPUT:    NONE                                               */
         /*                                                               */
         /*  OUTPUT:   THE PAGE HEADERS                                   */
         /*                                                               */
         /*******************************************************************/

    2     DCL PH_CLASS          CHARACTER    (11);
    3     DCL PH_SMSTR          CHARACTER    (11);
    4     DCL PH_SECTION        CHARACTER    (2);
```

```
5      DCL PH_#GRADES          FIXED DECIMAL (3);
6      DCL PH_PTS_PSBL (*)      FIXED DECIMAL (5);
7      DCL PH_TTL_PGM_PTS       FIXED DECIMAL (5);
8      DCL PH_TTL_TST_PTS       FIXED DECIMAL (5);


9      DCL PH_LINE_CNT          FIXED BINARY  (15);
10     DCL NDX                  FIXED BINARY  (15);


11            PUT PAGE EDIT ('CSCI ', PH_CLASS, 'SEC', PH_SECTION,
                    'GRADE ROLL', PH_SMSTR) (COL (40), A(5), A(5), A(4),
                    A(2), X(5), A(11), X(5), A(11));


12            PUT SKIP (3) EDIT ('LIST OF POINTS POSSIBLE')
                                            (COL (50), A(23));
13            PUT SKIP (2) EDIT ((PH_PTS_PSBL (NDX) DO NDX = 1 TO
                            PH_#GRADES)) (COL (35), (25)F(4));
14            PUT SKIP (3) EDIT
```

```
PL/I OPTIMIZING COMPILER          PGHDR:  PROC (PH_CLASS, PH_SMSTR,        PAGE  12
    STMT

                    ('TOTAL PROGRAM POINTS POSSIBLE', PH_TTL_PGM_PTS,
                     'TOTAL TEST POINTS POSSIBLE', PH_TTL_TST_PTS)
                     (COL (35), A(29), F(5), X(10), A(26), F(5));
15            PH_LINE_CNT = 1;
16            RETURN;
17      END PGHDR;
```

```
PL/I OPTIMIZING COMPILER          PGHDR:  PROC (PH_CLASS, PH_SMSTR,        PAGE  13
COMPILER DIAGNOSTIC MESSAGES
ERROR ID L   STMT    MESSAGE DESCRIPTION

COMPILER INFORMATORY MESSAGES

IEL0533I I         NO 'DECLARE' STATEMENT(S) FOR 'SYSPRINT'.
IEL0430I I   1     NO 'MAIN' OPTION ON EXTERNAL PROCEDURE.

END OF COMPILER DIAGNOSTIC MESSAGES


PL/I OPTIMIZING COMPILER  VERSION 1 RELEASE 5.1  TIME: 13.12.56  DATE: 10 MAR 87 PAGE  14
```

FIGURE 11–1 (continued)

```
OPTIONS SPECIFIED
GOSTMT;
*PROCESS;



PL/I OPTIMIZING COMPILER          AVGSRTN:  PROC (AR_STATUS, AR_GRADE,        PAGE  15
                   SOURCE LISTING
  STMT


   1     AVGSRTN:  PROC (AR_STATUS, AR_GRADE, AR_#GRADES,
                   AR_#PGMS, AR_TTL_PGM_PTS, AR_TTL_TST_PTS,
                   AR_FINAL_AVG, AR_PGM_AVG, AR_TST_AVG,
                   AR_PGM_PT_TTL, AR_TST_PT_TTL);
         /******************************************************************/
         /*                                                              */
         /*                    AVGSRTN SUBROUTINE                        */
         /*                                                              */
         /*  FUNCTION:  TO COMPUTE EACH STUDENT'S AVERAGE GRADE FOR      */
         /*             PROGRAMS, TESTS, AND THE COMBINED FINAL AVERAGE  */
         /*                                                              */
         /*                                                              */
         /*  PARAMETERS:  AR_GRADE                                       */
         /*               AR_#GRADES                                     */
         /*               AR_GRADE                                       */
         /*               AR_#GRADES                                     */
         /*               AR_#PGMS                                       */
         /*               AR_TTL_PGM_PTS                                 */
         /*               AR_TL_TST_PTS                                  */
         /*               AR_FINAL_AVG                                   */
         /*               AR_PGM_AVG                                     */
         /*               AR_TST_AVG                                     */
         /*               AR_PGM_PT_TTL                                  */
         /*               AR_TST_PT_TTL                                  */
         /*                                                              */
         /*  GLOBAL VARIABLES:   NONE                                    */
         /*                                                              */
         /*  SUBROUTINES CALLED:  NONE                                   */
         /*                                                              */
         /*  INPUT:    NONE                                              */
         /*                                                              */
```

```
/* OUTPUT:    NONE                                                    */
/*                                                                    */
/********************************************************************/

2      DCL AR_NAME             CHARACTER    (25);
3      DCL AR_SSN              CHARACTER    (9);

4      DCL AR_#PGMS            FIXED DECIMAL (3);
5      DCL AR_#TSTS            FIXED DECIMAL (3);
6      DCL AR_#GRADES          FIXED DECIMAL (3);
7      DCL AR_PGM_PT_TTL       FIXED DECIMAL (5);
8      DCL AR_TST_PT_TTL       FIXED DECIMAL (5);
9      DCL AR_STATUS           FIXED DECIMAL (1);
10     DCL AR_GRADE (*)        FIXED DECIMAL (3);
11     DCL AR_TTL_PGM_PTS      FIXED DECIMAL (5);
12     DCL AR_TTL_TST_PTS      FIXED DECIMAL (5);
13     DCL AR_PGM_AVG          FIXED DECIMAL (7,2);
14     DCL AR_TST_AVG          FIXED DECIMAL (7,2);
```

```
PL/I OPTIMIZING COMPILER          AVGSRTN:  PROC (AR_STATUS, AR_GRADE,      PAGE  16
    STMT
    15     DCL AR_FINAL_AVG        FIXED DECIMAL (7,2);

    16     DCL SUB                 FIXED BINARY  (15);
    17     DCL FLOAT               BUILTIN;
    18         AR_PGM_PT_TTL = 0;
    19         AR_TST_PT_TTL = 0;

    20         IF AR_STATUS = 0 THEN
        GET_GRD:    DO;

    21 RD_LP2:      DO SUB = 1 TO AR_#GRADES;
    22                  GET LIST (AR_GRADE (SUB));
    23              END RD_LP2;

    24 PGM_PTS:     DO SUB = 1 TO AR_#PGMS;
    25                  AR_PGM_PT_TTL = AR_PGM_PT_TTL +
                                        AR_GRADE (SUB);
    26              END PGM_PTS;
    27 TST_PTS:     DO SUB = AR_#PGMS + 1 TO AR_#GRADES;
    28                  AR_TST_PT_TTL = AR_TST_PT_TTL +
```

FIGURE 11-1 *(continued)*

```
                                    AR_GRADE (SUB);
29              END TST_PTS;
30            END GET_GRD;

31              AR_PGM_AVG = FLOAT (AR_PGM_PT_TTL) / FLOAT
                        (AR_TTL_PGM_PTS) * 100.0;
32              AR_TST_AVG = FLOAT (AR_TST_PT_TTL) / FLOAT
                        (AR_TTL_TST_PTS) * 100.0;
33              AR_FINAL_AVG = (AR_PGM_AVG * 4 + AR_TST_AVG * 6)
                        / 10.0;
34          RETURN;
35      END AVGSRTN;
```

```
PL/I OPTIMIZING COMPILER              AVGSRTN:  PROC (AR_STATUS, AR_GRADE,       PAGE  17
COMPILER DIAGNOSTIC MESSAGES
ERROR ID L   STMT    MESSAGE DESCRIPTION


COMPILER INFORMATORY MESSAGES


IEL0533I I          NO 'DECLARE' STATEMENT(S) FOR 'SYSIN'.
IEL0430I I    1     NO 'MAIN' OPTION ON EXTERNAL PROCEDURE.


END OF COMPILER DIAGNOSTIC MESSAGES
PL/I OPTIMIZING COMPILER  VERSION 1 RELEASE 5.1  TIME: 13.13.00  DATE: 10 MAR 87 PAGE  18
OPTIONS SPECIFIED
GOSTMT;
*PROCESS;


PL/I OPTIMIZING COMPILER              PRNTRTN:  PROC (PR_NAME, PR_GRADE, PR_SSN,   PAGE  19
                SOURCE LISTING
  STMT


    1    PRNTRTN:  PROC (PR_NAME, PR_GRADE, PR_SSN,
                    PR_STATUS, PR_FINAL_AVG, PR_PGM_AVG,
                    PR_TST_PT_TTL, PR_TST_AVG, PR_PGM_PT_TTL,
                    PR_#PGMS, PR_#GRADES, PR_LINE_CNT);
        /*****************************************************************/
        /*                                                             */
        /*                    PRNTRTN SUBROUTINE                       */
        /*                                                             */
        /*  FUNCTION:  TO PRINT THE DETAIL LINES FOR EACH STUDENT      */
```

```
      /*                                                       */
      /*  PARAMETERS:  PR_NAME                                 */
      /*               PR_GRADE                                */
      /*               PR_SSN                                  */
      /*               PR_STATUS                               */
      /*               PR_FINAL_AVG                            */
      /*               PR_PGM_AVG                              */
      /*               PR_TST_PT_TTL                           */
      /*               PR_TST_AVG                              */
      /*               PR_PGM_PT_TTL                           */
      /*               RPR_#PGMS                               */
      /*               PR_#GRADES                              */
      /*               PR_LINE_CNT                             */
      /*                                                       */
      /*  GLOBAL VARIABLES:  NONE                              */
      /*                                                       */
      /*  SUBROUTINES CALLED:  NONE                            */
      /*                                                       */
      /*  INPUT:   NONE                                        */
      /*                                                       */
      /*  OUTPUT:  DETAIL LINES FOR EACH STUDENT               */
      /*                                                       */
      /*********************************************************/

  2      DCL PR_SSN1            CHARACTER     (3);
  3      DCL PR_SSN2            CHARACTER     (2);
  4      DCL PR_SSN3            CHARACTER     (4);
  5      DCL PR_NAME            CHARACTER     (25);
  6      DCL PR_SSN             CHARACTER     (9);

  7      DCL PR_#PGMS           FIXED DECIMAL (3);
  8      DCL PR_#GRADES         FIXED DECIMAL (3);
  9      DCL PR_PGM_PT_TTL      FIXED DECIMAL (5);
 10      DCL PR_TST_PT_TTL      FIXED DECIMAL (5);
 11      DCL PR_STATUS          FIXED DECIMAL (1);
 12      DCL PR_GRADE (*)       FIXED DECIMAL (3);
 13      DCL PR_PGM_AVG         FIXED DECIMAL (7,2);
 14      DCL PR_TST_AVG         FIXED DECIMAL (7,2);
 15      DCL PR_FINAL_AVG       FIXED DECIMAL (7,2);
```

PL/I OPTIMIZING COMPILER   PRNTRTN: PROC (PR_NAME, PR_GRADE, PR_SSN,   PAGE 20

FIGURE 11–1 *(continued)*

```
STMT
  16    DCL PR_LINE_CNT          FIXED BINARY  (15);
  17    DCL NDX                  FIXED BINARY  (15);
  18    DCL SUBSTR               BUILTIN;


  19          PR_SSN1 = SUBSTR (PR_SSN, 1, 3);
  20          PR_SSN2 = SUBSTR (PR_SSN, 4, 2);
  21          PR_SSN3 = SUBSTR (PR_SSN, 6, 4);


  22          PUT SKIP (3) EDIT (PR_NAME,
                    PR_SSN1 || '-' || PR_SSN2 || '-' || PR_SSN3) (COL (5),
                       A(30), A(16));


  23            IF PR_STATUS ^= 0 THEN
                IF PR_STATUS = 2 THEN
                   PUT SKIP (0) EDIT
                           ('*** WITHDREW FAILING ***') (COL (50), A);
  24            ELSE

                IF PR_STATUS = 1 THEN
                   PUT SKIP (0) EDIT
                           ('*** WITHDREW PASSING ***') (COL (50), A);
  25                ELSE;
  26            ELSE
     PRINT:       DO;
  27                PUT SKIP (0) EDIT  ('***** FINAL AVG =',
                          PR_FINAL_AVG, 'PROGRAM AVG =', PR_PGM_AVG,
                          'TEST AVG =', PR_TST_AVG) (COL (50), A(17),
                           F(7,2), X(5), A(13), F(7,2), X(5), A(10),
                           F(7,2));

  28                PUT SKIP EDIT ('TOTAL PGM PTS =', PR_PGM_PT_TTL,
                          'PROGRAM GRADES =')
                             (COL (10), A(18), F(5), X(5), A(16));

  29 PT_LP2:        DO NDX = 1 TO PR_#PGMS;
  30                   PUT EDIT (PR_GRADE (NDX)) ((20)F(4));
  31                END PT_LP2;

  32                PUT SKIP EDIT ('TOTAL TEST POINTS =',
```

```
                              PR_TST_PT_TTL, 'TEST GRADES =')
                              (COL (9), A(19), F(5), X(8), A(13));

  33  PT_LP3:          DO NDX = NDX TO PR_#GRADES;
  34                      PUT EDIT (PR_GRADE (NDX)) ((5)F(4));
  35                   END PT_LP3;
  36               END PRINT;

  37               PR_LINE_CNT = PR_LINE_CNT + 6;
  38               RETURN;
  39           END PRNTRTN;
```

PL/I OPTIMIZING COMPILER   PRNTRTN: PROC (PR_NAME, PR_GRADE, PR_SSN,   PAGE 21
COMPILER DIAGNOSTIC MESSAGES
ERROR ID L   STMT   MESSAGE DESCRIPTION

COMPILER INFORMATORY MESSAGES

IEL0533I I   NO 'DECLARE' STATEMENT(S) FOR 'SYSPRINT'.
IEL0430I I   1   NO 'MAIN' OPTION ON EXTERNAL PROCEDURE.

END OF COMPILER DIAGNOSTIC MESSAGES

                    CSCI 260B SEC 2   GRADE ROLL   SPRING 1988

                              LIST OF POINTS POSSIBLE

                         25 50 75 50 100 100

                    TOTAL PROGRAM POINTS POSSIBLE 200   TOTAL TEST POINTS POSSIBLE 200

   ABLE, AMY   123-45-6789
                                      ***** FINAL AVG = 86.79   PROGRAM AVG = 89.49   TEST AVG = 84.9
         TOTAL PGM PTS =   179   PROGRAM GRADES = 20 45 70 44
         TOTAL TEST POINTS = 170   TEST GRADES = 95 75
```
```

FIGURE 11-1 (*continued*)

```
BAKER, BILL                 382-84-7291
                                          ***** FINAL AVG =  77.99      PROGRAM AVG =  79.49      TEST AVG =  76.99
        TOTAL PGM PTS =      159     PROGRAM GRADES =  25  35  60  39
        TOTAL TEST POINTS =  154        TEST GRADES =  81  73



CARTER, CARL                134-87-4673
                                          *** WITHDREW PASSING ***



DAWSON, DANA                837-29-3740
                                          *** WITHDREW FAILING ***



EARL, EARL                  573-93-9789
                                          ***** FINAL AVG =  78.99      PROGRAM AVG =  84.99      TEST AVG =  75.00
        TOTAL PGM PTS =      170     PROGRAM GRADES =  20  40  70  40
        TOTAL TEST POINTS =  150        TEST GRADES =  80  70



FUNK, FRIEDA                234-56-7890
                                          ***** FINAL AVG =  96.29      PROGRAM AVG =  95.99      TEST AVG =  96.49
        TOTAL PGM PTS =      192     PROGRAM GRADES =  23  48  73  48
        TOTAL TEST POINTS =  193        TEST GRADES =  98  95
```

11.3

Multidimensional Arrays

So far we have concentrated entirely on one-dimensional arrays, but arrays in PL/I can have up to 15 dimensions. Two-dimensional arrays are commonly used, but arrays with three or more dimensions are used only rarely. Certainly programmers in business data processing have little use for arrays with more than two dimensions.

11.3.1

Declaring and Referencing Multidimensional Arrays

The number of dimensions in an array is specified by the number of extents declared. The following examples illustrate the declaration of two-, three-, and four-dimensional arrays.

```
DCL TWO_DIM    (50, 5:25)            FIXED DEC (5,2);
DCL THREE_DIM (-100:0,75, 600:900) FIXED DEC (4);
DCL FOUR_DIM  (50, 50, 100, 100)    FIXED DEC (6);
```

When multidimensional arrays are declared in a subroutine using the asterisk form, an asterisk must be provided for each dimension. Thus, TWO_DIM
might be declared in a subroutine as

```
DCL SR_TWO_DIM (*,*) FIXED DEC (5,2);
```

In order to reference a specific element of a multidimensional array, you
must provide a subscript for each dimension:

```
TWO_DIM (45, 20)
THREE_DIM (I, J, K)
FOUR_DIM (I, J, K, L)
```

11.3.2

Two-Dimensional Arrays

Although arrays of three and more dimensions appear only rarely in general
data processing, two-dimensional arrays are commonly found. Before we
get into the techniques of manipulating them, we need to be sure to
understand some terms and have a conceptual understanding of two-
dimensional arrays. These arrays are usually described as being laid out in
rows and columns. A **row** is all the data on one line (i.e., rows run
horizontally across the page), and **column** is the same element in each row.
Thus, an array declared as

```
DCL TWO_DIM (5,3) CHAR (2);
```

and assumed to have been filled could be viewed *conceptually* like this:

```
        Columns
        1    2    3
Row 1:  AB   CD   EF
Row 2:  GH   IJ   KL
Row 3:  MN   OP   QR
Row 4:  ST   UV   WX
Row 5:  YZ   AZ   BY
```

As this conceptual diagram illustrates, the first dimension listed in the
declaration refers to rows and the second to columns.

Although the conceptual understanding gained from the diagram above is useful, it is also important to understand that computer memory is one-dimensional. To determine the order in which the data is stored in memory, the rule is this: The second (right) subscript in the input statement runs through its range before the first (left) subscript is incremented. For example, if we code (assuming enough input)

```
GET LIST (TWO_DIM);
```

the elements are filled in the following order: TWO_DIM(1,1) TWO_DIM(1,2) TWO_DIM(1,3) TWO_DIM(2,1) TWO_DIM(2,2) TWO_DIM(2,3) TWO_DIM(3,1) TWO_DIM(3,2) TWO_DIM(3,3) TWO_DIM(4,1) TWO_DIM(4,2) TWO_DIM(4,3) TWO_DIM(5,1) TWO_DIM(5,2) TWO_DIM(5,3). The values, then, are stored as AB CD EF GH IJ KL MN OP QR ST UV WX YZ AZ BY.

If we are to fill a two-dimensional array only partially (or print only some of its values), we can best do so by using nested iterative DO statements; for example,

```
DO ROW = 1 TO #_ROWS;
    DO COL = 1 TO #_COLS;
        GET EDIT (ALPHA (ROW, COL)(A(2));
    END;
END
```

or

```
GET EDIT (((ALPHA (ROW, COL) DO COL = 1 TO #_COLS) DO ROW =
        1 TO #_ROWS)) (A(2));
```

To print the array in row major order, use the same statement, with PUT replacing the GET. There might be times when to want to print the array in column major order. Assuming the array was filled with the GET EDIT above, you can print it in column major order with:

```
DO ROW = 1 TO #_ROWS;
    DO COL = 1 TO #_COLS;
        PUT EDIT (ALPHA (ROW, COL)) (A(5));
    END;
END;
```

or

```
PUT EDIT (((ALPHA (ROW, COL) DO ROW = 1 TO #_ROWS) DO COL =
    1 TO #_COLS)) (A(5));
```

To illustrate the use of a two-dimensional array we might suppose a teacher keeping class grades in such a format. Each row represents the grades for one student, and each column represents the grades given on one assignment. Because each element in the array has the same attribute as all others, it is impossible (if the grades are numeric) to include the student names, and it is undesirable to do so if the grades are characters because each element would have to be wide enough (20 columns?) to hold the name. There are two ways to solve this problem: The first possibility is that the teacher remembers which row represents each student. This would not be too difficult if the grades represented the students in alphabetized order. The teacher could look in her or his gradebook (presumably alphabetized also) and quickly see that row 7 contains the grades for HANSON, HILDA. To print Ms. Hanson's grade for the third assignment, the teacher merely codes

```
PUT EDIT (GRADES (7,3)) (F(3));
```

The second solution involves **parallel arrays**. In this case the teacher sets up a second (one-dimensional) array to hold the students' names, which are listed in the same order as the rows of grades in the two-dimensional array. The teacher searches the name array first, supplying a student's name as the search argument.

```
DO SUB = 1 TO #_ROWS WHILE (NAME_ARRAY (SUB) ^= INPUT_NAME);
    IF NAME_ARRAY (SUB) = INPUT_NAME THEN
        PUT EDIT (NAME_ARRAY (SUB)) (A);
END;
SUB = SUB - 1;
```

The value in SUB is also the value needed to identify the row containing the student's grades, and so the teacher can now code

```
PUT SKIP (0) EDIT (GRADES (SUB, 3)) (COL(22), F(3));
```

to list the student's third grade.

The two-dimensional array, then, is a very useful device in general data processing, and it will repay the time you spend mastering the techniques of manipulating it.

Multidimensional Arrays

As we noted earlier, arrays of more than two dimensions are rarely used in general data processing; therefore, we only note a few things about them. First, the terms "row" and "column" are not appropriate for arrays beyond two dimensions because the terms describe only two dimensions. Second, as more dimensions are added, the arrangement of the data becomes increasingly difficult to conceptualize. Perhaps the best way to try to conceptualize a three-dimensional array is to think literally of three dimensions, so that you have data on a vertical plane ("columns") and on two horizontal planes, one extending from right to left ("rows") and one extending from front to back. In other words, in extending an array from two to three dimensions, you can visualize each element in the two dimensions as having another element "behind" it. Of course, the data in a three- or more-dimensional array is actually stored in memory in the linear fashion we described for two-dimensional arrays. Third, it is easy to overlook the amount of memory being consumed by a multidimensional array. A five-dimensional array, for example, with each dimension having an extent of 10, contains 100,000 elements. If the array is declared as FIXED DEC (7), each element occupies four bytes, for a total of 400,000 bytes. It is very likely that your program will not be allocated that much memory, and consequently an execution error will result.

If you think your program requires a three-dimensional array, it might well be that you have not broken your problem into steps as simple as possible and desirable. Before using an array of more than two dimensions, then, you might want to give your algorithm a careful scrutiny to see whether there is a simpler approach. Parallel arrays, for example, might simplify matters in some situations.

11.4

Array BUILTIN Functions

PL/I provides a number of functions that are useful in manipulating arrays. We discuss DIM, HBOUND, LBOUND, PROD, and SUM, but there are others. Like all built-in functions, those used with arrays also each return a single value, not a value for each element in the array. Each also requires an array name (unsubscripted) for one (or the only) argument.

11.4.1

DIM

The syntax of the DIM function is DIM (x,y), where x is an array name and y is a value specifying a particular dimension of the array. The array, of course, must have at least as many dimensions as the value of y. DIM returns a value equal to the number of extents in the specified dimension, and the value is returned as a fixed binary with a precision of (15,0). Some examples are

```
DCL ONE_DIM (40)                            FIXED DEC (4),
    FOUR_DIM (-3:2, 11:15, -4:-7, 6:12) CHAR       (3),
    EXTENTS                             FIXED BIN (15);
EXTENTS = DIM (ONE_DIM, 1);      /* EXTENTS = 40 */
EXTENTS = DIM (FOUR_DIM, 4);     /* EXTENTS = 7  */
EXTENTS = DIM (FOUR_DIM, 3);     /* EXTENTS = 4  */
EXTENTS = DIM (FOUR_DIM, 2);     /* EXTENTS = 5  */
EXTENTS = DIM (FOUR_DIM, 1);     /* EXTENTS = 6  */
```

11.4.2

HBOUND and
LBOUND

The syntax for HBOUND and LBOUND are HBOUND (x,y) and LBOUND (x,y). The array name is again specified by x and the particular dimension by y. In both cases the value of the bound is returned in fixed binary with a precision of (15,0). Some examples:

```
DCL ONE_DIM (-19:20)      FIXED DEC (3),
    TWO_DIM (66:85, 35)   CHAR      (4),
    BOUND                 FIXED BIN (15);
BOUND = LBOUND (ONE_DIM, 1);      /* BOUND = -19 */
BOUND = HBOUND (ONE_DIM, 1);      /* BOUND = 20  */
BOUND = HBOUND (TWO_DIM, 2);      /* BOUND = 35  */
BOUND = LBOUND (TWO_DIM, 1);      /* BOUND = 66  */
```

11.4.3

PROD and SUM

The syntax for PROD and SUM are PROD (x) and SUM (x), where x is an array name in both cases. If the elements of the array are fixed-point integer values or strings, the result is a fixed-point integer value. In the case of strings, the values are converted to fixed decimal integers. Otherwise (i.e., if the elements contain noninteger fixed decimal, float or fixed binary, or float decimal values), the value is returned as a floating-point decimal. The value returned represents the product or sum of all elements in the array.

11.5

Array Processing—
Sort and
Search Routines

Perhaps the most common use of arrays is to allow for the sorting of data according to some preconceived order. One obvious example is simply to alphabetize a list of names. Sorting is such a pervasive activity in computing that an immense effort has been expended on arriving at efficient sorting routines. We do not go into the more complex of these algorithms; instead we present two relatively simple ones. The choice of a sorting algorithm depends in large measure on the size of the data set to be sorted, and these algorithms are satisfactory for the small arrays you sort as a student.

Once it has been sorted, a data set can be searched for a particular value much more efficiently. In Section 11.5.3 we present a very efficient technique for searching a sorted array.

11.5.1

Bubble Sort

The first step in sorting any data, regardless of the sort algorithm employed, is to determine whether the data is to be sorted in ascending or descending order. Suppose, for example, we have an array of 40 elements, and the value in each element represents the final average percentage grade for a class. The kindhearted teacher wants to grade on a curve, and this requires an ordered listing of the grades. The grades might be sorted in either ascending or descending order, of course, but probably most teachers prefer a descending order. Because CH11EX2 and its pseudocode near the end of the chapter demonstrate the implementation of the bubble sort, we discuss it here only on the conceptual level.

The bubble sort gets its name from the fact that data "bubbles" from the "bottom" of the array to its sorted position. In keeping with our hypothetical application above, we discuss the bubble sort as it sorts in descending order. This means, of course, that it is the larger values that "bubble up" from the bottom. The basic process is that Array (Subscript) is compared with Array (Subscript + 1). If Array (Subscript) is greater than Array (Subscript + 1), Subscript is incremented by 1, and the process is repeated. If Array (Subscript) is less than Array (Subscript + 1), however, Array (Subscript) is moved to a temporary save area; Array (Subscript + 1) is moved to Array (Subscript), and the contents of the temporary save area are then moved to Array (Subscript + 1). This set of actions results in a swap of values in the two elements. The process continues until Subscript + 1 is equal to the high bound of the array (or as long as Subscript is less than the high bound). At this point we say that we have completed one *pass*. The following is the pseudocode for a single pass.

```
Subscript ← Low Bound
DO WHILE (Subscript < High Bound)
    IF (Array (Subscript) < Array (Subscript + 1))
        Temporary Storage ← Array (Subscript)
        Array (Subscript) ← Array (Subscript + 1)
        Array (Subscript + 1) ← Temporary Storage
```

```
     ENDIF
     Subscript ← Subscript + 1
 ENDDO
```

When a single pass has been completed, we must set the subscript back to 1 and start over. This means, of course, that the logic requires a loop within a loop, but you might wonder how the higher-level loop is controlled. Typically, a flag is initialized to one of its two possible values, and the higher-level loop is executed while the value of the flag remains the same. Before the lower-level loop is entered, a second flag is set to one of its possible values. One statement in the DO-group following the IF-THEN statement in the lower level loop changes the flag to its other value. Following the END statement of the lower-level loop, the second flag is checked; if it contains its original value, it means that no values were swapped. This in turn means that the array is completely sorted, and so the first flag is changed to its second possible value. The complete algorithm for the bubble sort is as follows:

```
 Sorted-Flag ← False
 DO WHILE (Sorted-Flag = False)
    Swap-Flag ← False
    Subscript ← Low Bound
    DO WHILE (Subscript < High Bound)
       IF (Array (Subscript) < Array (Subscript + 1))
          Temporary Storage ← Array (Subscript)
          Array (Subscript) ← Array (Subscript + 1)
          Array (Subscript + 1) ← Temporary Storage
          Swap-flag ← True
       ENDIF
       Subscript ← Subscript + 1
    ENDDO
    IF (Swap-Flag = False)
       Sorted-Flag = True
    ENDIF
 ENDDO
```

To illustrate the action of the bubble sort we ran a simple program that sorted the array in descending order. We printed the array before the first pass and again after each pass. Here is the output:

```
ORIGINAL ARRAY:     19 16 12 22 10 23 44 38 83 55
AFTER PASS #1:      19 16 22 12 23 44 38 83 55 10
```

```
AFTER PASS #2:     19 22 16 23 44 38 83 55 12 10
AFTER PASS #3:     22 19 23 44 38 83 55 16 12 10
AFTER PASS #4:     22 23 44 38 83 55 19 16 12 10
AFTER PASS #5:     23 44 38 83 55 22 19 16 12 10
AFTER PASS #6:     44 38 83 55 23 22 19 16 12 10
AFTER PASS #7:     44 83 55 38 23 22 19 16 12 10
AFTER PASS #8:     83 55 44 38 23 22 19 16 12 10
AFTER PASS #9:     83 55 44 38 23 22 19 16 12 10
```

As you can see, on the first pass 22 and 12 were swapped; 10 was swapped with each succeeding number until it found its final location in Array (10). On the second pass 16 and 22 were swapped, and 12 was swapped with each number until it reached its sorted location of Array (9). On the third pass 22 and 19 were swapped, and 16 was swapped to its final destination of Array (8). By this time you no doubt see a pattern emerging: Upon completion of each pass the sorted portion of the array increases by one element. Notice, however, that the array elements are identical in pass 8 and pass 9. One extra pass has to be made to discover that no swaps were made. Thus, the efficiency of the algorithm could be increased slightly by including

```
High Bound ← High bound−1
```

after the inner loop ENDDO.

The bubble sort requires a relatively simple and easily conceptualized algorithm. For this reason, it is quite widely used in introducing students to the sorting process. In actual data processing situations, however, the size of the array can cause the inefficiencies of the bubble sort to outweigh the ease of its implementation, and other algorithms are used instead.

11.5.2

Insertion Sort

The insertion sort is similar to the bubble sort in that it, too, is satisfactory for small arrays. In its logic the insertion sort also requires a loop within a loop. The higher-level loop executes from Subscript (Lower Bound) to Subscript (Higher Bound − 1). The lower-level loop executes from Subscript (Lower Bound + 1) to Subscript (Higher Bound). On the first time through the lower-level loop, then, Subscript (Lower Bound) references Array (1) and Subscript (Lower Bound + 1) references Array (2). Within the lower-level loop Array (1) is compared to Array (2); if Array (1) is less than Array (2) (for a sort in descending sequence), the two are swapped. On successive executions of the lower level loop, Array (Subscript) is compared to Array

(Subscript + 1) and a swap made when necessary. Thus, after the first pass is complete, the highest value is in Array (1), which means the first element is sorted and need not be checked again. When the higher level DO WHILE executes to begin the second pass, its Subscript has the value of 2, and so the first element is not checked. On each successive pass, the unsorted portion of the array becomes smaller by one element. On the final pass it is necessary only to compare Higher Bound − 1 with Higher Bound and to make the swap if required.

As you can see from the results below, the insertion sort required the same number of passes as did the bubble sort to sort the same data. Nevertheless, the insertion sort is more efficient than the bubble sort because fewer comparisons are required.

```
ORIGINAL ARRAY:    19 16 12 22 10 23 44 38 83 55
AFTER PASS #1:     83 16 12 19 10 22 23 38 44 55
AFTER PASS #2:     83 55 12 16 10 19 22 23 38 44
AFTER PASS #3:     83 55 44 12 10 16 19 22 23 38
AFTER PASS #4:     83 55 44 38 10 12 16 19 22 23
AFTER PASS #5:     83 55 44 38 23 10 12 16 19 22
AFTER PASS #6:     83 55 44 38 23 22 10 12 16 19
AFTER PASS #7:     83 55 44 38 23 22 19 10 12 16
AFTER PASS #8:     83 55 44 38 23 22 19 16 10 12
AFTER PASS #9:     83 55 44 38 23 22 19 16 12 10
```

Keep in mind, however, that neither the bubble sort nor the insertion sort is particularly sophisticated and that neither is used to sort a large array. The two sorts come close to the manner in which you would sort a deck of cards and therefore provide a good introduction to the sorting process.

11.5.3

Binary Search

As we noted earlier, when the data in an array has been sorted in either ascending or descending order, a highly efficient search routine becomes possible. If the data is not sorted, the best that can be done is to carry out a sequential (linear) search. In a **sequential search** the array is searched starting with the first element and comparing its value with the argument value. The argument value is compared with each successive element until either a match is found or the end of the array is reached. Given an array of 100 elements, the worst case (i.e., a match on the last element or no match found) requires 100 comparisons, and on average a search requires 50 comparisons.

When the data in the array is sorted, however, it is possible to conduct a binary search. A **binary search** repeatedly divides the array in two halves and discards one half. This process continues until a match with the search argument is found or it is determined that no element matches the argument. The binary search, in fact, searches half of the (remaining) elements with a single seek. A **seek** is a single inspection of an element in the array, even if the inspection requires more than one comparison. Note that a binary search requires two comparisons (equal? larger? [or smaller?]), but the two comparisons constitute a single seek. A binary search is able to search half of the remaining array with one seek because (again assuming the data in descending order) if the middle element is less than the argument, all elements to the right are also less. The search routine can then go to the middle of the remaining left half of the original array and make another single seek. Once again it can discard half of the remaining elements. Given a worst-case scenario, the remaining unsearched elements are reduced to one, and either it matches the search argument or else there is no matching element in the table. Keep in mind that when the array (or the remaining portion) has an odd number of elements you will have to decide whether to truncate or round when you divide by two to find the middle element.

The pseudocode for a binary search is as follows:

```
Match ← No
Low ← Low Bound
High ← High Bound
DO WHILE (Match = No & Low <= High)
   Middle = (Low + High) / 2          [Round or truncate]
   IF (Argument = Array (Middle))
      Perform necessary functions on element
      Match ← Yes
   ELSE
      IF (Array (Middle) > Argument)
         Low = Middle + 1
      ELSE
         High = Middle - 1
   ENDIF
ENDDO
IF (Match = No)
   Perform 'No Match' functions
ENDIF
```

Notice, first of all, that this pseudocode represents the logic for the binary search of a single search argument; if you are searching for a series of arguments, the implementation of this logic must be placed within a higher-level loop.

The following table demonstrates the efficiency of the binary search over the sequential search. Notice that the sequential *average* is contrasted with the binary *worst case,* but also keep in mind that a sequential sort makes one comparison for each seek, while a binary search makes two comparisons per seek.

Number of elements	10	25	50	100	500	1000	5000
Sequential search (average)	5	12.5	25	50	250	500	2500
Binary search (worst case)	4	5	6	7	9	10	13

11.6
Example Program— Array Processing

Example program CH11EX2 (see Fig. 11–2) illustrates a situation common in programming: You write a program to solve a problem according to someone's specifications. The program is used for some time, and the user decides that changes need to be made. Usually the user wants to add certain features. CH11EX2 is a "new, improved" version of CH11EX1. We wrote CH11EX1, you recall, to report student grades in terms of average grade on programming projects, average grade on tests, and the two combined into a final average. After using the program for several classes, we noted certain shortcomings.

We discovered, first of all, that we had a tendency to make typing errors when we constructed the input records. We had designed the CH11EX1 output in such a way that the student names could be clipped off and the remainder posted, with student grades identified by social security numbers. Each student was asked to check her or his grades and notify us of any discrepancies. This process worked fine for cases in which we *underreported* the student's grades, but we found on occasion that a student was achieving above 100 percent on one or the other (or both) of the course components (programs and tests). We therefore decided to flag any instance of a student earning more program or test points than the total maximum. We decided to set flags in FIND_AVGSRTN and test them in PRNTRTN.

Second, some users wanted an easy method of identifying those students who failed the programming component or the test component, or both. We added a few lines of code to compare each student's average against the minimum passing grade of 68 percent. Again, we set these flags in AVGSRTN and test them in PRNTRTN.

Perhaps the most significant addition to CH11EX2 over CH11EX1 is the inclusion of SORTRTN and PRNTSSN. Students regularly ask to know their grade on the final examination and their letter grade for the course as soon as possible after the final examination has been taken. We wanted a report we could post for the students, and we felt it necessary to further ensure their anonymity. Since the social security numbers in CH11EX1 correspond to an alphabetized listing of the students in the class, it would not be too difficult to determine the name corresponding to a particular social security number. We therefore sorted the social security numbers in ascending order and, of course, kept SR_FINAL_AVG_TBL and SR_FINAL_TST_TBL parallel. All of this is handled in SORTRTN. PRNTSSN produces a report consisting of a blank line on which the teacher can enter the letter grade. The blank line is followed by the social security number, the student's grade on the final examination, and his or her course percentage average.

The minimum percentage grade for the various letter grades are officially 92, 84, 76, and 68 percent. Most teachers, however, want some leeway in determining letter grades. We therefore concluded that it would be helpful to sort and print the final percentage grades in descending order. This report allows the teacher to see at a glance what the grade spread is, where the larger gaps between grades fall, and where the grades are clustered. All of this, in turn, can make it easier to determine the cutoff points for the various letter grades. SRTAVG both sorts the final grades and produces the report.

Finally, you might want to note that we use list-directed input to facilitate the construction of the input records and edit-directed output to format the reports. Note also that we use the bubble sort in both sorting routines. The pseudocode for the MAIN and SORTAVG routines are as follows:

```
CH11EX2
    Call INHDR
    Call PGHDR
    Read a status-SSN-name record
    DO WHILE (more input)
        Call AVGSRTN
        IF (line counter > 40)
            Call PG-HDR-RTN
        ENDIF
        Call PRNTRTN
        Read a status-SSN-name record
    ENDDO
    Call SORTRTN
    Call PRNTSSN
    Call SRTAVG
STOP
SRTAVG
```

```
Last used element ← element counter - 1
Elements swapped flag ← off
DO WHILE (elements swapped flag = off and subscript > 2)
   Elements swapped flag ← on
   FOR (subscript - 2 to last used element by 1)
      Elements swapped flag ← off
      IF (final average array (subscript - 1) > final
                 average array (subscript))
         Save average ← final average array
                 (subscript - 1)
         Final average array (subscript - 1) ← final
                 average array (subscript)
         Final average array (subscript) ← save average
      ENDIF
   ENDFOR
   Print page header
   FOR (subscript = 1 to element counter - 1 by 1)
   IF (final average array (subscript) > -1)
      Print final average array (subscript)
   ENDIF
ENDFOR
EXIT
```

```
PL/I OPTIMIZING COMPILER  VERSION 1 RELEASE 5.1  TIME: 13.14.07  DATE: 10 MAR 87  PAGE   1
OPTIONS SPECIFIED
GOSTMT;


PL/I OPTIMIZING COMPILER         /* CH11EX2 PROCESSING ARRAYS */                 PAGE   2
                 SOURCE LISTING
   STMT

        /* CH11EX2 PROCESSING ARRAYS */
      1 CH11EX2:    PROCEDURE OPTIONS (MAIN);
        /*******************************************************************/
        /*                                                                 */
        /*   FUNCTION:  1.  TO COMPUTE STUDENTS' FINAL AVERAGES            */
        /*              2.  TO SORT FINAL AVERAGES IN DESCENDING ORDER     */
```

FIGURE 11–2
CH11EX2 (example of processing arrays).

FIGURE 11-2 *(continued)*

```
/*              3.  TO SORT SSN'S AND FINAL AVERAGES IN ASCENDING    */
/*                  ORDER ON SSN KEY                                 */
/*                                                                  */
/*    INPUT:    1.  THREE HEADER CARDS                              */
/*                  A.  'CLASS # (E.G. '465'), 'SECTION #' ('X'),   */
/*                      'SEMESTER' ('SPRING 19XX')                  */
/*                  B.  (1) NUMBER OF PROGRAM GRADES AND (2)        */
/*                      NUMBER OF TEST GRADES                       */
/*                  C.  MAXIMUM NUMBER OF POINTS FOR EACH PROGRAM   */
/*                      AND FOR EACH TEST                           */
/*              3.  TWO CARDS FOR EACH STUDENT                      */
/*                  A.  THE FIRST CONTAINING A STATUS,             */
/*                          0 = COMPLETED THE COURSE               */
/*                          1 = WITHDREW PASSING                   */
/*                          2 = WITHDREW FAILING                   */
/*                      STUDENT SSN                                 */
/*                      STUDENT NAME                                */
/*                  B.  THE SECOND CONTAINING THE STUDENT GRADES    */
/*                                                                  */
/*    OUTPUT:   1.  A REPORT LISTING                                */
/*                  A.  THE MAXIMUM POSSIBLE POINTS FOR EACH        */
/*                      PROGRAM AND TEST                            */
/*                  B.  THE MAXIMUM TOTAL POINTS FOR PROGRAMS AND   */
/*                      FOR TESTS                                   */
/*                  C.  FOR EACH STUDENT                            */
/*                      1.  NAME, SSN, AND                          */
/*                          IF (COURSE NOT COMPLETED) STATUS        */
/*                          ELSE                                    */
/*                          FINAL AVERAGE, PROGRAM AVERAGE, AND     */
/*                            TEST AVERAGE                          */
/*                      2.  THE TOTAL POINTS EARNED FOR PROGRAMS    */
/*                          AND THE POINTS EARNED FOR EACH PROGRAM  */
/*                      3.  THE TOTAL POINTS EARNED FOR TESTS AND   */
/*                          THE TOTAL TEST POINTS                   */
/*              2.  A REPORT LISTING ONLY SSN'S SORTED IN ASCENDING */
/*                  ORDER WITH CORRESPONDING FINAL TEST GRADES AND  */
/*                  FINAL AVERAGES                                  */
/*              3.  A REPORT LISTING THE FINAL AVERAGES SORTED IN   */
/*                  DESCENDING ORDER                                */
/*                                                                  */
```

```
        /*** NOTES:  1.  THE FINAL AVERAGE IS DETERMINED BY (PGM_AVG * 4     */
        /*              TST_AVG * 6) / 10.0 (PROGRAMS = 40%, TESTS = 60%      */
        /*          2.  STUDENTS WITH A FAILING AVERAGE ON EITHER PROGRAMS    */
        /*              OR TESTS OR BOTH ARE APPROPRIATELY FLAGGED            */
```

```
        /*          3.  IF PGM_PT_TTL > THAN TTL_PGM_PTS OR TST_PT_TTL >      */
        /*              TTL_TST_PTS OR BOTH THEN AN INPUT ERROR IS FLAGGED    */
        /*          4.  STUDENTS WHO WITHDREW ARE ASSIGNED FINAL AVERAGES     */
        /*              OF -1.0.  THEIR NAMES, SSN'S, AND FINAL AVERAGES      */
        /*              DO NOT APPEAR IN REPORTS 2 AND 3                       */
        /*                                                                    */
        /**********************************************************************/
```

```
        2    DCL CLASS              CHARACTER     (4);
        3    DCL SECTION            CHARACTER     (2);
        4    DCL SMSTR              CHARACTER     (11);
        5    DCL NAME               CHARACTER     (25);
        6    DCL SSN                CHARACTER     (9);
        7    DCL SSN_TBL (50)       CHARACTER     (9);

        8    DCL STATUS             FIXED DECIMAL (1);
        9    DCL #GRADES            FIXED DECIMAL (3);
       10    DCL #PGMS              FIXED DECIMAL (3);
       11    DCL #TSTS              FIXED DECIMAL (3);
       12    DCL GRADE (25)         FIXED DECIMAL (3);
       13    DCL FINAL_AVG_TBL(50)  FIXED DECIMAL (7,2);
       14    DCL FINAL_TST_TBL(50)  FIXED DECIMAL (3);
       15    DCL PTS_PSBL (25)      FIXED DECIMAL (5)  INIT (0);
       16    DCL TTL_PGM_PTS        FIXED DECIMAL (5);
       17    DCL TTL_TST_PTS        FIXED DECIMAL (5);
       18    DCL PGM_PT_TTL         FIXED DECIMAL (5)  INIT (0);
       19    DCL TST_PT_TTL         FIXED DECIMAL (5)  INIT (0);

       20    DCL PGM_AVG            FIXED DECIMAL (7,2);
       21    DCL TST_AVG            FIXED DECIMAL (7,2);
       22    DCL FINAL_AVG          FIXED DECIMAL (7,2);
```

FIGURE 11–2 *(continued)*

```
23      DCL PTR              FIXED BINARY  (15)  INIT (1);
24      DCL SUB              FIXED BINARY  (15);
25      DCL NDX              FIXED BINARY  (15);
26      DCL LINE_CNT         FIXED BINARY  (15);

27      DCL PGM_INPUT        BIT          (1);
28      DCL TST_INPUT        BIT          (1);
29      DCL FAIL_PGM         BIT          (1);
30      DCL FAIL_TST         BIT          (1);
31      DCL MORE_RECS        BIT          (1);
32      DCL    YES           BIT          (1)  INIT ('1'B);
33      DCL    NO            BIT          (1)  INIT ('0'B);
34      DCL SUBSTR           BUILTIN;
35      DCL INHDR            ENTRY;
36      DCL PGHDR            ENTRY;
37      DCL AVGSRTN          ENTRY;
38      DCL PRNTRTN          ENTRY;
39      DCL SRTAVG           ENTRY;
40      DCL PRNTSSN          ENTRY;
41      DCL SORTRTN          ENTRY;

42 MAIN:    MORE_RECS = YES;
43          ON ENDFILE (SYSIN) MORE_RECS = NO;

44          CALL INHDR (CLASS, SECTION, SMSTR, PTS_PSBL, #GRADES,
```

```
PL/I OPTIMIZING COMPILER        /* CH11EX2 PROCESSING ARRAYS */              PAGE   5
   STMT
                            #PGMS, #TSTS, TTL_PGM_PTS, TTL_TST_PTS);

45          CALL PGHDR (CLASS, SMSTR, SECTION, #GRADES, PTS_PSBL,
                        TTL_PGM_PTS, TTL_TST_PTS, LINE_CNT);
46          GET LIST (STATUS, SSN, NAME);

47 PROC_LP:  DO WHILE (MORE_RECS);
48              CALL AVGSRTN (STATUS, SSN, GRADE, #GRADES, #PGMS,
                        TTL_PGM_PTS, TTL_TST_PTS, PGM_AVG,
                        TST_AVG, PGM_PT_TTL, TST_PT_TTL,
                        FINAL_AVG_TBL, FINAL_TST_TBL, SSN_TBL,
```

```
                        FINAL_AVG, PGM_INPUT, TST_INPUT, FAIL_PGM,
                        FAIL_TST, PTR);

49              IF LINE_CNT > 40 THEN
                    CALL PGHDR (CLASS, SMSTR, SECTION, #GRADES,
                            PTS_PSBL, TTL_PGM_PTS, TTL_TST_PTS,
                            LINE_CNT);

50              CALL PRNTRTN (NAME, SSN, STATUS, GRADE, #GRADES,
                            #PGMS, PGM_PT_TTL, TST_PT_TTL, PGM_AVG,
                            TST_AVG, FINAL_AVG, PGM_INPUT, TST_INPUT,
                            FAIL_PGM, FAIL_TST, LINE_CNT);

51                  GET LIST (STATUS, SSN, NAME);
52              END PROC_LP;

53          CALL SORTRTN (SSN_TBL, FINAL_AVG_TBL, FINAL_TST_TBL, PTR);
54          CALL PRNTSSN (CLASS, SECTION, SMSTR, SSN_TBL, FINAL_TST_TBL,
                            FINAL_AVG_TBL, PTR);
55          CALL SRTAVG (CLASS, SECTION, SMSTR, FINAL_AVG_TBL, PTR);
56          RETURN;
57  END CH11EX2;
```

PL/I OPTIMIZING COMPILER   /* CH11EX2 PROCESSING ARRAYS */   PAGE   6
COMPILER DIAGNOSTIC MESSAGES
ERROR ID L   STMT   MESSAGE DESCRIPTION

COMPILER INFORMATORY MESSAGES

IEL0533I I   NO 'DECLARE' STATEMENT(S) FOR 'SYSIN'.

END OF COMPILER DIAGNOSTIC MESSAGES

********** We have omitted the INHDR and PGHDR routines to save space as they **********
********** are similar to the ones in CH11EX1.   **********

PL/I OPTIMIZING COMPILER VERSION 1 RELEASE 5.1 TIME: 13.14.34 DATE: 10 MAR 87 PAGE 15
OPTIONS SPECIFIED
GOSTMT;
*PROCESS;

FIGURE 11–2 *(continued)*

```
PL/I OPTIMIZING COMPILER          /*         AVGSRTN SUBROUTINE          */      PAGE  16
                   SOURCE LISTING
    STMT


        /*            AVGSRTN SUBROUTINE          */
    1      AVGSRTN:  PROC (AR_STATUS, AR_SSN, AR_GRADE, AR_#GRADES,
                      AR_#PGMS, AR_TTL_PGM_PTS, AR_TTL_TST_PTS,
                      AR_PGM_AVG, AR_TST_AVG, AR_PGM_PT_TTL,
                      AR_TST_PT_TTL, AR_FINAL_AVG_TBL, AR_FINAL_TST_TBL,
                      AR_SSN_TBL, AR_FINAL_AVG, AR_PGM_INPUT, AR_TST_INPUT,
                      AR_FAIL_PGM, AR_FAIL_TST, AR_PTR);
        /*****************************************************************/
        /*                                                              */
        /*                  AVGSRTN SUBROUTINE                          */
        /*                                                              */
        /*  FUNCTION:  TO COMPUTE EACH STUDENT'S AVERAGE GRADE FOR      */
        /*             PROGRAMS, TESTS, AND THE COMBINED FINAL AVERAGE  */
        /*                                                              */
        /*                                                              */
        /*  PARAMETERS:  AR_STATUS                                      */
        /*               AR_SSN                                         */
        /*               AR_GRADE                                       */
        /*               AR_#GRADES                                     */
        /*               AR_#PGMS                                       */
        /*               AR_TTL_PGM_PTS                                 */
        /*               AR_TTL_TST_PTS                                 */
        /*               AR_FINAL_AVG                                   */
        /*               AR_PGM_AVG                                     */
        /*               AR_TST_AVG                                     */
        /*               AR_PGM_PT_TTL                                  */
        /*               AR_TST_PT_TTL                                  */
        /*               AR_FINAL_AVG_TBL                               */
        /*               AR_FINAL_TST_TBL                               */
        /*               AR_SSN_TBL                                     */
        /*               AR_FAIL_PGM                                    */
        /*               AR_FAIL_TST                                    */
        /*               AR_PGM_INPUT                                   */
        /*               AR_TST_INPUT                                   */
        /*               AR_PTR                                         */
        /*                                                              */
```

```
    /*  GLOBAL VARIABLES:   NONE                                     */
    /*                                                               */
    /*  SUBROUTINES CALLED:  NONE                                    */
    /*                                                               */
    /*  INPUT:    NONE                                               */
    /*                                                               */
    /*  OUTPUT:   NONE                                               */
    /*                                                               */
    /*****************************************************************/

    2    DCL AR_NAME             CHARACTER     (25);
    3    DCL AR_SSN              CHARACTER     (9);
    4    DCL AR_SSN_TBL (*)      CHARACTER     (*);
```

```
    STMT
    5    DCL AR_#PGMS            FIXED DECIMAL (3);
    6    DCL AR_#GRADES          FIXED DECIMAL (3);
    7    DCL AR_PGM_PT_TTL       FIXED DECIMAL (5);
    8    DCL AR_TST_PT_TTL       FIXED DECIMAL (5);
    9    DCL AR_STATUS           FIXED DECIMAL (1);
   10    DCL AR_GRADE (*)        FIXED DECIMAL (3);
   11    DCL AR_FINAL_AVG_TBL(*) FIXED DECIMAL (7,2);
   12    DCL AR_FINAL_TST_TBL(*) FIXED DECIMAL (3);
   13    DCL AR_TTL_PGM_PTS      FIXED DECIMAL (5);
   14    DCL AR_TTL_TST_PTS      FIXED DECIMAL (5);
   15    DCL AR_PGM_AVG          FIXED DECIMAL (7,2);
   16    DCL AR_TST_AVG          FIXED DECIMAL (7,2);
   17    DCL AR_FINAL_AVG        FIXED DECIMAL (7,2);

   18    DCL AR_PGM_INPUT        BIT           (1);
   19    DCL AR_TST_INPUT        BIT           (1);
   20    DCL AR_FAIL_PGM         BIT           (1);
   21    DCL AR_FAIL_TST         BIT           (1);

   22    DCL AR_PTR              FIXED BINARY  (15);
   23    DCL SUB                 FIXED BINARY  (15);
   24    DCL FLOAT               BUILTIN;

   25        AR_PGM_INPUT, AR_TST_INPUT, AR_FAIL_PGM, AR_FAIL_TST = '0'B;
   26        AR_SSN_TBL (AR_PTR) = AR_SSN;
```

FIGURE 11–2 (*continued*)

```
27              AR_PGM_PT_TTL = 0;
28              AR_TST_PT_TTL = 0;

29              IF AR_STATUS ^= 0 THEN
    INIT:         DO;
30                  AR_FINAL_TST_TBL (AR_PTR) = -1;
31                  AR_FINAL_AVG_TBL (AR_PTR) = -1.0;
32                END INIT;
33              ELSE

    GRD_LP:     DO;
34  RD_LP2:       DO SUB = 1 TO AR_#GRADES;
35                  GET LIST (AR_GRADE (SUB));
36                END RD_LP2;

37  ADD_PP:       DO SUB = 1 TO AR_#PGMS;
38                  AR_PGM_PT_TTL = AR_PGM_PT_TTL + AR_GRADE (SUB);
39                END ADD_PP;
40              AR_FINAL_TST_TBL (AR_PTR) = AR_GRADE (AR_#GRADES);

41  ADD_TP:     DO SUB = AR_#PGMS + 1 TO AR_#GRADES;
42                  AR_TST_PT_TTL = AR_TST_PT_TTL + AR_GRADE (SUB);
43              END ADD_TP;

44              IF AR_PGM_PT_TTL > AR_TTL_PGM_PTS THEN
                    AR_PGM_INPUT = '1'B;
45              IF AR_TST_PT_TTL > AR_TTL_TST_PTS THEN
```

```
                    AR_TST_INPUT = '1'B;
46              AR_PGM_AVG = FLOAT (AR_PGM_PT_TTL) / FLOAT
                        (AR_TTL_PGM_PTS) * 100.0;
47              AR_TST_AVG = FLOAT (AR_TST_PT_TTL) / FLOAT
                        (AR_TTL_TST_PTS) * 100.0;
48              AR_FINAL_AVG = (AR_PGM_AVG * 4 + AR_TST_AVG * 6) / 10.0;
49              AR_FINAL_AVG_TBL (AR_PTR) = AR_FINAL_AVG;

50              IF AR_PGM_AVG < 68.0 THEN
```

```
                          AR_FAIL_PGM = '1'B;
      51              IF AR_TST_AVG < 68.0 THEN
                          AR_FAIL_TST = '1'B;
      52          END GRD_LP;

      53          AR_PTR = AR_PTR + 1;
      54          RETURN;
      55      END AVGSRTN;
```

PL/I OPTIMIZING COMPILER   /*   AVGSRTN SUBROUTINE   */   PAGE 19
COMPILER DIAGNOSTIC MESSAGES
ERROR ID L   STMT   MESSAGE DESCRIPTION

COMPILER INFORMATORY MESSAGES

IEL0533I I   NO 'DECLARE' STATEMENT(S) FOR 'SYSIN'.
IEL0430I I   1   NO 'MAIN' OPTION ON EXTERNAL PROCEDURE.

END OF COMPILER DIAGNOSTIC MESSAGES

PL/I OPTIMIZING COMPILER VERSION 1 RELEASE 5.1 TIME: 13.14.42 DATE: 10 MAR 87 PAGE 20
OPTIONS SPECIFIED
GOSTMT;
*PROCESS;
PL/I OPTIMIZING COMPILER   /*   PRNTRTN SUBROUTINE   */   PAGE 21
              SOURCE LISTING
    STMT

      /*   PRNTRTN SUBROUTINE   */
      1   PRNTRTN: PROC (PR_NAME, PR_SSN, PR_STATUS, PR_GRADE, PR_#GRADES,
                    PR_#PGMS, PR_PGM_PT_TTL, PR_TST_PT_TTL, PR_PGM_AVG,
                    PR_TST_AVG, PR_FINAL_AVG, PR_PGM_INPUT, PR_TST_INPUT,
                    PR_FAIL_PGM, PR_FAIL_TST, PR_LINE_CNT);
      /**/
      /*   */
      /*   PRNTRTN SUBROUTINE   */
      /*   */
      /* FUNCTION: TO PRINT THE DETAIL LINES FOR EACH STUDENT   */

FIGURE 11–2 *(continued)*

```
/*                                                                    */
/*                                                                    */
/*  PARAMETERS:  PR_NAME                                              */
/*               PR_SSN                                               */
/*               PR_STATUS                                            */
/*               PR_GRADE                                             */
/*               PR_#GRADES                                           */
/*               PR_#PGMS                                             */
/*               PR_PGM_PT_TTL                                        */
/*               PR_TST_PT_TTL                                        */
/*               PR_PGM_AVG                                           */
/*               PR_TST_AVG                                           */
/*               PR_FINAL_AVG                                         */
/*               PR_PGM_INPUT                                         */
/*               PR_TST_INPUT                                         */
/*               PR_LINE_CNT                                          */
/*                                                                    */
/*  GLOBAL VARIABLES:  NONE                                           */
/*                                                                    */
/*  SUBROUTINES CALLED:  NONE                                         */
/*                                                                    */
/*  INPUT:    NONE                                                    */
/*                                                                    */
/*  OUTPUT:  DETAIL LINES FOR EACH STUDENT                            */
/*                                                                    */
/**********************************************************************/

2     DCL PR_SSN1          CHARACTER     (3);
3     DCL PR_SSN2          CHARACTER     (2);
4     DCL PR_SSN3          CHARACTER     (4);
5     DCL PR_NAME          CHARACTER     (25);
6     DCL PR_SSN           CHARACTER     (9);

7     DCL PR_#PGMS         FIXED DECIMAL (3);
8     DCL PR_#GRADES       FIXED DECIMAL (3);
9     DCL PR_PGM_PT_TTL    FIXED DECIMAL (5);
10    DCL PR_TST_PT_TTL    FIXED DECIMAL (5);
11    DCL PR_STATUS        FIXED DECIMAL (1);
12    DCL PR_GRADE (*)     FIXED DECIMAL (3);
13    DCL PR_PGM_AVG       FIXED DECIMAL (7,2);
```

```
    STMT
     14     DCL PR_TST_AVG           FIXED DECIMAL (7,2);
     15     DCL PR_FINAL_AVG         FIXED DECIMAL (7,2);

     16     DCL PR_PGM_INPUT         BIT           (1);
     17     DCL PR_TST_INPUT         BIT           (1);
     18     DCL PR_FAIL_PGM          BIT           (1);
     19     DCL PR_FAIL_TST          BIT           (1);

     20     DCL PR_LINE_CNT          FIXED BINARY  (15);
     21     DCL NDX                  FIXED BINARY  (15);
     22     DCL SUBSTR               BUILTIN;

     23          PR_SSN1 = SUBSTR (PR_SSN, 1, 3);
     24          PR_SSN2 = SUBSTR (PR_SSN, 4, 2);
     25          PR_SSN3 = SUBSTR (PR_SSN, 6, 4);

     26          PUT SKIP (3) EDIT (PR_NAME,
                       PR_SSN1 || '-' || PR_SSN2 || '-' || PR_SSN3)
                       (COL(5), A (30), A(16));

     27 SLCT_S:   SELECT (PR_STATUS);
     28              WHEN (2)
                       PUT SKIP (0) EDIT
                           ('*** WITHDREW FAILING ***') (COL (50), A(25));
     29              WHEN (1)
                       PUT SKIP (0) EDIT
                           ('*** WITHDREW PASSING ***') (COL (50), A(25));
     30              OTHERWISE
        DETAIL:        DO;
     31                 PUT SKIP (0) EDIT  ('***** FINAL AVG =',
                           PR_FINAL_AVG, 'PROGRAM AVG =', PR_PGM_AVG,
                           'TEST AVG =', PR_TST_AVG) (COL (50), A(17),
                           F(7,2), X(5), A(13), F(7,2), X(5), A(10),
                           F(7,2));

     32 SLCT_E:        SELECT;
     33                  WHEN (PR_PGM_INPUT = '1'B & PR_TST_INPUT = '1'B)
                           PUT SKIP EDIT ('*** INPUT ERROR:   BOTH')
                                   (COL (2), A(25));
```

FIGURE 11-2 *(continued)*

```
34                      WHEN (PR_PGM_INPUT = '1'B)
                            PUT SKIP EDIT ('*** INPUT ERROR:  PROGRAMS')
                                     (COL (2), A(25));
35                      WHEN (PR_TST_INPUT = '1'B)
                            PUT SKIP EDIT ('*** INPUT ERROR:  TESTS')
                                     (COL (2), A(25));
36                   OTHERWISE
      CHK_F:             DO;

37  SLCT_F:            SELECT;
38                       WHEN (PR_FAIL_PGM = '1'B & PR_FAIL_TST = '1'B)
                             PUT SKIP EDIT ('*** PR_FAILED BOTH ***')
                                          (COL (2), A(30));
```

```
PL/I OPTIMIZING COMPILER       /*          PRNTRTN SUBROUTINE          */      PAGE  23
  STMT
  39                       WHEN (PR_FAIL_PGM = '1'B)
                             PUT SKIP EDIT ('*** FAILED PROGRAMS ***')
                                       (COL (2), A(30));
  40                       WHEN (PR_FAIL_TST = '1'B)
                             PUT SKIP EDIT ('*** FAILED TESTS ***')
                                       (COL (2), A(30));
  41                       OTHERWISE
                             PUT SKIP;
  42                     END SLCT_F;
  43                   END CHK_F;
  44                 END SLCT_E;

  45                 PUT SKIP EDIT ('TOTAL PGM PTS =' ,
                         PR_PGM_PT_TTL, 'PROGRAM GRADES =')
                           (COL (10), A(18), F(5), X(5), A(16));

  46  PT_LP2:        DO NDX = 1 TO PR_#PGMS;
  47                     PUT EDIT (PR_GRADE (NDX)) ((20)F(4));
  48                 END PT_LP2;

  49                 PUT SKIP EDIT ('TOTAL TEST POINTS =',
                         PR_TST_PT_TTL, 'TEST GRADES =')
                           (COL (9), A(19), F(5), X(8), A(13));
```

```
50  PT_LP3:        DO NDX = NDX TO PR_#GRADES;
51                    PUT EDIT (PR_GRADE (NDX)) ((5)F(4));
52                  END PT_LP3;
53                END DETAIL;
54            END SLCT_S;

55            PR_LINE_CNT = PR_LINE_CNT + 6;
56            RETURN;
57  END PRNTRTN;
```

PL/I OPTIMIZING COMPILER   /*   PRNTRTN SUBROUTINE   */   PAGE 24
COMPILER DIAGNOSTIC MESSAGES
ERROR ID L   STMT   MESSAGE DESCRIPTION

COMPILER INFORMATORY MESSAGES

IEL0533I I   NO 'DECLARE' STATEMENT(S) FOR 'SYSPRINT'.
IEL0430I I   1   NO 'MAIN' OPTION ON EXTERNAL PROCEDURE.

END OF COMPILER DIAGNOSTIC MESSAGES

PL/I OPTIMIZING COMPILER VERSION 1 RELEASE 5.1 TIME: 13.14.53 DATE: 10 MAR 87 PAGE 25
OPTIONS SPECIFIED
GOSTMT;
*PROCESS;

PL/I OPTIMIZING COMPILER   /*   SORTRTN SUBROUTINE   */   PAGE 26
             SOURCE LISTING
   STMT

      /*   SORTRTN SUBROUTINE   */
   1   SORTRTN: PROC (SR_SSN_TBL, SR_FINAL_AVG_TBL, SR_FINAL_TST_TBL,
               SR_PTR);
      /**/
      /*   */
      /*   SORTRTN SUBROUTINE   */
      /*   */
      /* FUNCTION: TO SORT SR_SSN_TBL IN ASCENDING ORDER ON   */
      /*   STUDENT SOCIAL SECURITY NUMBER KEEPING FINAL_ */

FIGURE 11–2 *(continued)*

```
    /*              AVG_TBL PARALLEL                              */
    /*                                                            */
    /*  PARAMETERS:  SR_SSN_TBL                                   */
    /*                 SR_FINAL_AVG_TBL                           */
    /*                 SR_PTR                                     */
    /*                                                            */
    /*     GLOBAL VARIABLES:  NONE                                */
    /*                                                            */
    /*     SUBROUTINES CALLED:  NONE                              */
    /*     INPUT:    NONE                                         */
    /*                                                            */
    /*     OUTPUT:   NONE                                         */
    /*                                                            */
    /***************************************************************/

    2     DCL SR_SAVE_SSN         CHARACTER      (9);
    3     DCL SR_SSN_TBL (*)      CHARACTER      (*);

    4     DCL SR_SAVE_AVG         FIXED DECIMAL (7,2);
    5     DCL SR_SAVE_TST         FIXED DECIMAL (7,2);
    6     DCL SR_FINAL_AVG_TBL (*) FIXED DECIMAL (7,2);
    7     DCL SR_FINAL_TST_TBL (*) FIXED DECIMAL (3);

    8     DCL SR_PTR              FIXED BINARY  (15);
    9     DCL NDX                 FIXED BINARY  (15);
   10     DCL SUB                 FIXED BINARY  (15);

   11     DCL SWAP            BIT        (1)  INIT ('0'B);
   12     DCL    ON           BIT        (1)  INIT ('1'B);
   13     DCL    OFF          BIT        (1)  INIT ('0'B);

   14          SUB = SR_PTR - 1;
   15 SORT_LP:  DO WHILE (SWAP = OFF : SUB > 2);
   16              SWAP = ON;

   17 SCAN:          DO NDX = 2 TO SUB;
   18                  SWAP = OFF;
   19                  IF SR_SSN_TBL (NDX-1) > SR_SSN_TBL (NDX) THEN
      SWAPS:              DO;
   20                        SR_SAVE_SSN = SR_SSN_TBL (NDX);
```

```
     21                           SR_SSN_TBL (NDX) = SR_SSN_TBL (NDX-1);
     22                           SR_SSN_TBL (NDX-1) = SR_SAVE_SSN;
```

PL/I OPTIMIZING COMPILER   /*   SORTRTN SUBROUTINE   */   PAGE 27
  STMT
```
     23                           SR_SAVE_TST = SR_FINAL_TST_TBL (NDX);
     24                           SR_FINAL_TST_TBL (NDX) = SR_FINAL_TST_TBL
                                  (NDX - 1);
     25                           SR_FINAL_TST_TBL (NDX - 1) = SR_SAVE_TST;
     26                           SR_SAVE_AVG = SR_FINAL_AVG_TBL (NDX);
     27                           SR_FINAL_AVG_TBL (NDX) = SR_FINAL_AVG_TBL
                                                        (NDX - 1);
     28                           SR_FINAL_AVG_TBL (NDX - 1) = SR_SAVE_AVG;
     29                        END SWAPS;
     30                      END SCAN;

     31                    SUB = SUB - 1;
     32                  END SORT_LP;
     33              RETURN;
     34   END SORTRTN;
```

PL/I OPTIMIZING COMPILER   /*   SORTRTN SUBROUTINE   */   PAGE 28
COMPILER DIAGNOSTIC MESSAGES
ERROR ID L   STMT   MESSAGE DESCRIPTION

COMPILER INFORMATORY MESSAGES

IEL0430I I   1   NO 'MAIN' OPTION ON EXTERNAL PROCEDURE.

END OF COMPILER DIAGNOSTIC MESSAGES

PL/I OPTIMIZING COMPILER VERSION 1 RELEASE 5.1 TIME: 13.14.57 DATE: 10 MAR 87 PAGE 29
OPTIONS SPECIFIED
GOSTMT;
*PROCESS;

PL/I OPTIMIZING COMPILER   /*   PRNTSSN SUBROUTINE   */   PAGE 30
                SOURCE LISTING

```
         /*          PRNTSSN SUBROUTINE          */
    1    PRNTSSN: PROC (PS_CLASS, PS_SECTION, PS_SMSTR, PS_SSN_TBL,
                        PS_FINAL_TST_TBL, PS_FINAL_AVG_TBL, PS_PTR);
         /*********************************************************************/
         /*                                                                 */
         /*                    PRNTSSN SUBROUTINE                           */
         /*                                                                 */
         /*  FUNCTION:  TO PRINT THE SORTED SSN'S, FINAL AVERAGE, AND       */
         /*             FINAL TEST GRADES WITH SPACE FOR LETTER GRADE        */
         /*                                                                 */
         /*  PARAMETERS:  PS_SSN_TBL                                        */
         /*               PS_FINAL_TST_TBL                                  */
         /*               PS_FINAL_AVG_TBL                                  */
         /*               PS_PNTR                                           */
         /*                                                                 */
         /*  GLOBAL VARIABLES: NONE                                         */
         /*                                                                 */
         /*  SUBROUTINES CALLED:  NONE                                      */
         /*                                                                 */
         /*  INPUT:   NONE                                                  */
         /*  OUTPUT:  FOR EACH STUDENT:  A 5-COLUMN LINE FOR LETTER GRADE,  */
         /*             SOCIAL SECURITY NUMBER AND GRADE ON LAST TEST        */
         /*                                                                 */
         /*********************************************************************/

    2    DCL PS_CLASS              CHARACTER     (*);
    3    DCL PS_SECTION            CHARACTER     (*);
    4    DCL PS_SMSTR              CHARACTER     (*);
    5    DCL PS_SSN_TBL (*)        CHARACTER     (*);
    6    DCL PS_SSN1               CHARACTER     (3);
    7    DCL PS_SSN2               CHARACTER     (2);
    8    DCL PS_SSN3               CHARACTER     (4);

    9    DCL PS_FINAL_TST_TBL (*) FIXED DECIMAL (3);
   10    DCL PS_FINAL_AVG_TBL (*) FIXED DECIMAL (7,2);

   11    DCL PS_PTR                FIXED BINARY  (15);
   12    DCL NDX                   FIXED BINARY  (15);
```

```
13    DCL SUB              FIXED BINARY (15);

14    DCL SUBSTR           BUILTIN;

15        PUT PAGE EDIT ('CSCI ', PS_CLASS, 'SEC', PS_SECTION,
                'GRADE ROLL', PS_SMSTR) (COL (40), A(5), A(5), A(4), A(2),
                X(5), A(11), X(5), A(11));
16            PUT SKIP (2) EDIT ('COURSE GRADE',
                   'TEST GRADE', 'COURSE AVG') (COL (35), A(31), A(15),
                   A(10));
17            PUT SKIP (0) EDIT ((12) '_', (10) '_',
```

PL/I OPTIMIZING COMPILER   /*   PRNTSSN SUBROUTINE   */   PAGE 31

PL/I OPTIMIZING COMPILER   /*   PRNTSSN SUBROUTINE   */   PAGE 31

```
   STMT
                        (10) '_') (COL (35), A(31), A(15), A(10));

18  WRT_LP:    DO NDX = 1 TO PS_PTR - 1;
19               IF PS_FINAL_AVG_TBL (NDX) > -1.0 THEN
    WRT_DTL:        DO;
20                     PS_SSN1 = SUBSTR (PS_SSN_TBL (NDX), 1, 3);
21                     PS_SSN2 = SUBSTR (PS_SSN_TBL (NDX), 4, 2);
22                     PS_SSN3 = SUBSTR (PS_SSN_TBL (NDX), 6, 4);

23                     PUT SKIP (2) EDIT ('___',
                          PS_SSN1 || '-' || PS_SSN2 || '-' || PS_SSN3,
                          PS_FINAL_TST_TBL (NDX), PS_FINAL_AVG_TBL (NDX))
                          (COL (41), A(8),A(20), F(3), X(10), F(7,2));
24                 END WRT_DTL;
25             END WRT_LP;
26             RETURN;
27  END PRNTSSN;
```

```
COMPILER DIAGNOSTIC MESSAGES
ERROR ID L   STMT   MESSAGE DESCRIPTION

COMPILER INFORMATORY MESSAGES

IEL0533I I          NO 'DECLARE' STATEMENT(S) FOR 'SYSPRINT'.
IEL0430I I   1      NO 'MAIN' OPTION ON EXTERNAL PROCEDURE.
```

FIGURE 11–2 *(continued)*

END OF COMPILER DIAGNOSTIC MESSAGES

PL/I OPTIMIZING COMPILER VERSION 1 RELEASE 5.1 TIME: 13.15.06 DATE: 10 MAR 87 PAGE 33
OPTIONS SPECIFIED
GOSTMT;
*PROCESS;

PL/I OPTIMIZING COMPILER   /*   SRTAVG SUBROUTINE   */   PAGE 34
                    SOURCE LISTING

    STMT


```
      /*          SRTAVG SUBROUTINE          */
  1     SRTAVG: PROC (SA_CLASS, SA_SECTION, SA_SMSTR, SA_FINAL_AVG_TBL,
                SA_PTR);
      /*****************************************************************/
      /*                                                             */
      /*                    SRTAVG SUBROUTINE                        */
      /*                                                             */
      /*  FUNCTION:  TO SORT FINAL AVERAGES IN DESCENDING ORDER AND  */
      /*             PRINT THEM                                      */
      /*                                                             */
      /*  PARAMETERS:  SA_CLASS                                      */
      /*               SA_SECTION                                    */
      /*               SA_SMSTR                                      */
      /*               SA_FINAL_AVG_TBL                              */
      /*               SA_PTR                                        */
      /*                                                             */
      /*  GLOBAL VARIABLES: NONE                                     */
      /*                                                             */
      /*  SUBROUTINES CALLED:  NONE                                  */
      /*                                                             */
      /*  INPUT:   NONE                                              */
      /*  OUTPUT:  THE CONTENTS OF THE SA_FINAL_AVG_TBL SORTED IN    */
      /*           DESCENDING ORDER                                  */
      /*                                                             */
      /*****************************************************************/

  2     DCL SA_CLASS          CHARACTER    (4);
```

```
    3       DCL SA_SECTION          CHARACTER   (2);
    4       DCL SA_SMSTR            CHARACTER   (11);


    5       DCL SA_FINAL AVG_TBL (*) FIXED DECIMAL (7,2);
    6       DCL SA_SAVE_AVG         FIXED DECIMAL (7,2);


    7       DCL SA_PTR              FIXED BINARY  (15);
    8       DCL SUB                 FIXED BINARY  (15);
    9       DCL NDX                 FIXED BINARY  (15);


   10       DCL SWAP            BIT      (1)  INIT ('0'B);
   11       DCL   ON            BIT      (1)  INIT ('1'B);
   12       DCL   OFF           BIT      (1)  INIT ('0'B);


   13           SUB = SA_PTR - 1;
   14           SWAP = OFF;
   15  SRT_AV:  DO WHILE (SWAP = OFF ! SUB > 2);
   16              SWAP = ON;


   17  SCN_AV:     DO NDX = 2 TO SUB;
   18                 SWAP = OFF;
   19                 IF SA_FINAL_AVG_TBL (NDX-1) < SA_FINAL_AVG_TBL
                         (NDX) THEN
       SWP_AV:            DO;
```

```
   STMT
   20                 SA_SAVE_AVG = SA_FINAL_AVG_TBL (NDX);
   21                 SA_FINAL_AVG_TBL (NDX) = SA_FINAL_AVG_TBL
                         (NDX -1);
   22                 SA_FINAL_AVG_TBL (NDX-1) = SA_SAVE_AVG;
   23              END SWP_AV;
   24           END SCN_AV;


   25           SUB = SUB - 1;
   26        END SRT_AV;


   27        PUT PAGE EDIT ('CSCI ', SA_CLASS, 'SEC', SA_SECTION,
                'GRADE ROLL', SA_SMSTR) (COL (40), A(5), A(5), A(4),
                A(4), X(5), A(11), X(5), A(11));
```

FIGURE 11-2 *(continued)*

```
28  WRT_AV:  DO NDX = 1 TO SA_PTR - 1;
29              IF SA_FINAL_AVG_TBL (NDX) > -1.0 THEN
                    PUT SKIP EDIT (SA_FINAL_AVG_TBL
                                      (NDX)) (COL (61), F(7,2));
30            END WRT_AV;
31  RETURN;
32  END SRTAVG;
```


PL/I OPTIMIZING COMPILER   /*   SRTAVG SUBROUTINE   */   PAGE 36
COMPILER DIAGNOSTIC MESSAGES
ERROR ID L   STMT   MESSAGE DESCRIPTION

COMPILER INFORMATORY MESSAGES

IEL0533I I   NO 'DECLARE' STATEMENT(S) FOR 'SYSPRINT'.
IEL0430I I   1   NO 'MAIN' OPTION ON EXTERNAL PROCEDURE.

END OF COMPILER DIAGNOSTIC MESSAGES

********** We have omitted the first pages of CH11EX2 output as it would be   **********
********** the same as in CH11EX1.   **********

                         CSCI 260B SEC 2   GRADE ROLL   SPRING 1988

                 COURSE GRADE   TEST GRADE   COURSE AVG

                    ___   111-11-6789   101   101.39

                    ___   111-22-2333   103   99.09

                    ___   112-23-3445   63   83.39

                    ___   123-45-6789   75   86.79

                    ___   135-79-2468   60   70.79

                    ___   192-83-7465   68   66.19

                    ___   223-34-4556   90   86.49

---	234-56-7890	95	96.29
---	321-65-4987	87	66.49
---	382-84-7291	73	77.99
---	573-93-9789	70	78.99

```
                    CSCI 260B SEC 2      GRADE ROLL      SPRING 1988
                               101.39
                                99.09
                                96.29
                                86.79
                                86.49
                                83.39
                                78.99
                                77.99
                                70.79
                                66.49
                                66.19
```

11.7

Summary

An array is one type of data aggregate, which is a group of data items that can be referred to either individually or collectively. An array is distinguished from other types of data aggregates by the fact that all its elements must have identical attributes. Arrays become necessary when you need to have a set of data items in memory at once; sorting and searching are common tasks requiring the array.

To declare an array you must specify the upper bound for each dimension; for example,

```
DCL AN_ARRAY (30) FIXED DEC (7,2);
```

If the lower bound is other than 1, you must include it also, separated from the upper bound by a colon:

```
DCL ARRAY_X (10:100) CHAR (5);
```

The lower and upper bounds are the numbers by which you reference the first and last elements of the array. You can express bounds as literals, variables, or expressions. An array also has an extent for each dimension; the extent is the number of elements in the array. Several rules apply to the declaration of arrays, and the following are the most important.

1. The dimension specification is enclosed in parentheses, and it must follow the declaration of the array name with nothing but optional blanks intervening.

2. One bound must be specified for each dimension.

3. The upper bound must be equal to or greater than the lower bound.

4. The default lower bound is 1.

5. When both lower and upper bounds are explicitly declared, they must be separated by a colon.

6. The array declaration can have the INITIAL attribute. You must provide a value for each element you initialized.

To reference an entire array, you need code only the array name. For example,

```
GET LIST (ARRAY_Y);
```

will read a value into each array element, assuming there are enough input data items. If there are too few input data items, an error condition exists. To reference an individual element in an array, you must use a subscript, which can be a literal, variable, or expression. You specify the subscript immediately following the array name and enclosed in parentheses. The subscript value cannot be less than the lower bound or greater than the upper bound.

The iterative DO is a particularly useful statement for manipulating arrays. The syntax of the iterative DO is

```
DO control variable = initial value TO limit value BY
     modification value;
```

The value in the control variable provides the subscript. PL/I automatically sets the control variable to the initial value before the loop executes the first time and modifies it by the modification value each time the loop executes. When the iterative DO appears in an I/O statement, it has no corresponding END statement.

```
GET LIST ((ARRAY_Q (SUB) DO SUB = 1 TO #_ITEMS));
```

In this example we do not provide modification value, and so it defaults to 1. Among the possible variations of the basic iterative DO statements are negative initial, limit, and modification values; fractional initial, limit, and modification values; multiple specifications; random values (in which case no modification value is specified); and the iterative DO with no modification value. In addition, you can append a WHILE (condition) or UNTIL (condition) to the iterative DO. When you choose these options, PL/I first checks the control variable for a value not greater than the limit value. If this test evaluates to true, PL/I tests the WHILE or UNTIL condition. It, too, must evaluate to true in order for the loop to execute. Iterative DO statements can also be nested. The higher-level control variable is held at each of its values while the lower-level control variable spins through each of its values.

PL/I offers two kinds of assignment operations for arrays: scalar-to-array and array-to-array. In the first, if the array name is unsubscripted, every element receives the scalar value. If the array name is subscripted, only the referenced element receives the scalar value. To use the array-to-array assignment, you must be certain that both arrays have the same bounds. Each element in the array named on the left receives the value in the corresponding element of the array named on the right. If an array expression is on the right hand side of the assignment symbol, the expression is evaluated for each element in the right-hand array, and the result is placed in the corresponding element of the left-hand array.

Arrays can appear in argument and parameter lists, but they must be unsubscripted in the parameter list. In the subroutine you can (PL/I) or must (PL/C) declare the dimensions of a parameter list array with an asterisk. Doing so allows the subroutine to receive arrays of different extents on different calls to it.

Single-dimensional arrays are used constantly in business data processing, and two-dimensional arrays are commonly used. Arrays of more than two dimensions are seldom encountered. You must declare bounds for each dimension, and you must provide a subscript for each dimension when you reference a particular element of an array.

The most useful of the BUILTIN functions for array manipulation are DIM, HBOUND, LBOUND, PROD, and SUM. The following table explains the use of each.

Two sort algorithms that are good for introducing beginners to sorting are the bubble sort and the insertion sort. The bubble sort compares Array (n) with Array (n + 1) and exchanges them as necessary. The insertion sort compares Array (n) successively with Array (n + 1), Array (n + 2), and so forth, until an exchange is necessary, and then it makes the exchange. When

Function	Returns
DIM	extent of specified dimension
HBOUND	value representing the higher bound
LBOUND	value representing the lower bound
PROD	product of all elements in the array
SUM	sum of all elements in the array

data in an array is unsorted, you can use only the sequential search to locate a value matching the search argument value. A sequential search requires extent/2 seeks for the average search and extent seeks for the worst-case search. If the array data is sorted, however, the binary search can eliminate one-half of (remainder of) the array after each seek. The larger the array, the greater the efficiency of the binary search over the sequential search.

11.8

Exercises

1. For each of the following array declarations give the number of dimensions, lower and upper bounds, and extent of each dimension.

   a.   `DCL ARRAY_1 (-6:0, 20) CHAR (3);`
   b.   `DCL ARRAY_2 (256) FIXED BIN (15);`
   c.   `DCL ARRAY_3 (20, 40:45, 30) FIXED DEC (2);`
   d.   `DCL ARRAY_4 (10.75:12.5) CHAR (2);`

2. Use the DECLARE statement to initialize the following as specified.

   a.   `DCL TABLE_1 (60) FIXED DEC (5);`
       Set all elements to 0.
   b.   `DCL TABLE_2 (16) FIXED DEC (2);`
       Set the first five elements to 99.
   c.   `DCL TABLE_3 (7) CHAR (3);`
       Set the following: `TABLE (2) ← 'AB'; TABLE (5)`
                               `← 'CD'; TABLE (7) ← 'EF'.`
   d.   `DCL TABLE_4 (6) CHAR (3);`
       Set the elements to the abbreviations of the first six months of the year.

3. For each of the arrays described below, give a possible declaration. If the array could be declared in several ways, give at least two declarations.

   a.   An array of twelve 5-character elements numbered 1 through 12.
   b.   An array of 50 fixed-decimal elements with four digits to the left and two digits to the right of the assumed decimal point.
   c.   An array numbered from −10 to 0, each element capable of containing a two-digit whole number.

d. An array numbered from 1900 to 1975, each element capable of holding a fixed decimal with six digits to the left and three to the right of the implied decimal point.

4. Code the input statement to read a value into each element of each array described in Exercise 3. Use list-directed input and the iterative DO within the input statement.

5. Using the iterative DO, write the code to accomplish each of the following. Assume the variable L_BND contains the lower bound and H_BND the higher bound.

a. `DCL ARRAY_5 (30) FIXED DEC (3);`
Increment the value in the lower-bound element and every fifth element after that by 15.

b. `DCL ARRAY_6 (3.5:5.75) CHAR (4);`
Starting with the higher bound add the contents to the element next to it. Add the contents of the higher bound to the lower bound.

c. `DCL ARRAY_7 (350) FIXED DEC (3,2);`
Assuming N contains an unknown value of <= 350, print the first N elements of the array.

d. `DCL ARRAY_8 (-39:+40) CHAR (2);`
Set the following elements (in the order listed) to XX: 0, 12, −16, 40, −37.

6. List the value in the control variable immediately after each iterative DO in Exercise 5 has finished executing (for item c list your answer as N + ?).

7. For each of the following, list the number of times the iterative DO will execute.

a. `DCL TABLE_5 (40) FIXED DEC (40);`
`DO SUB = 6 TO 40 BY 3;`

b. `DCL TABLE_6 (-5:5) CHAR (20);`
`L_BND = -5;`
`H_BND = +5;`
`DO SUB = 0 TO H_BND;`

c. `DCL TABLE_7 (50:62) FIXED DEC (4, 2);`
`NDX = 50;`
`DO SUB = NDX + 3 TO 60 BY 4;`

d. `DCL TABLE_8 (1.25:1.75) CHAR (6);`
`DO SUB = 0 TO 1.75 BY .25;`

8. Use the assignment statement to initialize the following array to the current year and successive years.

`DCL YR_ARRAY (10) FIXED DEC (4);`

9. Given the following two arrays, write the code to accomplish each task below.

```
DCL ARRAY_9  (25) FIXED DEC (3);
DCL ARRAY_10 (25) FIXED DEC (3);
```

    a.   Assign the values in ARRAY_10 to ARRAY_9.
    b.   Cause ARRAY_10 to have the ARRAY_9 values minus 5.
    c.   Cause ARRAY_9 to have the value for each element of the current value plus the value in the corresponding element of ARRAY_10.
    d.   Put the first N values of ARRAY_10 in the first N elements of ARRAY_9.

10. Given the following array and input data, show the contents of the array after the GET LIST statement has executed.

```
DCL ARRAY_11 (3,3) FIXED DEC (2);
```

    Input data: 10 12 14 16 18 20 22 24 26

```
GET LIST (((ARRAY_11 (ROW, COL) DO ROW = 1 TO 3) DO COL
          = 1 TO 3));
```

11. Use the following array declarations to list the value returned by each of the calls to BUILTIN functions.

```
DCL ARRAY_12 (6:8, -3:+3) FIXED DEC (2);
```

    a.   `VALUE = DIM (ARRAY_12, 2);`
    b.   `VALUE - HBOUND (ARRAY_12, 1);`
    c.   `VALUE = LBOUND (ARRAY_12, 2);`

12. Given that TABLE_9 contains the following data, what value is returned by the BUILTIN functions?

02	04	06	10

    a.   `VALUE = SUM (TABLE_9);`
    b.   `VALUE = PROD (TABLE_9);`

13. Given the following array, show its appearance after each pass of the bubble sort that is sorting in (a) ascending order and (b) descending order.

21	07	16	32	85

14. Use the array in Exercise 13 to show its appearance after each pass of an insertion sort that is sorting in (a) ascending order and (b) descending order.

15. Given the following array, how many seeks would a binary search have to make to find a match for each argument (or to determine no match is possible)? Assume we are truncating any fraction in locating the middle element.

10	11	12	13	14	15	16	17	18	19	20	21	22	23	24	25	26	27	28

a. ARG = 18;
b. ARG = 27;
c. ARG = 13;
d. ARG = 10;
e. ARG = 20;
f. ARG = 16;

16. Write a program that will read n numbers into an array. The value on the header record is n and is <= 100. Go through the array element by element and if the value is positive (or zero) move it to POS_ARRAY; if the value is negative, move it to NEG_ARRAY. Be sure you leave no empty elements in POS_#ARRAY and NEG-ARRAY. Print the portion of each array that has meaningful values in it.

17. Write a program to create a "look-up table" for a local information telephone operator. The first set of records will contain a seven-digit phone number and a 40-character name-address. The end of this set of records is indicated by a dummy record containing a phone number of 5555555. The next set of records contains only the name-address field and represents inquiries to the information operator. Read the phone numbers (excluding 5555555) into PHONE array and name- addresses into CUST_NAME_ADDR array. Using either the bubble or the insertion sort, perform a sort in ascending sequence on CUST_NAME_ADDR. Be sure you move the corresponding phone number each time you move a name- address. When the array is sorted, use a binary search to process the requests for telephone numbers. For each request, print the name-address given in the request and the proper phone number. If the name-address in the request is not in the directory, write an appropriate message following the name-address.

18. Telephone companies sometimes get legitimate requests (from the police, etc.) for the name-address corresponding to a known phone number. Rewrite the program of Exercise 17 to handle these requests. In this case, the input records following 5555555 will be phone numbers instead of name-addresses.

19. Write a program to compute the gross monthly sales of each of ten salespeople. The program will run at the end of each quarter of the fiscal year, and the header record will tell how many months are involved for this run (3 = first quarter; 12 = end of fiscal year). Your program should calculate the total sales to date for each salesperson and for all ten salespeople combined. It should then calculate the average sales to date for each salesperson and for all ten salespeople combined.

20. Write a program to keep track of the number of times each of 20 songs is played on a jukebox. The input will consist of two sets of records: first, a list of 20 songs. This set of records ends when a title of XXXXX is encountered. The second set of records consists of a series of numbers from 1 through 20. Each number represents one time the corresponding song was played. Read the song titles into one array and prepare a second array to hold the totals for each song. Then read the second set of records and update the times that song was played. Hint: Use the input number as the subscript into the times-played array. Print the names of the songs and the number of times each was played.

21. Add to the program of Question 20 a routine to sort the times-played array in ascending order (be sure to keep the song-title array in parallel order). Then print a bar graph showing how many times each song was played.

```
THE OLD RUGGED CROSS |$$$$$$$$$$$$$$$$$$$$$$
HOW GREAT THOU ART    |$$$$$$$$$$$$$$$$$$$$$$$$$$$$$
AMAZING GRACE         |$$$$$$$$$$$$$$$$$$$$$$$$$$$$$$$$$$$$$$$$$$$$$
```

C H A P T E R   1 2

STRUCTURES

In Chapter 11 we introduced arrays as the first of the data aggregates, and in this chapter we introduce the second, structures. A data aggregate, you remember, is a collection of data items that can be referred to either individually or collectively. Since both arrays and structures are data aggregates, perhaps the best way to introduce structures is to contrast them with arrays, and there are three important contrasts:

1. All elements of an array must have identical attributes, but the elements of a structure can have different attributes.

2. All elements of an array have the same name, and we reference the items collectively by simply referencing the array name. To reference a single item (element), on the other hand, we append a subscript to the array name. In a structure, however, each variable has, or can have, a unique name, and you will see in Section 12.1.1 how we refer to data items collectively.

3. In an array, no element can claim to be in any sense "more important" than any other; the only relationship among array elements is that each is an element in the array. A structure, however, creates hierarchical relationships among its parts; that is, necessary relationships are established, and each variable is not necessarily on an "equal" basis with every other.

12.1

Declaring Structures

Suppose we have input records consisting of a 5-digit customer number, a 25-character customer name, a 35-digit customer address, and a 5-digit amount due. Without using a structure, our first option for describing this record is

```
DCL CUST_REC        CHAR (70);
```

This declaration has two obvious drawbacks. The first is that we cannot reference any data items within the record; in other words, the entire record is a single data item. The second is that we should not perform any arithmetic on the amount-due field because we have declared it to be character data.

The second possibility for describing the record is

```
DCL CUST_ID         FIXED DEC (5);
DCL CUST_NAME       CHAR      (25);
DCL CUST_ADDRESS    CHAR      (35);
DCL AMNT_OWED       FIXED DEC (5,2);
```

Notice that this declaration allows us to access only the customer's entire name and entire address, unless we want to invoke the SUBSTR function. At the same time, we have given up the possibility of referencing the entire record with one variable name. To enable us to reference the various portions of the customer name and address, we might choose to describe the record as

```
DCL CUST_ID          FIXED DEC (5),
    CUST_LAST_NAME   CHAR      (15),
    CUST_FIRST_NAME  CHAR      (10),
    CUST_STREET      CHAR      (15),
    CUST_CITY_STATE  CHAR      (15),
    CUST_ZIP         CHAR      (5),
    AMNT_OWED        FIXED DEC (5,2);
```

Now we can reference the individual portions of CUST_NAME and CUST_ADDRESS, of course, but in exchange for that privilege we have given up the ability to reference the entire customer name and customer address with one variable each.

The ideal situation is to be able to reference

1. the entire record with one variable,

2. large units of the record with one variable, and

3. individual elements.

There is, in fact, a way that we can "have our cake and eat it, too," and it is by declaring the record to be a structure.

```
DCL 1  CUST_REC,
       2  CUST_ID             FIXED DEC (5),
       2  CUST_NAME,
          3  CUST_LAST_NAME    CHAR     (15),
          3  CUST_FIRST_NAME   CHAR     (10),
       2  CUST_ADDRESS,
          3  CUST_STREET       CHAR     (15),
          3  CUST_CITY_STATE   CHAR     (15),
       2  CUST_ZIP            CHAR     (5),
       2  AMNT_OWED           FIXED DEC (5,2);
```

With this declaration, a reference to CUST_REC is a reference to all of the elements listed below it. A reference to CUST_ID, however, is a reference to that field only, and a reference to CUST_NAME is the same as a reference to both CUST_LAST_NAME and CUST_FIRST_NAME. Thus, you can see that a structure declaration gives us a great deal more flexibility than we have when we declare each data item as discrete item; moreover, you can see that the structure does indeed create hierarchies: CUST_REC is "more important" than any other name because it refers to all the data in the record. CUST_ADDRESS is more important than CUST_STREET and CUST_CITY_STATE because it single-handedly references as much data as the two together do. Finally, remember that since the structure is an aggregate, it must be described in its entirety with a single DECLARE.

12.1.1

Level
Numbers

The hierarchical nature of the structure is revealed to PL/I by the use of **levels**, and the levels are specified to PL/I through **level numbers**. At the highest level is the variable name that refers to the entire data aggregate. This highest level is called a **major structure**, and the major structure name in our example is, of course, CUST_REC. There can be only one major structure in a structure, and it must be assigned the level number of 1. A variable at any level below the major structure, *if it is further subdivided*, is called a **minor structure**. CUST_NAME and CUST_ADDRESS are both examples of minor structures because they are subdivided. CUST_ID, CUST_ZIP, and AMNT_OWED, even though they are at the same level as CUST_NAME and CUST_ADDRESS, are not minor structures because they are not subdivided. In CUST_REC we assign the minor structure the level number of 2, but PL/I does not require that level numbers be consecutive, or even the same number throughout. The only rule PL/I has on level numbers for minor structures is that they be greater than the level number of the major structure. It would be quite possible to code

```
      .
   5   CUST_NAME,
      .
   8   CUST_ADDRESS,
      .
```

Variables (at whatever level below the major structure) that are not further subdivided are called **elements**, and the variable names are called **elementary names**. CUST_ID, CUST_LAST_NAME, CUST_FIRST_NAME, CUST_STREET, CUST_CITY_STATE, CUST_ZIP, and AMNT_OWED are all elementary names designating elements. Level numbers for elements that make up a minor structure must be greater than the largest used for a minor structure. This means in effect that the level numbers for minor structures must be less than the smallest used for minor structure elements, as well as larger than the one used for major structures. The use of different level numbers at the same level will probably only confuse a reader of your program, and we strongly urge you to avoid the practice.

There are a number of things to keep in mind when declaring and referencing structures. First, as we have said, both minor structures and elements can appear at any level below the major structure, and as our example illustrates, they can be at the same level. Second, PL/I does have one more rule regarding level numbers, and it is that one or more blanks must separate the level number from the variable name. Note that we leave two blanks between the level number and the variable, but we do so only to enhance our standard three-column indentation. As for the indentation itself, it is included only to make the record description more readable to human beings; the PL/I compiler pays no attention to it. The third point is that attributes are generally assigned only at the elementary level. This restriction is necessary for several reasons, the primary one being that typically there is more than one data type under the major structure or (perhaps less typically) under the minor structure. CUST_REC illustrates the first possibility, and for the second consider the following. A parts code might be comprised of, say, two letters followed by five digits. To aid in data verification, moreover, it might be necessary to carry out arithmetic operations on the numeric portion (if the part number is correct, the second and third digits equal the product of the fourth and fifth). This situation would require a declaration of

```
DCL 1  PART_RECORD,
       2  PART_CODE,
          3  ALPHA          CHAR      (2),
          3  NUMERIC,
             4  DIGIT_1      FIXED DEC (1),
             4  DIGITS_2_3   FIXED DEC (2),
             4  DIGIT_4      FIXED DEC (1),
             4  DIGIT_5      FIXED DEC (1)
       2   . . .
```

The INITIAL attribute, too, can be assigned only at the elementary level.

```
WRONG:    2  CUST_NAME INIT ('SMITH JOHN'),
CORRECT:  3  CUST_LAST_NAME CHAR (15) INIT ('SMITH'),
          3  CUST_FIRST_NAME CHAR (15) INIT ('JOHN'),
```

Earlier, we mentioned that each variable in a structure "has or can have" a unique name. As a matter of fact, the major structure *must* have a unique name. Any reference to a minor structure or element, moreover, must also provide a unique name, but it is possible to have nonunique names in the structure declaration itself. Consider, for example, the following portion of a payroll record:

```
2  HOURS,
   3  REGULAR     FIXED DEC (2),
   3  OVERTIME    FIXED DEC (2),
2  RATE,
   3  REGULAR     FIXED DEC (4,2),
   3  OVERTIME    FIXED DEC (4,2);
```

In order to reference REGULAR and OVERTIME in such a way that PL/I knows which element we are referring to, we must qualify the variable name. A **qualified name** is one that is concatenated with the variable name at one or more levels above it until it becomes unique. A period is used to join the names; for example,

```
REG_PAY = HOURS.REGULAR * RATE.REGULAR;
OT_PAY = HOURS.OVERTIME * RATE.OVERTIME;
```

Neither REGULAR nor OVERTIME is unique, but when they are qualified with HOURS and RATE, the resulting four names all are unique. A name need only be qualified with higher-level names until it becomes unique, and intermediate-level names can be omitted if they are not needed to provide uniqueness. The following highly abbreviated example illustrates this point.

```
DCL 1  A
         2  B
              3  C
                   4  D
                   4  E
         2  F
              3  C
                   4  D
                   4  E
```

The qualified names C.D and C.E are still ambiguous, but B.D, B.E, F.C, and F.D are unique. Of course, it is possible to include the intervening names: B.C.D, B.C.E, F.C.D, and F.C.E, but it is not required.

Finally, we should mention that you can factor elements in the structure if two or more adjacent elements have the same attributes. Suppose, for example, that instead of AMNT_OWED, CUST_REC had three identical fields of CURR_DUE, PAST_30_DUE, and PAST_60_DUE. These fields could then be declared as

```
DCL 1  CUST_REC,

         2  (CURR_DUE,
             PAST_30_DUE,
             PAST_60_DUE    FIXED DEC (5,2));
```

Level numbers, then, are the clues PL/I uses to create the hierarchy we have requested in our declaration. The level number is not included, however, in references elsewhere in the PL/I program to variables in a structure:

```
WRONG:    2  AMNT_OWED = 91.31;
CORRECT:  AMNT_OWED = 91.31;
```

*The PICTURE
Format*

The **picture** of a data item provides a character-by-character description of the data item by specifying exactly what data characters in the value can occupy that position. The syntax of the picture specification is PICTURE "specification." PICTURE can be abbreviated PIC, and the space between it and the specification is optional.

Because stream I/O provides for automatic data conversion, while record I/O does not, the PICTURE attribute is used primarily with record-oriented data transmission, especially on input. The PICTURE attribute, along with the P-format item, however, allows for more editing than we have presented up to this point and consequently is regularly used with stream-oriented output. It therefore seems appropriate to present both the PICTURE attribute and the P-format item at this time.

As you know by now, FIXED DECIMAL data is stored in a form which does not allow it to be printed in standard base 10 numbers without conversion to character form, and this conversion is automatic in stream-oriented data transmission. Arithmetic, on the other hand, should not be performed on data stored in character form, hence the value of the PICTURE attribute.

The first two picture characters we want to present are the **9** and the **V**. For example,

```
DCL AMNT_OWED PICTURE '999V99';
```

This declaration states that AMNT_OWED can receive numeric data of up to five digits (one for each 9 in the picture) and that two of the digits will be to the right of the implied decimal point. The implied decimal point is specified by the V. For stream-oriented data, the declaration is the same as

```
DCL AMNT_OWED FIXED DEC F(5,2);
```

However, we cannot declare a variable with both the PICTURE and the FIXED DECIMAL attributes. Other examples of the PICTURE attribute are

```
DCL NUM1 PIC 'V99';      /* NO WHOLE NUMBER DIGITS;
                            TWO FRACTIONAL DIGITS;
                            CF. F(2,2)                */
```

```
DCL NUM2 PIC '9999';      /* FOUR WHOLE-NUMBER DIGITS; NO
                             FRACTIONAL DIGITS; CF. F(4)    */
DCL NUM3 PIC '(4)9V99';   /* REPETITION FACTOR; SAME AS
                             9999V99. NOTE THAT THE 9
                             DOES NOT REPRESENT A FIFTH
                             DIGIT; CF. F(6,2)              */
```

The next set of picture characters allow you to specify signed values, and they are **S**, **+**, and **−**. Any one of these three characters can be the leftmost (first) or the rightmost (last) character in a PICTURE specification, but they cannot be embedded within other characters. The characters take up space in memory. An S specifies that the sign of the value is to appear as the leftmost or rightmost character of the value. If the value is less than 0, then the negative sign (−) is appended; otherwise, the positive sign (+) is appended. A + signifies that the first or last character is to be a + if the value is greater than or equal to 0; a blank appears for a negative number. A − specifies that the sign position contains a − if the value is less than 0, otherwise a blank. The following table illustrates the use and result of the sign characters. (We follow our usual practice of using a vertical rule (|) to indicate the implied decimal point.)

PICTURE Specification	Coded Arithmetic Form	Value	Result
S999V99	F (5,2)	−123.45	−123|45
999V99	F (5,2)	+123.45	123|45
+99V99	F (4,2)	+12.34	+12|34
+99V99	F (4,2)	−12.34	12|34
−999V9	F (4,10)	−123.4	−123|4
−999V9	F (4,1)	+123.4	123|4
999V9−	F (4,1)	−123.4	123|4−
S99V99	F (4,2)	0.0	+00|00

As we mentioned earlier, the PICTURE attribute describing fields in output records provides for editing facilities not otherwise available. Arithmetic can be done on values containing these editing characters, but it is most inefficient to do so, and you should avoid the practice. Instead, you should use the assignment statement to supply the value to a variable with editing characters in its picture. Suppose, for example, the input record contains fields named CURR_DUE, PAST_30, and PAST_60, and that we want to print the contents of each field and the total of the three fields.

```
BILL:  PROC OPTIONS (MAIN);
    DCL 1  INPUT_REC,
            2  CURR_DUE_IN        FIXED DEC (5,2),
            2  PAST_30_IN         FIXED DEC (5,2),
            2  PAST_60_IN         FIXED DEC (5,2),
            CURR_DUE_OUT          PIC '$$$9V.99',
            PAST_30_OUT           PIC '$$$9V.99',
            PAST_60_OUT           PIC '$$$9V.99',
            AMNT_OWED             PIC '$$$$9V.99' INIT (0.0);
            GET LIST (INPUT_REC);
            AMNT_OWED = CURR_DUE_IN + PAST_30_IN + PAST_60_IN;
            CURR_DUE_OUT = CURR_DUE_IN;
            PAST_30_OUT = PAST_30_IN;
            PAST_60_OUT = PAST_60_IN;
            PUT SKIP LIST ('CURRENTLY DUE '||CURR_DUE_OUT);
            PUT SKIP LIST ('30 DAYS PAST DUE '||PAST_30_OUT);
            PUT SKIP LIST ('60 DAYS PAST DUE '||PAST_60_OUT);
            PUT SKIP LIST ('TOTAL DUE '||CURR_DUE_OUT);
    END BILL;
```

The first of the editing characters are the **zero suppression** characters, and the name reveals their function. The first of the zero suppression characters is **Z**. If the value has a nonsignificant (leading) 0 (i.e., a zero to the left of any nonzero digit) in the position relative to a Z in the picture, that zero is suppressed, and a blank is printed instead. If the value has a nonzero digit or a significant zero (i.e., a zero anywhere to the right of the leftmost nonzero digit), that digit will print as usual. The second zero suppression character is the asterisk (*). When the asterisk is specified, leading zeros in those positions are replaced by asterisks. Three important rules to remember in using zero suppression characters follow.

1. All zero suppression characters must appear to the left of the leftmost 9.

2. If a zero suppression character is placed to the right of the implied decimal point, all remaining characters must also be zero suppression characters.

3. The Z and the * cannot be mixed in a single picture specification.

The following examples will help you in clarifying the use and results of zero suppression editing characters.

PICTURE Specification	Assigned Value	Character Value
ZZZ9V99	0012.34	bb12 34
ZZZ9V99	0001.23	bbb1 23
ZZZ9V99	0000.12	bbb0 12
ZZZZV99	0000.12	bbbb 12
ZZZ9V99	1002.34	1002 34
***9V99	0123.45	*123 45
***9V99	0000.12	***0 12
******	0000.00	**** **
****V**	1234.56	1234 56

Zero suppression is automatic in stream-oriented data transmission, of course, and consequently the zero suppression characters are seldom used in stream output. You might want to use the asterisk character, however, if your program is designed to write checks. Perhaps the most common use of the asterisk edit character is to make the amount of a check very difficult to alter.

The next group of editing characters are referred to as **insertion characters** because they neither suppress nor replace any characters in the value. Instead, they are inserted into the value at their relative positions. Note, also, that insertion characters *are* typically embedded within the nines and/or other editing characters. The first three insertion characters are the point (.), comma (,), and slash (/). Whichever of them is included in the picture is inserted into the value at the position corresponding to its position in the picture, *unless zero suppression occurs*. When zero suppression occurs, the insertion character is likewise suppressed. The point is most often used as a decimal point, of course, and it is important to realize that the point does not cause alignment. By the same token, a V does cause alignment, but it does not cause a point to be printed. Therefore, if you are printing a column of numbers that you want aligned on the point, you must include both a V and a point (V.) in your picture. Of course, the V can be used in conjunction with any of the other insertion characters, also, to cause alignment on that character. The final insertion character is the **B**, which causes a blank to be inserted into the value at the corresponding position. Again, the following table clarifies the foregoing discussion.

The next set of editing characters can be either **static** or **drifting**. When they are static, they must be either the leftmost or the rightmost character in the picture. We have already introduced two of these characters, the + and the −. The third is the dollar sign ($). Only one of these signs can be used in a single picture specification, but that one can be repeated as necessary.

PICTURE Specification	Assigned Value	Character Value
ZZ9V99	123.45	123|45
ZZ9V.99	123.45	123.45
Z,ZZ9	1234	1,234
Z,ZZZ,ZZZ	0012345	bb12345
99/99/99	060187	06/01/87
999B99	12345	123b45

Drifting edit characters, as a matter of fact, are created by repeating the edit character, and when the character is drifting, it causes zero suppression. The first occurrence of the character is static, and it prints immediately to the left of the leftmost nonsuppressed digit. Because the first occurrence is static, you must include one more occurrence of the character than the number of leading zeros you want suppressed. PIC '$$$$9V.99', for example, suppresses up to three (not four) leading zeros. These characters can be combined with insertion characters ($$,$$$,$$9V.99). Once again we present a series of examples to illustrate and clarify our discussion.

++++99	+00123	bb+123
++++99	−00123	bbb123
- - - - 9	−00012	bbb−12
- - - - 9	+00012	bbbb12
ZZZ99+	+00123	bb123+
ZZZ99−	−01234	b1234−
$ZZ9V.99	012.34	$b12.34
$$$9V.99	001.23	bb$1.23
$$,$$$,$$9V.99	0001234.56	bbb$1,234.56
$$,$$$,$$9V.99	0100123.45	b$100,123.45

The final group of editing characters is the **fixed insertion** characters and is comprised of **CR** and **DB**. Whichever character has been specified will appear in the rightmost two positions in the field *if the value is less than zero;* otherwise, blanks will occupy these positions. CR (Credit) and DB (Debit) cannot be used with any other sign values (S, +, −).

It is sometimes the case that we want either CR or DB to apply to a positive value. In making up a checking account statement, for example, we might want DB to indicate a check (withdrawal) and CR to indicate a deposit. If we assume that withdrawal values in the input stream contain a negative sign, while deposits do not (e.g., 500.00 −150.00 75.75 −42.50), we can use the following code to accomplish our goal.

PICTURE Specification	Assigned Value	Character Value
ZZZ9V.99CR	−1234.56	1234.56CR
ZZZ9V.99CR	+1234.56	1234.56bb
ZZZ9V.99DB	−0012.34	bb12.34DB
ZZZ9V.99DB	+0012.34	bb12.34bb

```
          DCL   BALANCE_IN          FIXED DECIMAL (6,2);
          DCL   AMNT_IN             FIXED DECIMAL (5,2);
          DCL   BALANCE_OUT         PIC 'ZZZ9V.99';
          DCL   CR_AMNT             PIC 'ZZ9V.99CR';
          DCL   DB_AMNT             PIC 'ZZ9V.99DB';
                GET LIST (BALANCE_IN, AMNT_IN);
PROC_LP:  DO WHILE (FILE_STATUS = MORE_RECS);
                IF AMNT_IN < 0 THEN
WDL:              DO;
                    DB_AMNT = AMNT_IN;
                    BALANCE_IN = BALANCE_IN + AMNT_IN;
                    BALANCE_OUT = BALANCE_IN;
                    PUT SKIP LIST (DB_AMNT, BALANCE_OUT);
                  END WDL;
                ELSE
DEPT:             DO;
                    CR_AMNT = -AMNT_IN;
                    BALANCE_IN = BALANCE_IN + AMNT_IN;
                    BALANCE_OUT = BALANCE_IN;
                    PUT SKIP LIST (CR_AMNT, BALANCE_OUT);
                  END DEPT;
                GET LIST (AMNT_IN);
          END PROC_LP;
```

By appending a negative sign to the value as it is assigned to CR_AMNT, we force PL/I to print the CR edit characters.

In addition to the decimal numeric picture specification of 9, PL/I also provides two character picture characters. An X signifies that this position can contain any of the 256 characters made possible by the combinations of bits in an eight-bit byte. An A signifies that this position can hold any alphabetic, national (@, #, $), or a blank character. The X is used primarily in the case in which the data is both character and numeric, mixed in an unpredictable manner, or for those rare occasions where special characters are included in the data stream. The A is used primarily for data verification. In other words, when valid data for a field can consist of only those characters allowed by the A, other characters cause the CONVERSION condition to be raised, thereby alerting the user to invalid data. Character data in a position described by a 9 produces the same result, of course, and therefore it also allows for data verification. Data verification, as a matter of fact, is a major reason for using the PICTURE attribute with stream-oriented input. Finally, A's, X's, and 9's can appear in the same picture specification.

```
DCL PROD-CODE PIC 'AAA(4)9' INIT ('DOM1933');
```

12.1.3

The P-Format
Item

The P-format item is used in the format list of edit-directed I/O statements. The item consists of a P followed by a picture specification enclosed in single quotation marks. Any picture or editing character allowed in the picture specification of the PICTURE attribute is allowed in the P-format item, and each editing character functions exactly the same in both instances. On input the P-format item must accurately describe the incoming data; otherwise, the CONVERSION condition is raised.

Input Value	P-Format Item	Internal Value
1234	P'99V99'	12⎪34
AB*1	P'XXXX'	AB*1
AB@4	P'AAA9'	AB@4

```
GET EDIT (FLD1, FLD2, FLD3) (COL(1), P'ZZZ9', P'999V99',
                            P'XXAA99');
```

On output the value is converted to the form specified by the picture specification before it is written.

Internal Value	P-Format Item	Printed Result
100.01	P'$$$9V.99'	$100.01
−5900	P'(4)9DB'	5900DB
695AC	P'ZZ99AA'	695AC
b#9!J	P'XXXXX'	b#9!J

```
PUT EDIT (GROSS, DSCNT, NET) (COL(20), P'$$$,$$9V.99',
                              P'$$$9V.99', P'$$$,$$9V.99');
```

12.2

Structures in Stream I/O

Although structures are used primarily with record-oriented data transmission, they can also be used with stream I/O. In fact, their use provides certain advantages in both list and edit I/O.

12.2.1

LIST I/O Revisited

The use of structures with list I/O creates no special problems. On input, you must be sure, of course, that the input data items are listed in the order specified in the structure declaration, and the items must be in list format (separated by a comma and/or space(s); character items delimited by single quotation marks). The advantage a structure offers over a declaration of the same variables discretely is that the declaration of a structure requires almost no more typing, but the data list in the I/O statement is reduced to a single variable, the major structure name. Consider, for example,

```
DCL 1 STDNT_REC,
      2  SSN        CHAR (9),
      2  NAME       CHAR (20),
      2  MAJOR      CHAR (4),
      2  CLASS      CHAR (2);
```

With this declaration we code

```
GET LIST (STDNT_REC);
```

instead of the

```
GET LIST (SSN, NAME, MAJOR, CLASS);
```

that is required if we declare each variable discretely. Output can be handled the same way, but PL/I still treats each elementary item as a separate data item. Therefore, the statement

```
PUT LIST (STDNT_REC);
```

causes the SSN value to start in column 1, the NAME value in column 25, the MAJOR value in column 49, and the CLASS value in column 73. Of course, minor structure and elementary names can also be used in list-directed I/O statements.

EDIT I/O
Revisited

Structures in edit-directed I/O can also be referenced by major or minor structure names, or by elementary item names. However, you must provide a data item in the format list for each elementary item that is implicitly referenced when you reference a major or minor structure name.

```
DCL 1   CLASS_REC,
        2  CLASS_ID,
            3  CLASS_NAME      CHAR (4),
            3  CLASS_#         CHAR (3),
        2  TEACHER            CHAR (20),
        2  MEET_IN,
            3  BLDG            CHAR (2),
            3  ROOM            CHAR (3),
        2  TIME,
            3  START_HR        CHAR (5),
            3  END_HR          CHAR (5),
            3  DAYS            CHAR (3);
```

To read the record, we must code

```
GET EDIT (CLASS_REC) (COL(1), A(4), A(3), A(20), A(2), A(3),
                      A(5), A(5), A(3));
```

Some possible PUT EDIT statements are (some fields are larger than minimum to create spaces between output items):

```
PUT SKIP EDIT (TIME) (COL(10), A(6), A(11), A);
              /* PRINTS START_HR, END_HR, DAYS */
PUT SKIP EDIT (CLASS_ID) (COL(25), A(5), A);
              /* PRINTS CLASS_NAME, CLASS_#    */
PUT SKIP EDIT (TEACHER, MEET_IN) (COL(45), A(25), A(5), A);
              /* PRINTS TEACHER, BLDG, ROOM    */
```

Example Program—
Structures and I/O

The CH12EX1 program (see Fig. 12–1) is designed to produce a salary report for the work-study program at the university. For each student in the program, we need a social security number, name, the employing university unit, the worker's job title, hours worked, and gross salary. The federal government provides 80% of the funds for the work-study program, and the university provides 20%. The government (we are assuming) requires this salary breakdown for each employee. In addition, it requires the total gross salary for each employing unit, and the government and university dollar contributions. Finally, by way of summary, the government requires the final totals, again broken down into university and government contributions. The logic employed in constructing the program is the by-now-familiar report logic, and we need not dwell on that.

From a pedagogical point of view the main purpose of the program is to illustrate the use of structures in stream I/O. The header record, input data records, and the output records are all declared as structures. These structure declarations necessitate the series of assignment statements (Statements 32–36).

Other matters of interest are, first of all, the use of the PICTURE to declare the variables SSN, START_PAY, and END_PAY. Although bad input to these fields would not cause execution to terminate, the picture does act as a check on the data. The variables UNIT_TOTAL, UNIV_SUBTTL, and GOVT_SUBTTL (Statements 48–50), as well as UNIV_TOTAL, GOVT_TOTAL, and GRAND_TOTAL in the output are described with the P-format item in order to take advantage of its extra editing capabilities. Finally, note that the employer unit, job title, and pay rate are stored in arrays and that their respective codes serve as subscripts into the arrays. This practice results in a substantial decrease in the length of the input records. The pseudocode is as follows:

```
CH12EX1
    Read the header record
    Read an input record
    DO WHILE (more input)
        Save code ← unit code
        Print-record unit ← unit array (unit code)
        Print page, column headers
        Unit total, university subtotal, government subtotal
                ← 0
        DO WHILE (save code = unit code and more input)
            Print-record SSN ← student SSN
            Print-record name ← student name
            Print-record title ← title array (title code)
            Print-record hours ← hours worked
            Print-record rate ← rate array (rate code)
```

                    Print-record gross ← hours worked · print-record
                        rate
                    Print-record university share ← print-record
                        groos · .20
                    Print-record government share ← print-record
                        gross – print-record university share
                    Unit total ← university subtotal + print-record
                        gross
                    University subtotal ← university subtotal +
                        print-record university share
                    Government subtotal ← government subtotal +
                        print-record government share
                    Print a record
                    Read a record
                ENDDO
                University total ← university total + university
                    subtotal
                Government total ← government total + government
                    subtotal
                Print labels, unit total, university subtotal,
                    government subtotal
            ENDDO
            Grand total ← university total + government total
            Print summary-page header
            Print labels, university total, government total, grand
                total
        STOP

PL/I OPTIMIZING COMPILER VERSION 1 RELEASE 5.1 TIME: 13.15.16 DATE: 10 MAR 87 PAGE   1
OPTIONS SPECIFIED
GOSTMT;

PL/I OPTIMIZING COMPILER   /* CH12EX1 USING STRUCTURES */   PAGE   2
                    SOURCE LISTING
    STMT

        /* CH12EX1 USING STRUCTURES */
    1 CH12EX1:   PROCEDURE OPTIONS (MAIN);

FIGURE 12–1
CH12EX1 (the use of structures).

FIGURE 12–1 *(continued)*

```
/*********************************************************************/
/*                                                                 */
/*    FUNCTION:   TO PRODUCE A PAYROLL REPORT FOR WORK-STUDY        */
/*                STUDENTS                                          */
/*                                                                 */
/*    INPUT:      A FILE CONTAINING                                 */
/*                1.  A HEADER RECORD GIVING THE STARTING AND       */
/*                    ENDING DATE OF A TWO-WEEK PAY PERIOD          */
/*                2.  A RECORD FOR EACH WORK-STUDY STUDENT          */
/*                    CONTAINING A SOCIAL SECURITY NUMBER, NAME,    */
/*                    EMPLOYER-UNIT CODE, JOB-TITLE CODE, PAY-      */
/*                    RATE CODE, AND HOURS WORKED.                  */
/*                THE INPUT IS IN LIST-DIRECTED FORMAT AND IS       */
/*                SORTED ON EMPLOYER-UNIT CODE                      */
/*                                                                 */
/*    OUTPUT:     FOR EACH STUDENT A DETAIL LINE LISTING SOCIAL     */
/*                SECURITY NUMBER, NAME, EMPLOYER UNIT, JOB TITLE,  */
/*                PAY RATE, GROSS SALARY, UNIVERSITY CONTRIBUTION,  */
/*                AND GOVERNMENT CONTRIBUTION.                      */
/*                SUBTOTALS FOR EACH EMPLOYER UNIT                  */
/*                TOTALS FOR THE WORK-STUDY PROGRAM                 */
/*                                                                 */
/*    NOTES:      THE EMPLOYER-UNIT CODES ARE                       */
/*                    1--COMPUTER CENTER                           */
/*                    2--COMPUTER SCIENCE DEPT.                    */
/*                    3--LIBRARY                                   */
/*                    4--STUDENT CENTER                            */
/*                                                                 */
/*                THE JOB-TITLE CODES ARE                          */
/*                    1--OPERATOR ASSISTANT                        */
/*                    2--TEACHING ASSISTANT                        */
/*                    3--CLERK                                     */
/*                    4--DESK ATTENDANT                            */
/*                                                                 */
/*                THE PAY-RATE CODES ARE                           */
/*                    1--$3.45/HOUR                                */
/*                    2--$3.65/HOUR                                */
/*                    3--$3.80/HOUR                                */
/*                    4--$6.75/HOUR                                */
/*                                                                 */
/*********************************************************************/
```

```
2     DCL UNIT_ARRAY (4)        CHARACTER (25)   INIT
              ('COMPUTER CENTER', 'COMPUTER SCIENCE DEPT',
              'LIBRARY', 'STUDENT CENTER');
3     DCL TITLE_ARRAY (4)       CHARACTER (20)   INIT
              ('OPERATOR ASSISTANT', 'TEACHING ASSISTANT',
```

```
              'CLERK', 'DESK ATTENDANT');
4     DCL RATE_ARRAY (4)        FIXED DEC (3,2)  INIT
              (3.45, 3.65, 3.80, 6.75);

5     DCL 1 HDR_REC,
        2 START_PAY        PIC'99X99X99',
        2 END_PAY          PIC'99X99X99';

6     DCL 1 INPUT_REC,
        2 SSN              PIC'999X99X9999',
        2 NAME             CHARACTER (20),
        2 UNIT_CODE        FIXED DEC (1),
        2 TITLE_CODE       FIXED DEC (1),
        2 RATE_CODE        FIXED DEC (1),
        2 HOURS_WORKED     FIXED DEC (2);

7     DCL 1 PRNT_REC,
        2 PRNT_SSN         CHARACTER (11),
        2 PRNT_NAME        CHARACTER (20),
        2 PRNT_UNIT        CHARACTER (25),
        2 PRNT_TITLE       CHARACTER (20),
        2 PRNT_HOURS       FIXED DEC (2),
        2 PRNT_RATE        FIXED DEC (4,2),
        2 PRNT_GROSS       FIXED DEC (5,2),
        2 PRNT_UNIV_SHR    FIXED DEC (5,2),
        2 PRNT_GOVT_SHR    FIXED DEC (5,2);

8     DCL SAVE_CODE            FIXED DEC (1);
9     DCL GOVT_SUBTTL          FIXED DEC (6,2);
10    DCL UNIV_SUBTTL          FIXED DEC (6,2);
11    DCL UNIT_TOTAL           FIXED DEC (6,2);
12    DCL UNIV_TOTAL           FIXED DEC (7,2)  INIT (0);
13    DCL GOVT_TOTAL           FIXED DEC (7,2)  INIT (0);
```

FIGURE 12–1 *(continued)*

```
14      DCL GRAND_TOTAL              FIXED DEC (8,2)  INIT (0);

15      DCL MORE_RECS               BIT      (1);
16      DCL    YES                  BIT      (1)   INIT ('1'B);
17      DCL    NO                   BIT      (1)   INIT ('0'B);

18      DCL ROUND                   BUILTIN;
19          MORE_RECS = YES;
20          ON ENDFILE (SYSIN) MORE_RECS = NO;

21          GET LIST (HDR_REC);
22          GET LIST (INPUT_REC);

23  PROC_LP: DO WHILE (MORE_RECS);
24          SAVE_CODE = UNIT_CODE;
25          PRNT_UNIT = UNIT_ARRAY (UNIT_CODE);
26          PUT PAGE EDIT ('WORK-STUDY SALARY REPORT')
                        (COL (50), A);
27          PUT SKIP (3) EDIT ('PAY PERIOD STARTING ', START_PAY,
                    ' ENDING ', END_PAY, 'EMPLOYER UNIT:  ',
                    PRNT_UNIT) (COL(11), A, A(13), A, A, X(15), A, A);
```

STMT

```
28          PUT SKIP (4) EDIT ('SSN', 'NAME', 'EMPLOYER UNIT',
                    'JOB TITLE', 'HOURS', 'RATE', 'GROSS', 'UNIV',
                    'GOVT')
                    (COL(4), A, COL(17), A, COL(35), A, COL(60), A,
                     COL(80), A, COL(87), A, COL(95), A, COL(104), A,
                     COL(113), A);

29          PUT SKIP EDIT ('FUNDS', 'FUNDS') (COL(104), A, COL(113),
                            A);
30          UNIT_TOTAL, UNIV_SUBTTL, GOVT_SUBTTL = 0;

31  UNIT_LP:    DO WHILE (SAVE_CODE = UNIT_CODE & MORE_RECS);
32              PRNT_SSN = SSN;
33              PRNT_NAME = NAME;
34              PRNT_TITLE = TITLE_ARRAY (TITLE_CODE);
```

```
35              PRNT_HOURS = HOURS_WORKED;
36              PRNT_RATE = RATE_ARRAY (RATE_CODE);
37              PRNT_GROSS = ROUND (HOURS_WORKED * PRNT_RATE, 2);
38              PRNT_UNIV_SHR = ROUND (PRNT_GROSS * .20, 2);
39              PRNT_GOVT_SHR = PRNT_GROSS - PRNT_UNIV_SHR;
40              UNIT_TOTAL = UNIT_TOTAL + PRNT_GROSS;
41              UNIV_SUBTTL = UNIV_SUBTTL + PRNT_UNIV_SHR;
42              GOVT_SUBTTL = GOVT_SUBTTL + PRNT_GOVT_SHR;

43              PUT SKIP (2) EDIT (PRNT_REC) (COL(1), A(14), A, A, A,
                        X(1), F(2), X(2), F(6,2), F(9,2), F(9,2),
                        F(9,2));

44                GET LIST (INPUT_REC);
45            END UNIT_LP;

46              UNIV_TOTAL = UNIV_TOTAL + UNIV_SUBTTL;
47              GOVT_TOTAL = GOVT_TOTAL + GOVT_SUBTTL;

48              PUT SKIP (3) EDIT ('THE UNIT TOTAL IS', UNIT_TOTAL,
                                ' FOR ', PRNT_UNIT) (COL(53), A,
                                P'$$$$$9V.99', A, A);
49              PUT SKIP (2) EDIT ('THE UNIVERSITY CONTRIBUTION IS ',
                                UNIV_SUBTTL)
                        (COL(40), A, P'$$$$9V.99');
50              PUT SKIP (2) EDIT ('THE GOVERNMENT CONTRIBUTION IS ',
                        GOVT_SUBTTL) (COL(40), A, P'$$$$9V.99');
51            END PROC_LP;

52          GRAND_TOTAL = UNIV_TOTAL + GOVT_TOTAL;
53          PUT PAGE;
54          PUT SKIP (6) EDIT ('WORK-STUDY SALARY SUMMARY')
                        (COL(55), A);
55          PUT SKIP(5) EDIT ('TOTAL UNIVERSITY CONTRIBUTION', UNIV_TOTAL)
                        (COL(40), A, P'$$$$$9V.99');
56          PUT SKIP(2) EDIT ('TOTAL GOVERNMENT CONTRIBUTION', GOVT_TOTAL)
```

```
                                (COL(40), A, P'$$$$$9V.99');
57          PUT SKIP (4) EDIT
```

FIGURE 12–1 *(continued)*

```
                    ('TOTAL WORK-STUDY EXPENDITURES FOR THE PAY PERIOD',
                      GRAND_TOTAL) (COL(19), A, P'$$$$,$$9V.99');
        58  END CH12EX1;


PL/I OPTIMIZING COMPILER        /* CH12EX1 USING STRUCTURES */            PAGE   6
COMPILER DIAGNOSTIC MESSAGES
ERROR ID L   STMT    MESSAGE DESCRIPTION

COMPILER INFORMATORY MESSAGES

IEL0533I I          NO 'DECLARE' STATEMENT(S) FOR 'SYSIN','SYSPRINT'.

END OF COMPILER DIAGNOSTIC MESSAGES
```

```
                            WORK-STUDY SALARY REPORT


      PAY PERIOD STARTING 05/05/87     ENDING 05/19/87        EMPLOYER UNIT:  COMPUTER CENTER
```

SSN	NAME	EMPLOYER UNIT	JOB TITLE	HOURS	RATE	GROSS	UNIV FUNDS	GOVT FUNDS
101-01-0101	ADAMS, AMY	COMPUTER CENTER	CLERK	20	3.45	69.00	13.80	55.20
111-10-1000	JONES, HARRY	COMPUTER CENTER	OPERATOR ASSISTANT	20	3.45	69.00	13.80	55.20
110-10-1010	ZIEGLER, J. K.	COMPUTER CENTER	DESK ATTENDANT	15	3.65	54.75	10.95	43.80

```
                    THE UNIT TOTAL IS  $192.75 FOR COMPUTER CENTER

                THE UNIVERSITY CONTRIBUTION IS    $38.55

                THE GOVERNMENT CONTRIBUTION IS  $154.20
```

WORK-STUDY SALARY REPORT

PAY PERIOD STARTING 05/05/87   ENDING 05/19/87   EMPLOYER UNIT: COMPUTER SCIENCE DEPT

SSN	NAME	EMPLOYER UNIT	JOB TITLE	HOURS	RATE	GROSS	UNIV FUNDS	GOVT FUNDS
112-11-1100	BAKER, BETTY	COMPUTER SCIENCE DEPT	TEACHING ASSISTANT	15	6.75	101.25	20.25	81.00
121-12-1211	CARLSON, CARL	COMPUTER SCIENCE DEPT	CLERK	18	3.80	68.40	13.68	54.72
212-21-2121	DAWKINS, DANA	COMPUTER SCIENCE DEPT	TEACHING ASSISTANT	14	6.75	94.50	18.90	75.60
211-12-1122	MASON, TERRY	COMPUTER SCIENCE DEPT	TEACHING ASSISTANT	19	6.75	128.25	25.65	102.60

THE UNIT TOTAL IS $392.40 FOR COMPUTER SCIENCE DEPT

THE UNIVERSITY CONTRIBUTION IS $78.48

THE GOVERNMENT CONTRIBUTION IS $313.92

WORK-STUDY SALARY REPORT

PAY PERIOD STARTING 05/05/87   ENDING 05/19/87   EMPLOYER UNIT: LIBRARY

SSN	NAME	EMPLOYER UNIT	JOB TITLE	HOURS	RATE	GROSS	UNIV FUNDS	GOVT FUNDS
331-13-1323	CRABBE, CORA	LIBRARY	CLERK	17	3.80	64.60	12.92	51.68
303-03-0303	KRUPP, TOM	LIBRARY	TEACHING ASSISTANT	17	6.75	114.75	22.95	91.80
313-13-1313	LARSON, LARS	LIBRARY	CLERK	20	3.80	76.00	15.20	60.80
321-12-3123	NEARS, ANITA	LIBRARY	DESK ATTENDANT	12	3.45	41.40	8.28	33.12

FIGURE 12–1 *(continued)*

```
                        THE UNIT TOTAL IS  $296.75 FOR LIBRARY

              THE UNIVERSITY CONTRIBUTION IS   $59.35

              THE GOVERNMENT CONTRIBUTION IS  $237.40
```

```
                              WORK-STUDY SALARY REPORT


         PAY PERIOD STARTING 05/05/87      ENDING 05/19/87          EMPLOYER UNIT:  STUDENT CENTER
```

SSN	NAME	EMPLOYER UNIT	JOB TITLE	HOURS	RATE	GROSS	UNIV FUNDS	GOVT FUNDS
424-42-2442	ELEFSON, ELMER	STUDENT CENTER	DESK ATTENDANT	20	6.75	135.00	27.00	108.00
666-66-6666	FRAME, FREDA	STUDENT CENTER	CLERK	11	3.80	41.80	8.36	33.44
696-69-6969	QANTAS, JERI	STUDENT CENTER	OPERATOR ASSISTANT	16	6.75	108.00	21.60	86.40

```
                          THE UNIT TOTAL IS  $284.80 FOR STUDENT CENTER

              THE UNIVERSITY CONTRIBUTION IS   $56.96

              THE GOVERNMENT CONTRIBUTION IS  $227.84
```

```
                              WORK-STUDY SALARY SUMMARY



              TOTAL UNIVERSITY CONTRIBUTION  $233.34

              TOTAL GOVERNMENT CONTRIBUTION  $933.36


       TOTAL WORK-STUDY EXPENDITURES FOR THE PAY PERIOD  $1,166.70
```

12.3

Using
Structures

In the following sections we explain several aspects of structures that enable you to take fuller advantage of them. We describe several assignment possibilities and the use of structures as parameters.

12.3.1

Structure
Assignment

Two types of structure assignments are possible. In the first, called **structure to structure**, a structure or portion of a structure is assigned to a
second structure or portion of a structure. The only restrictions on structure-
to-structure assignment is that both must be identically declared in the sense
that they have the same **relative structure**—that is, the same minor
structuring with the same number of elementary items. If arrays are contained within the structure, their bounds must be the same. As long as these
two restrictions are observed, you can assign a minor structure to a major
structure or a major structure to a minor structure; the difference in level
numbers is of no concern, and the attributes can differ so long as they are
compatible. Of course, major-to-major and minor-to-minor structure
assignments are also possible, but the same restrictions apply.

```
DCL 1  MASTER_REC,
       2  SSN               CHAR (9),
       2  NAME              CHAR (20),
       2  HISTORY,
          3  LAST_VISIT     CHAR (6),
          3  PHYSICIAN      CHAR (20),
          3  DIAGNOSIS      CHAR (40),
    1  UPDATE_REC,
       2  NEW_VISIT         CHAR (6),
       2  NEW_PHYSICIAN     CHAR (20),
       2  NEW_DIAGNOSIS     CHAR (40);
HISTORY = UPDATE_REC;
```

In this example the corresponding elementary items have identical attributes, but identical attributes are not required. The receiving field can be of
different length than the sending field. However, if the receiving field is
shorter, the possibility of truncation exists. If the data types differ, conversion (if possible) will automatically occur. Keep in mind that elementary
items correspond *positionally;* that is, the first elementary item in the receiving structure will be assigned the value in the first elementary item in the

sending structure. What in fact happens (invisibly to the programmer) is that the assignment statement (e.g., HISTORY = UPDATE_REC) is expanded to generate a separate assignment statement for each elementary item. Thus the assignment statement above expands to generate

```
LAST_VISIT = NEW_VISIT;
PHYSICIAN = NEW_PHYSICIAN;
DIAGNOSIS = NEW_DIAGNOSIS;
```

The second type of assignment is called **scalar to structure**. A scalar is a single data item or elementary item, and it can be assigned to a structure; for example, the following causes the entire structure to be set to blanks.

```
DCL 1  CLASS_DATA,
        2  DEPT        CHARACTER (4),
        2  NUMBER      CHARACTER (4),
        2  TEACHER     CHARACTER (20);
CLASS_DATA = ' ';
```

12.3.2

BY NAME Assignment

The BY NAME option (used for structure-to-structure assignments) frees us from the restriction that both structures have the same number of elementary items, but it imposes a restriction of its own. When we assign a structure BY NAME, only those elementary items having the same name in both structes are moved. If the elementary items have minor structures over them, the minor structure names must also be identical in both structures. The BY NAME option can be used only in an assignment statement, and it must be preceded by a comma.

```
DCL 1  INVENTORY,
        2  BK_ID,
            3  BK_NUM      FIXED DEC (12),
            3  AUTHOR      CHARACTER (20),
            3  TITLE       CHARACTER (30),
        2  #_IN_STOCK      FIXED DEC (3),
        2  UNIT_COST       FIXED DEC (4,2);
     1  OUTPUT_REC,
        2  BK_ID,
            3  AUTHOR      CHARACTER (20),
            3  TITLE       CHARACTER (30),
        2  #_IN_STOCK      FIXED DEC (3),
```

```
                    2  INV_VALUE              PIC '$$$9V.99';
            OUTPUT_REC = INVENTORY, BY NAME;
```

Notice that you will have to use qualified names to refer to those elementary items that also appear in OUTPUT REC. You will have to decide whether the advantage of BY NAME in a particular instance outweighs the disadvantage of qualified names.

*The STRING
Function*

Conspicuous by its absence (we hope) in the foregoing discussion of structure assignment was any mention of structure-to-scalar assignment. A structure, as a matter of fact, *cannot* be assigned to a scalar, although there are instances when we would like to be able to do so. The STRING built-in function solves the problem by concatenating the specified elements of a data aggregate (array or structure) into a single element. The resulting value can then be assigned to a scalar.

```
DCL 1   INPUT_REC,
        2   NAME_ADDR,
            3   NAME              CHAR (20),
            3   STREET            CHAR (20),
            3   CITY              CHAR (15),
            3   STATE             CHAR (2),
            3   ZIP               CHAR (5),
        NM_ADR                    CHAR (62);
    NM_ADR = NAME_ADDR;               /* ILLEGAL ASSIGNMENT */
    NM_ADR = STRING (NAME_ADDR);      /* LEGAL ASSIGNMENT   */
```

The second assignment, using the STRING function, is legal because before the assignment is made, STRING has concatenated the elementary items into a scalar.

This problem can be solved other ways; concatenation, for example, serves the same purpose, but the statement is likely to be long and awkward. The DEFINED attribute, introduced in the following chapter, might be a more efficient solution than the use of STRING. The STRING built-in function, however, is a simple means of solving the problem of assigning a structure to a scalar.

*Structures as
Parameters*

In coding internal procedures with parameters, it is easy to get the idea that the calling procedure is the "boss" and that the procedure parameters must conform to the arguments the calling routine sends it. We get this idea because we probably code the calling routine first. As a matter of fact, however, most procedures in industry are external procedures and can be called by many calling routines. Many of these calling routines have yet to be written; therefore, we should think of the procedure as the "boss," even though it is under control of the calling routine. In other words, you determine what parameters the procedure needs to carry out its function, and it is up to the various calling routines to conform by passing the proper arguments. All of this is by way of explaining the emphasis on parameters as opposed to arguments.

The use of structures as parameters poses no real problems beyond those you have been introduced to. The following example illustrates some possibilities of parameters and arguments and provides the basis for our discussion.

```
PLITST: PROCEDURE OPTIONS (MAIN);
    DCL 1  RECORD,
           2  NAME,
              3  F_NAME        CHAR (10) INIT ('KATHY'),
              3  L_NAME        CHAR (10) INIT ('GLENN'),
           2  ADDRESS,
              3  STREET        CHAR (20) INIT ('110 TAYLOR ST.'),
              3  CITY          CHAR (10) INIT ('DEKALB');
        CALL SUB1 (RECORD);
        F_NAME = 'KYLA';
        L_NAME = 'DOMINA';
        CALL SUB2 (NAME);
        STREET = '414 GAYLE AVE.';
        CALL SUB3 (STREET);
    END PLITST;
SUB1:  PROCEDURE (S1_RECORD);
    DCL 1  S1_RECORD,
           2  S1_NAME,
              3  S1_F_NAME     CHAR (*),
              3  S1_L_NAME     CHAR (*),
           2  S1_ADDRESS,
              3  S1_STREET     CHAR (*),
              3  S1_CITY       CHAR (*);
        PUT SKIP LIST (S1_RECORD);
        RETURN;
```

```
END SUB1;
SUB2:   PROCEDURE (S2_NAME);
    DCL 1   S2_NAME,
            2   S2_F_NAME           CHAR (*),
            2   S2_L_NAME           CHAR (*);
        PUT SKIP LIST (S2_NAME);
        RETURN;
END SUB2;
SUB3:   PROCEDURE (S3_STREET);
    DCL S3-STREET                   CHAR (*);
        PUTSKIP LIST (S3_STREET);
        RETURN;
END SUB3;
```

As you can see from a glance at the example, the rule that the argument must match the parameter holds true when the parameter is a structure. SUB1 has what we might call the entire structure for its parameter, and the argument passed to it must also be the "full" structure. SUB2 illustrates three items of interest:

1. The parameter structure must be a major structure (i.e., it must have a 1-level name).

2. The level numbers of the argument need not match those of the parameter, so long as they each contain the same number of elementary items and these items possess the same attributes.

3. The argument to the parameter can be a minor structure, but as we mentioned, the parameter must be a major structure. The parameter of SUB3 is not even a structure, of course, but we included it to show that it is possible to pass an elementary item that is part of a structure to an elementary item parameter that is not part of a structure.

If you keep these simple rules in mind, you should have no problem using structures as parameters (and arguments).

12.4

Structures and Arrays

A structure can contain one or more arrays as elements. By the same token, structures can be the elements of an array. It is even possible, in fact, for an array to consist of structures which have arrays within them. These matters are the subject of the following two sections.

Arrays in Structures

An array within a structure is declared at the elementary level. Consider, for example, the following declaration.

```
DCL 1  CLASS_REC,
       2  SSN              CHARACTER (9),
       2  NAME             CHARACTER (20),
       2  STANDING         CHARACTER (2),
       2  GRADES (6)       FIXED DEC (3);
```

The declaration provides 9 bytes of storage for SSN, of course, 20 bytes for NAME, and 2 bytes for STANDING. It then sets aside space for six elements of GRADES, each capable of holding a three-digit number. The elements are referenced in the usual manner; for example,

```
DO SUB = 1 TO 6 WHILE (GRADES (SUB) > 0);
   GRD_TTL = GRD_TTL + GRADES (SUB);
END;
#_GRDS = SUB - 1;
```

Remember, however, that to use:

```
GET LIST (CLASS_REC);
```

you must provide six grades in each input record. In our example, we are assuming that any grade not yet recorded is listed as a 0 in the input record.

Arrays of Structures

An array of structures, as you would expect, is an array whose elements are structures. The same rules apply when the array elements are structures as when they are scalar elements, and this is primarily the fact that each structure must be identical in format to every other structure. To declare an array of structures, you provide the dimension attribute at the major structure level or at a minor structure level (however, PL/C does not support a declaration at the minor structure level).

```
DCL 1  CLASS_ARRAY (35),
       2  SSN                CHARACTER (9),
       2  NAME,
```

```
                3   LAST              CHARACTER (10),
                3   FIRST             CHARACTER (10),
            2   CURR_AVG              FIXED DEC (5,2);
```

This declaration provides for 35 instances of the structure. With this array of structures you can access a corresponding SSN, NAME, and CURR_AVG through CLASS_ARRAY (SUB). NAME (SUB) references FIRST (SUB) and LAST (SUB). In other words, you can reference the structure at any level in its hierarchy.

```
DCL  1   HISTORY,
            2   SSN               CHARACTER (9),
            2   NAME              CHARACTER (20),
            2   CLASSES (50),
                3   CLASS         CHARACTER (4),
                3   SMSTR         CHARACTER (1),
                3   YEAR          CHARACTER (2),
                3   GRADE         FIXED DEC (1);
```

HISTORY sets aside storage for 50 instances of the minor structure CLASSES. Again, you can reference the array at the level of CLASSES or any elementary item under it. Minor structure level arrays are not available to PL/C.

Finally, you can combine arrays in structures with structures of arrays. Let's return to CLASS_REC and modify its declaration to read

```
DCL  1   CLASS_REC (45),
            2   SSN               CHARACTER (9),
            2   NAME              CHARACTER (20),
            2   STANDING          CHARACTER (2),
            2   GRADES (6)        FIXED DEC (3);
```

With this declaration we have set aside storage for 45 instances of CLASS_REC, and for each of those instances, space is reserved for 6 instances of GRADES. You can reference CLASS_REC and any elementary item except GRADES with a single subscript: CLASS_REC (SUB) or STANDING (SUB). You must use two subscripts, however, to reference GRADES: GRADES (SUB, NDX). Another method of referencing GRADES is CLASS_REC (SUB).GRADES (NDX). In other words, GRADES is a two-dimensional array.

Example Program—
Arrays and Structures

The problem we solve in program CH12EX2 (see Fig. 12–2) is simple enough, at least on the surface. The program was requested by Records and Registration because each semester a number of students request permission to carry an academic overload for the coming semester. We are to devise a program that will determine overload eligibility based on a cumulative grade point average (GPA) of 3.00. The students' records are maintained in such a form that the GPA for each semester is included. This format, of course, requires an array within the structure, and if you look at the declaration of OVLD_REC, you will see that the array has 14 elements. Experience has taught the Records and Registration officials that students rarely enroll in more than that many semesters before they either graduate or drop out.

Records and Registration also wants two distinct reports: one listing students (including their cumulative GPA) who are eligible to carry an overload, and a second report for those who are not eligible. To meet this request requires us to set up two arrays of structures: one containing eligible student records and the other containing the records of the ineligible students. The officials also assured us that no more than 50 students apply for overload permission for a particular semester. We therefore gave each array 50 elements because of the possibility that all applicants will be either eligible or ineligible. Finally, Records and Registration told us they place zeros in the semester GPA array elements that do not contain valid GPA's. Because Records and Registration wants the report to reflect the appropriate semester and year, we chose to include a header record containing that information.

Turning to the program, you will notice that because the input records have data items for all the OR_GPA_TAB elements, we can code a simple GET LIST (OVLD_REC); (Statements 19 and 38). Within the main loop we first determine the cumulative GPA. The purpose of Statement 26 (#_SMSTRS = SUB − 1;) is that SUB is declared as a binary field. If we were to code in Statement 27

```
AVG_GPA = ROUND (TTL_GPA / (SUB - 1));
```

it would force TTL_GPA to be converted to binary. As it converts from a decimal to a binary fraction and back again, it might give inaccurate results; therefore, we convert the binary SUB − 1 to the decimal #_SMSTRS. Since we are now dividing a decimal by a decimal, no conversion is required.

Once the cumulative GPA has been determined, we assign the record to either NELG_TAB or ELG_TAB, depending on its GPA relative to the 3.00 cutoff. Notice that each array is controlled by a separate subscript (N_NDX, E_NDX); this is necessary to prevent any empty elements in either array.

The final steps are to call the page header routine (HDR_RTN), print the list of eligible students, and print the identifying message at the end of the list. We go through the same steps for the ineligible report, and end the program. The pseudocode is as follows:

```
CH12EX2
    Read the header record
    Read a data record
    Eligible subscript, not-eligible subscript ← 1
    DO WHILE (more input)
        FOR (subscript = 1 to 14 by 1 WHILE (GPA array
                (subscript) > 0))
            Total GPA ← total GPA + GPA array (subscript)
        ENDFOR
        Number of semesters ← subscript - 1
        Average GPA ← total GPA / number of semesters
        IF (average GPA < 3.00)
            Not-eligible student data (not-eligible subscript)
                ← student data
            Not-eligible cumulative GPA (not-eligible
                subscript) ← average GPA
            Not-eligible subscript ← not-eligible subscript
                + 1
        ELSE
            Eligible student data (eligible subscript) ←
                student data
            Eligible cumulative GPA (eligible subscript) ←
                average GPA
            Eligible subscript ← eligible subscript + 1
        ENDIF
        Read a record
    ENDDO
    Call HDR_RTN
    FOR (subscript = 1 to eligible subscript - 1 by 1)
        Print eligible array (subscript)
    ENDFOR
    Print 'permission granted' message
    Call HDR_RTN
    FOR (subscript = 1 to not-eligible subscript - 1 by 1)
        Print not-eligible array (subscript)
    ENDFOR
    Print 'permission denied' message
```

```
                    STOP
                    HDR_RTN
                       Print page header
                    EXIT
```

PL/I OPTIMIZING COMPILER VERSION 1 RELEASE 5.1 TIME: 13.16.05 DATE: 10 MAR 87 PAGE 1
OPTIONS SPECIFIED
GOSTMT;

```
   STMT

       /* CH12EX2 USING STRUCTURES, ARRAYS */
     1 CH12EX2:    PROCEDURE OPTIONS (MAIN);
       /******************************************************************/
       /*                                                              */
       /* FUNCTION:  TO PROCESS A FILE OF STUDENTS REQUESTING          */
       /*            PERMISSION TO CARRY AN ACADEMIC OVERLOAD.         */
       /*                                                              */
       /* INPUT:     A FILE CONTAINING A HEADER RECORD LISTING THE    */
       /*            YEAR AND SEMESTER, AND A DATA RECORD FOR EACH    */
       /*            STUDENT REQUESTING AN OVERLOAD.  EACH DATA RECORD */
       /*            CONTAINS A SOCIAL SECURITY NUMBER, NAME, MAJOR,  */
       /*            AND 10 SEMESTER GRADE POINT AVERAGES             */
       /*                                                              */
       /* OUTPUT:    TWO REPORTS:  THE FIRST LISTING STUDENTS GRANTED */
       /*            PERMISSSION TO CARRY AN OVERLOAD, THE SECOND     */
       /*            LISTING THE STUDENTS DENIED PERMISSION.          */
       /*                                                              */
       /* NOTES:     STUDENTS HAVE A GPA OF 0 FOR SEMESTERS THEY HAVE */
       /*            YET ENROLLED IN.  TO GAIN PERMISSION FOR AN OVER- */
       /*            LOAD THE STUDENTS MUST HAVE A CUMULATIVE GPA OF  */
       /*            3.00.                                            */
       /*                                                              */
       /******************************************************************/
```

FIGURE 12–2
CH12EX2 (both structures and arrays).

```
2      DCL 1  HDR_REC,
              2  YEAR              CHARACTER (4),
              2  SMSTR             CHARACTER (6);


3      DCL 1  OVLD_REC,
              2  OR_STDNT_DATA,
                 3  OR_SSN         CHARACTER (9),
                 3  OR_NAME        CHARACTER (20),
              2  OR_GPA_TAB (14)   FIXED DEC (3,2);


4      DCL 1  ELG_TAB (50),
              2  ET_STDNT_DATA,
                 3  SSN            CHARACTER (9),
                 3  NAME           CHARACTER (20),
              2  ET_CUM_GPA        FIXED DEC (3,2);


5      DCL 1  NELG_TAB (50),
              2  NT_STDNT_DATA,
                 3  SSN            CHARACTER (9),
                 3  NAME           CHARACTER (20),
              2  NT_CUM_GPA        FIXED DEC (3,2);
6      DCL TTL_GPA                 FIXED DEC (5,2);
7      DCL AVG_GPA                 FIXED DEC (3,2);
8      DCL #_SMSTRS                FIXED DEC (2);
```

```
9      DCL SUB                     FIXED BIN (15);
10     DCL E_NDX                   FIXED BIN (15);
11     DCL N_NDX                   FIXED BIN (15);

12     DCL MORE_RECS               BIT      (1);
13     DCL    YES                  BIT      (1)   INIT ('1'B);
14     DCL    NO                   BIT      (1)   INIT ('0'B);

15     DCL    ROUND                BUILTIN;
16            MORE_RECS = YES;
17            ON ENDFILE (SYSIN) MORE_RECS = NO;

18            GET LIST (HDR_REC);
```

FIGURE 12–2 *(continued)*

```
19              GET LIST (OVLD_REC);
20              E_NDX, N_NDX = 1;

21  MAIN_LP: DO WHILE (MORE_RECS);
22                  TTL_GPA = 0;

23  GPA_LP:     DO SUB = 1 TO 14 WHILE (OR_GPA_TAB (SUB) > 0.0);
24                  TTL_GPA = TTL_GPA + OR_GPA_TAB (SUB);
25              END GPA_LP;
26              #_SMSTRS = SUB - 1;
27              AVG_GPA = ROUND (TTL_GPA / #_SMSTRS, 2);

28              IF AVG_GPA < 3.00 THEN
    NT_ELG:     DO;
29                  NT_STDNT_DATA (N_NDX) = OR_STDNT_DATA;
30                  NT_CUM_GPA (N_NDX) = AVG_GPA;
31                  N_NDX = N_NDX + 1;
32              END NT_ELG;
33              ELSE

    ELIG:       DO;
34                  ET_STDNT_DATA (E_NDX) = OR_STDNT_DATA;
35                  ET_CUM_GPA (E_NDX) = AVG_GPA;
36                  E_NDX = E_NDX + 1;
37              END ELIG;
38              GET SKIP LIST (OVLD_REC);
39          END MAIN_LP;

40          CALL HDR_RTN (HDR_REC);

41  PRNT_EL: DO SUB =  1 TO E_NDX - 1;
42              PUT SKIP (2) EDIT (ELG_TAB (SUB)) (COL(38), A(14), A(25),
                                        F(4,2));
43          END PRNT_EL;
44          PUT SKIP (5) EDIT ('***** PERMISSION GRANTED *****')
                            (COL (40), A);
45          CALL HDR_RTN (HDR_REC);

46  PRNT_NL: DO SUB =  1 TO N_NDX - 1;
47              PUT SKIP (2) EDIT (NELG_TAB (SUB)) (COL(38), A(14), A(25),
                                        F(4,2));
```

STMT
```
    48          END PRNT_NL;
    49          PUT SKIP (5) EDIT ('***** PERMISSION DENIED *****')
                        (COL(40), A);
```

STMT
```
    50 HDR_RTN:  PROC (HR_HDR_REC);
       /*******************************************************************/
       /*                                                               */
       /* FUNCTION:  TO PRINT PAGE AND COLUMN HEADERS                    */
       /*                                                               */
       /* PARAMETERS:  HR_HDR_REC  THE HEADER RECORD                     */
       /*                                                               */
       /* GLOBAL VARIABLES:  NONE                                        */
       /*                                                               */
       /* INPUT:     NONE                                                */
       /*                                                               */
       /* OUTPUT:    PAGE AND COLUMN HEADERS                             */
       /*                                                               */
       /*******************************************************************/

    51     DCL 1  HR_HDR_REC,
                   2  HR_YEAR              CHARACTER (4),
                   2  HR_SMSTR             CHARACTER (6);

    52          PUT PAGE;
    53          PUT EDIT ('ACADEMIC OVERLOAD REQUEST REPORT FOR ', HR_HDR_REC)
                         (COL(35), A, A(5), A);
    54          PUT SKIP (4) EDIT ('SSN', 'NAME', 'CUM GPA') (COL(42), A,
                                 X(10), A, X(16), A);
    55          RETURN;
    56 END HDR_RTN;
    57 END CH12EX2;
```

COMPILER DIAGNOSTIC MESSAGES
ERROR ID L   STMT   MESSAGE DESCRIPTION

COMPILER INFORMATORY MESSAGES

FIGURE 12-2 *(continued)*

```
IEL0533I I          NO 'DECLARE' STATEMENT(S) FOR 'SYSIN','SYSPRINT'.
END OF COMPILER DIAGNOSTIC MESSAGES
```

```
                    ACADEMIC OVERLOAD REQUEST REPORT FOR 1988 SUMMER


                  SSN           NAME            CUM GPA

              101011010     BEARD, BETTY          3.50

              212122121     DAWN, DELTA           3.50

              414414141     KELP, KEVIN           3.57




                    ***** PERMISSION GRANTED *****
------------------------------------------------------------------------------
                    ACADEMIC OVERLOAD REQUEST REPORT FOR 1988 SUMMER


                  SSN           NAME            CUM GPA

              691136913     CARSON, CINDY         2.73

              987654321     EARL, EARL            2.94

              555511515     FUNK, FRIEDA          2.66

              777137713     HANSEN, HANS          2.85

              333445555     JONES, JERRY          2.95




                    ***** PERMISSION DENIED *****
```

12.5

**Example Program—
Structures**

The problem we solve in CH12EX3 (see Fig. 12–3) is one familiar to every college student and teacher: providing timely and accurate grade rolls at the beginning of the semester. In writing our program we have supposed the following situation. There are two distinct registration periods; the first is a "regular" period before the semester opens, perhaps a mail registration, and the second is "add/drop" registration occurring during the first week of the semester. Students who registered during the regular period have their data entered into a data set; these students are identified with an 'R'. When this registration period ends, a dummy record is entered to separate the 'R' students from those who will register during drop/add. As students drop or add the course during drop/add registration, their data is entered following the dummy record.

```
PL/I OPTIMIZING COMPILER  VERSION 1 RELEASE 5.1  TIME: 11.05.45  DATE: 13 MAR 87 PAGE   1
OPTIONS SPECIFIED
GOSTMT;


PL/I OPTIMIZING COMPILER        /* CH12EX3 USING STRUCTURES */              PAGE   2
                SOURCE LISTING
   STMT


      /* CH12EX3 USING STRUCTURES */
   1  CH12EX3:    PROCEDURE OPTIONS (MAIN);
      /****************************************************************/
      /*                                                          */
      /*   FUNCTION:  TO PRODUCE A CLASS ROLL FOR A CLASS FROM    */
      /*              ENROLLMENT RECORDS.                         */
      /*                                                          */
      /*   INPUT:     A FILE CONTAINING                           */
      /*              1.  A HEADER RECORD LISTING THE CLASS NAME AND */
      /*                  NUMBER, SECTION, SEMESTER, AND YEAR.    */
      /*              2.  A RECORD FOR EACH STUDENT WHO ENROLLED IN THE */
      /*                  CLASS DURING REGULAR RESGISTRATION.  THESE  */
      /*                  RECORDS CONTAIN  A STATUS-BYTE, SSN, NAME,  */
      /*                  STREET, CITY-STATE-ZIP, COLLEGE OF MAJOR, AND */
      /*                  TOTAL NUMBER OF HOURS ENROLLED IN.  THE END  */
```

FIGURE 12–3
CH12EX3 (another example of structures).

FIGURE 12–3 *(continued)*

```
/*                       OF THIS GROUP OF RECORDS IS DELIMITED BY ALL  */
/*                       X'S IN CHARACTER FIELDS AND ALL 9'S IN        */
/*                       NUMERIC FIELDS.                               */
/*                    3. A RECORD CONTAINING THE SAME DATA AS THE      */
/*                       RECORDS ABOVE FOR EACH STUDENT WHO ADDED OR   */
/*                       DROPPED THE CLASS DURING LATE REGISTRATION.   */
/*                                                                     */
/*   OUTPUT:          1. A CLASS ROLL FOR THE STUDENTS STILL ENROLLED  */
/*                       IN THE CLASS.                                 */
/*                    2. A LISTING OF THOSE STUDENTS WHO DROPPED THE   */
/*                       CLASS.                                        */
/*                                                                     */
/*   NOTES:           1. THE STATUS-BYTE CAN HAVE THE VALUES OF        */
/*                       R--ENROLLED DURING REGULAR REGISTRATION       */
/*                       A--ADDED THE CLASS DURING THE DROP/ADD PERIOD */
/*                       D--DROPPED THE CLASS DURING DROP/ADD          */
/*                    2. THE REGULAR RESGISTRATION RECORDS ARE READ    */
/*                       INTO ONE ARRAY AND THE DROP/ADD RECORDS INTO  */
/*                       A SECOND ONE.                                 */
/*                    3. THE ADD RECORDS ARE MOVED TO THE FIRST ARRAY, */
/*                       AND THAT ARRAY IS 'SORTED' USING SUBSCRIPTED  */
/*                       SUBSCRIPTS.                                   */
/*                    4. THE DROP RECORDS ARE PROCESSED BY SEARCHING   */
/*                       THE FIRST ARRAY (THROUGH THE SUBSCRIPTED      */
/*                       SUBSCRIPTS) AND ITS STATUS-BYTE CHANGED TO    */
/*                       'D'.                                          */
/*                    5. THE TWO REPORTS ARE PRINTED.                  */
/*                                                                     */
/***********************************************************************/


  2     DCL 1 HDR_REC,
              2 CLASS               CHARACTER (9),
              2 SECTION             CHARACTER (2),
              2 CLS_HRS             FIXED DEC (1),
              2 TERM                CHARACTER (7),
              2 YEAR                CHARACTER (4);
```

PL/I OPTIMIZING COMPILER   /* CH12EX3 USING STRUCTURES */   PAGE   3
    STMT

```
3      DCL 1  REG_ARRAY (50),
              2  R_STATUS_BYTE        CHARACTER (1),
              2  REG_DATA,
                 3  R_SSN             CHARACTER (9),
                 3  R_NAME            CHARACTER (20),
                 3  R_ADDRESS,
                    4  R_STREET       CHARACTER (15),
                    4  R_CITY_ST_ZIP  CHARACTER (20),
                 3  R_MAJOR_COLL      CHARACTER (4),
                 3  R_CREDITS         FIXED DEC (2);

4      DCL 1  DA_ARRAY (50),
              2  DA_STATUS_BYTE       CHARACTER (1),
              2  DA_DATA,
                 3  DA_SSN            CHARACTER (9),
                 3  DA_NAME           CHARACTER (20),
                 3  DA_ADDRESS,
                    4  DA_STREET      CHARACTER (15),
                    4  DA_CITY_ST_ZIP CHARACTER (20),
                 3  DA_MAJOR_COL      CHARACTER (4),
                 3  DA_CREDITS        FIXED DEC (2);

5      DCL PNTRS (50)               FIXED BIN (15);
6      DCL SUB                      FIXED BIN (15);
7      DCL NDX                      FIXED BIN (15);
8      DCL REG_CNTR                 FIXED BIN (15);
9      DCL DA_CNTR                  FIXED BIN (15);

10     DCL MORE_RECS                BIT    (1);
11     DCL   YES                    BIT    (1)  INIT ('1'B);
12     DCL   NO                     BIT    (1)  INIT ('0'B);

13 MAIN:   MORE_RECS = YES;
14         ON ENDFILE (SYSIN) MORE_RECS = NO;

15         CALL READ_RTN (HDR_REC, REG_ARRAY, DA_ARRAY);
16         CALL PROC_ADD_RTN (REG_ARRAY, DA_ARRAY, CLS_HRS);
17         CALL PROC_DROP_RTN (REG_ARRAY, DA_ARRAY, CLS_HRS);
18         CALL SORT_PNTRS_RTN (REG_ARRAY, PNTRS);
19         CALL PRNT_RTN (HDR_REC, REG_ARRAY, PNTRS);
```

FIGURE 12–3 *(continued)*

```
PL/I OPTIMIZING COMPILER        /* CH12EX3 USING STRUCTURES */                    PAGE   4
  STMT
    20   READ_RTN: PROC (RR_HDR_REC, RR_REG_ARRAY,RR_DA_ARRAY);
         /***********************************************************************/
         /*                                                                   */
         /*  FUNCTION:  TO READ DATA INTO REG_ARRAY UNTIL DUMMY RECORD        */
         /*             READ, THEN TO READ REMAINING DATA INTO DA_ARRAY       */
         /*                                                                   */
         /*  PARAMETERS:  RR_HDR_REC    THE HEADER RECORD                     */
         /*               RR_REG_ARRAY  THE ARRAY FOR REGULAR REGISTRANTS     */
         /*               RR_DA_ARRAY   THE ARRAY FOR DROP/ADDS               */
         /*                                                                   */
         /*  GLOBAL VARIABLES:  NONE                                         */
         /*                                                                   */
         /*  SUBROUTINES CALLED:  NONE                                       */
         /*                                                                   */
         /*  INPUT:   THE HEADER AND DATA RECORDS                            */
         /*                                                                   */
         /*  OUTPUT: NONE                                                    */
         /*                                                                   */
         /***********************************************************************/
    21   DCL 1 RR_HDR_REC,
               2 RR_CLASS              CHARACTER (9),
               2 RR_SECTION            CHARACTER (2),
               2 RR_CLS_HRS            FIXED DEC (1),
               2 RR_TERM               CHARACTER (7),
               2 RR_YEAR               CHARACTER (4);

    22   DCL 1 IN_REC,
               2 IN_STATUS_BYTE        CHARACTER (1),
               2 IN_DATA,
                 3 IN_SSN              CHARACTER (9),
                 3 IN_NAME             CHARACTER (20),
                 3 IN_ADDRESS,
                   4 IN_STREET         CHARACTER (15),
                   4 IN_CITY_ST_ZIP    CHARACTER (20),
                 3 IN_MAJOR_COLL       CHARACTER (4),
                 3 IN_CREDITS          FIXED DEC (2);

    23   DCL 1 RR_REG_ARRAY (*),
```

```
              2  RR_R_STATUS_BYTE          CHARACTER (1),
              2  RR_REG_DATA,
                 3  RR_R_SSN              CHARACTER (9),
                 3  RR_R_NAME             CHARACTER (20),
                 3  RR_R_ADDRESS,
                    4  RR_R_STREET        CHARACTER (15),
                    4  RR_R_CITY_ST_ZIP   CHARACTER (20),
                 3  RR_R_MAJOR_COLL       CHARACTER (4),
                 3  RR_R_CREDITS          FIXED DEC (2);

   24    DCL 1  RR_DA_ARRAY (*),
              2  RR_DA_STATUS_BYTE         CHARACTER (1),
              2  RR_DA_DATA,
                 3  RR_DA_SSN             CHARACTER (9),
```

```
                 3  RR_DA_NAME            CHARACTER (20),
                 3  RR_DA_ADDRESS,
                    4  RR_DA_STREET       CHARACTER (15),
                    4  RR_DA_CITY_ST_ZIP  CHARACTER (20),
                 3  RR_DA_MAJOR_COL       CHARACTER (4),
                 3  RR_DA_CREDITS         FIXED DEC (2);

   25         GET LIST (HDR_REC);
   26         GET LIST (IN_REC);

   27 REG_LP: DO SUB = 1 TO 50 WHILE (MORE_RECS & IN_STATUS_BYTE < 'X');
   28            IF IN_STATUS_BYTE < 'X' THEN
                     RR_REG_ARRAY (SUB) = IN_REC;
   29            GET LIST (IN_REC);
   30         END REG_LP;

   31         REG_CNTR = SUB;
   32         SUB = 1;
   33         GET LIST (IN_REC);

   34 DA_LP:  DO WHILE (MORE_RECS);
   35            RR_DA_ARRAY (SUB) = IN_REC;
   36            GET LIST (IN_REC);
   37            SUB = SUB + 1;
```

FIGURE 12–3 *(continued)*

```
 38          END DA_LP;

 39          DA_CNTR = SUB - 1;
 40          RETURN;
 41 END READ_RTN;
```

```
PL/I OPTIMIZING COMPILER        /* CH12EX3 USING STRUCTURES */                PAGE   6
  STMT
    42   PROC_ADD_RTN:  PROC (PA_REG_ARRAY, PA_DA_ARRAY, PA_CLS_HRS);
         /*******************************************************************/
         /*                                                               */
         /*  FUNCTION:  TO ADD THOSE STUDENTS WHO ADDED THE COURSE TO     */
         /*             PA_REG_ARRAY AND TO COMPUTE NEW TOTAL HOURS       */
         /*             ENROLLED FOR                                      */
         /*                                                               */
         /*  PARAMETERS:  PA_REG_ARRAY  ARRAY OF THOSE REGISTERED FOR THE */
         /*                             COURSE (REGULARS + ADDS)          */
         /*               PA_DA_ARRAY   ARRAY OF THOSE WHO HAVE ADDED OR  */
         /*                             DROPPED THE COURSE                */
         /*               PA_CLS_HRS    NUMBER OF CREDIT HOURS OF THE     */
         /*                             COURSE                           */
         /*                                                               */
         /*  GLOBAL VARIABLES:  NONE                                     */
         /*                                                               */
         /*  SUBROUTINES CALLED:  NONE                                   */
         /*                                                               */
         /*  INPUT:  NONE                                                */
         /*                                                               */
         /*  OUTPUT:  NONE                                               */
         /*                                                               */
         /*******************************************************************/
    43   DCL 1  PA_REG_ARRAY (*),
               2  PA_R_STATUS_BYTE       CHARACTER (1),
               2  PA_REG_DATA,
                  3  PA_R_SSN            CHARACTER (9),
                  3  PA_R_NAME           CHARACTER (20),
                  3  PA_R_ADDRESS,
                     4  PA_R_STREET      CHARACTER (15),
                     4  PA_R_CITY_ST_ZIP CHARACTER (20),
```

```
                    3  PA_R_MAJOR_COLL        CHARACTER (4),
                    3  PA_R_CREDITS           FIXED DEC (2);


     44     DCL 1  PA_DA_ARRAY (*),
                    2  PA_DA_STATUS_BYTE      CHARACTER (1),
                    2  PA_DA_DATA,
                       3  PA_DA_SSN           CHARACTER (9),
                       3  PA_DA_NAME          CHARACTER (20),
                       3  PA_DA_ADDRESS,
                          4  PA_DA_STREET     CHARACTER (15),
                          4  PA_DA_CITY_ST_ZIP CHARACTER (20),
                       3  PA_DA_MAJOR_COL     CHARACTER (4),
                       3  PA_DA_CREDITS       FIXED DEC (2);


     45     DCL PA_CLS_HRS                    FIXED DEC (1);
     46  ADD_LP:  DO SUB = 1 TO DA_CNTR;
     47             IF PA_DA_STATUS_BYTE (SUB) = 'A' THEN

     MOVE_LP:         DO;
     48                 PA_REG_ARRAY (REG_CNTR) = PA_DA_ARRAY (SUB);
     49                 PA_R_CREDITS (REG_CNTR) = PA_R_CREDITS (REG_CNTR) +
                                          PA_CLS_HRS;
```

```
     50                 REG_CNTR = REG_CNTR + 1;
     51               END MOVE_LP;
     52           END ADD_LP;
     53         RETURN;
     54  END PROC_ADD_RTN;
```

```
     55  PROC_DROP_RTN:  PROC (PD_REG_ARRAY, PD_DA_ARRAY, PD_CLS_HRS);
         /****************************************************************/
         /*                                                              */
         /*  FUNCTION:    TO LOGICALLY DELETE RECORDS IN PD_REG_ARRAY OF  */
         /*               THOSE STUDENTS WHO DROPPED THE COURSE.         */
         /*               PD_STATUS_BYTE IS SET TO 'D'.                  */
         /*                                                              */
         /****************************************************************/
```

FIGURE 12–3 *(continued)*

```
56    DCL 1  PD_REG_ARRAY (*),
               2  PD_R_STATUS_BYTE        CHARACTER (1),
               2  PD_REG_DATA,
                  3  PD_R_SSN             CHARACTER (9),
                  3  PD_R_NAME            CHARACTER (20),
                  3  PD_R_ADDRESS,
                     4  PD_R_STREET       CHARACTER (15),
                     4  PD_R_CITY_ST_ZIP  CHARACTER (20),
                  3  PD_R_MAJOR_COLL      CHARACTER (4),
                  3  PD_R_CREDITS         FIXED DEC (2);

57    DCL 1  PD_DA_ARRAY (*),
               2  PD_DA_STATUS_BYTE       CHARACTER (1),
               2  PD_DA_DATA,
                  3  PD_DA_SSN            CHARACTER (9),
                  3  PD_DA_NAME           CHARACTER (20),
                  3  PD_DA_ADDRESS,
                     4  PD_DA_STREET      CHARACTER (15),
                     4  PD_DA_CITY_ST_ZIP CHARACTER (20),
                  3  PD_DA_MAJOR_COL      CHARACTER (4),
                  3  PD_DA_CREDITS        FIXED DEC (2);

58    DCL PD_CLS_HRS                      FIXED DEC (1);
59    DCL NOT_FOUND                       BIT      (1);
60    DCL    TRUE                         BIT      (1)   INIT ('1'B);
61    DCL    FALSE                        BIT      (1)   INIT ('0'B);

62 DRP_LP:  DO SUB = 1 TO DA_CNTR;
63              IF PD_DA_STATUS_BYTE (SUB) = 'D' THEN
   DRP_1:        DO;
64                  NOT_FOUND = TRUE;

65 SRCH_LP:        DO NDX = 1 TO 50 WHILE (NOT_FOUND);
66                    IF PD_DA_NAME (SUB) = PD_R_NAME (NDX) THEN
   DROP:             DO;
67                      PD_R_STATUS_BYTE (NDX) = 'D';
68                      PD_R_CREDITS (NDX) = PD_R_CREDITS (NDX) -
                                            PD_CLS_HRS;
69                      NOT_FOUND = FALSE;
70                    END DROP;
```

```
71                   END SRCH_LP;
72                END DRP_1;
73           END DRP_LP;
74    .      RETURN;
75 END PROC_DROP_RTN;
```

```
   STMT
    76  SORT_PNTRS_RTN: PROC (SP_REG_ARRAY, SP_PNTRS);
        /*****************************************************************/
        /*                                                             */
        /*  FUNCTION:  TO "SORT" SR_REG_ARRAY IN ASCENDING ORDER USING */
        /*             'TAG-SORT'.  PNTRS ARRAY IS ACTUALLY SORTED.     */
        /*                                                             */
        /*  PARAMETERS:  SP_REG_ARRAY  ARRAY OF STUDENTS ENROLLED IN THE*/
        /*                             COURSE (REGULARS + ADDS)         */
        /*               SP_PNTRS      ARRAY OF SUBSCRIPTS INTO SP_REG_ */
        /*                             ARRAY                            */
        /*  GLOBAL VARIABLES:                                           */
        /*      REG_CNTR  NUMBER OF SP_REG_ARRAY ELEMENTS HAVING VALID */
        /*                DATA.  SET BY PROC_ADD_RTN, IT CANNOT BE      */
        /*                CHANGED.                                      */
        /*                                                             */
        /*  SUBROUTINES CALLED:  NONE                                   */
        /*                                                             */
        /*  INPUT:  NONE                                                */
        /*                                                             */
        /*  OUTPUT:  NONE                                               */
        /*                                                             */
        /*****************************************************************/
    77  DCL 1  SP_REG_ARRAY (*),
            2  SP_R_STATUS_BYTE       CHARACTER (1),
            2  SP_REG_DATA,
               3  SP_R_SSN            CHARACTER (9),
               3  SP_R_NAME           CHARACTER (20),
               3  SP_R_ADDRESS,
                  4  SP_R_STREET      CHARACTER (15),
                  4  SP_R_CITY_ST_ZIP CHARACTER (20),
               3  SP_R_MAJOR_COLL     CHARACTER (4),
               3  SP_R_CREDITS        FIXED DEC (2);
```

FIGURE 12–3 *(continued)*

```
78    DCL SP_PNTRS (*)              FIXED BIN (15);
79    DCL TEMP                      FIXED BIN (15);

80    DCL SWAP                      BIT     (1);
81    DCL    OFF                    BIT     (1)   INIT ('0'B);
82    DCL    ON                     BIT     (1)   INIT ('1'B);
83    DCL MORE_TO_SORT              BIT     (1);
84    DCL    MORE                   BIT     (1)   INIT ('1'B);
85    DCL    NO_MORE                BIT     (1)   INIT ('0'B);

86 LD_PNTR: DO SUB = 1 TO 50;
87             SP_PNTRS (SUB) = SUB;
88          END LD_PNTR;

89          MORE_TO_SORT = MORE;
90 SORT_LP: DO WHILE (MORE_TO_SORT);
91             SWAP = OFF;
92             SUB = 1;
```

PL/I OPTIMIZING COMPILER   /* CH12EX3 USING STRUCTURES */   PAGE 10
   STMT

```
  93 SWAP_LP:    DO WHILE (SUB < REG_CNTR -1);
  94              IF SP_R_NAME (SP_PNTRS (SUB)) > SP_R_NAME
                                (SP_PNTRS (SUB + 1)) THEN

     SWAP_1:          DO;
  95                    TEMP = SP_PNTRS (SUB);
  96                    SP_PNTRS (SUB) = SP_PNTRS (SUB + 1);
  97                    SP_PNTRS (SUB + 1) = TEMP;
  98                    SWAP = ON;
  99                  END SWAP_1;
 100                  SUB = SUB + 1;
 101              END SWAP_LP;

 102         IF SWAP = OFF THEN
                 MORE_TO_SORT = NO_MORE;
 103         END SORT_LP;
 104         RETURN;
 105 END SORT_PNTRS_RTN;
```

STMT

```
106   PRNT_RTN:  PROC (PR_HDR_REC, PR_REG_ARRAY, PR_PNTRS);
      /*********************************************************************/
      /*                                                                 */
      /*  FUNCTION:    TO PRINT 'GRADE ROLL' AND 'WITHDRAWAL' REPORTS     */
      /*                                                                 */
      /*  PARAMETERS:  PR_HDR_REC    THE HEADER RECORD                   */
      /*               PR_REG_ARRAY  THE ARRAY OF STUDENTS WHO HAVE       */
      /*                             ENROLLED, WITH DROPS LOGICALLY       */
      /*                             DELETED (PR_STATUS_BYTE = 'D')       */
      /*               PR_PNTRS      SUBSCRIPT ARRAY (NOW POINTING AT     */
      /*                             PR_REG_ARRAY IN ALPHABETICAL ORDER   */
      /*  GLOBAL VARIABLES:                                              */
      /*       REG_CNTR              THE NUMBER OF ELEMENTS IN PR_REG-     */
      /*                             ARRAY HAVING VALID DATA              */
      /*                                                                 */
      /*  SUBROUTINES CALLED:  NONE                                      */
      /*                                                                 */
      /*  INPUT:   NONE                                                  */
      /*                                                                 */
      /*  OUTPUT:  THE 'GRADE ROLL' AND 'WITHDRAWAL' REPORTS             */
      /*                                                                 */
      /*********************************************************************/
107   DCL 1 PR_HDR_REC,
            2 PR_CLASS              CHARACTER (9),
            2 PR_SECTION            CHARACTER (2),
            2 PR_CLS_HRS            FIXED DEC (1),
            2 PR_TERM               CHARACTER (7),
            2 PR_YEAR               CHARACTER (4);

108   DCL 1 PR_REG_ARRAY (*),
            2 PR_R_STATUS_BYTE      CHARACTER (1),
            2 PR_REG_DATA,
              3 PR_R_SSN            CHARACTER (9),
              3 PR_R_NAME           CHARACTER (20),
              3 PR_R_ADDRESS,
                4 PR_R_STREET       CHARACTER (15),
                4 PR_R_CITY_ST_ZIP  CHARACTER (20),
              3 PR_R_MAJOR_COLL     CHARACTER (4),
              3 PR_R_CREDITS        FIXED DEC (2);
```

FIGURE 12–3 *(continued)*

```
109      DCL PR_PNTRS (*)                    FIXED BIN (15);
110          PUT PAGE EDIT ('GRADE ROLL') (COL(55), A);
111          PUT SKIP (3) EDIT ('CLASS', 'SECTION', 'HOURS',
                             'SEMESTER', 'YEAR') (COL(35), A(14),
                                  A(12), A(10), A(13), A);
112          PUT SKIP EDIT (PR_CLASS, PR_SECTION, PR_CLS_HRS, PR_TERM,
                             PR_YEAR) (COL(33), A(18), A(9), F(2), X(9),
                                  A(12), A);
113          PUT SKIP (3);

114  ADD_LST: DO SUB = 1 TO REG_CNTR - 1;
115              IF PR_R_STATUS_BYTE (PNTRS (SUB)) ^= 'D' THEN
```

```
PL/I OPTIMIZING COMPILER        /* CH12EX3 USING STRUCTURES */            PAGE  12
  STMT
                    PUT SKIP EDIT (PR_REG_DATA (PNTRS (SUB)))
                             (COL(20), A(14), A(25), A(20), A(25),
                                  A(9), F(2));
116          END ADD_LST;

117          PUT PAGE EDIT ('WITHDRAWALS FROM ',PR_CLASS, ' SECTION ',
                             PR_SECTION) (COL(42), A, A, A, A);
118          PUT SKIP (3) EDIT ('CLASS', 'SECTION', 'HOURS',
                             'SEMESTER', 'YEAR') (COL(35), A(14),
                                  A(12), A(10), A(13), A);
119          PUT SKIP EDIT (PR_CLASS, PR_SECTION, PR_CLS_HRS, PR_TERM,
                             PR_YEAR) (COL(33), A(18), A(9), F(2), X(9),
                                  A(12), A);
120          PUT SKIP (3);

121  DRP_LST: DO SUB = 1 TO REG_CNTR - 1;
122              IF PR_R_STATUS_BYTE(PNTRS (SUB)) = 'D' THEN
                    PUT SKIP EDIT (PR_REG_DATA (PNTRS (SUB)))
                             (COL(20), A(14), A(25), A(20), A(25),
                                  A(9), F(2));
123          END DRP_LST;
124          RETURN;
125  END PRNT_RTN;
126  END CH12EX3;
```

COMPILER INFORMATORY MESSAGES

IEL0533I I   NO 'DECLARE' STATEMENT(S) FOR 'SYSIN','SYSPRINT'.

END OF COMPILER DIAGNOSTIC MESSAGES

GRADE ROLL

CLASS	SECTION	HOURS	SEMESTER	YEAR
CSCI 300	5	5	SUMMER	1987

100101000	ARNOLD, ARNOLD	101 N. 1ST ST	DEKALB, IL 60115	LA&S	15
666559999	BAKER, BONNIE	LINCOLN HALL	CAMPUS	EDUC	9
102030405	COLE, ROBERT	COLE HALL	CAMPUS	MUS	16
654456543	HARPER, WILLY	DOUGLAS HALL	CAMPUS	LA&S	19
908070605	LISLE, LYLE	LINCOLN HALL	CAMPUS	BUSI	18
829137465	MARX, MARY	LINCOLN HALL	CAMPUS	PFST	10
889977669	STEVENS, L.ROBERT	GRANT TOWERS	CAMPUS	LA&S	12
405060708	THOMAS, RITA	216 RIO AVE.	DE KALB, IL 60115	EDUC	11
807060504	VERI, VERA	NEPTUNE HALL	CAMPUS	PFST	19

WITHDRAWALS FROM CSCI 300 SECTION 5

CLASS	SECTION	HOURS	SEMESTER	YEAR
CSCI 300	5	5	SUMMER	1987

304050607	DIRKS, DORIS	GRANT TOWERS	CAMPUS	LA&S	20
203040506	PETERS, ZELDA	24 85TH ST	MALTA, IL 60116	PART	13
747747447	ZANDER, XAVIER	21 GREEN DR.	SYCAMORE, IL 60114	BUSI	7

When the drop/add period has ended, the input data is ready to be processed by our program. The READ_RTN reads those records ahead of the dummy record into IN_REC, and each is assigned to REG_ARRAY. We cannot read the input directly into the array because then there is no way of keeping the dummy record from also being assigned to the array. As it is, we test each record in IN_REC to see if it is the dummy (Statement 28). We then read the remaining records into DA_ARRAY in a similar fashion.

Notice that in PROC-ADD_RTN and PROC_DROP_RTN we scan DA_ARRAY twice, once looking for the Add code and once looking for the Drop code. PROC_ADD_RTN assigns the records of those students who have added the course to REG_ARRAY; it now holds all of those who have enrolled in the course. The function of DROP_RTN, then, is to logically delete those who have subsequently dropped the course. In order to use a binary search, we might have sorted REG_ARRAY before calling PROC_DROP_RTN, but we had a reason for doing otherwise. Therefore, we use a linear search to "logically delete" those who dropped by changing their status byte to 'D'.

The function of SORT_RTN is to allow for an alphabetized listing of those enrolled and those who dropped. The obvious solution is simply to sort REG-ARRAY; that method, however, involves a considerable amount of data movement. To reduce the data movement, we chose to use a "tag-sort," which involves subscripted subscripts. In other words, the subscripts into REG_ARRAY are themselves the contents of elements of an array (PNTRS). It is these subscripts, rather than the contents of REG_ARRAY, that are sorted. To give you a better grasp of how tag-sorting works, here is what the two arrays look like when the LD_PNTRS loop in SORT_RTN finishes executing. We use the student's last name to represent the entire element of REG-ARRAY, and we only go as far into the arrays as we have valid data for REG-ARRAY. The numbers in parentheses represent the subscript for each element.

REG ARRAY		PNTRS	
ARNOLD	(1)	1	(1)
ZANDER	(2)	2	(2)
BAKER	(3)	3	(3)
STEVENS	(4)	4	(4)
MARX	(5)	5	(5)
COLE	(6)	6	(6)
PETERS	(7)	7	(7)
HARPER	(8)	8	(8)
DIRKS	(9)	9	(9)
THOMAS	(10)	10	(10)
LISLE	(11)	11	(11)
VERI	(12)	12	(12)

As you can see, the values in the elements of PNTRS can serve as subscripts into REG_ARRAY. Contrast this with the following, which shows the two when SORT_RTN finishes sorting.

```
REG ARRAY                        PNTRS
---------                        -----
┌─────────┐                      ┌────┐
│ ARNOLD  │ (1)                  │  1 │ (1)
├─────────┤                      ├────┤
│ ZANDER  │ (2)                  │  3 │ (2)
├─────────┤                      ├────┤
│ BAKER   │ (3)                  │  6 │ (3)
├─────────┤                      ├────┤
│ STEVENS │ (4)                  │  9 │ (4)
├─────────┤                      ├────┤
│ MARX    │ (5)                  │  8 │ (5)
├─────────┤                      ├────┤
│ COLE    │ (6)                  │ 11 │ (6)
├─────────┤                      ├────┤
│ PETERS  │ (7)                  │  5 │ (7)
├─────────┤                      ├────┤
│ HARPER  │ (8)                  │  7 │ (8)
├─────────┤                      ├────┤
│ DIRKS   │ (9)                  │  4 │ (9)
├─────────┤                      ├────┤
│ THOMAS  │ (10)                 │ 10 │ (10)
├─────────┤                      ├────┤
│ LISLE   │ (11)                 │ 12 │ (11)
├─────────┤                      ├────┤
│ VERI    │ (12)                 │  2 │ (12)
└─────────┘                      └────┘
```

If you now look at the contents of PNTRS in relation to the subscripts of both PNTRS and REG_ARRAY, you will notice that the contents of PNTRS, taken in ascending order, point to the elements of REG_ARRAY in alphabetical order: PNTRS (1) contains a 1, and REG_ARRAY (1) (ARNOLD) is first in an alphabetized list. PNTRS (2) contains a 3, which is the subscript for BAKER, and PNTR (3) contains a 6—the subscript for COLE, and so on. In Statement 94 you can see that the comparison is between the names in successive elements in SP_REG_ARRAY, but Statements 95–97 show you that it is the contents of SP_PNTRS that are being moved. Another look at Statement 94 will verify our claim that REG_ARRAY must be referenced through subscripted subscripts—(SP_PNTRS (SUB))—and a check of Statements 115 and 122 (notice both the IF and the PUT) further illustrate the point. The result is an elegant operation in which a minimum of data movement occurs; yet it produces the desired result. The flowchart is shown in Figure A–19.

12.6

Summary

Structures give you the advantage of being able to address entire records, segments of records, or individual data items with a single variable for each. A structure is a data aggregate in which some variables are subordinate to others. At the highest level is the major structure, which is designated by a level number of 1. A structure can have only one major structure. The next lower level is the minor structure. A variable names a minor structure *only if it is further divided.* It must have a level number greater than 1. If a variable has no levels below it, it is an elementary item, and an elementary item must have a level number greater than the level number of the minor structure (if any) it is under. Elementary items and minor structures can appear at the same level. In all cases, the level number must be separated from its variable by at least one blank. As a general rule, attributes are assigned only at the elementary level.

A major structure must have a unique name, but minor structures and elementary items need not. We can achieve a unique name by concatenating the minor structure or elementary name with one or more higher-level names, each separated from the other by a period. Intermediate names can be omitted if they are not required for uniqueness.

The PICTURE declaration is used primarily with record I/O, especially on input because stream-directed data transmissions provide automatic data conversion. On output, however, the picture provides editing features not otherwise available; consequently, pictures are used regularly with stream-directed output. A second reason for using the picture specification in some cases is that it provides data verification. The PICTURE (which can be abbreviated PIC) specification characters determine exactly what data type can occupy that relative position in the field. PL/I provides a specification character for alphabetic (A–Z, national, blank) data (A), for any of the 256 possible characters (X), and for numeric data (9). The possibilities for numeric, as you might expect, are the most numerous. One digit can occupy the position occupied by each 9 in the picture. Other numeric characters include the V, which indicates the location of the implied point.

The next group of picture characters are those that designate signed numbers. Only one can be used in a picture, and it must be either the leftmost or the rightmost character in the picture. An S specifies that the sign of the value is to appear in the field in the position the S appears in. A − specifies that a negative sign is to appear if the value is less than 0; otherwise a blank occupies the position. A + specifies that a positive sign is to appear if the number is 0 or greater; otherwise a blank is appended. Each of these characters takes up space in memory. The zero suppression characters are Z and *. A Z is replaced by a blank if the field contains a leading zero in the same relative position, and an * causes the * to replace the leading zero. These characters must appear to the left of the leftmost 9.

Insertion characters do not affect the contents of the field. Instead, they are added to the contents of the field at their relative positions unless they are embedded in zero suppression or drifting characters. The three insertion characters are the point (.), the comma (,), and the (B), which inserts a blank in its relative position within the field. Keep in mind that a point does not cause alignment on the point and that a V does not cause a point to print. If you want to print a point and align on it, you must code both (V.). Picture characters that can be either static or drifting are the negative sign (−), the positive sign (+), and the dollar sign ($). If any of these characters is repeated, the first (leftmost) instance is static, and all others are drifting. The second and successive character will cause leading zeros to be suppressed and the character to be placed immediately to the left of the leftmost significant digit. Remember that the first character is static; therefore, you must include one more repetition than the number of leading zeros you want to suppress.

The final numeric picture characters are fixed insertion characters and consist of CR and DB. These are placed as the rightmost two characters in the field *when the value is negative; otherwise, those positions are filled with blanks.*

In edit I/O the P-format item can be used to produce the same results as the picture attribute. The P-format item consists of a P followed by the picture enclosed in single quotation marks. The advantage of the PICTURE or P-format item in list- or edit-directed input is that it allows you to verify the input data, it allows for some extra editing on output.

When a structure is assigned to a structure, the only restrictions are that both must be declared with the same relative structure and that, if they contain arrays, the bounds must be the same in both. If these conditions are met, you can assign a major structure to a minor structure or vice versa. A scalar can also be assigned to a structure; in this case, each variable in the structure takes on the value of the scalar. The BY NAME assignment is a means of avoiding the restriction that both structures have identical relative structuring, but it imposes the restriction that elementary items and perhaps minor structures must have identical names. Only those items having identical names are affected by the assignment statement. The BY NAME option is valid only in an assignment statement, and it must be preceded by a comma. A structure can serve as a parameter, but it *must be a major structure,* even if it corresponds to a minor structure in the argument. When the parameter is a structure, the argument, of course, must also be a structure.

A structure can contain one or more arrays as elementary items. To declare an array within a structure, simply code the bounds attribute after the elementary name. Arrays, likewise, can be composed of structures, provided that all structures are identical in format. To declare an array of structures, code the bounds attribute after the major structure or (in PL/I, but not PL/C) a minor structure name. You can then reference any portion of the structure by appending a subscript. It is also possible to have an array of structures that contain arrays. In this case, the portion of the structure that is not an array is referenced in the normal way, but the portion that is an array requires two subscripts.

12.7

Exercises

1. Locate the seven errors in the following structure.

```
DCL  ASTRUC
         5  X            CHARACTER (4),
         5  Y            FIXED DEC (6),
             2  Z        FIXED DEC (2),
             2  A        FIXED DEC (4,2),
          3  B,
          4  C           CHARACTER (5)
```

2. Change the following declaration into a structure that allows you to reference TEMP, REL_HUM, and WIND, as well as each item given below.

```
DCL  T_HIGH          FIXED DEC (4,1),
     T_LOW           FIXED DEC (4,1),
     RH_HIGH         FIXED DEC (3),
     RH_LOW          FIXED DEC (3),
     WIND_DIR        CHARACTER (3),
     WIND_VEL        FIXED DEC (3);
```

3. Declare a structure for the following description of an input record (a field with no columns indicates a unit to be referenced).

Social security number
first three digits   3 characters
second two digits   2 characters
last four digits   4 characters
Professor name
last name   10 characters
first name   10 characters
Status
rank   9 characters
department   4 characters
tenured   1 character
salary   99999.99 (9 = any digit)

4. Show the output results for each of the following.

	Picture	Internal Value
a.	999V.99	100.01
b.	S999V.99	001.12
c.	+++9V.99	000.88
d.	99999−	−25
e.	−ZZZ99	−00621
f.	$$$$9V.99	0000.99
g.	$$,$$9V.99	1827.58
h.	$$,$$9V.99	0066.13
i.	99B99B99	010288
j.	ZZZZZV.99DB	−00900.69
k.	ZZZZZV.99CR	−90000.96
l.	****99	000666

5. Given below are internal values and the resulting printed output. Code the output P-format item for each item. Always assume leading zeros, and always cause one digit to print in front of the decimal point.

	Internal Value	Printed Result
a.	43.97	**43.97
b.	0019.45	$19.47
c.	0007.16	$***7.16
d.	00028.40	28.40

6. Code the structure declaration for Exercise 3, this time using the PICTURE attribute.

7. Code the edit-directed output statement to print the structure in Exercise 2. Put a five-column blank field between each elementary item under a minor structure and a ten-column blank field separating minor structures.

8. Code the structure in Exercise 3 so that "salary" becomes an array capable of holding the teacher's salary for each of the past five years.

9. Code the structure of Exercise 2 as an array of structures containing the data for each day of the month.

10. Given the following declaration, show how you would reference the various elementary items.

```
DCL 1  PRSNL_REC,
       2  FACULTY,
          3  P_DATA
             4  SSN           CHARACTER (9),
             4  NAME          CHARACTER (20),
             4  TITLE         CHARACTER (15),
             4  SALARY        FIXED DEC (7,2),
       2  STAFF,
          3  P_DATA
             4  SSN           CHARACTER (9),
             4  NAME          CHARACTER (20),
             4  TITLE         CHARACTER (15),
             4  SALARY        FIXED DEC (7,2),
```

11. Code the statement to assign all of the data in the structure you created for Exercise 2 to a scalar named WEATHER_DATA.

12. Declare an output structure according to the specifications of Exercise 7. You want to be able to use the BY NAME option to assign the input structure to the output structure.

13. Using edit-directed input with the data declared as a structure, and the P-format item for output, code an accounts receivable program. Input records each contain

Customer number	Cols 1–5
Customer name	6–20
Invoice number	21–24
Invoice amount	25–33 (use dollars and cents format)

In addition, use the DATE function to print the current date as mm/dd/yy in the page header. Your program should print, in addition to the input, the total of accounts receivable. Be sure your output is edited according to standard practice.

14. Write a program to produce a listing of patrons of a library and the call number of each book currently charged to each patron. The input consists of:

Patron number   Cols 1–4
Patron name   5–24
Patron address
     Street   25–44
     City-state-zip   45–64

An array of ten elements, each containing ten characters for the call number of the borrowed books.

The end of the valid data in the array is signaled by a call number of XXX000.

Use either list- or edit-directed input as your teacher directs. The output should have a page header and column headers for all items in the input. The patron's number, name, and address should appear once with the first call number. Subsequent call numbers should then appear it the appropriate column, with all other columns blank.

NUMBER	NAME	ADDRESS	CALL #
1234	JOHN SMITH	408 WOOD AVE.	PZ91L821C5
			PR4K911777
			PQ811D41

15. Use the structure declaration of Exercise 9 as the basis for a program that will print the input, as well as the average temperature, relative humidity and wind velocity for each day. The array must be suitable, of course, for any month, and a header record should give the actual month. The days of the month should print as the leftmost field. Be sure, also, that all DO loops on the array can handle from 28 to 31 elements of valid data.

C H A P T E R   1 3

RECORD

I/O

Record I/O is the norm in commercial data processing and is a considerably more efficient means of data transmission (in terms of execution time) than is stream I/O. We began with stream I/O because the fact that a number of things are carried out automatically makes it easier for the beginner to use.

13.1

Record vs. Stream I/O

We have already suggested that one major difference between record and stream I/O is that the former executes more rapidly. To appreciate the significance of this fact, you must be able to think in terms of commercial data processing, where

1. huge programs involve dozens of routines.

2. huge data sets contain thousands of records and therefore involve massive amounts of data transmission.

3. these programs process these data sets on, perhaps, a daily basis.

Clearly, even small gains in execution efficiency can become extremely significant in this situation. In the following two sections we discuss some other significant differences between the two I/O formats.

Stream I/O

Throughout our discussion of stream-oriented data transmission, we have spoken of "records." We warned you at the outset, however, that the term is not technically correct when applied to stream I/O. In stream I/O the entire data set is a single record, which is therefore a single data stream. Thus, we must use various combinations of GET (or PUT) SKIP (or LINE, COLUMN), and data lists to create the divisions we want within this continuous data stream. The final result is that with one I/O statement we can read in or send to the printer as much or as little of the data stream as we wish.

The second attribute of stream I/O has a direct bearing on its decreased execution efficiency compared to record I/O: the fact that data conversion in stream I/O is automatic. Therefore, you have no problem in reading numeric data in character format into a field that has been declared to be FIXED DECIMAL.

13.1.2

Record I/O

In contrast to the entire file being a single record in stream I/O, a file defined as RECORD is a set of discrete records, and this fact has two major corollaries.

1. You must define an entire record in your declaration, accounting for every byte in the record (even those unused).

2. You cannot transmit more or less than a total record with a single I/O statement.

The output record need not be identical with the input record, of course, but having defined a record, it is the unit of transfer. All of this helps to explain why structures are typically used with record I/O, and we give more explanation later on.

A second major quality of record files is that their data is not converted during I/O. All instream input data is necessarily in character format. With stream input PL/I has no difficulty (as we noted above) in converting this data to the base, scale, and precision (or length, in the case of character data) specified in the variable declaration. With record I/O, however, no such automatic conversion occurs; the data is stored internally in exactly the format in which it appeared on the input file. The same is true on output: Unless the program includes a statement that causes the data to be converted, the output will be in exactly the same format as that in which it was stored. This lack of automatic data conversion, which seems like a drawback when we think in terms of instream input and output to the printer, is in fact an advantage in many situations. Data in FIXED DECIMAL format, for example, takes up less room than the same data stored in character format (for an explanation, see Section 13.3.1). Because record-oriented data transmission on output can write to a disk or tape in FIXED DECIMAL format, the records can occupy significantly less space. By the same token, record-oriented input can read in data that has been stored in FIXED DECIMAL format, while stream input is restricted to data in character format. Still another advantage of record I/O is that it can take advantage of disk storage formats that again are unavailable to stream I/O.

13.2

Declaring Files

We have been able to spare you the details of declaring files because we have been using files that conform to the default values for the file attributes. For the input file, the default values are

File name	SYSIN
Input or output	INPUT
Data transmission	STREAM
Record format	F(ixed)
Blocksize	80

Record format indicates that the records are of fixed length (i.e., all the same length), and blocksize tells PL/I that the records are 80 bytes long. There are other attributes, but these are all we need to concern ourselves with.

The default values for the output record are

File name	SYSPRINT
Input or output	OUTPUT
Data transmission	STREAM
Record format	F(ixed)
Blocksize	121 (usually, but can vary from one installation to another)

13.2.1

Declaring Stream Files

Why, you ask, since we have gotten along so nicely thus far without declaring any files, would we want to declare a stream file? There are several reasons, especially for output data sets. First of all, however, there are two reasons that apply equally to both input and output files:

1. You might want to rid your program of the compiler informatory message, NO 'DECLARE' STATEMENT(S) FOR 'SYSIN', 'SYSPRINT'.

2. You might want to name your files something other than SYSIN and SYSPRINT.

Before you change the file names, however, you must know that doing so has consequences for the Job Control Language, which we don't discuss here but which you must understand.

It might be especially desirable to declare two output files, both directed to the printer but using SYSPRINT to handle messages and your debugging output (PUT DATA or CHECK) and the other to print the regular output from your program. At any rate, the syntax for declaring a stream file is

```
DCL filename FILE attributes;
```

A specific example of an input file declaration is

```
DCL SYSIN FILE INPUT STREAM ENV (F BLKSIZE (80));
```

The FILE options are INPUT, STREAM, and ENV(IRONMENT). The EN-VIRONMENT has its options, too, and since we are specifying more than one, we must enclose the list in parentheses. The BLKSIZE value (80) must be enclosed in yet another pair of parentheses. The FILE options must be separated by one or more blanks, and the same is true of the ENVIRON-MENT options. The options for each of FILE and ENVIRONMENT can be listed in any order, but the ENVIRONMENT options must immediately follow the keyword, wherever it appears among the FILE options. A specific example of an output stream FILE declaration is

```
DCL SYSPRINT FILE OUTPUT STREAM PRINT ENV (F BLKSIZE (133));
```

The FILE option PRINT is required if you are using carriage control options such as LINE, PAGE, and SKIP. The BLKSIZE of 133 gives you 132 columns for your output data; the first column is reserved for the carriage control character, which is placed there by LINE, PAGE, and SKIP. Finally, we should point out that the FILE declaration we illustrate is for IBM OS; if you are working under a different operating system, you probably will have to include other options.

Once you have declared your stream file, it is advisable to both open and close the file. PL/I will perform both tasks for you, but it is often helpful to someone reading the program to see at what point the files are opened and closed. A second reason for explicitly opening a file is that you can provide attributes of the file other than the defaults. For example, you can override both the default number of columns per line (if the default is less than the maximum), and you can also override the default number of lines per page (remembering always that blank lines are also "lines"). Finally, most professional programmers would advise you to get into the habit of explicitly opening each file (and closing it), for the sake of completeness and balance. For our purposes, then, the OPEN statement has the basic syntax of

```
OPEN FILE (filename);
OPEN FILE (SYSIN);
```

If you have more than one file to open, you can open them with a single OPEN:

```
OPEN FILE (INFILE), FILE (OUTFILE);
```

The comma separating the two files is required. You can also use a separate OPEN for each file:

```
OPEN FILE (INFILE);
OPEN FILE (OUTFILE);
```

If the file has been declared to have the PRINT attribute, you can, as we noted above, override the number of print columns per line if the default is less than the maximum. It is not uncommon, for example, to find a default of 121 print columns per line, even though the printer (and paper) can handle 133 columns. In this situation, if you wish to take advantage of all 133 columns, you can code the following:

```
OPEN FILE (SYSPRINT) LINESIZE (133);
```

The maximum number of print lines per page is generally 55–60. In certain situations you might want to change to fewer lines per page, and you can accomplish that by coding

```
OPEN FILE (SYSPRINT) PAGESIZE (48);
```

The ENDPAGE condition will then be raised when you direct the printer to print a line beginning at line 49 or greater. The standard action of the ENDPAGE condition being raised is to skip to a new page and continue execution. PL/I maintains a **current line counter** for the use of ENDPAGE. The raising of the ENDPAGE condition *does not*, however, reset that counter. To reset the counter, a PAGE option or format item or a LINE option or format item that specifies a line less than the current line number must execute. In this case, the ENDPAGE condition is not raised. To use this facility, you might code

```
ON ENDPAGE (SYSPRINT)
   CALL PG_HDR_RTN;
```

Except for the keyword, the CLOSE statement is the same as the OPEN:

```
CLOSE FILE (INFILE),
      FILE (OUTFILE);
```

or

```
CLOSE FILE (INFILE);
CLOSE FILE (OUTFILE);
```

Also, having declared your stream files, it is worth while for you to reference the appropriate file in each GET or PUT statement. As a matter of fact, you are required to do so unless you are using the default files of SYSIN and SYSPRINT.

```
GET FILE (INFILE) LIST (DATA);
PUT FILE (SYSPRINT) EDIT (DATA) (COL(20), A);
```

13.2.2

Declaring Record Sequential Files

Data sets can have any one of several organizations; the **organization** of the data set refers to the manner or order in which records are stored in the data set. The only organization we are concerned with in this text is physical sequential. **Physical sequential** organization simply means that the records are stored physically in the same sequence in which they are presented to the data set. Clearly, an instream input data set has physical sequential organization and, just as clearly, so does a data set sent to the printer. Sequential (here the same as physical sequential) is the default data set organization for record files (the option is not applicable to stream files);

therefore, we do not have to include it as a FILE option, although we can if we wish.

The only required change between a declaration for record and stream files is that STREAM must become RECORD. Also, if the output stream file declaration contains the PRINT option, it must be deleted in a record file declaration because it is not compatible with record output. A specific example of an input record file declaration is

```
DCL SYSIN FILE SEQL INPUT RECORD ENV (F BLKSIZE (80));
```

SEQL is the acceptable abbreviation for SEQUENTIAL. A specific example of an output record file declaration is

```
DCL SYSPRINT OUTPUT RECORD ENV (F BLKSIZE (128));
```

Everything we said about the OPEN and CLOSE statements and referencing the file name in I/O statements applies equally to record-oriented data transmission, and the syntax of the OPEN and CLOSE statements is the same.

13.3

The Mechanics of Record I/O

In the following sections first of all we present you with an explanation of why (as we noted in Chapter 12) the PICTURE attribute is so desirable in record I/O. Then we go on to describe in detail how record I/O statements are coded, and in the process complete our discussion of structures.

13.3.1

The PICTURE Attribute Revisited

When you use the terminal to create records containing numeric data (such as instream input files), the data is in character format. For example, a field containing 1234567 is stored internally (that is, with no data conversion) as follows:

```
F1 F2 F3 F4 F5 F6 F7
```

The same data in FIXED DECIMAL format, on the other hand, is stored in this fashion (in some circumstances the final character will be A, B, C, D, or E, instead of F):

```
12 34 56 7F
```

This second format is called **packed decimal** because each byte except the last has two digits "packed" into it. You cannot create packed decimal data at the typical (ASCII) terminal, nor can you print it in base 10 format. In PL/I (but not PL/C) numeric input can be described as CHARACTER; the code the compiler generates to convert CHARACTER data to FIXED DECIMAL (packed decimal) is considerably less efficient than the code it generates to convert PICTURE data to packed decimal because when the compiler encounters a PICTURE field, it knows from the outset that the data is intended for arithmetic purposes (that is, the compiler assumes it will be asked to convert the data to packed decimal). Numeric data that is not involved in any arithmetic operations is generally described as CHARACTER. If the contents of a PICTURE field are involved in arithmetic operations in several statements in the program, the data will have to be converted anew for each statement. In this case it is generally more efficient to assign the PICTURE field to a FIXED DECIMAL field, perform the various arithmetic operations using the FIXED DECIMAL variable name, then assign it back to the PICTURE field (or perhaps a second PICTURE field containing editing characters) before printing it.

The READ
Statement

Now that we are involved with record I/O, we can no longer use the keywords GET and PUT to begin the I/O statements. Instead, for the input statement we must begin with the keyword READ. The syntax of the READ statement is

```
READ FILE (filename) INTO (input area);
```

The keyword FILE is required, as is the file name (which can be no more than seven characters). The keyword INTO is not strictly required; however, we do not go into those circumstances in which it is not required. Therefore, for purposes of this text, you can assume the INTO is required. The parentheses surrounding both the file name and the input area name are also required. The input area is a variable you have included in your DECLARE statements, and it must be the same length as the input record, even if much of the input record contains no data. After the READ statement executes, the input data is available in this area and can be referenced with its variable name. The following is an example of the code necessary to read in a record. (Even though the OPEN statement is not strictly required, we always use it.)

```
DCL SYSIN FILE INPUT RECORD ENV (F BLKSIZE (80));
    FACULTY_IN      CHARACTER (80);
```

```
OPEN FILE (SYSIN);
READ FILE (SYSIN) INTO (FACULTY_IN);
```

The declaration of FACULTY_IN is, of course, a minimal definition of the input area. It makes possible a reference to the entire input record but not a reference to any smaller unit. The only way we can reference a smaller unit of the input record in record input is to define the input area as a structure. You will find that there are occasions when you need only reference the entire record, and on those occasions a declaration of the input area as a scalar will suffice. The far more common situation, however, requires you to reference various portions of the input record and to give those portions different attributes. Consequently, the structure is indispensable to record I/O, and a more typical declaration of an input area is something like

```
DCL 1   FACULTY_IN,
        2   SSN                 CHARACTER (9),
        2   NAME                CHARACTER (20),
        2   DEPT                CHARACTER (4),
        2   RANK                CHARACTER (9),
        2   TENURED             CHARACTER (1),
        2   YRS_SERVICE         PIC '99',
        2   SALARY              PIC '99999V99',
        2   BLANK               CHARACTER (28);
```

The final elementary item, BLANK, is required in order to extend FACULTY_IN to the 80-byte length that we declared SYSIN to be. It is important to remember that in record I/O the record declaration must account for every byte in the record, whether or not those bytes contain data.

13.3.3

The WRITE Statement

Just as we cannot use GET for record input, so we cannot use PUT for record output. Instead, we must use WRITE, and its syntax is

```
WRITE FILE (filename) FROM (output area);
```

and a specific example is

```
WRITE FILE (SYSPRINT) FROM (FACULTY_OUT);
```

As you can see, the WRITE statement is very similar to the READ statement, the only differences being that the WRITE replaces READ and FROM replaces INTO. Again, you are required to enclose the file name and output area name in parentheses.

In its action, the WRITE statement is the mirror image of the READ statement; that is, the contents of the output area are written to the specified file. Record output directed to the printer involves considerably more coding of declarations and assignment statements than does stream output; however, record output executes enough more efficiently that it is used in those instances in which a program produces a significant amount of printed output, especially if the program is executed on a regular basis over a long period of time. Output to disk or tape, moreover, is regularly in record format. It is quite possible, certainly, to declare SYSIN as RECORD, while defining SYSPRINT (or any other output file) as STREAM, and the reverse is equally possible.

It is again possible to define the output area as a scalar; however, it is more common (especially if the output is directed to the printer) to define the output area as a structure. The following example uses FACULTY_IN as the source data, and we format the output as it might need to be for a report. (Note: SYSPRINT has been declared to contain 133 bytes.)

```
DCL 1   FACULTY_OUT,
        2  BLANK1              CHARACTER (29) INIT (' '),
        2  SSN                 CHARACTER (9),
        2  BLANK2              CHARACTER (5)  INIT (' '),
        2  NAME                CHARACTER (20),
        2  BLANK3              CHARACTER (5)  INIT (' '),
        2  DEPT                CHARACTER (4),
        2  BLANK4              CHARACTER (5)  INIT (' '),
        2  RANK                CHARACTER (9),
        2  BLANK5              CHARACTER (5)  INIT (' '),
        2  TENURED             CHARACTER (1),
        2  BLANK6              CHARACTER (5)  INIT (' '),
        2  YRS-SERVICE         PIC 'Z9',
        2  BLANK7              CHARACTER (5)  INIT (' '),
        2  SALARY              PIC '$$$,$$9V.99',
        2  BLANK8              CHARACTER (23) INIT (' ');
```

Perhaps it has already occurred to you that the declaration could be shortened: BLANK2-6 could be eliminated by adding five to the length of each of SSN, NAME, DEPT, and RANK. BLANK7, however, is required (unless you want to add five more $'s to SALARY), and BLANK8 is also required because data in numeric fields right-justify. As with record input, we must once again account for every byte of the declared length of the output file. A second detail to notice regarding the declaration of FACULTY_OUT vis-à-vis FACULTY_IN is that identical names are used for the data fields that appear in both; thus, we can build FACULTY_OUT with the statement

```
FACULTY_OUT = FACULTY_IN, BY NAME;
```

Keep it in mind that if you are using record-oriented data transmission to the printer, any page or column headers must be similarly declared, with each byte of declared length of the output file accounted for. The example program in Section 13.5 illustrates the number of items it might be necessary to declare when record output is sent to the printer.

Record output to the printer poses one more problem that you are unfamiliar with: It does not contain anything comparable to the ENDPAGE condition that occurs in stream output. In stream output the ENDPAGE condition arises when the printer has covered as many lines as you have indicated in the PAGESIZE option of a stream file declaration, or as many lines as the default at your installation. If no provision is made, record output prints across the perforation of the paper, a situation that is not very desirable. The standard method of overcoming this problem is to declare a variable to be used as a line counter. Each time a line is printed the line counter is incremented by the number of lines involved in the printing (remember, blank lines also must be accounted for). You then test the line counter against your predetermined number of lines per page, and when the printer has reached that figure, write an instruction to cause a skip to a new page. This instruction might very well be a call to a page header routine; the control field of the page header might be initialized to '1', which, of course, would cause a skip to a new page before printing the page and column headers. Suppose, for example, that the page header routine prints a page header, double spaces, and prints column headers. As the last statement in the routine, you might then code

```
LINE_CNT = 3;
```

The first thing you would then do in the loop that prints the detail lines is test the value in LINE_CNT:

```
IF LINE_CNT >= 55 THEN
   CALL PG_HDR_RTN;
```

and, of course, you increment LINE_CNT each time you print a detail line.

*Carriage
Controls*

The LINE, PAGE, and SKIP options that you have been using with stream I/O are not available to record I/O; in their place is a somewhat more limited set of options. If you are using an IBM mainframe, you have two options available to you for controlling the printer carriage, and you specify your choice as an ENVIRONMENT option. The first possibility is CTLASA (ConTroL American Standards Association; the ASA is now known as the American National Standards Institute). To use this option, your file declaration must resemble the following:

```
DCL SYSPRINT FILE OUTPUT RECORD ENV (CTLASA F BLKSIZE (121));
```

Because we have specified a carriage control character, we have room for 120 columns of output data. CTLASA is the most widely used carriage control character set, mainly because the specified action occurs *before* the line is printed. Adjusting the carriage before printing is consonant with PAGE, LINE, and SKIP, of course, and to standard typewriters; therefore, it seems more natural to most programmers. The following table shows the character codes and the resulting actions.

Character	*Action*
Blank	Advance one line; then print (single space)
0	Advance two lines; then print (double space)
—	Advance three lines; then print (triple space)
1	Advance to top of page; then print
+	Do not advance carriage; overprint

It is the responsibility of the programmer to place a valid carriage control character in the *first* column of each record, but for single spacing all you need do is avoid putting any character in the first column (i.e., be sure it is left blank). To facilitate the placement of the carriage control character, the first elementary item name in the output area declaration is often a one-byte variable:

```
DCL 1  FACULTY_OUT,
       2  CARR_CNTRL          CHARACTER (1),
       2  . . .
```

The second option, available to users of the IBM 360 or 370 series, is called CTL360, but we do not describe it here. You can find the list of character codes in the *System/370 Reference Summary,* published by IBM.

If you specify CTLASA as an ENVIRONMENT option on an input file, you must be certain that each input record contains a valid ANSI character code (including blank) in the first byte.

13.4

The DEFINED Attribute

In our discussion of structures in assignment statements in Chapter 12 (Section 12.3.1), we pointed out that it is possible to assign a scalar to a structure, and the example we used is that of setting an entire structure to blanks.

```
DCL 1  CLASS_DATA,
       2  DEPT              CHARACTER (4),
       2  NUMBER            CHARACTER (4),
       2  TEACHER           CHARACTER (20);
   CLASS_DATA = ' ';
```

The assignment statement actually generates the following three statements.

```
DEPT = ' ';
NUMBER = ' ';
TEACHER = ' ';
```

PL/I provides another means of setting the entire structure to blanks, and it generates only the statement you code. This method makes use of the DEFINED attribute. Assume CLASS_DATA is declared as above. We now add to that declaration

```
DCL CLS_DATA CHARACTER (28) DEFINED CLASS_DATA;
```

In this example, CLASS_DATA is called the **base variable**, and CLS_DATA is called the **defined variable**. CLS_DATA refers to exactly the same area in memory as CLASS_DATA; in other words, the defined variable **redefines** the area occupied by the base variable. CLS_DATA, however, "sees" that memory area as a scalar, whereas CLASS_DATA "sees" it as a structure. Because to CLS_DATA the area is a scalar, we can code

```
CLS_DATA = ' ';
```

and this single statement suffices to blank out the entire area in memory, without generating any additional statements.

You will also recall that in Section 12.3.3 of Chapter 12 we invoked the STRING function to assign a structure to a scalar. We also pointed out that the DEFINED attribute is usually a more efficient means of assigning a structure to a scalar. The code in the STRING function must execute, of course, whenever the function is invoked; with the DEFINED attribute, however, no extra code is executed because the variable is a scalar, even though it is "overlaying" a structure. Therefore, we are simply assigning a scalar to a scalar.

It is also possible to attach the DEFINED attribute to a structure; in other words, we can redefine a structure as a second structure. Suppose, for example, that we have input records from which to construct a class roll and that the format of the records depends on whether the student is a major or nonmajor. The declaration and code to handle this situation follows:

```
DCL 1   NON_MAJORS,
        2   CODE            CHARACTER (1),
        2   SSN             CHARACTER (9),
        2   NAME            CHARACTER (20),
        2   MJR_DEPT        CHARACTER (4),
        2   ADVISOR         CHARACTER (20),
        2   BLANK           CHARACTER (26),
     1  MAJORS DEFINED NON_MAJORS,
        2   CODE            CHARACTER (1),
        2   SSN             CHARACTER (9),
        2   NAME            CHARACTER (20),
        2   STANDING        CHARACTER (2),
        2   EMPHASIS        CHARACTER (1);
    OPEN FILE (SYSIN);
    READ FILE (SYSIN) INTO (NON_MAJORS);
    IF NON_MAJORS.CODE = 'N' THEN
        CALL N_MJR_RTN (NON_MAJORS);
    ELSE
        CALL MAJOR_RTN (MAJORS);
```

You should note several details in this example:

1. The defined variable (MAJORS) does not have to redefine an area as large as the base variable (NON_MAJORS), but the defined variable can never define an area *larger* than the base variable.

2. The fields in the two definitions need not begin and end in the same relative locations.

3. All of the fields are CHARACTER.

It is possible to redefine structures containing the PICTURE attribute by overlaying it with one containing CHARACTER fields, but to do so requires a more sophisticated understanding of numeric data storage than we present in this text.

Two other restrictions you must be aware of in using the DEFINED attribute are as follows:

1. Although the base variable can be redefined as many times as necessary, it cannot itself be DEFINED:

```
Legal:     DCL  1  NM_SSN-ADDR,
                2   NAME      CHARACTER (20),
                2   SSN       CHARACTER (9),
                2   ADDR      CHARACTER (35),
             1  SSN_NM_ADDR DEFINED NM_SSN_ADDR,
                2   SSN       CHARACTER (9),
                2   NAME      CHARACTER (20),
                2   ADDR      CHARACTER (35),
             1  NM_ADDR DEFINED NM_SSN_ADDR,
                2   NAME      CHARACTER (20),
                2   ADDR      CHARACTER (35);
Illegal:   DCL  1  NM_SSN-ADDR,
                2   NAME      CHARACTER (20),
                2   SSN       CHARACTER (9),
                2   ADDR      CHARACTER (35),
             1  SSN_NM_ADDR DEFINED NM_SSN_ADDR,
                2   SSN       CHARACTER (9),
                2   NAME      CHARACTER (20),
                2   ADDR      CHARACTER (35),
             1  NM_ADDR DEFINED SSN_NM_ADDR,
                2   NAME      CHARACTER (20),
                2   ADDR      CHARACTER (35);
```

It is quite legal, of course, for SSN_NM_ADDR to redefine NM_SSN_ADDR; what is illegal is for NM_ADDR to redefine SSN_NM_ADDR because SSN_NM_ADDR itself contains the DEFINED attribute.

2. The INITIAL attribute cannot be attached to the defined variable.

Remember, both the base and the defined variables refer to the same area in memory, and two values cannot be at the same location at the same time. PL/I allows us to intitialize the base variable but not the defined variable.

Assuming that you have given the defined variable the same length as the input or output file, under PL/I (but not PL/C) you can read records into or write records from the defined variable.

13.5

Example Program— Record I/O

Of the various types of stream I/O, record I/O most closely resembles edit-directed stream I/O. In order to allow you to compare record- with edit-directed stream I/O, we have rewritten CH12EX1 to use record I/O. We deliberately chose an edit-directed I/O example containing several different output lines (e.g., page header, column header, detail line) to give you a good idea of record output directed to the printer, as well as record input. By way of reminder, the problem the two programs solve is that of producing a report on work-study students. The programs must determine the gross pay, university contribution, and government contribution for each employee. They must also provide these three totals for each employer unit and for the entire work-study program.

In comparing the CH13EX1 program (see Fig. 13–1) with CH12EX1, perhaps the first thing you will notice is, as we mentioned, that record output used for output to the printer requires a good deal more coding of declarations. The DECLARE of CH12EX1 covers 40 lines of code, whereas the DECLARE of CH13EX1 (not counting the FILE declaration) includes 130 lines. Notice, too, that the PICTURE data, such as HOURS_WORKED, is assigned to a FIXED DECIMAL field (HRS_WRKD) in Statement 55 before arithmetic operations are performed on it. Other FIXED DECIMAL fields, such as GROSS, UNIV_SHR, and GOVT_SHR, are explicitly assigned to PICTURE fields (Statements 62-64) before printing. We go to this trouble, of course, to avoid conversions in each arithmetic statement.

Finally, let us point out that since the logic of CH13EX1 is identical to that of CH12EX1, we do not repeat the pseudocode here.

PL/I OPTIMIZING COMPILER   VERSION 1 RELEASE 5.1   TIME: 13.17.55   DATE: 10 MAR 87 PAGE   1
OPTIONS SPECIFIED
GOSTMT;

FIGURE 13–1
CH13EX1 (the use of record I/O).

FIGURE 13-1 *(continued)*

```
PL/I OPTIMIZING COMPILER          /* CH13EX1 USING RECORD I/O */                    PAGE   2
                    SOURCE LISTING

    STMT


        /* CH13EX1 USING RECORD I/O */
     1  CH13EX1:     PROCEDURE OPTIONS (MAIN);
        /********************************************************************/
        /*                                                                */
        /*   FUNCTION:  TO PRODUCE A PAYROLL REPORT FOR WORK-STUDY        */
        /*              STUDENTS                                          */
        /*                                                                */
        /*   INPUT:     A FILE CONTAINING                                 */
        /*              1.  A HEADER RECORD GIVING THE STARTING AND       */
        /*                  ENDING DATE OF A TWO-WEEK PAY PERIOD          */
        /*              2.  A RECORD FOR EACH WORK-STUDY STUDENT          */
        /*                  CONTAINING A SOCIAL SECURITY NUMBER, NAME,    */
        /*                  EMPLOYER-UNIT CODE, JOB-TITLE CODE, PAY-      */
        /*                  RATE CODE, AND HOURS WORKED.                  */
        /*              THE INPUT IS IN LIST-DIRECTED FORMAT AND IS       */
        /*              SORTED ON EMPLOYER-UNIT CODE                      */
        /*                                                                */
        /*   OUTPUT:    FOR EACH STUDENT A DETAIL LINE LISTING SOCIAL     */
        /*              SECURITY NUMBER, NAME, EMPLOYER UNIT, JOB TITLE,   */
        /*              PAY RATE, GROSS SALARY, UNIVERSITY CONTRIBUTION,   */
        /*              AND GOVERNMENT CONTRIBUTION.                      */
        /*              SUBTOTALS FOR EACH EMPLOYER UNIT                  */
        /*              TOTALS FOR THE WORK-STUDY PROGRAM                 */
        /*                                                                */
        /*   NOTES:     THE EMPLOYER-UNIT CODES ARE                      */
        /*                  1--COMPUTER CENTER                           */
        /*                  2--COMPUTER SCIENCE DEPT.                    */
        /*                  3--LIBRARY                                   */
        /*                  4--STUDENT CENTER                           */
        /*                                                                */
        /*              THE JOB-TITLE CODES ARE                         */
        /*                  1--OPERATOR ASSISTANT                       */
        /*                  2--TEACHING ASSISTANT                       */
        /*                  3--CLERK                                    */
        /*                  4--DESK ATTENDANT                           */
        /*                                                                */
```

```
/*          THE PAY-RATE CODES ARE                         */
/*               1--$3.45/HOUR                             */
/*               2--$3.65/HOUR                             */
/*               3--$3.80/HOUR                             */
/*               4--$6.75/IIOUR                            */
/*                                                         */
/***********************************************************/
```

```
STMT
   2      DCL UNIT_ARRAY (4)        CHARACTER (25)   INIT
                ('COMPUTER CENTER', 'COMPUTER SCIENCE DEPT',
                 'LIBRARY', 'STUDENT CENTER');

   3      DCL TITLE_ARRAY (4)       CHARACTER (20)   INIT
                ('OPERATOR ASSISTANT', 'TEACHING ASSISTANT',
                 'CLERK', 'DESK ATTENDANT');

   4      DCL RATE_ARRAY (4)        FIXED DEC (3,2)  INIT
                (3.45, 3.65, 3.80, 6.75);

   5      DCL 1  HDR_REC,
              2 START_PAY        PIC'99X99X99',
              2 END_PAY          PIC'99X99X99',
              2 HR_BLANK         CHARACTER (64);

   6      DCL 1  INPUT_REC,
              2 SSN              PIC'999X99X9999',
              2 NAME             CHARACTER (20),
              2 UNIT_CODE        PIC '9',
              2 TITLE_CODE       PIC '9',
              2 RATE_CODE        PIC '9',
              2 HOURS_WORKED     PIC '99',
              2 IR_BLANK         CHARACTER (44);

   7      DCL 1  PG_HDR1,
              2 PH1_CC           CHARACTER (1)    INIT ('1'),
              2 PH_BLANK         CHARACTER (55)   INIT (' '),
              2 HEADER           CHARACTER (77)   INIT
                                 ('WORK-STUDY SALARY REPORT');
```

FIGURE 13–1 (continued)

```
  8      DCL 1  PG_HDR2,
                2  PH2_CC            CHARACTER (1)    INIT ('-'),
                2  START_LABEL       CHARACTER (20)   INIT
                                     ('PAY PERIOD STARTING'),
                2  START_DATE        CHARACTER (15),
                2  END_LABEL         CHARACTER (7)    INIT ('ENDING'),
                2  END_DATE          CHARACTER (36),
                2  EMP_UNTI          CHARACTER (16)   INIT ('EMPLOYER UNIT'),
                2  HDR_E_UNIT        CHARACTER (38);


  9      DCL 1  COL_HDR1,
                2  CNTRL_CARR        CHARACTER (1)    INIT ('0'),
                2  CH1_BLANK         CHARACTER (10)   INIT (' '),
                2  CH_SSN            CHARACTER (16)   INIT ('SSN'),
                2  CH_NAME           CHARACTER (20)   INIT ('NAME'),
                2  CH_EMP            CHARACTER (25)   INIT ('EMPLOYER UNIT'),
                2  CH_TITLE          CHARACTER (15)   INIT ('JOB TITLE'),
                2  CH_HOURS          CHARACTER (8)    INIT ('HOURS'),
                2  CH_RATE           CHARACTER (10)   INIT ('RATE'),
                2  CH_GROSS          CHARACTER (11)   INIT ('GROSS'),
                2  CH_UNIV           CHARACTER (11)   INIT ('UNIV'),
                2  CH_GOVT           CHARACTER (6)    INIT ('GOVT');
```

PL/I OPTIMIZING COMPILER   /* CH13EX1 USING RECORD I/O */   PAGE   4
  STMT

```
 10      DCL 1  COL_HDR2,
                2  CH2_BLANK         CHARACTER (116)  INIT (' '),
                2  U_SHR             CHARACTER (11)   INIT ('FUNDS'),
                2  G_SHR             CHARACTER (6)    INIT ('FUNDS');


 11      DCL 1  PRNT_REC,
                2  PR_CC             CHARACTER (1)    INIT ('0'),
                2  PR_BLANK1         CHARACTER (5)    INIT (' '),
                2  PRNT_SSN          CHARACTER (16),
                2  PRNT_NAME         CHARACTER (20),
                2  PRNT_UNIT         CHARACTER (25),
                2  PRNT_TITLE        CHARACTER (20),
                2  PRNT_HOURS        PIC '99',
                2  PR_BLANK2         CHARACTER (5)    INIT (' '),
```

('WORK-STUDY SALARY SUMMARY');

```
                2  PRNT_RATE        PIC 'Z9V.99',
                2  PR_BLANK3        CHARACTER (5)    INIT (' '),
                2  PRNT_GROSS       PIC 'ZZ9V.99',
                2  PR_BLANK4        CHARACTER (5)    INIT (' '),
                2  PRNT_UNIV_SHR    PIC 'ZZ9V.99',
                2  PR_BLANK5        CHARACTER (5)    INIT (' '),
                2  PRNT_GOVT_SHR    PIC 'ZZ9V.99',
                2  PR_BLANK6        CHARACTER (1)    INIT (' ');

 12    DCL 1  UNIT_FTNG1,
                2  UF1_CC           CHARACTER (1)    INIT ('-'),
                2  UF1_BLANK        CHARACTER (62)   INIT (' '),
                2  UF1_LABEL1       CHARACTER (18)   INIT
                                   ('THE UNIT TOTAL IS '),
                2  UF_UNIT_TTL      PIC '$$$$9V.99',
                2  UF1_LABEL2       CHARACTER (5)    INIT (' FOR'),
                2  UF_UNIT          CHARACTER (39)   INIT (' ');

 13    DCL 1  UNIT_FTNG2,
                2  UF2_CC           CHARACTER (1)    INIT ('0'),
                2  UF2_BLANK1       CHARACTER (39)   INIT (' '),
                2  UF2_LABEL        CHARACTER (30)   INIT
                                   ('THE UNIVERSITY CONTRIBUTION IS'),
                2  UNIV_AMNT        PIC '$$$$9V.99',
                2  UF2_BLANK2       CHARACTER (55)   INIT (' ');

 14    DCL 1  UNIT_FTNG3,
                2  UF3_CC           CHARACTER (1)    INIT ('0'),
                2  UF3_BLANK1       CHARACTER (39)   INIT (' '),
                2  UF3_LABEL        CHARACTER (30)   INIT
                                   ('THE GOVERNMENT CONTRIBUTION IS'),
                2  GOVT_AMNT        PIC '$$$$9V.99',
                2  UF3_BLANK2       CHARACTER (55)   INIT (' ');

 15    DCL 1  SUM_PG_HDR,
                2  SPH_CC           CHARACTER (1)    INIT ('1'),
                2  SPH_BLANK        CHARACTER (59)   INIT (' '),
                2  SPH_LABEL        CHARACTER (73)   INIT
```

FIGURE 13–1 *(continued)*

```
STMT
  16     DCL 1  U_SUM_LINE,
                2  GSL_CC          CHARACTER (1)    INIT ('-'),
                2  GSL_BLANK1      CHARACTER (39)   INIT (' '),
                2  GSL_LABEL       CHARACTER (29)   INIT
                                     ('TOTAL UNIVERSITY CONTRIBUTION'),
                2  GSL_TTL         PIC '$$,$$9V.99',
                2  GSL_BLANK2      CHARACTER (55)   INIT (' ');

  17     DCL 1  G_SUM_LINE,
                2  USL_CC          CHARACTER (1)    INIT ('-'),
                2  USL_BLANK1      CHARACTER (39)   INIT (' '),
                2  USL_LABEL       CHARACTER (29)   INIT
                                     ('TOTAL UNIVERSITY CONTRIBUTION'),
                2  USL_TTL         PIC '$$,$$9V.99',
                2  USL_BLANK2      CHARACTER (55)   INIT (' ');

  18     DCL 1  T_SUM_LINE,
                2  TSL_CC          CHARACTER (1)    INIT ('-'),
                2  SL_BLANK1       CHARACTER (28)   INIT (' '),
                2  SH_LABEL        CHARACTER (48)   INIT
                        ('TOTAL WORK-STUDY EXPENDITURES FOR THE PAY PERIOD'),
                2  SH_GRND_TTL     PIC '$$$$,$$9V.99',
                2  SL_BLANK2       CHARACTER (45)   INIT (' ');

  19     DCL SAVE_CODE            FIXED DEC (1);
  20     DCL HRS_WRKD             FIXED DEC (2);
  21     DCL GROSS                FIXED DEC (5,2);
  22     DCL UNIV_SHR             FIXED DEC (5,2);
  23     DCL GOVT_SHR             FIXED DEC (5,2);
  24     DCL GOVT_SUBTTL          FIXED DEC (6,2)  INIT (0);
  25     DCL UNIV_SUBTTL          FIXED DEC (6,2)  INIT (0);
  26     DCL UNIT_TOTAL           FIXED DEC (6,2)  INIT (0);
  27     DCL UNIV_TOTAL           FIXED DEC (7,2)  INIT (0);
  28     DCL GOVT_TOTAL           FIXED DEC (7,2)  INIT (0);

  29     DCL MORE_RECS            BIT     (1);
  30     DCL    YES               BIT     (1)   INIT ('1'B);
  31     DCL    NO                BIT     (1)   INIT ('0'B);
```

```
32      DCL ROUND                    BUILTIN;
33      DECLARE
            SYSIN FILE INPUT RECORD ENV (F BLKSIZE (80)),
            SYSPRINT FILE OUTPUT RECORD ENV (F BLKSIZE (133) CTLASA);

34          OPEN FILE (SYSIN),
                 FILE (SYSPRINT);
35          MORE_RECS = YES;
36          ON ENDFILE (SYSIN) MORE_RECS = NO;

37          READ FILE (SYSIN) INTO (HDR_REC);
38          READ FILE (SYSIN) INTO (INPUT_REC);
39          START_DATE = START_PAY;
```

```
   STMT
    40      END_DATE = END_PAY;

    41  PROC_LP: DO WHILE (MORE_RECS);
    42          HDR_E_UNIT = UNIT_ARRAY (UNIT_CODE);
    43          PRNT_UNIT = UNIT_ARRAY (UNIT_CODE);
    44          SAVE_CODE = UNIT_CODE;

    45          WRITE FILE (SYSPRINT) FROM (PG_HDR1);
    46          WRITE FILE (SYSPRINT) FROM (PG_HDR2);
    47          WRITE FILE (SYSPRINT) FROM (COL_HDR1);
    48          WRITE FILE (SYSPRINT) FROM (COL_HDR2);

    49  UNIT_LP:    DO WHILE (SAVE_CODE = UNIT_CODE & MORE_RECS);
    50              PRNT_SSN = SSN;
    51              PRNT_NAME = NAME;
    52              PRNT_TITLE = TITLE_ARRAY (TITLE_CODE);
    53              PRNT_HOURS = HOURS_WORKED;
    54              PRNT_RATE = RATE_ARRAY (RATE_CODE);
    55              HRS_WRKD = HOURS_WORKED;
    56              GROSS = ROUND (HRS_WRKD * RATE_ARRAY (RATE_CODE), 2);
    57              UNIV_SHR = ROUND (GROSS * .20, 2);
    58              GOVT_SHR = GROSS - UNIV_SHR;
    59              UNIT_TOTAL = UNIT_TOTAL + GROSS;
```

FIGURE 13-1 *(continued)*

```
60                    UNIV_SUBTTL = UNIV_SUBTTL + UNIV_SHR;
61                    GOVT_SUBTTL = GOVT_SUBTTL + GOVT_SHR;
62                    PRNT_UNIV_SHR = UNIV_SHR;
63                    PRNT_GOVT_SHR = GOVT_SHR;
64                    PRNT_GROSS = GROSS;

65                    WRITE FILE (SYSPRINT) FROM (PRNT_REC);
66                    READ FILE (SYSIN) INTO (INPUT_REC);
67                  END UNIT_LP;

68                  UNIV_TOTAL = UNIV_TOTAL + UNIV_SUBTTL;
69                  GOVT_TOTAL = GOVT_TOTAL + GOVT_SUBTTL;
70                  UF_UNIT_TTL = UNIT_TOTAL;
71                  UNIV_AMNT = UNIV_SUBTTL;
72                  GOVT_AMNT = GOVT_SUBTTL;
73                  UF_UNIT = UNIT_ARRAY (SAVE_CODE);

74                  WRITE FILE (SYSPRINT) FROM (UNIT_FTNG1);
75                  WRITE FILE (SYSPRINT) FROM (UNIT_FTNG2);
76                  WRITE FILE (SYSPRINT) FROM (UNIT_FTNG3);
77                END PROC_LP;

78                USL_TTL = UNIV_TOTAL;
79                GSL_TTL = GOVT_TOTAL;
80                SH_GRND_TTL = UNIV_TOTAL + GOVT_TOTAL;
81                WRITE FILE (SYSPRINT) FROM (SUM_PG_HDR);
82                WRITE FILE (SYSPRINT) FROM (U_SUM_LINE);
```

```
PL/I OPTIMIZING COMPILER        /* CH13EX1 USING RECORD I/O */          PAGE   7
    STMT
    83                WRITE FILE (SYSPRINT) FROM (G_SUM_LINE);
    84                WRITE FILE (SYSPRINT) FROM (T_SUM_LINE);
    85                CLOSE FILE (SYSIN),
                          FILE (SYSPRINT);
    86  END CH13EX1;


PL/I OPTIMIZING COMPILER        /* CH13EX1 USING RECORD I/O */          PAGE   8
COMPILER DIAGNOSTIC MESSAGES
ERROR ID L   STMT    MESSAGE DESCRIPTION
```

```
COMPILER INFORMATORY MESSAGES

IEL0906I I    70, 71, 72, 78, 79, 80    DATA CONVERSION WILL BE DONE BY SUBROUTINE CALL.

END OF COMPILER DIAGNOSTIC MESSAGES
```

13.6

Summary

Record I/O, as opposed to stream I/O, is standard practice in commercial data processing centers, partly because it executes more efficiently than does stream I/O. There are several significant contrasts between stream- and record-oriented data transmission.

1. Since the default is STREAM, RECORD files must be explicitly declared.

2. Record I/O involves "true" records as opposed to the single-record "stream" of data in stream I/O. This difference creates two major corollaries for record I/O.

    a. The declaration of a record must account for every byte declared in the FILE declaration, even if some bytes contain only blanks (i.e., are unused).

    b. Neither more nor less than a single record can be read or written with a single I/O statement. In stream I/O you can read or write as little or as much data as you please.

3. No data conversion occurs when record-format data is read or written; this fact renders the PICTURE attribute nearly essential to numeric data in record I/O.

4. In contrast to the GET and PUT statements of stream I/O, record I/O uses READ and WRITE statements.

5. Carriage control options such as LINE, PAGE, and SKIP are not available to record I/O. In their place CTLASA (or CTL360) carriage control characters must be used.

The syntax for declaring a file is

```
DCL filename FILE attributes;
```

and a typical input file declaration resembles the following:

```
DCL SYSIN FILE INPUT RECORD ENV (F BLKSIZE (80));
```

With this declaration you tell PL/I that the file's name is SYSIN; it is an input file in RECORD format; and the file description (ENVIRONMENT) is that all records are the same length (F) of 80 bytes (BLKSIZE (80)). A typical output file declaration is

```
DCL SYSPRINT FILE OUTPUT RECORD ENV (CTLASA F BLKSIZE (133));
```

Here you are informing PL/I that the file named SYSPRINT is an output file in RECORD format, that the first byte of each record in the file is reserved for an ANSI (formerly ASA) control character, and that all records in the file are of the same length (F), which is 133 bytes (BLKSIZE).

If you fail to open and close your files, the system will do it for you, but it is standard practice for the programmer to explicitly open and close files. The syntax for the OPEN and CLOSE statements is

```
OPEN FILE (filename);
OPEN FILE (SYSIN),
     FILE (SYSPRINT);
CLOSE FILE (filename);
CLOSE FILE (SYSIN),
      FILE (SYSPRINT);
```

The syntax of the record-oriented input statement is

```
READ FILE (filename) INTO (input area);
```

The keywords READ FILE are required, as is INTO in any of the situations we discuss in this text. Filename, of course, is the name of the file, and input area is a variable (scalar or structure), which you have declared to have a length equal to the BLKSIZE value in the FILE declaration. A specific example is

```
READ FILE (SYSIN) INTO (INPUT_AREA);
```

After the READ statement executes, its contents are available in INPUT_ AREA. The syntax of the WRITE statement is

```
WRITE FILE (filename) FROM (output area);
```

In this case the data in output area is written to the file specified by filename. After the WRITE statement executes, the data is still available in the output area. Specific examples are:

```
WRITE FILE (SYSPRINT) FROM (PG_HDR);
WRITE FILE (SYSPRINT) FROM (COL_HDR);
WRITE FILE (SYSPRINT) FROM (STDNT_DATA);
```

CTLASA is the more commonly used carriage control character set because the specified action occurs before the line is printed. A blank specifies "advance one line" (single space); a zero (0) specifies double space; a hyphen (-), triple space; a plus sign (+), do not advance carriage; and a one (1), advance to a new page. It is the programmer's responsibility to place the correct character in the first byte of each record.

The DEFINED attribute is generally a more efficient means of assigning a scalar to a structure and a structure to a scalar. The first variable is called the base variable, and the second (containing the keyword DEFINED) is called the defined variable. When the base variable is a structure, the defined variable can also be a structure. There are several things you need to remember if you are to use the DEFINED attribute successfully:

1. Both the base and the defined variables address the same area in memory.

2. The defined variable can redefine an area shorter than the area defined by the base variable, but the defined variable can never define an area longer than that defined by the base variable.

3. The INITIAL attribute can be used with the base variable, but not with the defined variable.

4. The base variable can be redefined as many times as necessary, but it cannot itself be a defined variable (the base variable cannot contain the DEFINED attribute).

5. If both the base and defined variables are declared to have the same length as a file, I/O operations can reference either variable (in PL/I; PL/C must reference the base variable).

6. You must use extreme care in redefining a base variable that contains numeric PICTURE data with a CHARACTER defined variable. To do so requires an understanding of the storage of PICTURE numeric items that goes beyond that presented in this text.

13.7

Exercises

1. List five significant differences between stream and record I/O.

2. What is the significance for the programmer using record I/O that she or he is involved with "true" records?

3. Write the declaration for each of the following files:

	Name	I/O	Data Type	Record Length
a.	SYSIN	Input	Stream	40
b.	PRNTR	Output	Record	133
c.	STDNT	Input	Record	75
d.	TCHR	Output	Stream	121

4. Write OPEN and CLOSE statements to (a) open (then close) each file of Exercise 3 in a separate statement, and (b) open all files in one statement and close them all in another single statement.

5. Declare as a scalar the input or output area for each file in Exercise 3.

6. Based on the following description, define the input area for STDNT file in Exercise 3.

Columns   1–9   Social security number
           10–29   Name
           30–34   Tuition (dollars and cents)
           35–40   Dorm fees (dollars and cents)
           Remainder unused

7. Suppose PRNTR (see Exercise 3) is the output file to the printer. Declare the output area so that the input from Exercise 6, plus a 'total-due' field is to be printed. Leave five blank columns between each field and approximately the same number of blank columns at either margin.

8. Declare an output area named COL_HDRS that will place column headers over the detail lines in Exercise 7.

9. Write an input statement to read a record from Exercise 6 and output statements for Exercises 7 and 8.

10. List six important things to keep in mind when using the DEFINED attribute.

11. Suppose now that STDNT file has two record formats: for dorm students the records are in the format of Exercise 6, and for off-campus students the following format.

Columns   1–10   Last name
           11–29   Local address
           30–34   Tuition

Write the appropriate statement to redefine the input area you defined in Exercise 6. Redefine only as much as is needed.

12. Write a program to produce a credit report. The input file records have the following format.

Columns 2–8   Customer ID
         9–28   Customer name
         30–35   Balance owed (dollars and cents)
         40–45   Credit limit (dollars and cents)
         75–80   Date (yymmdd) of last record update

The program should print a page header naming the credit company (BREAK-LEGS CREDIT CORP.) and the date of the report (use DATE function). The detail lines should list the input and the amount of credit still available or the amount the customer is over the limit. The starting column for each output field is

Column   10   Customer ID
         22   Customer name
         45   Balance owed
         55   Credit limit
         65   Credit available
         75   Overlimit
         85   Date (mm/dd/yy)

If the customer has credit available, the overlimit field should be blank, and if he or she is over the limit, the credit available field should be blank. (Note: This requires you to blank out the print line after each printing.) Provide column headers, and declare the output file as length 121. Use CTLASA characters for appropriate vertical spacing.

13. Write a program to produce invoices for customer purchases. If the same invoice number appears on successive records, those purchases belong on the same invoice, but that customer might appear again later, which requires a new invoice. The 80-byte input records have the format of:

Columns 1–5   Customer number
         6–25   Customer name
         26–30   Invoice number
         31–36   Invoice date (yymmdd)
         37–42   Invoice amount (dollars and cents)

The input records are sorted in ascending order on invoice number. If an invoice number is found out of sequence, that record should be printed with an appropriate error message. The output should print each invoice on a separate page in the following format.

Column   20   Invoice number
          30   Invoice date
          40   Customer number
          50   Customer name
          75   Invoice amount

A summary page should specify the number of records out of sequence and the total of these invoice amounts. The page should then list the total of all invoice amounts from the records in proper sequence.

14. Write a program to produce a book checkout report for the university library. Books can be checked out by either students or faculty, and the format of the input record depends on which. The student checkout record format is

Column   1   Patron code ('S' = student; 'F' = faculty)
          2-13   Call number
          14-23   Author
          24-43   Title
          44-50   Student ID
          51-70   Student name
          71-80   Dorm name

The faculty records have the same format through Title, but then

          44-52   Social security number
          53-72   Faculty name
          73-76   Department code

Use the DATE built-in function to get the current date to compute the due date. For students the due date is the same day of the following month, and for faculty it is the same day of the month six months later. Use your own discretion in formatting the output, but be sure your report is easily readable and attractive.

C H A P T E R   1 4

USER

DEFINED

FUNCTIONS

By way of review, let us remind you that there are three kinds of procedures in PL/I. The first of these is the main procedure, identified by the OPTIONS (MAIN) specification on the procedure statement. The second and third are the subroutine procedure and the function procedure. Both subroutines and functions can be either internal or external. If they are internal, they are compiled along with the main procedure of which they are a part, and they are available only to that main procedure. If they are external, on the other hand, they are compiled separately from any main procedure, and they are available to any procedure at that installation. We described the subroutine procedure in Chapters 6 and 10; the function procedure is the subject of this chapter.

14.1
User
Functions

Since early in the book, you have been using BUILTIN functions. These functions, which are a part of the PL/I compiler, consist of routines that are likely to be used by large numbers of programmers wherever PL/I is used. User functions, however, are written for those situations for which there are no built-in functions. User functions are typically written in two circumstances:

1. A single program might need to include the same block of code at several points. Since one of the attributes of a function is that it can be invoked from anywhere in the program, it seems to make sense to write that code once (as a function) and to invoke it from various points in the program at which the routine or block of code needs to be executed.

2. An even more common situation is that all or most of the programmers at a given installation execute the same block of code repeatedly.

One example of this second situation is that illustrated by the two versions of our example program in Section 14.5. It is much easier to compute the number of days between two dates if the dates are presented in Julian format (yyddd); therefore, dates are often stored in Julian format. Most people, however, are not comfortable with Julian dates; they prefer the standard mmddyyyy format. An installation might therefore allow the standard format on input records and write a function to convert that date to Julian, which is what our example does. The installation would very likely have a second function to convert a Julian date back to standard format. At any rate, rather than forcing each programmer to write the code for this routine each time she or he needs it in a program, the code is written once and stored on disk in such a manner that it is then available to every programmer at the installation simply by including the PL/I statement that invokes the function. Obviously, external functions are more commonly used than internal functions.

Our presentation of user functions proceeds by first comparing and contrasting them with subroutine procedures and then comparing and contrasting them with built-in functions. We then discuss "standard" function-arguments, dummy arguments, the RETURN statement, the RETURNS option and attribute, and the declaration of functions.

14.1.1

User Functions vs. Subroutines

There are, first of all, several similarities between user functions and subroutines:

1. Both can be invoked from anywhere in the program.

2. Both can contain parameters that refer to the same storage areas as the corresponding arguments included in the invoking statement.

3. Both can be internal to or external from the invoking program.

4. Both have the same PROCEDURE statement, namely procname: PROCEDURE (parameters).

5. In both cases control passes to the procedure from the invoking statement, and when the procedure code has executed, control is returned to the point of invocation.

At the same time, there are significant differences between subroutine and function procedures:

1. Subroutines are not declared as such, but it is sometimes necessary to declare a function in the invoking program.

2. A subroutine is invoked through a CALL statement, while a function is invoked simply by supplying the function name (and an argument list, if required).

3. As item 2 suggests, if you want to reference the value returned by the function, you must provide a variable to receive that value.

4. A function, generally speaking, returns a single value; that is, the result of the execution of the code in the function logically replaces the function name in the invoking statement. A subroutine, on the other hand, does not return any values.

14.1.2

User Functions vs. Built-In Functions

The primary similarities between built-in and user functions are that both are invoked in the same manner and both return their value in the same manner.

The differences between built-in and user functions are, first of all, that built-in functions are included in the PL/I compiler. Second, several built-in functions are capable of returning arrays, but a user function can return only an element. Many built-in functions, moreover, generate "in-line" code; therefore, there is no branch to a procedure.

14.2

**Arguments
to Functions**

14.2.1

*Arguments
Revisited*

Throughout this section we will suppose that we are invoking a function called CTEMP, which computes a Celsius temperature from a Fahrenheit temperature passed to it. In general, the most common form of an argument passed to a function is simply a variable. If we suppose our main procedure reads Fahrenheit temperatures into a variable named F_TEMP and prints each as a Celsius temperature from a variable named C_TEMP, then the statement to invoke the function is

```
C_TEMP = CTEMP (F_TEMP);
```

or, of course, we might simply code

```
PUT LIST ('CELSIUS TEMPERATURE = ', CTEMP (F_TEMP));
```

A second commonly-used argument format is a literal:

```
C_TEMP = CTEMP (85);
```

where 85 is the temperature expressed in Fahrenheit. A third possibility for the format of the argument is that it be in the form of an arithmetic expression:

```
AVG_C_TEMP = CTEMP ((H_F_TEMP + L_F_TEMP) / 2.0);
```

or

```
AVG_C_TEMP = CTEMP ((85 + 68) / 2.0);
```

In this case, the arithmetic expression is evaluated first, and the result of the evaluation is the argument, which is passed to CTEMP.

The argument to a user function can also contain a BUILTIN function, and these built-in functions are handled differently (you will recall from Chapter 10) depending on whether or not the built-in function itself has arguments. Let us breifly review. Built-in functions *with* arguments are invoked *before* the user function is invoked. Suppose, for example, that we have a Fahrenheit thermometer so finely calibrated that it gives the tempera-

ture to three decimal places; suppose, in addition, that we do not want that
much precision in our Celsius equivalent.

```
C_TEMP = CTEMP (ROUND ((H_F_TEMP + L_F_TEMP) / 2.0, 2));
```

The arithmetic expression is evaluated first, and the result is the argument to
the function ROUND. ROUND is then invoked, and the value it returns
is the argument to CTEMP. Built-in functions *without* arguments are not
invoked before the user function is invoked; instead, the function name
itself is passed to the user function. Suppose the user function requires the
Fahrenheit temperature and the time.

```
C_TEMP = CTEMP (F_TEMP, TIME);
```

It is the function name (TIME) that is passed to CTEMP, not the value of
TIME. If you want to invoke a built-in function with no arguments *before* the
user function is invoked, you can place an extra set of parentheses around
the built-in function name:

```
C_TEMP = CTEMP (F_TEMP, (TIME));
```

Now it is the value returned by TIME that is the second argument to CTEMP.

14.2.2

Dummy Arguments

To understand the concept of dummy arguments, you must keep in mind
that it is the *location* (address) of the argument, rather than the value in the
location, that is passed to a function or subroutine. When the argument is a
variable, the variable name is a symbolic representation of the numeric
address or location of the value in memory. Therefore, a variable as an
argument poses no problem. A literal or the result of the evaluation of an
arithmetic expression, however, has no name. In this case the compiler must
choose a name for the address of the literal and pass this name to the
function or subroutine. These compiler generated names are called **dummy
arguments**. The first thing to understand is that dummy arguments are *not*
available to the programmer, even if the programmer could determine what
name the compiler had assigned to the literal. The second thing to understand about a dummy argument is how it is assigned its attributes. First,
the compiler acquires storage belonging to the invoking program. If the
function is internal or if it is external and a parameter-descriptor list is
provided, the compiler will use the parameter attributes to convert the

dummy argument value and assign it to the acquired storage. A reference to this storage area is then sent to the function as a dummy argument. Any change to the value of the dummy argument by the function changes the value in the acquired storage, of course, but it does not change the value of the literal in the invoking program. For external functions with no parameter-descriptor list, the process of acquiring storage for the dummy argument value is the same; however, the dummy argument derives its attributes from the value because the parameter attributes are unavailable to it. Thus, the literal 85 gives the dummy argument the attributes of FIXED DECIMAL (2), and 3.1416 gives the dummy argument the attributes of FIXED DECIMAL (5,4). Remember that the attributes of an argument must match exactly those of the corresponding parameter; if they do not match, an error condition exists. You can avoid this error condition in three ways.

1. Express the literal with a precision that matches the precision declared for the parameter.

2. Assign the literal to a variable whose attributes match those of the corresponding parameter; then use the variable as the argument when you invoke the function.

3. Provide a parameter-descriptor list along with the ENTRY attribute statement of the declaration of the function.

14.3

The RETURN Statement

As we have already pointed out, both the argument list and control of execution are passed to a function when that function is invoked. The RETURN statement, on the other hand, returns both the specified value and control of execution to the point of invocation. The syntax of the RETURN statement (when used in a function) is

```
RETURN (expression);
```

Notice the difference between the use of RETURN here, where it is followed by an expression in parentheses, and RETURN as it is used in a subroutine (RETURN;). In its use in a subroutine, it simply returns control of execution to the point of invocation. As it is used in a function, however, it returns the result of the evaluation of the expression, as well as control of execution. Suppose we want a function that will find the difference between two numbers received as parameters. We will suppose that the first parameter is always the minuend and the second the subtrahend.

```
DIFF:  PROCEDURE (MINU, SUBTRA);
    DCL MINU     FIXED DECIMAL (5);
    DCL SUBTRA  FIXED DECIMAL (5);
    RETURN (MINU - SUBTRA);
END DIFF;
```

It is also possible for a function to contain more than one RETURN statement. Using the same basic function as above, let us now suppose that the larger value of the two parameters is always to be the minuend. Our function might now appear as

```
DIFF:  PROCEDURE (NUM1, NUM2);
    DCL NUM1    FIXED DECIMAL (5);
    DCL NUM2    FIXED DECIMAL (5);
    IF NUM1 > NUM2 THEN
        RETURN (NUM1 - NUM2);
    ELSE
        RETURN (NUM2 - NUM1);
END DIFF;
```

Our examples show an arithmetic expression in the RETURN statement, but that need not be the case. We could, for example, assign the result of our computations to a variable and return that variable:

```
DIFF:  PROCEDURE (NUM1, NUM2);
    DCL (NUM1,
         NUM2,
         RSLT)    FIXED DECIMAL (5);
    IF NUM1 > NUM2 THEN
        RSLT = NUM1 - NUM2;
    ELSE
        RSLT = NUM2 - NUM1;
    RETURN (RSLT);
END DIFF;
```

You will recall that one of the attributes of structured programming (as we understand the term) is "one entrance, one exit." Notice that multiple RETURN statements violate this requirement; therefore, we recommend the second of the formats above.

14.4

The RETURNS Attribute and Option

The compiler must know the attributes of the value the function is returning to the invoking procedure so that, if necessary, it can generate the code to convert the value to the correct attributes. The attributes of the value being returned must match those of the variable on the left-hand side of the assignment statement which invokes the function because in user functions no conversion occurs in the invoking procedure. The attributes of the value being returned can be declared in two ways.

1. By default according to the first character of the function name. If the function name begins with one of the characters A through H or O through Z, the attributes of the value being returned are FLOAT DECIMAL (6) because these are the default attributes of names beginning with those characters. If the function name begins with one of the characters I through N, the attributes of the value being returned are FIXED BINARY (15). These defaults place too great a restriction on the attributes of the values being returned by the function. They do not allow, for example, for a user function that returns a character string.

2. To avoid the restrictions imposed by the defaults we can explicitly declare the attributes of the returned value, and the declaration depends on whether the function is internal or external. If the function is internal, we must declare these attributes in the function only. The keyword is RETURNS (note the S), and in the function it is referred to as the RETURNS **option**. The RETURNS option is attached to the PROCEDURE statement.

14.5

Declaring Functions

When a function is coded internally, both the invoking procedure and the function procedure are compiled together. The compiler therefore is aware of the arguments, the parameters, and the value being returned. Because the compiler has this awareness, we need only tell it to generate the code necessary to convert the value being returned to the specified attributes, and we do this by coding the RETURNS option on the function's PROCEDURE statement. If any dummy arguments are passed, the compiler creates the dummy argument with the attributes of the corresponding parameter.

When the function is external, the situation is different, however. As the compiler is compiling the invoking procedure, it can have no direct knowledge of the attributes of the corresponding parameters. If a dummy argument is created, the compiler has no choice but to give the dummy argument the attributes of its value. The way around this limitation is to provide a **parameter-descriptor list** in the declaration of the function, as well as indicating the attributes of the value being returned. The syntax of the declaration, then, is

```
DCL function name ENTRY (parameter-descriptor list) RETURNS
    (attributes);
```

and a specific example is

```
DCL DIFF ENTRY (FIXED DEC (2), FIXED DEC (2)) RETURNS
    (FIXED DEC (5));
```

The ENTRY attribute tells PL/I that the function is external and that it is to be entered for execution at the entry point (usually the PROCEDURE statement) designated by the name following ENTRY. The parameter-descriptor list follows the ENTRY attribute and describes the parameter attributes. The relationship of the items in the descriptor list and the parameters is positional. If one or more arguments will always have the same attributes as its corresponding parameter, the parameter(s) need not be described, but its position must be marked by an extra comma. For example,

```
DCL AFUNC ENTRY (, FIXED DEC (5),, CHAR(8));
```

specifies *four* parameters: The first is specified by the leading comma, the second by FIXED DEC (5), the third by the second of the consecutive commas, and the fourth by CHAR (8). It follows that if all of the arguments always match the parameters no parameter-descriptor list is required. Parameter-descriptor lists are used primarily to assure that dummy arguments will match their parameters, although certainly no harm is done by describing parameters whose arguments agree with them.

The RETURNS option is specified in the function the same way it is when the function is internal. In the invoking procedure, we speak of the RETURNS **attribute**, and it forms a part of the declaration of the function.

Using the RETURNS option and if necessary the parameter-descriptor list, then, we can free ourselves from the constraint imposed by the implicit declaration of the attributes of the returned value, and we can guarantee that dummy arguments will match their parameters.

14.6

Example Program— User Functions

As we mentioned earlier, it is much easier to add and subtract dates if they are in Julian format (yyddd), where January 1 is day 001, and the days are numbered consecutively throughout the year. Most people, however, do not want to take the trouble of determining the Julian date from the "standard" date (mmddyyyy). Therefore, it is a common practice to allow input records to contain the date in standard format and to provide a function to carry out the conversion algorithm. In practice, such a function would almost certainly be an external function. We have chosen to present it as both an internal and an external function, however, and for this reason: Practices in introductory PL/I classes vary. In some schools, students are taught to write external functions and subroutines. In other schools, only internal procedures are taught in the introductory course.

The logic of the main procedure of programs CH14EX1 and CH14EX2 (see Figs. 14-1 and 14-2) is extremely simple. It prints page and column headers, reads the input, calls the function, and prints the detail lines.

The problem posed by the function JULIAN is a bit more complex. It has to allow for leap years, and typically we say any year evenly divisible by four is a leap year. As a matter of fact, however, the last year of any century (1700, 1800, 1900, 2000) is evenly divisible by four, but only those evenly divisible by 400 are actually leap years.

To find the day of the Julian date (apart from the possible February 29), we use an array with each entry initialized to the highest possible Julian day for the preceding month (the first element, January, has no preceding month; therefore, it is initialized to 0). This allows us to use the J_MONTH parameter value as the subscript into the array. Thus, if the standard month and day are 0201, the 02 will reference the second element, and when that value (31) is added to the J_DAY parameter value (01), the result is 32, and February 1, of course, is the thirty-second day of the year. We then check to see if the date is later than February 28 because it is only at that point in the year that a leap year will make any difference. If the date we are converting is later than February 28, we use the MOD built-in function to check for leap year. We first make the '400 check' and then the '4 check'. If the year is evenly divisible by either 400 or 4, we add one to the day calculated by adding the J_DAY value to the appropriate element in the MONTH_TAB. Notice that Statement 36 of CH14EX1 (Statement 14 of JULIAN procedure in CH14EX2) is a null ELSE; the failure to notice it could cause you to puzzle over the logic.

In Statement 37 of CH14EX1 (Statement 15 of JULIAN procedure in CH14EX2) we perform the necessary arithmetic to get our two-part date into a single number as demanded by the Julian format. Then, in Statement 38 (16) we return both the Julian date and control to the invoking procedure.

It was not necessary for us to provide a parameter-descriptor list in

CH14EX2 because we are not sending any dummy arguments. We wanted to illustrate parameter-descriptor lists, however, and the list does provide excellent documentation for the main procedure.

Finally, note the information message printed following JULIAN procedure in CH14EX2.

```
NO 'MAIN' OPTION ON PROCEDURE.
```

This is merely an information message, and PL/I does *not* mean that you should change the procedure statement to.

```
JULIAN:  PROCEDURE OPTION (MAIN);
```

The pseudocode is as follows:

```
CH14EX1/CH14EX2
    Print page, column headers
    Read a record
    DO WHILE (more input)
        Standard year ← century * 100 + decade
        Call JULIAN
        Print a detail line
        Read a record
    ENDDO
STOP
JULIAN
    Julian year ← century * 100 + decade
    Julian date ← day + month array (month)
    IF (Julian date > 59)
        IF (remainder of Julian year / 400 = 0)
            Julian date ← Julian date + 1
        ELSE
            IF (remainder of Julian year / 4 = 0)
                Julian date ← Julian date + 1
            ENDIF
        ENDIF
    ELSE
        Julian date ← decade * 1000 + Julian year
    ENDIF
EXIT
```

FIGURE 14-1
CH14EX1 (user-defined internal functions).

```
PL/I OPTIMIZING COMPILER  VERSION 1 RELEASE 5.1  TIME: 11.06.03  DATE: 13 MAR 87 PAGE   1
OPTIONS SPECIFIED
GOSTMT;



PL/I OPTIMIZING COMPILER         /* CH14EX1 USING USER FUNCTIONS (INTERNAL) */   PAGE   2
                    SOURCE LISTING
     STMT


          /* CH14EX1 USING USER FUNCTIONS (INTERNAL) */
     1  CH14EX1:    PROCEDURE OPTIONS (MAIN);
          /*****************************************************************/
          /*                                                             */
          /*  FUNCTION:  TO DETERMINE A JULIAN DATE FROM A STANDARD DATE  */
          /*                                                             */
          /*  INPUT:     A FILE WHOSE RECORDS CONTAIN A STANDARD DATE IN  */
          /*             THE FORMAT OF MMDDYYYY (IN EDIT-DIRECTED FORMAT)  */
          /*                                                             */
          /*  OUTPUT:    THE INPUT DATES IN BOTH STANDARD AND JULIAN      */
          /*             FORMAT                                           */
          /*                                                             */
          /*  NOTES:     THE USER-CODED FUNCTION JULIAN MAKES THE DATE    */
          /*             CONVERSION FROM A STANDARD TO A JULIAN DATE       */
          /*                                                             */
          /*****************************************************************/
     2      DCL MONTH              FIXED DECIMAL (3);
     3      DCL DAY                FIXED DECIMAL (3);
     4      DCL CENTURY            FIXED DECIMAL (3);
     5      DCL DECADE             FIXED DECIMAL (3);
     6      DCL J_DATE             FIXED DECIMAL (5);
     7      DCL STD_YEAR           FIXED DECIMAL (5);

     8      DCL MORE_RECS          BIT       (1);
     9      DCL    YES             BIT       (1)    INIT ('1'B);
    10      DCL    NO              BIT       (1)    INIT ('0'B);

    11      DCL MOD                BUILTIN;

    12  MAIN:    MORE_RECS = YES;
    13           ON ENDFILE (SYSIN) MORE_RECS = NO;
```

```
14              PUT EDIT ('STANDARD TO JULIAN DATE CONVERSION')
                         (COL(45), A);
15              PUT SKIP (3) EDIT ('STANDARD DATE', 'JULIAN DATE')
                                 (COL(45), A(20), A);
16              PUT SKIP (3);

17              GET EDIT (MONTH, DAY, CENTURY, DECADE) ((4)F(2));
18   PROC_LP: DO WHILE (MORE_RECS);
19              STD_YEAR = CENTURY * 100 + DECADE;
20              J_DATE = JULIAN (MONTH, DAY, CENTURY, DECADE);
21              PUT SKIP (2) EDIT (MONTH, '/', DAY, '/', STD_YEAR, J_DATE)
                                 (COL(46), F(2), A, F(2), A, F(4),
                                  COL(68), F(5));
22              GET SKIP EDIT (MONTH, DAY, CENTURY, DECADE) ((4)F(2));
23            END PROC_LP;
```

PL/I OPTIMIZING COMPILER   /* CH14EX1 USING USER FUNCTIONS (INTERNAL) */   PAGE   3

```
   STMT
   24  JULIAN:  PROCEDURE (J_MONTH, J_DAY, J_CENTURY, J_DECADE)
                          RETURNS (FIXED DECIMAL (5));
       /*******************************************************************/
       /*                                                                 */
       /*  FUNCTION:  TO COMPUTE A JULIAN DATE FROM A 'STANDARD' DATE      */
       /*                                                                 */
       /*  INPUT:     NONE                                                */
       /*                                                                 */
       /*  OUTPUT:    NONE                                                */
       /*                                                                 */
       /*  PARAMETERS:                                                    */
       /*     J-MONTH          THE MONTH                                  */
       /*     J_DAY            THE DAY                                    */
       /*     J-CENTURY        THE CENTURY                                */
       /*     J_DECADE         THE DECADE                                 */
       /*                                                                 */
       /*  PROCEDURES CALLED:  NONE                                       */
       /*                                                                 */
       /*******************************************************************/
   25     DCL J_MONTH           FIXED DECIMAL (3);
   26     DCL J_DAY             FIXED DECIMAL (3);
   27     DCL J_CENTURY         FIXED DECIMAL (3);
   28     DCL J_DECADE          FIXED DECIMAL (3);
```

FIGURE 14–1 *(continued)*

```
29    DCL J_YEAR              FIXED DECIMAL (5);
30    DCL JUL_DATE            FIXED DECIMAL (5);
31    DCL MONTH_TAB (12)      FIXED DECIMAL (3)      INIT
                              (0,31,59,90,120,151,181,212,242,273,303,334);

32        J_YEAR = J_CENTURY * 100 + J_DECADE;
33        JUL_DATE = J_DAY + MONTH_TAB (J_MONTH);

34        IF JUL_DATE > 59 THEN
              IF MOD (J_YEAR, 400) = 0 THEN
                 JUL_DATE = JUL_DATE + 1;
35            ELSE
              IF MOD (J_YEAR, 4) = 0 THEN
                 JUL_DATE = JUL_DATE + 1;
36        ELSE;

37        JUL_DATE = J_DECADE * 1000 + JUL_DATE;
38        RETURN (JUL_DATE);
39        END JULIAN;
40    END CH14EX1;


PL/I OPTIMIZING COMPILER       /* CH14EX1 USING USER FUNCTIONS (INTERNAL) */   PAGE   4
COMPILER DIAGNOSTIC MESSAGES
ERROR ID L   STMT    MESSAGE DESCRIPTION
COMPILER INFORMATORY MESSAGES

IEL0533I I          NO 'DECLARE' STATEMENT(S) FOR 'SYSIN','SYSPRINT'.

END OF COMPILER DIAGNOSTIC MESSAGES

                              STANDARD TO JULIAN DATE CONVERSION


                              STANDARD DATE       JULIAN DATE


                              1/15/1985           85015

                              3/ 1/1944           44061
```

11/12/1933	33315
6/15/1956	56167
1/24/1938	38024
12/31/1988	88366
10/ 1/1961	61274
9/ 6/1932	32249
5/30/1963	63150
2/ 3/1968	68034

PL/I OPTIMIZING COMPILER   VERSION 1 RELEASE 5.1 TIME: 13.19.31 DATE: 10 MAR 87 PAGE   1
OPTIONS SPECIFIED
GOSTMT;

PL/I OPTIMIZING COMPILER   /* CH14EX2 USING USER FUNCTIONS (EXTERNAL) */   PAGE   2
                  SOURCE LISTING

    STMT

        /* CH14EX2 USING USER FUNCTIONS (EXTERNAL) */
    1 CH14EX2:   PROCEDURE OPTIONS (MAIN);
        /**/
        /*   */
        /*   FUNCTION: TO DETERMINE A JULIAN DATE FROM A STANDARD DATE */
        /*   */
        /*   INPUT: A FILE WHOSE RECORDS CONTAIN A STANDARD DATE IN */
        /*   THE FORMAT OF MMDDYYYY (IN EDIT-DIRECTED FORMAT) */
        /*   */
        /*   OUTPUT: THE INPUT DATES IN BOTH STANDARD AND JULIAN   */
        /*   FORMAT   */
        /*   */
        /*   NOTES: THE USER-CODED FUNCTION JULIAN MAKES THE DATE   */
        /*   CONVERSION FROM A STANDARD TO A JULIAN DATE   */

FIGURE 14–2
CH14EX2 (user-defined external functions).

FIGURE 14–2 *(continued)*

```
     /*                                                               */
     /*****************************************************************/


PL/I OPTIMIZING COMPILER        /* CH14EX2 USING USER FUNCTIONS (EXTERNAL) */   PAGE   3
    STMT
      2       DCL MONTH              FIXED DECIMAL (3);
      3       DCL DAY                FIXED DECIMAL (3);
      4       DCL CENTURY            FIXED DECIMAL (3);
      5       DCL DECADE             FIXED DECIMAL (3);
      6       DCL J_DATE             FIXED DECIMAL (5);
      7       DCL STD_YEAR           FIXED DECIMAL (5);

      8       DCL MORE_RECS          BIT        (1);
      9       DCL   YES              BIT        (1)    INIT ('1'B);
     10       DCL   NO               BIT        (1)    INIT ('0'B);

     11       DCL MOD                BUILTIN,
                  JULIAN ENTRY (FIXED DEC (3), FIXED DEC (3), FIXED DEC (3),
                               FIXED DEC (3)) RETURNS (FIXED DEC (5));

     12  MAIN:    MORE_RECS = YES;
     13           ON ENDFILE (SYSIN) MORE_RECS = NO;
     14           PUT EDIT ('STANDARD TO JULIAN DATE CONVERSION')
                          (COL(45), A);
     15           PUT SKIP (3) EDIT ('STANDARD DATE', 'JULIAN DATE')
                                 (COL(45), A(20), A);
     16           PUT SKIP (3);

     17           GET EDIT (MONTH, DAY, CENTURY, DECADE) ((4)F(2));
     18  PROC_LP: DO WHILE (MORE_RECS);
     19              STD_YEAR = CENTURY * 100 + DECADE;
     20              J_DATE = JULIAN (MONTH, DAY, CENTURY, DECADE);
     21              PUT SKIP (2) EDIT (MONTH, '/', DAY, '/', STD_YEAR, J_DATE)
                                    (COL(46), F(2), A, F(2), A, F(4),
                                    COL(68), F(5));
     22              GET SKIP EDIT (MONTH, DAY, CENTURY, DECADE) ((4)F(2));
     23           END PROC_LP;
     24  END CH14EX2;
```

COMPILER DIAGNOSTIC MESSAGES
ERROR ID L   STMT   MESSAGE DESCRIPTION

COMPILER INFORMATORY MESSAGES

IEL0533I I   NO 'DECLARE' STATEMENT(S) FOR 'SYSIN','SYSPRINT'.

END OF COMPILER DIAGNOSTIC MESSAGES

PL/I OPTIMIZING COMPILER VERSION 1 RELEASE 5.1 TIME: 13.19.38 DATE: 10 MAR 87 PAGE   5
OPTIONS SPECIFIED
GOSTMT;
*PROCESS;

                SOURCE LISTING
    STMT

        /* CH14EX2 USING USER FUNCTIONS (EXTERNAL) */

    STMT
    1 JULIAN: PROCEDURE (J_MONTH, J_DAY, J_CENTURY, J_DECADE)
                       RETURNS (FIXED DECIMAL (5));
      /**/
      /*   */
      /* FUNCTION: TO COMPUTE A JULIAN DATE FROM A 'STANDARD' DATE   */
      /*   */
      /* INPUT: NONE   */
      /*   */
      /* OUTPUT: NONE   */
      /*   */
      /* PARAMETERS: */
      /*   J-MONTH   THE MONTH   */
      /*   J_DAY   THE DAY   */
      /*   J-CENTURY   THE CENTURY   */
      /*   J_DECADE   THE DECADE   */
      /*   */

FIGURE 14–2 *(continued)*

```
     /*  PROCEDURES CALLED:  NONE                                     */
     /*                                                               */
     /*****************************************************************/
  2     DCL J_MONTH            FIXED DECIMAL (3);
  3     DCL J_DAY              FIXED DECIMAL (3);
  4     DCL J_CENTURY          FIXED DECIMAL (3);
  5     DCL J_DECADE           FIXED DECIMAL (3);
  6     DCL J_YEAR             FIXED DECIMAL (5);
  7     DCL JUL_DATE           FIXED DECIMAL (5);
  8     DCL MONTH_TAB (12)     FIXED DECIMAL (3)     INIT
                       (0,31,59,90,120,151,181,212,242,273,303,334);

  9     DCL MOD                BUILTIN;

 10         J_YEAR = J_CENTURY * 100 + J_DECADE;
 11         JUL_DATE = J_DAY + MONTH_TAB (J_MONTH);

 12         IF JUL_DATE > 59 THEN
                IF MOD (J_YEAR, 400) = 0 THEN
                    JUL_DATE = JUL_DATE + 1;
 13             ELSE
                    IF MOD (J_YEAR, 4) = 0 THEN
                        JUL_DATE = JUL_DATE + 1;
 14         ELSE;

 15         JUL_DATE = J_DECADE * 1000 + JUL_DATE;
 16         RETURN (JUL_DATE);
 17 END JULIAN;


PL/I OPTIMIZING COMPILER        /* CH14EX2 USING USER FUNCTIONS (EXTERNAL) */   PAGE   8
COMPILER DIAGNOSTIC MESSAGES
ERROR ID L   STMT    MESSAGE DESCRIPTION

COMPILER INFORMATORY MESSAGES

IEL0430I I   1      NO 'MAIN' OPTION ON EXTERNAL PROCEDURE.

END OF COMPILER DIAGNOSTIC MESSAGES
```

Summary

PL/I allows for three kinds of procedures: main procedures, subroutine procedures, and function procedures. Function procedures are distinguishable from subroutine procedures by the manner in which they are invoked. The use of the function name, followed by an argument list (if required) is all that is necessary to invoke a function. The function generally returns a single value, and the value logically replaces the function name. In contrast, a subroutine returns no values.

Functions, in turn, are of two kinds: built-in and user-defined. The built-in functions are extensions of the PL/I language itself, while user-defined functions are coded by various PL/I programmers at individual installations. In general, functions are coded for routines that are commonly used, and the built-in functions cover the most widely used routines in both scientific and business applications. User functions are coded for routines widely used at a particular installation. Still a third difference between built-in and user-defined functions is the fact that some built-in functions return an array, but user functions are restricted to returning a single element.

The attributes of an argument must match those of the corresponding parameter. When an argument is a literal or the result of evaluating an operational expression, the compiler has no variable name (hence no attributes) to associate with the memory location at which the value resides. The compiler must therefore create a variable name, and these variables are referred to as dummy arguments. These dummy arguments are not available to the programmer, and the dummy argument derives its attributes from the attributes of the corresponding parameter, if the parameter is available to the compiler. Otherwise, it derives its attributes from the value to be placed in it. The parameter attributes are always available to the compiler when the function is internal; if you are sending a dummy argument to an external function, however, you might find it easiest to include a parameter-descriptor list on the function declaration in the invoking routine. Another method of assuring the necessary consistency is to assign the value to a variable having the required attributes and send the variable as the argument.

The RETURN statement is coded in the user function to return control to the invoking procedure and to indicate the value to be returned to the invoking statement. Its syntax is

```
RETURN (expression);
```

Functions can be internal or external, but in either case the compiler needs to know the attributes of the value being returned; it is supplied with this information either by defaults based on the starting character of the function name or by the RETURNS option and attribute. When used in the function, the keyword is called the RETURNS option, and this RETURNS statement is all that is required when the procedure is internal. The RETURNS option forms a part of the function's PROCEDURE statement. When used in the main procedure, the keyword is referred to as the RETURNS attribute, and it forms a portion of the declaration of the function. The function must be so declared if it is external to the invoking procedure, and the syntax is

```
DCL function name ENTRY (parameter-descriptor list)
        RETURNS (attributes);
```

If all arguments match their corresponding parameters, the parameter-descriptor list can be omitted.

14.8

Exercises

1. Name three kinds of procedures available to PL/I and briefly describe the differences among them.

2. Briefly describe a situation in which a programmer might code an internal function.

3. In a short paragraph, explain why a user function is more likely to be external than internal.

4. List five similarities between subroutine and function procedures.

5. List four significant differences between subroutines and functions.

6. Explain under what circumstances dummy arguments are sent to subroutines and functions and why it is necessary that they be sent.

7. How does the compiler determine the attributes to assign to a dummy argument? Consider more than one situation.

8. Write the DECLARE and invoking statements in the invoking procedures, based on the code given for the functions. Use recognizably similar, but not identical, variable names.
   a.
```
   RECAREA:   PROC (LENGTH, WIDTH)
                        RETURNS (FIXED DEC (9,2));
              DCL  LENGTH    FIXED DEC (6,2),
                   WIDTH     FIXED DEC (5,2),
                   AREA      FIXED DEC (9,2);
```

```
b.   TRIAREA:   PROC (BASE, ALT) RETURNS (FIXED DEC (8,1));
          DCL   BASE        FIXED DEC (3),
                ALT         FIXED DEC (3),
                AREA        FIXED DEC (8,1);
c.   PYTHAG:   PROC (LEG1, LEG2) RETURNS FIXED DEC (5));
          DCL   LEG1        FIXED DEC (2),
                LEG2        FIXED DEC (2),
                HYPOT       FIXED DEC (5);
```

9. Assume the functions in problem 8 are external. Code the function declaration in the invoking routine under the assumption that dummy arguments will be passed in all cases.

10. In our example program we coded a function to compute a Julian date from a standard date. Write a function (internal or external as your teacher directs) to accept a Julian date (yyddd) and return a standard date (mmddyy). Assume all are twentieth-century dates. Code a main procedure to read and print the dates and invoke the function.

11. Write a main procedure that will read temperatures in either Celsius or Fahrenheit and invoke a function to convert the temperature to the opposite scale. The input will have the format of

```
F103
C027
etc.
```

The output should list both temperatures in two appropriately named columns.

12. Write a main procedure that will read in miles or kilometers; code two separate functions: one to convert miles to kilometers, the other to convert kilometers to miles. Print both the miles and the kilometers. The input will have the format of

```
K4390
M1277
etc.
```

C H A P T E R   1 5

LIST

PROCESSING

Pl/I is very powerful in performing the functions that are associated with list processing; therefore, we include this chapter on list processing. However, due to the nature of this book, we cannot include all there is to know about list processing. We present a "polite" coverage of this topic. There are some good books discussing list processing; please refer to these sources in the area of data structures to gain the knowledge which you need to complete your understanding of this topic.

By "lists" we mean linear lists, trees, stacks, and queues. We try to define the elementary concepts of lists for those of you who do not have any background in list processing. We first begin by defining and giving examples of each type of list. Then we show you the PL/I code necessary to define and process the lists. And, as usual, example programs utilizing the list processing concepts contained in this chapter are presented.

15.1

Tables

A table is a linear list defined as an array. We presented arrays in Chapter 11. The elements of the list (array) are stored consecutively in memory and accessed via subscripts within the array name (e.g., TABLE (INDEX)). The table in Figure 15–1 shows just such a linear list of courses in which a student is enrolled.

CSCI 200 A
CSCI 304
CSCI 350
CSCI 367
CSCI 375

FIGURE 15–1
A sequential table.

15.2

Linked Linear Lists

A linked linear list differs from Figure 15–1 in that the elements do not occupy consecutive areas of memory but rather are "linked" together through the use of pointers. Let us first illustrate this idea with the diagram in Figure 15–2 showing the elements as they would appear within a linked linear list.

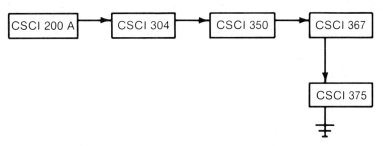

FIGURE 15–2
A linked list.

As you can tell from Figure 15–2, each element in a linked list consists of a structure containing at least two fields. The first is the data portion (which can also be a structure). The second field contains the address of (hence, "points to") the next element in the linked list. The pointer field within the last element contains what is called a NULL value. The NULL

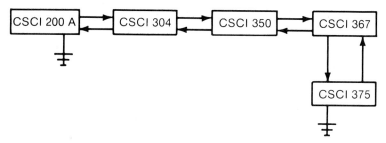

FIGURE 15–3
A doubly linked list.

value is usually used to denote the end of the list (the pointer contains no value at all).

Figure 15–2 shows a **singly linked** list. There is only one pointer in each element of the list. The list can be processed in only one direction—from beginning to end. Other types of linked lists are **doubly linked** lists, in which each element points not only to the next element in the list but also to its predecessor, as shown in Figure 15–3. You will notice that a doubly linked list can be processed either from beginning to end or from end to beginning.

Another type of linked list you might run across in your programming career is the **circularly** linked list. In the circularly linked list, the last element of the linked list points to the first element, rather than containing a NULL value. You can create singly linked circular lists or even doubly linked circular lists, in which the first element also points to the last element as its predecessor. (In reality, the circular list has no beginning and no ending element.) A singly linked circular list is shown in Figure 15–4.

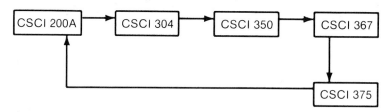

FIGURE 15–4
A circularly linked list.

15.3

Processing Linked Linear Lists

The processing that is usually performed upon linked lists is inserting an element, deleting an element, sorting the list, and searching the list. Naturally, this is the same processing that would take place upon tables. However, as you may have already discovered, there are advantages to linked lists over tables. In case you have not, we describe the advantages now.

1. Insertion into a linked list is simply the creation of an element and changing the necessary pointers to reflect the new element's position in the list. This is unlike tables, in which the contents of the elements would have to be moved in order to insert a new element in an element other than following the currently last element with valid data.

2. Deletion of an element from a linked list is a matter of resetting the pointers, skipping the element that is to be deleted. Thus, although the element remains physically present, it has been logically deleted because we can no longer access it. When an element is no longer needed that is somewhere in the middle of a table, one of two things may happen. First, the element could be simply set to all blanks to denote "not in use" and left at its position in the table (this could cause problems in processing). Or, second, the element could be removed (deleted) by moving all the elements lower in the table up one position. This causes a lot of movement of the data during processing and is not an ideal situation.

15.3.1

Insertion in a Linked List

Let's now look at an example of inserting an element in a singly linked list. Using the diagram in Figure 15–4, showing an ordered list of courses in which a student is enrolled, let's insert the course CSCI 324 in its proper position in the list.

Step 1: Start at the beginning of the list comparing the names of the courses with CSCI 324 until one is found that is greater (CSCI 350 in this case).

Step 2: We know that CSCI 324 should be inserted prior to the current element in the list. In order to be able to find and change the appropriate pointers, we will maintain two extra pointers as we are going through the list: CURR_ELEM_PTR and PREV_ELEM_PTR.

Step 3: Set the pointer in CSCI 324 to the value in CURR_ELEM_PTR.

Step 4: Set the pointer in the element pointed to by PREV_ELEM_PTR to point to CSCI 324 element.

Step 5: You have now inserted the CSCI 324 element.

Follow the above steps while studying the diagram in Figure 15–5, where we insert an element containing "CSCI 324" into the linked list.

As an exercise see if you can insert an element in a circular doubly linked list. Remember to *use* the values within the pointers *before* you change them to perform the insertion. (Otherwise, you may destroy the values of the pointers and not be able to correctly link in the new element.)

15.3.2

Deleting an Element from a Linked List

As you may have guessed, the processing to delete an element from a linked list is almost the opposite of the insertion process (see Fig. 15–6). Before you read the following algorithm, try to delete the CSCI 324 element that we added in the previous section. Then read the algorithm below to see whether you were correct.

Step 1: Find the element to delete, maintaining two pointers: CURR_ELEM_PTR and PREV_ELEM_PTR. The CURR_ELEM_PTR will point to the element to be deleted.

Step 2: Change the NEXT_ELEM_PTR in the element pointed to by PREV_ELEM_PTR to contain the pointer value in the NEXT_ELEM_PTR of the element pointed to by CURR_ELEM_PTR. Here we are setting the pointer in the previous element to skip over the element to be deleted and point to the next element in the list.

Step 3: Change the NEXT_ELEM_PTR in the element to be deleted to NULL.

Step 4: We have now removed (deleted) the element from the list.

15.4

Linking the Nodes of Trees

A tree is a special type of linked list in which each of its elements, called **nodes**, has at least two pointers. A node is a **parent** if it points to at least one other node in the tree. A **child** node is one that is pointed to by a parent node, and a child node may also be a parent. Think of your family tree. Each member of your tree will be a child of someone, and might be a parent to someone else. The very top node of the tree, the one without any parent, is called the **root** of the tree. Actually, every node in a tree is a child node except one, the root node at the top of the tree. Figure 15–7 shows a binary tree (binary trees are restricted in that each node may have at most only two children). Notice, also, that each node can have one, and only one, parent in this tree structure.

Step 1 and Step 2:

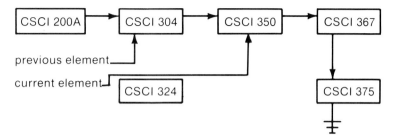

Step 3:

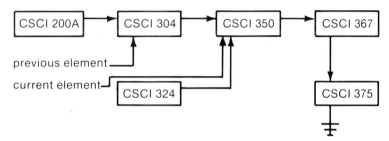

Step 4:

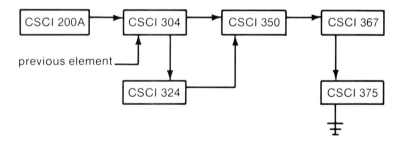

Step 5:

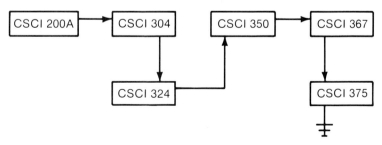

FIGURE 15–5
Inserting an element in a linked list.

Step 1:

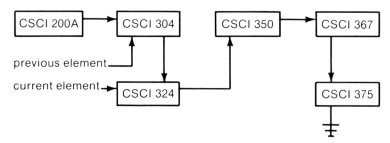

Step 2:

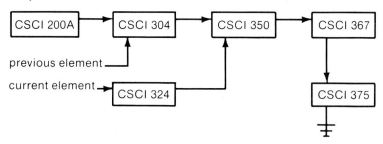

Step 3:

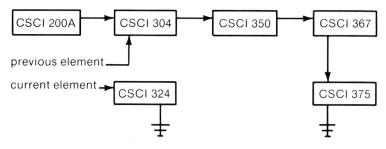

Step 4:

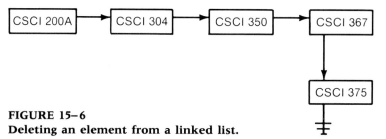

FIGURE 15–6
Deleting an element from a linked list.

The trees can be set up in many ways. One possibility is to say that the data portion of each parent node must be alphabetically greater than any of the nodes that can be reached by following the pointer to the left child and alphabetically less than any of the nodes reached by following the right child pointer.

The processing performed on trees is quite similar to the processing that is performed on linked lists in general: inserting nodes in the tree, deleting nodes from the tree, searching the tree for a specific node, and perhaps even sorting the nodes of the tree if they are not built in the order needed.

Before you continue with the next sections of the book see if you can perform an insertion and deletion using the tree shown above. Then compare your logic with the logic shown in the following sections.

15.4.1

Inserting a Node in a Binary Tree

Depending upon the ordering of the nodes within the tree, the insertion of a node into a tree structure is a matter of finding the correct location for the new node. This may require following certain rules such as this one for a binary tree: No node can have more than two children. Examine the

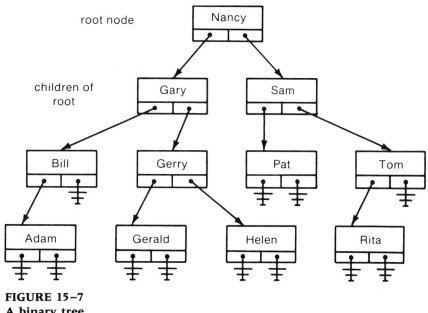

FIGURE 15–7
A binary tree.

algorithm below and for the insertion of a node into the binary tree of Figure 15–7. The insertion rules that apply to this binary tree are as follows:

1. If the name in the node to be inserted is less than the name in the node being looked at, follow the left subtree.

2. If the name in the node to be inserted is greater than the name in the node being looked at, follow the right subtree.

The following algorithm is used to insert the node.

```
Obtain the new node to be inserted
Set P and Q to point to the root of the tree
DO WHILE (name in new_node not = name in tree_node (P)
          AND Q not = NULL)
     P = Q
     IF name in new_node less than name in tree_node(P)
          Q = left_pointer (P)
     ELSE
          Q = right_pointer (P)
     ENDIF
ENDDO
IF name in new_node = name in tree_node (P)
     Process duplicate name
ELSE
     IF name in new_node less than name in tree_node(P)
          Set the left pointer of tree_node(P)
     ELSE
          Set the right pointer of tree_node(P)
     ENDIF
ENDIF
```

15.4.2

Deleting a Node from a Binary Tree

The deletion of a node is a matter of first finding the node to be deleted, deleting the node, and resetting the pointers to show that the tree has been modified. We will leave this algorithm for an exercise at the end of the chapter.

15.4.3

Traversing a Binary Tree

There are three basic methods of traversing a binary tree. The methods differ according to which node is actually processed first. The three methods are called **preorder**, **inorder**, and **postorder**. The name of each states when the root node of the tree (or subtree) is to be processed.

The traversing of a binary tree always begins at the root. Then, the tree is divided into subtrees with a "new" root being assigned. Traversing binary trees using any of the three methods requires a recursive algorithm (a procedure is **recursive** if it invokes itself) because each subtree created is processed as if it were a new tree entirely. Let's examine the preorder traversing method using the binary tree in Figure 15–7.

Using preorder traversing means that the root of each subtree is processed first, then a new subtree is created using the left child as the new root. After the left subtree is completely processed in preorder, the right child is used as the root of a new subtree. The new subtree is then processed in preorder, also. The recursive algorithm for processing a binary tree in preorder is

```
Preorder: (Pointer-to-the-root)
     IF the Pointer-to-the-root is not NULL
          CALL Process-the-root (Pointer-to-the-root)
          CALL Preorder (Pointer-to-the-left-child)
          CALL Preorder (Pointer-to-the-right-child)
     ENDIF
END Preorder.
```

Notice that calling the Preorder procedure using the Pointer-to-the-left-child is simply making this pointer the Pointer-to-the-root as the Preorder procedure is entered. Thus we have subdivided the tree into a subtree with the left child as the root.

If you follow the preorder algorithm using the data in the nodes of the binary tree in 15.7., you will see that the following would be printed. (Assuming that the procedure "Process-the-root" prints the root pointed to by the argument.)

```
NANCY, GARY, BILL, ADAM, GERRY, GERALD, HELEN, SAM, PAT,
RITA, TOM
```

The inorder and postorder algorithms are very similar. The only difference is the location of the "CALL Process-the-root (Pointer-to-the-root)" statement. In inorder it is located between processing the left and right children. In postorder it is after processing both the children. We will leave these algorithms as an exercise at the end of the chapter.

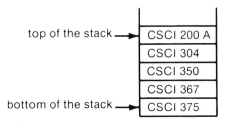

top of the stack → CSCI 200 A
CSCI 304
CSCI 350
CSCI 367
bottom of the stack → CSCI 375

FIGURE 15–8
A stack.

15.5

Stacks

A **stack** is a special type of data structure in that the processing done on the items is done at only one end called the "top" item in the stack. Data may be inserted into the stack, which is called "pushed" onto the stack, and data may be deleted from the stack which is called "popped" off the stack. Let's examine the diagram in Figure 15–8, which shows a stack. The element (item) available for processing in the stack of Figure 15–8 is "CSCI 200 A" and is located at the top of the stack.

In Figure 15–8 we show the stack with the "top" up. However, the stack is just a conceptual idea. We could view it from the side, with the top on the left (or right). We could also view it with the top down and still have the same stack as long as the elements were shown in order from the top of the stack to the bottom of the stack.

A stack is sometimes called a **Last-Out**, **First-In** (LIFO) queue because of the nature of the processing. A stack may also be known as a **pushdown queue** because the process of inserting an element is known as pushing.

Stacks can be implemented in several of the methods described above. We can create a stack as a table, allowing only the first element to be processed. A stack can be a linked list of nodes, when a pointer is set up to the top of the stack (and possibly a pointer to the bottom also). PL/I also offers a storage class that can be used to implement stacks. If you remember, we have discussed two storages classes, AUTOMATIC and STATIC in Chapter 6. Another is called CONTROLLED and may be used in processing stacks (discussed later in this chapter).

15.5.1

Popping an Element off a Stack

Deleting an element from a stack is called **popping** an element off a stack. The process includes removing the top element and redoing the stack so that the second element becomes the new top element (see Fig. 15–9).

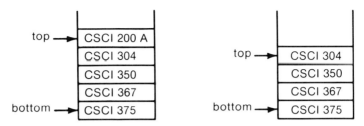

FIGURE 15–9
Popping an element off a stack.

The algorithm for popping an element off a stack must first make certain that the stack contains at least one element. The pseudocode for popping an element is as follows:

```
Pop-stack:  (Stack, Popped-element, Empty-flag)
      IF (Stack is EMPTY)
            Set Empty-flag to show empty stack
      ELSE
            Popped-element ← Stack (TOP)
            TOP ← TOP – 1
      ENDIF
END Pop-stack.
```

We do not go into the process that determines whether the stack is empty because that process depends upon how the stack is being implemented. The stack is empty if the first element of the table is NULL, or a stack is empty if the TOP pointer is NULL. You can study the example program on stacks, shown later, to learn how it determined whether or not the stack was empty.

15.5.2

*Pushing an Element
onto a Stack*

The process of inserting an element onto a stack is called **pushing** an element. Pushing an element is creating a new element and setting the top of the stack to the new element as shown in Figure 15–10, where the element containing "CSCI 250" is pushed on to the stack.

The algorithm for pushing an element onto a stack should check to see whether there is enough space available to insert another element. If you examine the "Push" algorithm below, you will see that, again, we have left this part vague because determining whether or not any space is left in the stack is based upon how the stack is implemented.

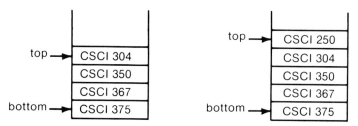

FIGURE 15–10
Pushing an element onto a stack.

```
Push-stack:  (Stack, Element-to-push, Full-flag)
     IF (Stack is full)
          Set the Full-flag to show that the stack is full
     ELSE
          TOP ← TOP + 1
          Stack(TOP) ← Element-to-push
     ENDIF
END Push-stack.
```

15.5.3

Using Stacks

Stacks are used any time a data structure is needed in which the inserting and deleting are done only at one end. In other words, you should choose to use a stack when the stack processing fits the problem at hand. In Figure 15–11 we show a stack that is "in motion"—that is, one in which many elements have been inserted and deleted during processing.

The processing that took place on the stack in Figure 15–11 is described below using the procedures Pop-stack and Push-stack.

```
CALL Push-stack (Stack, 14, Full-flag)
CALL Push-stack (Stack, 15, Full-flag)
CALL Push-stack (Stack, 16, Full-flag)
CALL Pop-stack (Stack, Popped-element, Empty-flag)
CALL Pop-stack (Stack, Popped-element, Empty-flag)
CALL Push-stack (Stack, 17, Full-flag)
CALL Push-stack (Stack, 18, Full-flag)
```

15.6

Queues

A **queue** is a special data structure containing a set of elements that are inserted at the end of the list (called the "rear" of the queue) and deleted

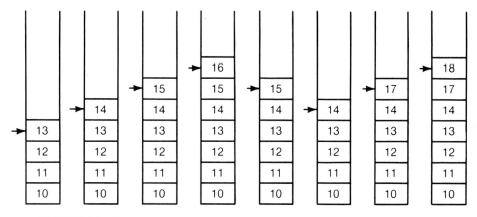

FIGURE 15–11
A stack in motion.

from the beginning of the list (called the "front" of the queue). This type of queue is sometimes called a "First-In, First-Out" (FIFO) queue. You can think of the checkout line at the grocery store as a queue. The first person in line is processed first and removed from the line as he or she pays and leaves the store, while new customers are added to the rear of the line. Figure 15–12 shows a queue.

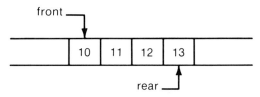

FIGURE 15–12
A queue.

As with stacks, queues can be implemented in various ways. A table could be set up as a queue, as could a linked list. Actually a queue could be a specialized linked list, with pointers to the front and the rear.

15.6.1

Inserting an Element in a Queue

An element is inserted at the rear of the queue by moving the pointer to the rear and inserting the data (Fig. 15–13). However, as with stacks, you

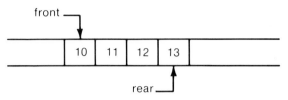

insert an element containing 14 into the queue

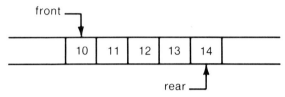

FIGURE 15-13
Inserting an element into a queue.

should first check to be sure there is enough storage left in the queue to add a new element.

The procedure Insert-Queue shows the algorithm necessary to insert an element in a queue, as follows:

```
Insert-queue:   (Queue, Element-to-be-inserted, Full-flag)
      IF (the Queue is full)
            Set the Full-flag to show Queue is full
      ELSE
            REAR ← REAR + 1
            Queue (REAR) ← Element-to-be-inserted
      ENDIF
   END Insert-queue.
```

15.6.2

Deleting from a Queue

The deletion of an element must remove the element that was inserted first in order to meet the definition of a queue. Therefore, the deletion takes place at the front of the queue. Figure 15-14 shows the element at the front (containing 10) being deleted from the queue in Figure 15-13.

The algorithm shown below to delete an element from a queue first checks to be sure that the queue is not empty. Notice that if the pointers to the rear (REAR) and the front (FRONT) point to the same element, there is only one element in the queue.

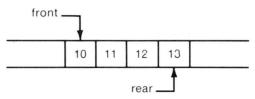

delete the first element in the queue

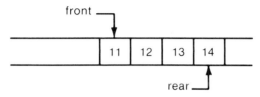

FIGURE 15–14
Deleting an element from a queue.

```
Delete-queue:   (Queue, Element-deleted, Empty-flag)
      IF (the queue is empty)
            Set the Empty-flag to show the queue is empty
      ELSE
            Element-deleted ← Queue (FRONT)
            FRONT ← FRONT + 1
      ENDIF
END Delete-queue.
```

15.6.3

Using
Queues

Queues are used when the data being processed is needed in the order in which it is inserted into the queue. For example, if a procedure is processing messages that have been sent to it from another procedure, the message-processing procedure would probably want to receive the messages in the order in which they were sent. This is a good problem for a queue. The messages would be inserted into the queue when they are sent by the message-sending procedure. Then as the message-processing procedure is ready to handle a new message, it simply removes the message at the front of the queue and processes it.

Figure 15–15 shows a message-processing queue in motion, where messages are inserted into the queue when sent and deleted from the queue when received. The processing involved in Figure 15–15 is shown below using the Insert-queue and Delete-queue procedures.

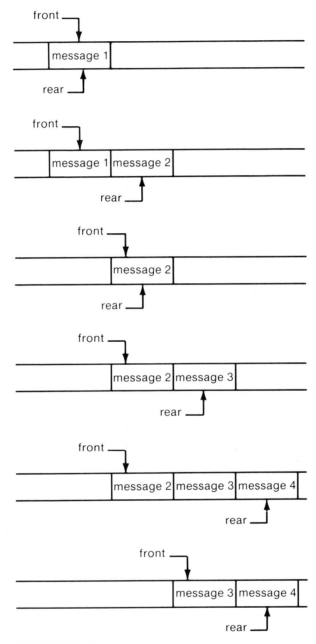

FIGURE 15–15
A queue in motion.

```
CALL Insert-queue (Queue, "message 2", Full-Flag)
CALL Delete-queue (Queue, Message-received, Empty-flag)
CALL Insert-queue (Queue, "message 3", Full-flag)
CALL Insert-queue (Queue, "message 4", Full-flag)
CALL Delete-queue (Queue, Message-received, Empty-flag)
```

15.7

Implementing Stacks Using the CONTROLLED Storage Class

The CONTROLLED storage class differs from both AUTOMATIC and STATIC storage classes in that memory is not reserved for the variable being declared at the time the procedure is entered. For CONTROLLED storage, memory must be specifically requested when needed. The request is made by using the ALLOCATE statement. As you probably have guessed, you declare a variable as CONTROLLED within the DECLARE statement, such as

```
DECLARE MESSAGE  CHAR (25) CONTROLLED;
DECLARE AMOUNT FIXED DEC (5,2) CONTROLLED;
```

When you need memory for the controlled variable, you issue the ALLOCATE statement as follows:

```
ALLOCATE MESSAGE;
ALLOCATE AMOUNT;
```

You may request storage for several variables within the same ALLOCATE statement as

```
ALLOCATE MESSAGE, AMOUNT;
```

Once the storage is allocated for a controlled variable, you can use the variable as you would a variable of any other storage class we have discussed. The controlled variable can be used in an assignment statement, a GET or PUT statement, or the like.

Now you might be wondering what happens to the storage when the procedure in which the controlled variables are declared is exited. Nothing. The storage remains allocated to the variable until you specifically request PL/I to "free" it. This is done by using the FREE statement. Examples:

```
FREE MESSAGE;
FREE AMOUNT;
```

Again, you may request the freeing of storage for several variables within the same FREE statement as

```
FREE MESSAGE, AMOUNT;
```

15.7.1

Errors That Can Occur While Using CONTROLLED Storage

If you try to reference a CONTROLLED variable before you have issued an ALLOCATE statement for that variable, PL/I gives you an error. If you try to FREE a variable for which no ALLOCATE statement was executed, no error occurs (actually no action at all is taken). And, naturally, you must declare the variable to be CONTROLLED in order to be able to issue the ALLOCATE statement using the variable name.

Another error that could occur when you are using the ALLOCATE statement is that you have run out of memory for your program. As you know, when you run your program it is assigned a specific amount of memory. If you try to allocate too much memory, you may exceed the maximum amount assigned to your program. If this happens, PL/I issues an error message.

One thing we must mention, and this is a good point at which to do it, is that PL/C *does not support the CONTROLLED storage class* or any of the keywords (ALLOCATE, FREE) that are used with CONTROLLED storage. Therefore, any programs that you write using the CONTROLLED storage class *must* be written using a full PL/I compiler.

15.7.2

Special Characteristics of the CONTROLLED Storage Class

As we described above, storage is allocated to the variable declared as CONTROLLED when an ALLOCATE statement is executed. This allows us to specify the length of the varible at execution time rather than at compile time. Examine the following DECLARE statement.

```
DECLARE MESSAGE CHAR (N + 1) CONTROLLED;
```

Now if we set N to 20 and issue an ALLOCATE statement, MESSAGE is allocated storage of length 21.

```
N = 20;
ALLOCATE MESSAGE;
```

You may use a similar expression for the bounds of an array. Look at the following PL/I statements.

```
DECLARE COURSES (X) CHAR (12) CONTROLLED;
X = 6;
ALLOCATE COURSES;
```

Now that we have shown you arrays used as controlled variables, we need to examine them further. If you use the array name, as we did above (i.e., COURSES) in the ALLOCATE statement, the entire array is allocated storage. However, if you specify a part of the array in the ALLOCATE statement, only that piece is allocated storage.

```
DECLARE MATRIX (6,7) FIXED DEC (2) CONTROLLED;
ALLOCATE MATRIX;          /* ALLOCATE ENTIRE 6x7 MATRIX */
ALLOCATE MATRIX (2,4);    /* CREATES A 2X4 MATRIX */
```

When the last ALLOCATE is executed, the bounds overrides the bounds used in the DECLARE statement, and a matrix of 2 × 4 is allocated storage.

A characteristic of the CONTROLLED storage class that allows us to implement stacks easily is that when ALLOCATE statements are issued consecutively without the corresponding FREE statements, new "generations" of the controlled variable are created. Let's look at the following example

```
DECLARE MESSAGE CHAR (25) CONTROLLED;
ALLOCATE MESSAGE;          /* CREATE FIRST MESSAGE */
MESSAGE = "THIS IS THE FIRST MESSAGE";
ALLOCATE MESSAGE;          /* CREATE SECOND MESSAGE */
MESSAGE = "THIS IS THE SECOND MESSAGE";
ALLOCATE MESSAGE;          /* CREATE THIRD MESSAGE */
MESSAGE = "THIS IS THE THIRD MESSAGE";
```

Now we have a stack of three messages. When we issue a FREE statement, the *last* message allocated will be freed.

```
FREE MESSAGE;          /* FREE THE THIRD MESSAGE */
FREE MESSAGE;          /* FREE THE SECOND MESSAGE */
FREE MESSAGE;          /* FREE THE FIRST MESSAGE */
```

Using CONTROLLED
Storage to
Implement Stacks

Let us reexamine the stack-processing procedures described above. Here we will show you how to implement the PUSH_STACK and POP_STACK procedures using variables declared as CONTROLLED. When you read the PL/I code, please notice that there is one difference between the PUSH_STACK procedure and the pseudocode shown previously. The difference pertains to the use of the ALLOCATE statement. When you attempt to allocate more memory than you are allowed within your program, an error occurs. Therefore, we do not need to decide, in this case, whether there is enough memory to allocate another element of the stack.

```
MAIN:  PROCEDURES OPTIONS (MAIN);
DECLARE STACK           CHAR (25) CONTROLLED,
        INPUT_ELEMENT   CHAR (25),
        STACK_FLAG      BIT(1);
GET LIST (INPUT_ELEMENT);
CALL PUSH_STACK (STACK, INPUT_ELEMENT);
    .
    .
    .
CALL POP_STACK (STACK, POPPED_ELEMENT, STACK_FLAG);
IF STACK_FLAG = '0'B THEN
    PUT SKIP LIST (POPPED_ELEMENT);
    .
    .
RETURN;
END MAIN;

PUSH_STACK:  PROCEDURE (STACK, ELEMENT_TO_PUSH);
DECLARE STACK           CHAR (*) CONTROLLED,
        ELEMENT_TO_PUSH CHAR (*);
ALLOCATE (STACK);
STACK = ELEMENT_TO_PUSH;
RETURN;
END PUSH_STACK;

POP_STACK:  PROCEDURE (STACK, POPPED_ELEMENT, EMPTY_FLAG);
DECLARE  STACK           CHAR (*) CONTROLLED,
         POPPED_ELEMENT  CHAR(*),
         EMPTY_FLAG      BIT(1);
IF ALLOCATION (STACK) = 0 THEN
```

```
                    EMPTY_FLAG = '1'B;
        ELSE
        DO;
                    POPPED_ELEMENT = STACK;
                    FREE STACK,
        END;
        RETURN;
        END POP_STACK;
```

There is a built-in function within PL/I called ALLOCATION. The function returns a value depending upon whether or not storage was actually allocated for the controlled variable, in our case the stack. A FIXED BINARY (15) number is returned that gives the number of allocations that are available to the active task. Therefore, a zero is returned when no storage was allocated.

15.7.4

Example Program—
The CONTROLLED
Storage Class

The CH15EX1 program (see Fig. 15–16) was written especially to demonstrate the use of stacks in PL/I. The main function of the program is to calculate and print the grade point average (GPA) for an unknown number of students. In order to utilize stacks in this process, the course credit hours and grade are pushed onto a stack until all the courses for a single student have been read. At that time, the courses are popped off the stack one at a time, and the GPA is calculated.

```
PL/I OPTIMIZING COMPILER  VERSION 1 RELEASE 5.1  TIME: 15.34.08  DATE: 10 MAR 87 PAGE   1
OPTIONS USED


PL/I OPTIMIZING COMPILER          /* CH15EX1 -- EXAMPLE PROGRAM -- STACKS */    PAGE   2
                  SOURCE LISTING
   STMT

        /* CH15EX1 -- EXAMPLE PROGRAM -- STACKS */
   1  CH15EX1: PROCEDURE OPTIONS (MAIN);
        /***********************************************************/
        /*                                                       */
        /* FUNCTION: THIS PROGRAM ILLUSTRATES THE USE OF STACKS BY */
```

FIGURE 15–16
CH15EX1 (stacks [CONTROLLED storage])

FIGURE 15-16 *(continued)*

```
/*          READING A STUDENT'S SSN, CREDIT HOURS, AND GRADE*/
/*          FOR EACH COURSE A STUDENT HAS TAKEN.  THESE      */
/*          COURSES ARE PUSHED ONTO A STACK UNTIL ALL THE    */
/*          COURSES FOR A SINGLE STUDENT HAVE BEEN OBTAINED.*/
/*          THE GPA FOR THE STUDENT IS CALCULATED BY POPPING*/
/*          THE COURSES OFF THE STACK ONE AT A TIME AND      */
/*          CALCULATING THE HONOR POINTS AND TOTAL CREDIT    */
/*          HOURS.  WHEN THE STACK IS EMPTY, THE GPA IS      */
/*          PRINTED FOR THE STUDENT.                         */
/*                                                           */
/*  INPUT:  AN UNKNOWN UNUMBER OF LINES EACH CONTAINING      */
/*                STUDENT'S SSN                              */
/*                CREDIT HOURS FOR A COURSE                  */
/*                GRADE RECEIVED IN THE COURSE               */
/*                                                           */
/*  OUTPUT: FOR EACH STUDENT PROCESSED:                      */
/*                STUDENT'S SSN                              */
/*                THE CALCULATED GPA                         */
/*                                                           */
/*************************************************************/


2  DECLARE
     STUDENT_SSN               CHAR     (9),
     STUDENT_CREDIT_HOURS      FIXED DEC (3),
     STUDENT_GRADE             CHAR     (1),

     ALLOCATION                BUILTIN,

     1 POPPED_ELEMENT,
       2 POPPED_CREDIT_HOURS   FIXED DEC (3),
       2 POPPED_GRADE          FIXED DEC (1),

     MORE_ELEMENTS_TO_POP      BIT (1) INIT ('1'B),

     OLD_SSN                   CHAR     (9),

     SYSIN FILE INPUT STREAM,
     SYSPRINT FILE OUTPUT STREAM,

     END_OF_FILE               BIT      (1) INIT ('0'B),
```

```
                NO                    BIT      (1) INIT ('0'B),
                YES                   BIT      (1) INIT ('1'B),
        1   STUDENT_COURSES_STACK     CONTROLLED,
            2   STACK_CREDIT_HOURS    FIXED DEC (3),
            2   STACK_GRADE           FIXED DEC (1),
```

```
            HONOR_POINTS              FIXED DEC (5),
            TOTAL_CREDIT_HOURS        FIXED DEC (5),
            STUDENT_GPA               FIXED DEC (4,3);

    3   ON ENDFILE(SYSIN) END_OF_FILE = YES ;

    4   GET  EDIT (STUDENT_SSN,STUDENT_CREDIT_HOURS,STUDENT_GRADE)
            (COL(1),A(9),COL(10),F(3),A(1));


    5   DO WHILE (END_OF_FILE = NO);
    6       OLD_SSN = STUDENT_SSN;
    7       DO WHILE (OLD_SSN = STUDENT_SSN & END_OF_FILE = NO);

    8           CALL PUSH_STACK;
    9           GET EDIT (STUDENT_SSN,STUDENT_CREDIT_HOURS,STUDENT_GRADE)
                    (COL(1),A(9),COL(10),F(3),A(1));

   10       END;

   11       CALL CALCULATE_GPA;
   12       CALL PRINT_STUDENT;
   13   END;

   14   PUT  SKIP EDIT ('*** ALL STUDENTS PROCESSED ***')
                        (COL(1),A);

   15   RETURN;
```

FIGURE 15–16 *(continued)*

```
16  PUSH_STACK:  PROCEDURE;

    /*****************************************************************/
    /*                                                             */
    /*                    PUSH_STACK                               */
    /*                                                             */
    /* FUNCTION: TO ADD THE NEW COURSE'S INFORMATION TO THE TOP    */
    /*           OF THE STACK.                                     */
    /*                                                             */
    /* GLOBAL VARIABLES:   STUDENT_COURSES_STACK                  */
    /*                                                             */
    /* PARAMETERS:  NONE                                          */
    /*                                                             */
    /* SUBROUTINES CALLED:  NONE                                  */
    /*                                                             */
    /* INPUT:    NONE                                             */
    /*                                                             */
    /* OUTPUT:   NONE                                             */
    /*                                                             */
    /*****************************************************************/

17  ALLOCATE  STUDENT_COURSES_STACK;

18  STACK_CREDIT_HOURS = STUDENT_CREDIT_HOURS;

19  SELECT (STUDENT_GRADE);
20      WHEN ('A') STACK_GRADE = 4;
21      WHEN ('B') STACK_GRADE = 3;
22      WHEN ('C') STACK_GRADE = 2;
23      WHEN ('D') STACK_GRADE = 1;
24      WHEN ('F') STACK_GRADE = 0;
25      OTHERWISE
            PUT SKIP EDIT ('***** ERROR GRADE NOT VALID ****')
                            (COL(1),A);
26  END;

27  RETURN;
28  END PUSH_STACK;
```

    STMT

```
 29  CALCULATE_GPA: PROCEDURE;

     /****************************************************************/
     /*                                                            */
     /*                   CALCULATE_GPA                            */
     /*                                                            */
     /*  FUNCTION: TO CALCULATE THE STUDENT'S GPA FROM THE COURSE  */
     /*            INFORMATION THAT IS IN THE STACK.               */
     /*                                                            */
     /*  GLOBAL VARIABLES:   HONOR_POINTS                          */
     /*                      TOTAL_CREDIT_HOURS                     */
     /*                      MORE_ELEMENTS_TO_POP                   */
     /*                      POPPED_ELEMENT                         */
     /*                                                            */
     /*  PARAMETERS:  NONE                                         */
     /*                                                            */
     /*  SUBROUTINES CALLED:  POP_STACK                            */
     /*                                                            */
     /*  INPUT:    NONE                                            */
     /*                                                            */
     /*  OUTPUT:   NONE                                            */
     /*                                                            */
     /****************************************************************/

 30  HONOR_POINTS = 0;
 31  TOTAL_CREDIT_HOURS =0;
 32  MORE_ELEMENTS_TO_POP = YES;
 33  CALL POP_STACK;

 34  DO WHILE (MORE_ELEMENTS_TO_POP = YES);
 35      HONOR_POINTS = HONOR_POINTS + POPPED_CREDIT_HOURS * POPPED_GRADE;
 36      TOTAL_CREDIT_HOURS = TOTAL_CREDIT_HOURS + POPPED_CREDIT_HOURS;

 37      CALL POP_STACK;
 38  END;

 39  STUDENT_GPA = HONOR_POINTS / TOTAL_CREDIT_HOURS;
 40  RETURN;
```

FIGURE 15–16 *(continued)*

```
PL/I OPTIMIZING COMPILER        /* CH15EX1 -- EXAMPLE PROGRAM -- STACKS */      PAGE   6
   STMT
    41  POP_STACK:  PROCEDURE;


       /**************************************************************/
       /*                                                          */
       /*                   POP_STACK                              */
       /*                                                          */
       /*  FUNCTION:  TO SEE IF THERE IS ANY MORE ELEMENTS ON THE   */
       /*             STACK, AND IF THERE IS REMOVE IT.            */
       /*                                                          */
       /*  GLOBAL VARIABLES:   POPPED_ELEMENT                       */
       /*                      STUDENT_COURSE_STACK                */
       /*                                                          */
       /*  PARAMETERS:  NONE                                        */
       /*                                                          */
       /*  SUBROUTINES CALLED:  NONE                                */
       /*                                                          */
       /*  INPUT:    NONE                                           */
       /*                                                          */
       /*  OUTPUT:   NONE                                           */
       /*                                                          */
       /**************************************************************/
    42  IF ALLOCATION (STUDENT_COURSES_STACK) = 0 THEN
           MORE_ELEMENTS_TO_POP  = NO;
    43  ELSE
           DO ;
    44          POPPED_ELEMENT = STUDENT_COURSES_STACK;
    45          FREE STUDENT_COURSES_STACK;
    46      END;

    47  RETURN;
    48  END POP_STACK;
    49  END CALCULATE_GPA;



PL/I OPTIMIZING COMPILER        /* CH15EX1 -- EXAMPLE PROGRAM -- STACKS */      PAGE   7
   STMT
    50  PRINT_STUDENT:  PROCEDURE;
```

```
     /****************************************************************/
     /*                                                          */
     /*                 PRINT_STUDENT                            */
     /*                                                          */
     /*  FUNCTION: TO PRINT THE STUDENT INFORMATION.             */
     /*                                                          */
     /*  GLOBAL VARIABLES:    OLD_SSN                            */
     /*                       TOTAL_CREDIT_HOURS                 */
     /*                       HONOR_POINTS                       */
     /*                       STUDENT_GPA                        */
     /*                                                          */
     /*  PARAMETERS:  NONE                                       */
     /*                                                          */
     /*  SUBROUTINES CALLED:  NONE                               */
     /*                                                          */
     /*  INPUT:    NONE                                          */
     /*                                                          */
     /*  OUTPUT:   THE STUDENT SSN AND GPA ARE PRINTED.          */
     /*                                                          */
     /****************************************************************/

51   DECLARE
         LINE_COUNT          FIXED DEC (3)  INIT (60) STATIC,
         PAGE_COUNT          FIXED DEC (5)  INIT (0) STATIC;
52   DECLARE
         REPEAT              BUILTIN;

53   IF LINE_COUNT > 55 THEN
         DO;
54           PAGE_COUNT = PAGE_COUNT + 1;
55           PUT FILE (SYSPRINT) PAGE EDIT ('STUDENT GPA REPORT')
                         (COL (21), A);
56           PUT FILE (SYSPRINT) SKIP(2) EDIT
             ('STUDENT SSN', 'TOTAL CREDIT HOURS','HONOR POINT','GPA')
                     (COL(1),A,COL(20),A,COL(43),A,COL(58),A);
57           PUT FILE (SYSPRINT) SKIP(0) EDIT
             (REPEAT('_',11),REPEAT('_',18),REPEAT('_',11),REPEAT('_',3))
             (COL(1),A,COL(20),A,COL(43),A,COL(58),A);

58           PUT FILE (SYSPRINT) SKIP(3);
59           LINE_COUNT = 5;
60       END;
```

FIGURE 15–16 *(continued)*

```
61  PUT FILE (SYSPRINT) EDIT
        (OLD_SSN,TOTAL_CREDIT_HOURS,HONOR_POINTS,STUDENT_GPA)
        (COL(1),A,COL(23),F(5),COL(46),F(5),COL(58),F(5,3));
62  LINE_COUNT = LINE_COUNT + 2;
63  RETURN;
64  END PRINT_STUDENT;
65  END CH15EX1;
```

```
PL/I OPTIMIZING COMPILER       /* CH15EX1 -- EXAMPLE PROGRAM -- STACKS */    PAGE   8
NO MESSAGES PRODUCED FOR THIS COMPILATION
COMPILE TIME    0.00 MINS        SPILL FILE:    0 RECORDS, SIZE  4051
```

                          STUDENT GPA REPORT

STUDENT·SSN	TOTAL CREDIT HOURS	HONOR POINT	GPA
273621893	13	48	3.692
345676543	11	27	2.454
365478906	3	12	4.000
456543219	17	65	3.823

*** ALL STUDENTS PROCESSED ***

15.8

Implementing Queues Using the BASED Storage Class

As we described above, a queue can be implemented using the linked list data structure. When this data structure is used, pointers are needed, pointers between the elements within the linked list or elements of the queue. In this section we describe the BASED storage class and the concept of pointers within PL/I. We use the queue data structure as the example throughout.

15.8.1

POINTER Variables

The POINTER variable is used to hold the address of another variable within PL/I. Such a variable is declared as

```
DECLARE P POINTER;
DECLARE ELEMENT_TO_BE_PUSHED POINTER;
DECLARE ITEM POINTER;
```

You can abbreviate POINTER to PTR if you wish.

A pointer variable is assigned the address of another variable with the use of the ADDR built-in function. The ADDR built-in function returns the location of the variable specified.

```
P = ADDR (MESSAGE);
ELEMENT_TO_BE_PUSHED = ADDR (MESSAGE);
```

The value in P, above, is the address of MESSAGE, *not the value in MESSAGE*. In order to get to the value in MESSAGE through the pointer P we need a new symbol. Examine the PL/I code

```
P→MESSAGE = "Message value";
```

The element of MESSAGE that is located at the address in P is assigned the value "Message value". We will come back to this idea of "qualifying" a variable in a moment.

15.8.2

The BASED
Storage Class

At this point you might be wondering why a pointer is needed. Pointers come in very handy, and are a necessity, when using BASED variables. Variables declared as BASED are not allocated any memory at the time the procedure in which they are declared is entered. Study the DECLARE statement below.

```
DECLARE MESSAGE CHAR (25) BASED (P);
DECLARE P POINTER;
DECLARE INPUT_AREA CHAR (80);
P = ADDR(INPUT_AREA);
PUT LIST (MESSAGE);
```

This DECLARE statement says that PL/I is to use the variable name MESSAGE as an outline (template) for the storage at the location given in P. The code above would print out the first 25 characters of the input area because that is where P is pointing.

In the example above, we declared an actual storage area, INPUT_AREA, and set P to point at it. This is not always necessary. The ALLOCATE statement can be used to assign storage to a BASED variable and return the new storage address in P. Consider the following PL/I code.

```
DECLARE MESSAGE CHAR (25) BASED (P),
        (P,Q,R) POINTER;
ALLOCATE MESSAGE;
MESSAGE = "••• FIRST MESSAGE •••";
Q = P;                          /• SET Q TO THE FIRST MESS •/
ALLOCATE MESSAGE;
MESSAGE = "••• SECOND MESSAGE •••";
R = P;                          /• SET R TO THE SECOND MESS •/
ALLOCATE MESSAGE;
MESSAGE = "••• THIRD MESSAGE •••";
PUT EDIT (Q→MESSAGE, R→MESSAGE, P→MESSAGE)
        (COL(1), A, SKIP, COL(1), A, SKIP, COL(1), A);
```

The following would be the output printed.

```
••• FIRST MESSAGE •••
••• SECOND MESSAGE •••
••• THIRD MESSAGE •••
```

There are a few items we should mention about the code above. First, when an ALLOCATE statement is issued, the storage is assigned and the address is returned in the pointer variable specified in the DECLARE statement. Next, when the BASED variable is used without the pointer, the generation accessed is the one whose address is in the pointer variable. And last, a BASED variable may be qualified by any pointer variable. (Notice, the PUT statement above contains MESSAGE qualified three times by three different pointer variables.)

The FREE statement is used, as with the CONTROLLED variable, to FREE the storage assigned to a BASED variable via an ALLOCATE statement. The FREE statement takes on two forms:

```
FREE MESSAGE;
FREE Q→MESSAGE;
```

The first FREE statement frees the storage pointed to by the pointer variable assigned (in the case above, P). The second FREE statement frees the storage pointed to by Q.

15.8.3

Using BASED Storage with Queues

To illustrate BASED storage used with queues, the first thing we need to do is define an element of the queue that we use in the example. Let's use the message queue described previously as a singly linked list. In this case, each element of the queue is declared as:

```
DECLARE  1  QUEUE ELEMENT BASED(ELEM_PTR),
            2  QUEUE_MESSAGE CHAR (25),
            2  QUEUE_NEXT_PTR PTR;
```

Also needed are the pointers to the front and rear of the queue. The following DECLARE statement is added for these pointers.

```
DECLARE  (FRONT_PTR, REAR_PTR)  POINTER;
```

We now can actually code the procedures for inserting an element in a queue and deleting an element from a queue. The pseudocode shown in the previous section is used.

```
MAIN:  PROCEDURE OPTIONS(MAIN);
DECLARE  1  QUEUE BASED (ELEM_PTR),
            2  QUEUED_MESSAGE    CHAR (25),
            2  NEXT_ELEM_PTR     PTR,
         (ELEM_PTR, FRONT_PTR, REAR_PTR)  PTR;
DECLARE  INPUT_MESSAGE  CHAR(25),
         QUEUE_FLAG       BIT(1),
         MESSAGE_DELETED CHAR (25);
GET LIST (INPUT_MESSAGE);
CALL INSERT_QUEUE (REAR_PTR, INPUT_MESSAGE);
  .
  .
  .

CALL DELETE_QUEUE (FRONT_PTR, MESSAGE_DELETED, QUEUE_FLAG);
PUT SKIP LIST (MESSAGE-DELETED);
  .
  .

RETURN;
INSERT_QUEUE:  PROCEDURE (REAR_PTR, ELEMENT_TO_BE_INSERTED);
DECLARE  REAR_PTR                   POINTER,
         ELEMENT_TO_BE_INSERTED    CHAR (*);
ALLOCATE QUEUE;
MESSAGE = ELEMENT_TO_BE_INSERTED;
NEXT_ELEM_PTR = NULL;
REAR_PTR→NEXT_ELEM_PTR = ELEM_PTR;  /* INSERT ELEMENT */
REAR_PTR = ELEM_PTR;                /* SET REAR POINTER TO LAST */
                                    /* ELEMENT                 */
RETURN;
END INSERT_QUEUE;
```

```
DELETE_QUEUE:    PROCEDURE    (FRONT_PTR,    DELETED_ELEM_PTR,
                              EMPTY_FLAG);
DECLARE    DELETED_ELEM                CHAR (*),
           FRONT_PTR                   POINTER,
           EMPTY_FLAG                  BIT (1);


IF ALLOCATION(QUEUE) = 0 THEN
    EMPTY_FLAG = '1'B;
ELSE
    DO;
        DELETED_ELEM = FRONT_PTR→MESSAGE;
        ELEM_PTR = FRONT_PTR;
        FRONT_PTR = ELEM_PTR→NEXT_ELEM_PTR;
        FREE QUEUE;
    END;
RETURN;
END DELETE_QUEUE;
```

We are sure you noticed that there is a difference between the pseudocode and the actual PL/I code. This is due to the fact that the pseudocode was
meant to describe the conceptual idea of insertion and deletion. However,
the PL/I code describes an actual implementation using BASED variables.
This is not the only way to insert or delete from a queue implemented as a
linked list. It is just one way.

15.8.4

*Implementing Trees
Using BASED Storage*

The binary tree shown in Figure 15–7 can be implemented using a BASED
storage class variable. Examine the following PL/I code.

```
DECLARE   1 TREE_NODE BASED (NODE_PTR),
          2 TREE_NAME       CHAR (10),
          2 TREE_LEFT_CHILD POINTER,
          2 TREE_RIGHT_CHILD POINTER;
```

A node in the tree can be allocated and the appropriate pointers set
when an insertion is needed. The deletion simply sets the appropriate
pointer to NULL. The following PL/I code implements the preorder traversal
routine shown above. You will notice that the PREORDER procedure is
RECURSIVE.

```
PREORDER:  PROCEDURE (ROOT_PTR) RECURSIVE;
DECLARE    ROOT_PTR POINTER;
DECLARE  1 TREE_NODE BASED (ROOT_PTR),
            2 TREE_NAME        CHAR (10),
            2 TREE_LEFT_CHILD POINTER,
            2 TREE_RIGHT_CHILD POINTER;
IF ROOT_PTR ^= NULL THEN
    DO;
         CALL PROCESS_THE_ROOT (ROOT_PTR);
         CALL PREORDER (TREE_LEFT_CHILD);
         CALL PREORDER (TREE_RIGHT_CHILD);
    END;
RETURN;
END PREORDER;
```

15.8.5

*Example Program—
BASED Storage*

The CH15EX2 program (see Fig. 15–17) uses BASED storage in producing a course request report for a university. Every semester the departments need to know how many students are going to request their courses prior to the actual registration. Therefore, a course request report is produced. The courses are read into an array. Attached to each element of the array are two linked lists. The first is the list of students actually registered in the course. The second is the list of the students who requested the course but because the course was full could not be registered in it. This is a "wait queue." The students will be added to the wait queue in the order in which the request came through, since that is the proper order for registration. After all the requests are processed, a report is produced showing the students actually registered and the students, in order, in the wait queue.

```
PL/I OPTIMIZING COMPILER  VERSION 1 RELEASE 5.1  TIME: 11.04.52  DATE: 13 MAR 87 PAGE   1
OPTIONS SPECIFIED


PL/I OPTIMIZING COMPILER  /*  CH15EX2 -- LINKED LISTS AND BASED STORAGE   */   PAGE   2
                  SOURCE LISTING
    STMT

        /*  CH15EX2 -- LINKED LISTS AND BASED STORAGE    */
```

FIGURE 15–17
CH15EX2 (queues [BASED storage])

FIGURE 15–17 *(continued)*

```
1  CH15EX2:  PROCEDURE OPTIONS (MAIN);
   /*****************************************************************/
   /*                                                             */
   /* FUNCTION:   THIS PROGRAM REGISTERS STUDENTS IN CLASSES.     */
   /*             THE COURSES ARE READ INTO AN ARRAY.  THE STUDENTS*/
   /*             REQUEST THE COURSE THEY WISH THROUGH COURSE      */
   /*             REQUEST RECORDS.  WHEN A COURSE REQUEST RECORD IS*/
   /*             PROCESSED, ONE OF TWO ACTIONS WILL TAKE PLACE.   */
   /*             EITHER A STUDENT IS REQUESTING TO ADD A COURSE   */
   /*             OR TO DROP A COURSE.  IF A STUDENT IS ASKING TO  */
   /*             ADD A COURSE AND THE COURSE IS NOT FULL, THE     */
   /*             STUDENT IS ADDED TO THE COURSE'S REGISTERED LIST.*/
   /*             IF THE COURSE IS FULL, THE STUDENT IS ADDED TO   */
   /*             COURSE'S WAIT QUEUE.                             */
   /*             WHEN A DROP REQUEST COMES THROUGH, A STUDENT     */
   /*             IS REMOVED FROM THE REGISTERED LIST.   IF THERE  */
   /*             ARE ANY STUDENTS AWAITNG REGISTRATION, THE FIRST */
   /*             ONE IN THE WAIT QUEUE IS MOVED TO THE            */
   /*             REGISTRATION LIST.                               */
   /*                                                             */
   /* SUBROUTINES CALLED:   COMPLETE_COURSE_FILE                  */
   /*                       READ_REQUEST_FILE                     */
   /*                       PROCESS_ADD_REQUEST                   */
   /*                       PROCESS_DROP_REQUEST                  */
   /*                       WRITE_REQUEST_REPORT                  */
   /*                                                             */
   /* INPUT:   A FILE CONTAINING ONE RECORD FOR EACH COURSE       */
   /*              THAT IS BEING OFFERED                          */
   /*          A REQUEST FILE CONTAINING ONE RECORD FOR EACH      */
   /*              ADD OR DROP REQUEST SUBMITTED BY A STUDENT      */
   /*                                                             */
   /* OUTPUT:   A REPORT SHOWING THE NUMBER OF STUDENTS THAT      */
   /*              HAVE REQUESTED A COURSE                        */
   /*              HAVE BEEN REGISTERED IN THE COURSE             */
   /*              HAVE BEEN ADDED TO THE WAITING QUEUE           */
   /*                                                             */
   /*****************************************************************/

2  DECLARE
       1 COURSE_ARRAY (200),
```

```
                2 COURSE_ID                  CHAR (8),
                2 COURSE_TITLE               CHAR (30),
                2 COURSE_CREDIT_HOURS        CHAR (4),
                2 COURSE_LIMIT               PIC '999',
                2 COURSE_NUMBER_REGISTERED   PIC '999',
                2 COURSE_STUDENT_REGISTERED  POINTER,
                2 COURSE_NUMBER_WAITING      PIC '999',
                2 COURSE_STUDENT_WAIT_QUEUE,
                    3 COURSE_WAIT_REAR       POINTER,
                    3 COURSE_WAIT_FRONT      POINTER;
```

```
    3  DECLARE
            1 STUDENT_COURSE_REQUEST,
               2 STUDENT_REQUEST_INDICATOR CHAR(1),
               2 STUDENT_SSN               CHAR(9),
               2 STUDENT_NAME              CHAR(30),
               2 STUDENT_REQUESTED_COURSE  CHAR(8),
               2 STUDENT_CLASSIFICATION    CHAR(1);


    4  DECLARE
            QUEUE_PTR                       POINTER,
            1 COURSE_QUEUE_ELEMENT BASED(QUEUE_PTR),
               2 COURSE_QUEUE_STUDENT_SSN CHAR(9),
               2 COURSE_QUEUE_STUDENT_NAME CHAR(30),
               2 COURSE_QUEUE_STUDENT_CLASS CHAR(1),
               2 COURSE_QUEUE_ELEMENT_NEXT POINTER;


    5  DECLARE
            NUM_OF_COURSES                  FIXED DEC (3) INIT (0),
            REQUEST_FILE_EOF_FLAG           BIT(1) INIT ('0'B),
               MORE_RECORDS                 BIT(1) INIT ('0'B);

    6  DECLARE
            (NULL)                          BUILTIN;


    7  CALL COMPLETE_COURSE_FILE;

    8  CALL READ_REQUEST_FILE;
```

FIGURE 15–17 *(continued)*

```
 9  DO WHILE (REQUEST_FILE_EOF_FLAG = MORE_RECORDS);

10     IF STUDENT_REQUEST_INDICATOR = 'A' THEN
            CALL PROCESS_ADD_REQUEST;
11     ELSE
            CALL PROCESS_DROP_REQUEST;

12     CALL READ_REQUEST_FILE;
13  END;

14  CALL WRITE_REQUEST_REPORT;

15  RETURN;
```

```
PL/I OPTIMIZING COMPILER  /*  CH15EX2 -- LINKED LISTS AND BASED STORAGE    */   PAGE   4
   STMT
   16  COMPLETE_COURSE_FILE:  PROCEDURE;
       /****************************************************************/
       /*                                                            */
       /*                  COMPLETE_COURSE_FILE                       */
       /*                                                            */
       /* FUNCTION: TO COMPLETE THE COURSE ARRAY SO THAT THE COURSE */
       /*           REQUEST REPORT CAN BE PROCESSED.  A RECORD IS     */
       /*           READ FOR EACH COURSE AND ADDED TO THE ARRAY.      */
       /*                                                            */
       /* PARAMETERS: NONE                                            */
       /*                                                            */
       /* GLOBAL VARIABLES:   COURSE_ARRAY                            */
       /*                     NUM_OF_COURSES                          */
       /*                                                            */
       /* SUBROUTINES CALLED:   NONE                                  */
       /*                                                            */
       /* INPUT:    THE COURSE RECORDS                                */
       /*                                                            */
       /* OUTPUT:   NONE                                              */
       /*                                                            */
       /****************************************************************/

   17  DECLARE
```

```
          I                         FIXED BIN (15),
        CRSEFILE FILE INPUT RECORD,
        CRSE_EOF_FLAG               BIT(1) INIT ('0'B),
            END_OF_FILE             BIT(1) INIT ('1'D),
            MORE_RECORDS            BIT(1) INIT ('0'B);

 18  DECLARE
        1 COURSE_INPUT_RECORD,
          2 COURSE_INPUT_ID     CHAR(8),
          2 COURSE_INPUT_TITLE   CHAR(30),
          2 COURSE_INPUT_HOURS   CHAR(4),
          2 COURSE_INPUT_LIMIT   PIC '999';

 19  DECLARE
        STRING                  BUILTIN;

 20  ON ENDFILE (CRSEFILE)
        CRSE_EOF_FLAG = END_OF_FILE;

 21  READ FILE (CRSEFILE) INTO (COURSE_INPUT_RECORD);

 22  DO I = 1 TO 200 WHILE (CRSE_EOF_FLAG = MORE_RECORDS);

 23      COURSE_ID (I) = COURSE_INPUT_ID;
 24      COURSE_TITLE (I) = COURSE_INPUT_TITLE;
 25      COURSE_CREDIT_HOURS (I) = COURSE_INPUT_HOURS;
 26      COURSE_LIMIT (I) = COURSE_INPUT_LIMIT;
 27      COURSE_NUMBER_REGISTERED (I) = 0;
 28      COURSE_STUDENT_REGISTERED (I) = NULL;
 29      COURSE_NUMBER_WAITING (I) = 0;
 30      COURSE_STUDENT_WAIT_QUEUE (I) = NULL;
```

```
    STMT
 31      READ FILE (CRSEFILE) INTO (COURSE_INPUT_RECORD);

 32  END;

 33  NUM_OF_COURSES = I - 1; /* SUBSCRIPT INCREMENTED BEFORE EOF CHECKED */
                                                    /* <-- TN */

 34  RETURN;
 35  END COMPLETE_COURSE_FILE;
```

FIGURE 15–17 *(continued)*

```
PL/I OPTIMIZING COMPILER  /*  CH15EX2 -- LINKED LISTS AND BASED STORAGE    */   PAGE   6
   STMT
     36  READ_REQUEST_FILE: PROCEDURE;

          /**************************************************************/
          /*                                                          */
          /*                    READ_REQUEST_FILE                     */
          /*                                                          */
          /*  FUNCTION: TO READ THE STUDENT'S COURSE REQUEST RECORDS. */
          /*                                                          */
          /*  PARAMETERS:  NONE                                       */
          /*                                                          */
          /*  GLOBAL VARIABLES:   STUDENT_REQUEST                     */
          /*                       REQUEST_EOF_FLAG                    */
          /*                                                          */
          /*  SUBROUTINES CALLED:   NONE                              */
          /*                                                          */
          /*  INPUT:    THE STUDENT COURSE REQUEST RECORDS            */
          /*                                                          */
          /*  OUTPUT:  NONE                                           */
          /*                                                          */
          /**************************************************************/

     37  DECLARE
             MORE_RECORDS                BIT(1) INIT ('0'B),
             END_OF_FILE                 BIT(1) INIT ('1'B),
             REQFILE FILE INPUT RECORD;

     38  ON ENDFILE (REQFILE)
             REQUEST_FILE_EOF_FLAG = END_OF_FILE;

     39  READ FILE (REQFILE) INTO (STUDENT_COURSE_REQUEST);

     40  RETURN;
     41  END READ_REQUEST_FILE;


PL/I OPTIMIZING COMPILER  /*  CH15EX2 -- LINKED LISTS AND BASED STORAGE    */   PAGE   7
   STMT
     42  PROCESS_ADD_REQUEST:  PROCEDURE;
```

```
/******************************************************************/
/*                                                                */
/*                    PROCESS_ADD_REQUEST                         */
/*                                                                */
/*  FUNCTION: TO PROCESS A STUDENT REQUEST RECORD THAT IS         */
/*            REQUESTING TO ADD A COURSE.                         */
/*                                                                */
/*  PARAMETERS:    NONE                                           */
/*                                                                */
/*  GLOBAL VARIABLES:    STUDENT_COURSE_REQUEST                   */
/*                       COURSE_ARRAY                             */
/*                       NUM_OF_COURSES                           */
/*                                                                */
/*  SUBROUTINES CALLED:    PROCESS_ADD_WAIT                       */
/*                         PROCESS_ADD_REGISTERED                 */
/*                                                                */
/*  INPUT:    NONE                                                */
/*                                                                */
/*  OUTPUT:    AN ERROR MESSAGE IF THE COURSE ID CANNOT BE        */
/*             FOUND IN THE COURSE ARRAY                          */
/*                                                                */
/******************************************************************/

43  DECLARE
       I                        FIXED BIN (15);


44  DO I = 1 TO NUM_OF_COURSES
          WHILE (COURSE_ID (I) ^= STUDENT_REQUESTED_COURSE);
45  END;

46  IF COURSE_ID (I) ^= STUDENT_REQUESTED_COURSE THEN
       PUT SKIP LIST ('**** ERROR COURSE ID INVALID ***',
              STUDENT_REQUESTED_COURSE);
47  ELSE
       DO;
48        IF COURSE_NUMBER_REGISTERED (I) = COURSE_LIMIT(I) THEN
              CALL PROCESS_ADD_WAIT;
49        ELSE
              CALL PROCESS_ADD_REGISTERED;
50     END;
```

FIGURE 15-17 *(continued)*

```
    51  RETURN;


PL/I OPTIMIZING COMPILER   /*  CH15EX2 -- LINKED LISTS AND BASED STORAGE    */   PAGE   8
    STMT
    52  PROCESS_ADD_REGISTERED:  PROCEDURE;


        /*****************************************************************/
        /*                                                             */
        /*                 PROCESS_ADD_REGISTERED                      */
        /*                                                             */
        /* FUNCTION: TO REGISTER A STUDENT IN THE COURSE BY ADDING     */
        /*           THE STUDENT'S INFORMATION TO THE COURSE ARRAY.    */
        /*                                                             */
        /* PARAMETERS:   NONE                                          */
        /*                                                             */
        /* GLOBAL VARIABLES:   COURSE_ELEMENT                          */
        /*                     STUDENT_COURSE_REQUEST                  */
        /*                                                             */
        /* SUBROUTINES CALLED:   NONE                                  */
        /*                                                             */
        /* INPUT:   NONE                                               */
        /*                                                             */
        /* OUTPUT:  NONE                                               */
        /*                                                             */
        /*****************************************************************/


    53  DECLARE
            (Q,R)                         POINTER;

    54  ALLOCATE COURSE_QUEUE_ELEMENT;

    55  COURSE_QUEUE_STUDENT_SSN = STUDENT_SSN;
    56  COURSE_QUEUE_STUDENT_NAME = STUDENT_NAME;
    57  COURSE_QUEUE_STUDENT_CLASS = STUDENT_CLASSIFICATION;
    58  COURSE_QUEUE_ELEMENT_NEXT = NULL;

    59  IF COURSE_STUDENT_REGISTERED (I) = NULL THEN
            COURSE_STUDENT_REGISTERED (I) = QUEUE_PTR;
    60  ELSE
```

```
          DO;
 61            Q = COURSE_STUDENT_REGISTERED (I);

 62            DO WHILE (Q ^= NULL);
 63                R = Q;
 64                Q = Q->COURSE_QUEUE_ELEMENT_NEXT;
 65            END;

 66            R->COURSE_QUEUE_ELEMENT_NEXT = QUEUE_PTR;
 67        END;
 68 COURSE_NUMBER_REGISTERED (I) =
            COURSE_NUMBER_REGISTERED (I) + 1;
 69 RETURN;
 70 END PROCESS_ADD_REGISTERED;
```

```
 71 PROCESS_ADD_WAIT:  PROCEDURE;
    /****************************************************************/
    /*                                                            */
    /*                 PROCESS_ADD_WAIT                           */
    /*                                                            */
    /* FUNCTION: TO ADD A STUDENT'S INFORMATION TO THE WAIT        */
    /*           QUEUE FOR THE REQUESTED COURSE.                   */
    /*                                                            */
    /* PARAMETERS:    NONE                                        */
    /*                                                            */
    /* GLOBAL VARIABLES:   COURSE_ELEMENT                         */
    /*                     STUDENT_COURSE_REQUEST                 */
    /*                                                            */
    /* SUBROUTINES CALLED:    NONE                               */
    /*                                                            */
    /* INPUT:    NONE                                             */
    /*                                                            */
    /* OUTPUT:   NONE                                             */
    /*                                                            */
    /****************************************************************/

 72 ALLOCATE COURSE_QUEUE_ELEMENT;

 73 COURSE_QUEUE_STUDENT_SSN = STUDENT_SSN;
```

FIGURE 15–17 *(continued)*

```
74  COURSE_QUEUE_STUDENT_NAME = STUDENT_NAME;
75  COURSE_QUEUE_STUDENT_CLASS = STUDENT_CLASSIFICATION;
76  COURSE_QUEUE_ELEMENT_NEXT = NULL;

77  IF COURSE_WAIT_REAR (I) ^= NULL THEN
        COURSE_WAIT_REAR(I)->COURSE_QUEUE_ELEMENT_NEXT=QUEUE_PTR;
78  COURSE_WAIT_REAR (I) = QUEUE_PTR;

79  IF COURSE_WAIT_FRONT (I) = NULL THEN
        COURSE_WAIT_FRONT (I) = QUEUE_PTR;
80  COURSE_NUMBER_WAITING (I) = COURSE_NUMBER_WAITING (I) + 1;
81  RETURN;
82  END PROCESS_ADD_WAIT;


83  END PROCESS_ADD_REQUEST;


PL/I OPTIMIZING COMPILER  /*  CH15EX2 -- LINKED LISTS AND BASED STORAGE    */   PAGE  10
  STMT
    84  PROCESS_DROP_REQUEST:  PROCEDURE;
        /***************************************************************/
        /*                                                           */
        /*                     PROCESS_DROP_REQUEST                   */
        /*                                                           */
        /*  FUNCTION: TO PROCESS A STUDENT'S REQUEST TO DROP A COURSE.*/
        /*                                                           */
        /*  PARAMETERS:    NONE                                      */
        /*                                                           */
        /*  GLOBAL VARIABLES:    STUDENT_COURSE_REQUEST              */
        /*                       COURSE_ARRAY                        */
        /*                       NUM_OF_COURSES                      */
        /*                                                           */
        /*  SUBROUTINES CALLED:   DELETE_LINK_LIST                   */
        /*                                                           */
        /*  INPUT:    NONE                                           */
        /*                                                           */
        /*  OUTPUT:   AN ERROR MESSAGE WHEN THE COURSE ID IN NOT IN  */
        /*                THE COURSE ARRAY.                          */
        /*                                                           */
        /***************************************************************/
```

```
85  DECLARE
        (P,Q)                           POINTER,
        I                               FIXED BIN (15);
86  DO I = 1 TO NUM OF COURSES
            WHILE (COURSE_ID (I) ^= STUDENT_REQUESTED_COURSE);
87  END;

88  IF COURSE_ID (I) ^= STUDENT_REQUESTED_COURSE THEN
        PUT FILE (SYSPRINT) SKIP LIST
          ('**** ERROR COURSE ID INVALID ***', STUDENT_REQUESTED_COURSE);
89  ELSE                                                        /* <-- TN */
        DO;
90          QUEUE_PTR = COURSE_STUDENT_REGISTERED (I);
91          Q = NULL;
92          DO WHILE (COURSE_QUEUE_ELEMENT_NEXT ^= NULL &
                COURSE_QUEUE_STUDENT_SSN ^= STUDENT_SSN);

93              Q = QUEUE_PTR;
94              QUEUE_PTR = COURSE_QUEUE_ELEMENT_NEXT;
95          END;

96          IF COURSE_QUEUE_STUDENT_SSN = STUDENT_SSN THEN
                CALL DELETE_LINK_LIST;
97          ELSE
                PUT FILE (SYSPRINT) SKIP LIST
                    ('*** STUDENT NOT REGISTERED - DROPPING ***');

98      END;
99  RETURN;


PL/I OPTIMIZING COMPILER /* CH15EX2 -- LINKED LISTS AND BASED STORAGE   */  PAGE  11
 STMT
 100 DELETE_LINK_LIST:  PROCEDURE;
     /****************************************************************/
     /*                                                            */
     /*                 DELETE_LINK_LIST                           */
     /*                                                            */
     /* FUNCTION: TO DELETE A STUDENT'S INFORMATION FROM THE       */
     /*           LIST OF STUDENTS REGISTERED FOR A COURSE.        */
     /*                                                            */
     /* PARAMETERS:    NONE  FILL IN                               */
```

FIGURE 15–17 *(continued)*

```
      /*                                                         */
      /*  GLOBAL VARIABLES:   COURSE_ELEMENT                     */
      /*                      STUDENT_COURSE_REQUEST             */
      /*                                                         */
      /*  SUBROUTINES CALLED:   NONE                             */
      /*                                                         */
      /*  INPUT:   NONE                                          */
      /*                                                         */
      /*  OUTPUT:   NONE                                         */
      /*                                                         */
      /***********************************************************/

 101  DECLARE
          (P, R)                        POINTER;
 102  IF Q ^= NULL THEN
          Q -> COURSE_QUEUE_ELEMENT_NEXT = COURSE_QUEUE_ELEMENT_NEXT;
 103  ELSE
          COURSE_STUDENT_REGISTERED (I) = COURSE_QUEUE_ELEMENT_NEXT;
 104  COURSE_NUMBER_REGISTERED (I) = COURSE_NUMBER_REGISTERED (I) - 1;
 105  FREE COURSE_QUEUE_ELEMENT;

 106  IF COURSE_WAIT_FRONT (I) ^= NULL THEN
          DO;
 107          P = COURSE_WAIT_FRONT (I);
 108          COURSE_WAIT_FRONT (I) = P -> COURSE_QUEUE_ELEMENT_NEXT;

 109          IF COURSE_WAIT_FRONT (I) = NULL THEN
                  COURSE_WAIT_REAR (I) = NULL;
 110          COURSE_NUMBER_WAITING (I) = COURSE_NUMBER_WAITING (I) - 1;

 111          IF COURSE_STUDENT_REGISTERED (I) = NULL THEN
                  COURSE_STUDENT_REGISTERED (I) = P;
 112          ELSE
                  DO;
 113                  Q = COURSE_STUDENT_REGISTERED (I);
 114                  DO WHILE (Q ^= NULL);
```

PL/I OPTIMIZING COMPILER /* CH15EX2 -- LINKED LISTS AND BASED STORAGE   */   PAGE 12
  STMT

```
115                  R = Q;
116                  Q = Q -> COURSE_QUEUE_ELEMENT_NEXT;
117              END;
118              R -> COURSE_QUEUE_ELEMENT_NEXT = P;
119          END;

120      P -> COURSE_QUEUE_ELEMENT_NEXT = NULL;
121      COURSE_NUMBER_REGISTERED (I) =
                 COURSE_NUMBER_REGISTERED (I) + 1;
122    END;
123 END DELETE_LINK_LIST;
124 END PROCESS_DROP_REQUEST;
```

```
  STMT
  125 WRITE_REQUEST_REPORT:    PROCEDURE;


      /****************************************************************/
      /*                                                            */
      /*                  WRITE_REQUEST_REPORT                      */
      /*                                                            */
      /* FUNCTION: TO PRINT THE REPORT THAT SHOWS ALL THE COURSES   */
      /*           AND THE STATUS OF ALL THE STUDENTS IN THE        */
      /*           COURSES.  THE NUMBER OF STUDENTS REGISTERED AND  */
      /*           THEIR NAMES ALONG WITH THE NUMBER IN THE WAIT    */
      /*           QUEUE AND THEIR NAMES.                           */
      /*                                                            */
      /* PARAMETERS:    NONE                                        */
      /*                                                            */
      /* GLOBAL VARIABLES:   COURSE_ARRAY                           */
      /*                     COURSE_QUEUE_ELEMENT                   */
      /*                                                            */
      /* SUBROUTINES CALLED:   PRINT_HEADINGS                       */
      /*                       PRINT_STUDENTS                       */
      /*                                                            */
      /* INPUT:    NONE                                             */
      /*                                                            */
      /* OUTPUT:   THE COURSE AND STUDENT INFORMATION               */
      /*                                                            */
      /****************************************************************/
```

FIGURE 15–17 *(continued)*

```
126  DECLARE
        I                              FIXED BIN(15);


127  DO I = 1 TO NUM_OF_COURSES;

128     CALL PRINT_HEADINGS;

129     IF COURSE_STUDENT_REGISTERED (I) = NULL THEN
            PUT SKIP(2) LIST ('*** NO STUDENTS REGISTERED ***');
130     ELSE
           DO;
131             PUT EDIT ('LIST OF STUDENTS THAT ARE REGISTERED')
                   (SKIP(2),COL(1),A);

132             QUEUE_PTR = COURSE_STUDENT_REGISTERED (I);
133             CALL PRINT_STUDENTS;
134         END;

135     PUT SKIP(2);
136     IF COURSE_WAIT_FRONT (I) = NULL THEN
            PUT EDIT ('*** NO STUDENTS WAITING ***')
                   (COL(1),A);
137     ELSE
           DO;
138             PUT EDIT ('LIST OF STUDENTS IN THE WAIT QUEUE')
                   (SKIP(2),COL(1),A);
```

```
PL/I OPTIMIZING COMPILER  /*  CH15EX2 -- LINKED LISTS AND BASED STORAGE    */   PAGE  14
   STMT
   139             QUEUE_PTR = COURSE_WAIT_FRONT (I);
   140             CALL PRINT_STUDENTS;
   141         END;
   142  END;
   143  RETURN;
```

```
PL/I OPTIMIZING COMPILER  /*  CH15EX2 -- LINKED LISTS AND BASED STORAGE    */   PAGE  15
   STMT
```

```
144  PRINT_HEADINGS:  PROCEDURE;
     /*****************************************************************/
     /*                                                            */
     /*                   PRINT_HEADINGS                           */
     /*                                                            */
     /* FUNCTION:  TO PRINT THE HEADINGS FOR THE COURSE REQUEST    */
     /*            REPORT.                                         */
     /*                                                            */
     /* PARAMETERS:   NONE                                        */
     /*                                                            */
     /* GLOBAL VARIABLES:   NONE                                  */
     /*                                                            */
     /* SUBROUTINES CALLED:   NONE                                */
     /*                                                            */
     /* INPUT:   NONE                                             */
     /*                                                            */
     /* OUTPUT:   THE HEADINGS ARE PRINTED ON THE REPORT.         */
     /*                                                            */
     /*****************************************************************/

145  DECLARE
         PAGE_COUNT                    FIXED DEC (5) INIT (0) STATIC;

146  PAGE_COUNT = PAGE_COUNT + 1;

147  PUT PAGE EDIT ('STUDENT COURSE REQUEST REPORT')
                   (COL(45),A)
                   ('PAGE: ',PAGE_COUNT) (COL(1),A,F(5))
                   ('ID','TITLE','HOURS','LIMIT','REGISTERED','WAITING')
                   (SKIP(2),COL(1),A,COL(10),A,COL(45),A,COL(55),A,COL(65),
                       A,COL(80),A);                          /* <-- TN */

148  PUT EDIT (COURSE_ID (I), COURSE_TITLE (I),
         COURSE_CREDIT_HOURS (I), COURSE_LIMIT (I),
         COURSE_NUMBER_REGISTERED (I), COURSE_NUMBER_WAITING (I))
         (SKIP(2),COL(1),A,COL(10),A,COL(45),A,COL(55),P'ZZ9',
             COL(65),P'ZZ9',COL(80),P'ZZ9');

149  PUT EDIT ('STUDENT SSN','STUDENT NAME','STUDENT CLASS')
             (COL(1),A,COL(25),A,COL(60),A);

150  RETURN;
```

FIGURE 15-17 *(continued)*

```
151  END PRINT_HEADINGS;
```

```
PL/I OPTIMIZING COMPILER  /*  CH15EX2 -- LINKED LISTS AND BASED STORAGE    */   PAGE  16
   STMT
   152  PRINT_STUDENTS:  PROCEDURE;
        /****************************************************************/
        /*                                                            */
        /*                    PRINT_STUDENTS                          */
        /*                                                            */
        /*  FUNCTION: TO PERFORM THE PRINTING OF THE STUDENT          */
        /*            INFORMATION.                                    */
        /*                                                            */
        /*  PARAMETERS:   NONE                                        */
        /*                                                            */
        /*  GLOBAL VARIABLES:   COURSE_QUEUE_ELEMENT                  */
        /*                                                            */
        /*  SUBROUTINES CALLED:   NONE                                */
        /*                                                            */
        /*  INPUT:   NONE                                             */
        /*                                                            */
        /*  OUTPUT:   THE STUDENT INFORMATION IS PRINTED.             */
        /*                                                            */
        /****************************************************************/
   153  DO WHILE (QUEUE_PTR ^= NULL);
   154     PUT SKIP EDIT (COURSE_QUEUE_STUDENT_SSN,
                           COURSE_QUEUE_STUDENT_NAME,
                           COURSE_QUEUE_STUDENT_CLASS)
                          (COL(1),A,COL(25),A,COL(60),A);
   155     QUEUE_PTR = COURSE_QUEUE_ELEMENT_NEXT;
   156  END;
   157  RETURN;
   158  END PRINT_STUDENTS;
   159  END WRITE_REQUEST_REPORT;
   160  END CH15EX2;
```

```
PL/I OPTIMIZING COMPILER  /*  CH15EX2 -- LINKED LISTS AND BASED STORAGE    */   PAGE  17
COMPILER DIAGNOSTIC MESSAGES
ERROR ID L   STMT    MESSAGE DESCRIPTION
```

WARNING DIAGNOSTIC MESSAGES

IEL0966I W 1   COMPILER RESTRICTION.   EXTERNAL NAME 'CRSEFILE' TOO LONG.   NAME SHORTENED TO 7 CHARACTERS.

COMPILER INFORMATORY MESSAGES

IEL0533I I   NO 'DECLARE' STATEMENT(S) FOR 'SYSPRINT'.

END OF COMPILER DIAGNOSTIC MESSAGES


```
                              STUDENT COURSE REQUEST REPORT
PAGE:    1

ID     TITLE                   HOURS    LIMIT    REGISTERED    WAITING

ART 200  ART APRECIATION        3.0      10        1             0
STUDENT SSN          STUDENT NAME                  STUDENT CLASS


LIST OF STUDENTS THAT ARE REGISTERED
234565438            ANNA SMITH                       S
*** NO STUDENTS WAITING ***
---------------------------------------------------------------------------
                              STUDENT COURSE REQUEST REPORT
PAGE:    2

ID     TITLE                   HOURS    LIMIT    REGISTERED    WAITING

ART 301  ADVANCED SCULPTURE     4.0       3        3             2
STUDENT SSN          STUDENT NAME                  STUDENT CLASS

LIST OF STUDENTS THAT ARE REGISTERED
245655435            ROBERT JOHNSON                   J
345467649            SAMATHA ANDERSON                 J
565743452            KARA JONES                       R

LIST OF STUDENTS IN THE WAIT QUEUE
579849483            HARRY NEWTON                     J
786665567            JULIE THOMPSON                   J
---------------------------------------------------------------------------
```

FIGURE 15–17 *(continued)*

```
                                STUDENT COURSE REQUEST REPORT
PAGE:     3

ID       TITLE                          HOURS    LIMIT    REGISTERED    WAITING

ENGL100A FRESHMAN ENGLISH                3.0      120         0            0
STUDENT SSN              STUDENT NAME                    STUDENT CLASS

*** NO STUDENTS REGISTERED ***

*** NO STUDENTS WAITING ***
```

15.9

Summary

A table is a linear list that is implemented as an array. A linked list is a list in which the elements are linked together via pointers. A tree is a special kind of linked list in which each node points to its children. The top node of the tree is called the root and has no parent. Each node of the tree must have at most one parent. A binary tree is a special kind of tree in which each node can have at most two children. A stack is a special list in which all processing takes place at one end. Elements are inserted into a stack by pushing onto the stack. Deletion of an element of a stack is called popping. Queues are linked lists in which insertion takes place at the rear and deletion at the front. A queue is analogous to a checkout line in a grocery store.

Tables, trees, stacks, and queues are implemented in PL/I using either arrays, CONTROLLED storage, or BASED storage. Arrays are fine for tables but not very efficient for the other types of data structures. The CONTROLLED storage class is excellent for implementing stacks. The BASED storage class is used whenever pointers are needed between the elements of the data structure, such as in trees and queues. Linked lists are implemented with the BASED storage class.

When using the CONTROLLED storage class, memory is requested and assigned to a variable via the ALLOCATE statement. The allocated memory can then be returned to PL/I via the FREE statement. Several ALLOCATE statements issued without subsequent FREE statements create generations of the controlled variable. Any access to the variable name is to the most recently acquired element of the controlled variable. Once memory has been allocated, variables declared using the CONTROLLED storage class can be used in the same way as any other variable.

When using a variable declared in the BASED storage class, a pointer variable must be used. The ALLOCATE statement is also used to assign memory to a based variable. When a based variable is allocated, however, the ALLOCATE statement returns the address in memory assigned in the pointer variable. The FREE statement returns the memory pointed to by the pointer variable to the system. A based variable may have several pointer variables used to access it. However, only one pointer variable is actually used in the allocation and freeing of storage. This is the variable shown in the declaration of the based variable. A based variable may be qualified by using a pointer

```
P→MESSAGE = 'THIS IS THE MESSAGE';
```

In this case the variable MESSAGE, whose address is contained in P, is assigned the message in quotes. Another way to obtain the address of a variable is via the ADDR built-in function.

```
P = ADDR (MESSAGE);
```

Another built-in function that may be used with the ALLOCATE statement is ALLOCATION. The ALLOCATION built-in function returns a 0 or "false" value if no memory is currently allocated. A 1 or "true" value is returned if memory is allocated. The ALLOCATION function is used to determine whether stacks or queues are empty.

15.10

Exercises

1. Write the pseudocode to insert and delete an element from a doubly linked list.

2. Write the pseudocode to perform an inorder and postorder traversal of a binary tree.

3. Write the pseudocode to perform the "push" and "pop" functions on a stack that is implemented as an array.

4. Modify the Insert-queue and Delete-queue pseudocode to perform the same functions on a queue implemented as a circularly doubly linked list.

5. Write a program that will read in a set of people's names and build a binary tree similar to the tree in Figure 15–6. After the tree is built, print it in preorder, inorder, and postorder using recursive functions. Last, read in a set of names, and delete the corresponding nodes from the binary tree.

6. Write the pseudocode for the PROCESS_DROP_REQUEST of CH15EX2 (see Fig. 15–17).

A P P E N D I X   A

FLOWCHARTS FOR

THE EXAMPLE

PROGRAMS

Appendix A contains the flowcharts for most of the example programs throughout the text. Even though we highly discourage the use of on-page/off-page connectors for your flowcharts, we had to use them in these examples in order to have them typeset to fit within the page size of this book. Please continue to avoid connectors whenever possible.

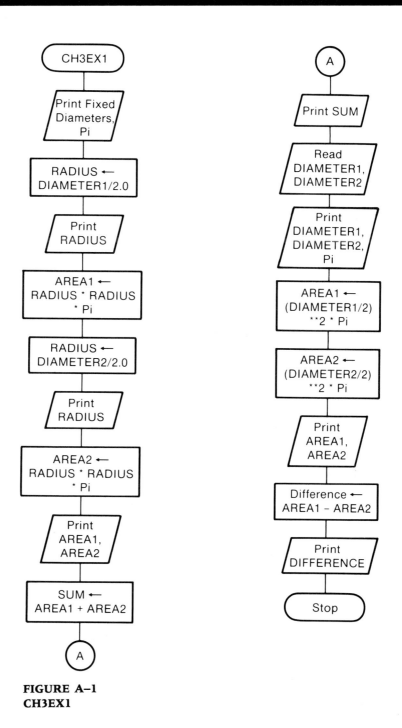

FIGURE A–1
CH3EX1

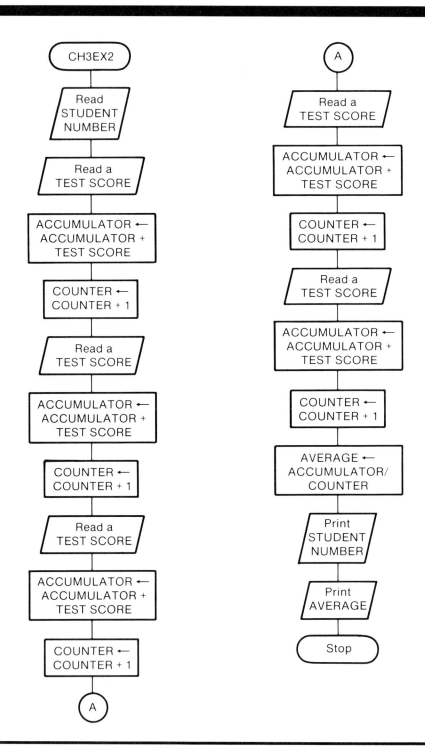

FIGURE A–2
CH3EX2

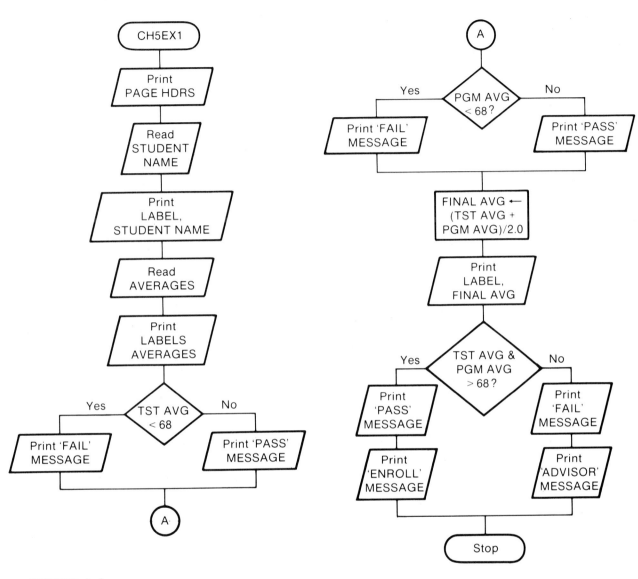

FIGURE A–3
CH5EX1

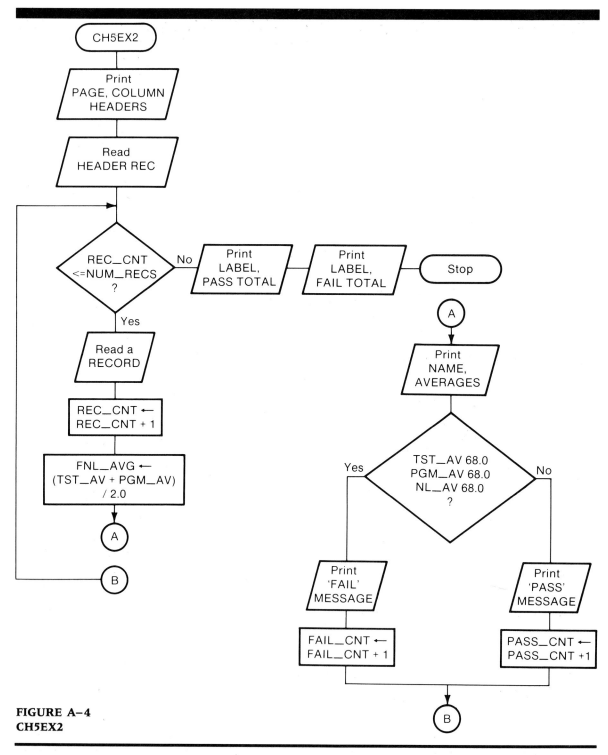

FIGURE A–4
CH5EX2

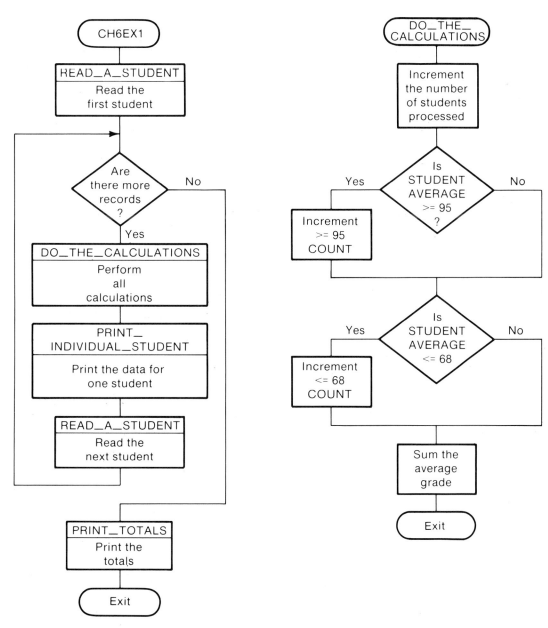

FIGURE A–5
CH6EX1

FIGURE A–5 *(continued)*

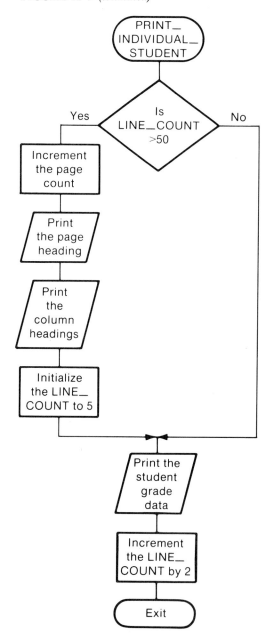

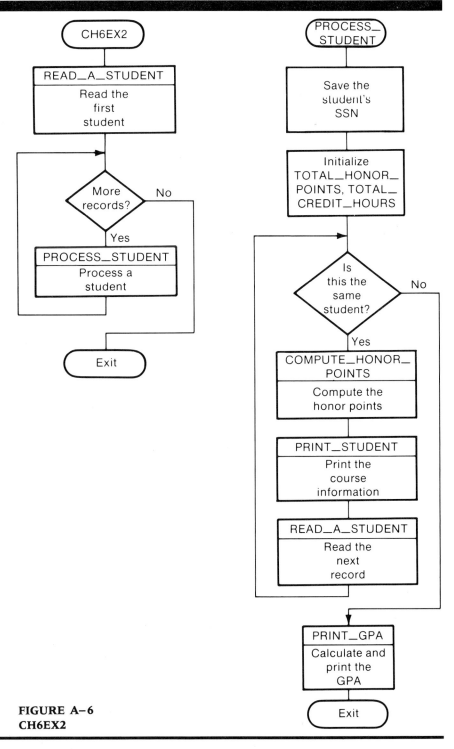

FIGURE A-6
CH6EX2

FIGURE A–6 *(continued)*

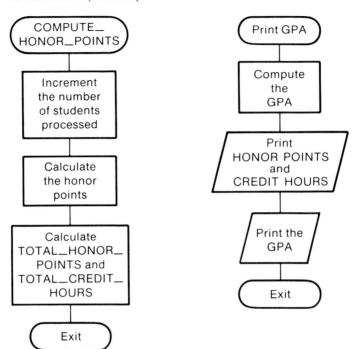

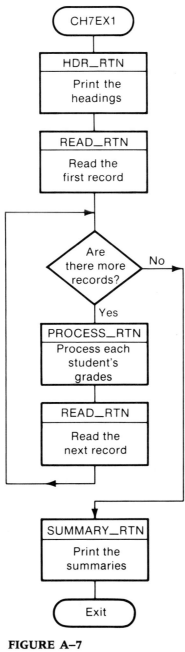

FIGURE A–7
CH7EX1

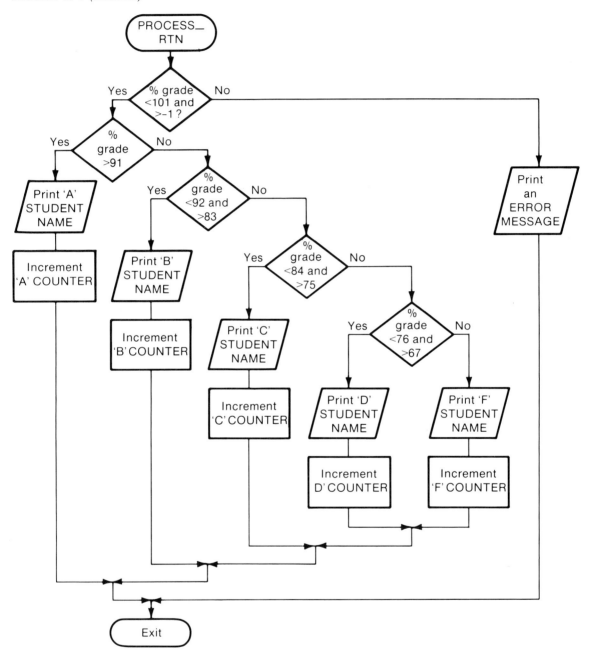

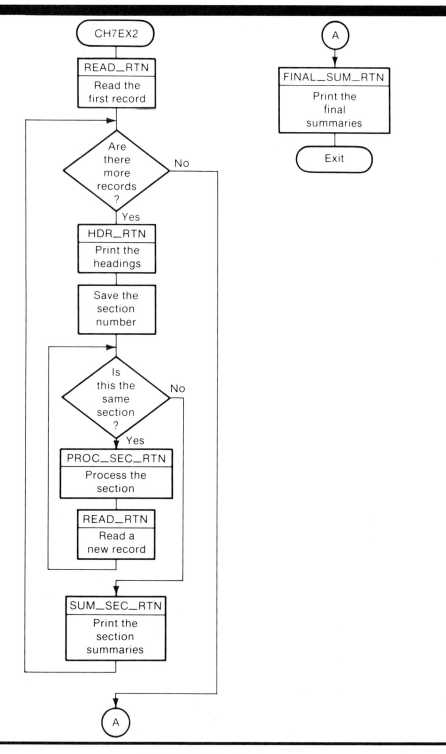

FIGURE A–8
CH7EX2

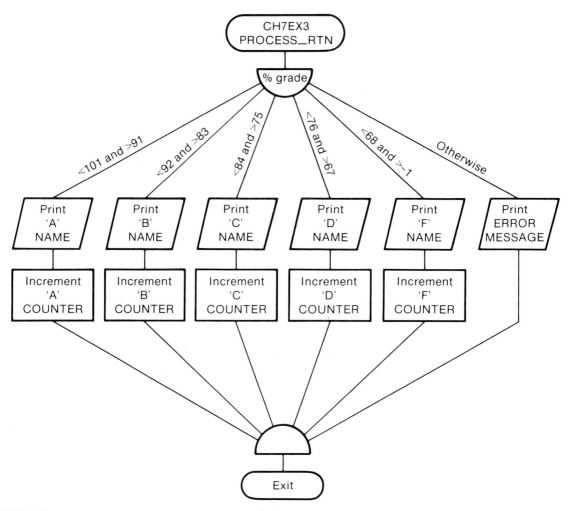

FIGURE A–9
CH7EX3 (PROCESS_RTN)

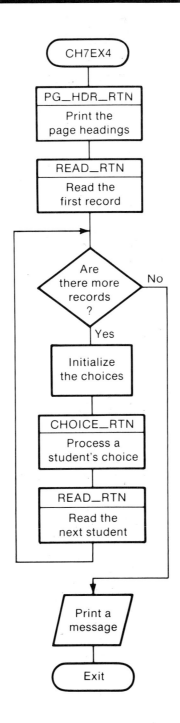

FIGURE A–10
CH7EX4

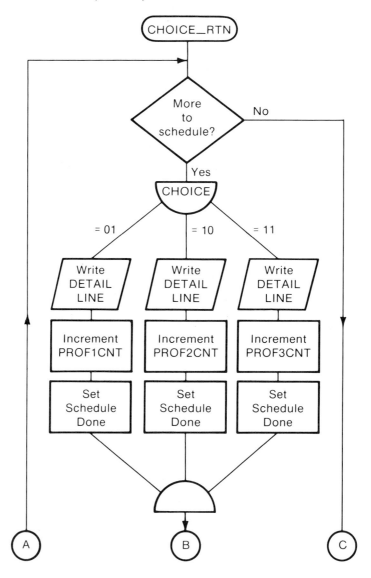

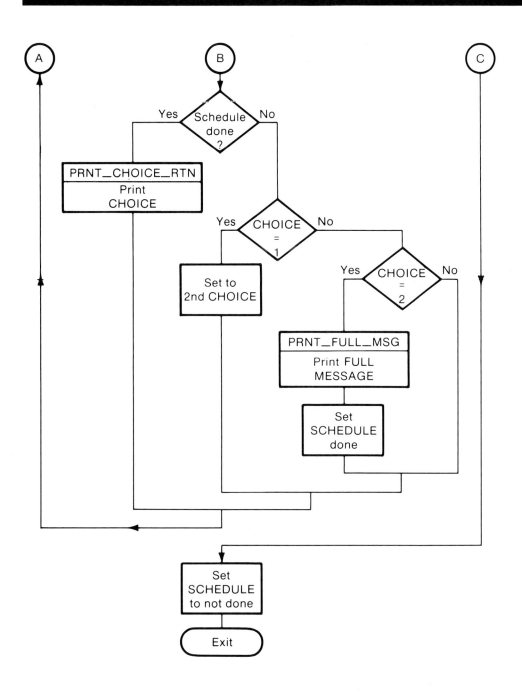

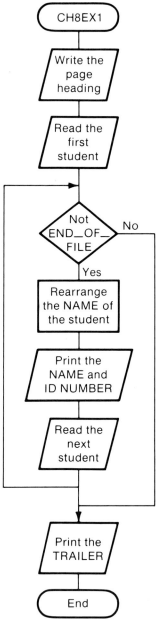

FIGURE A–11
CH8EX1

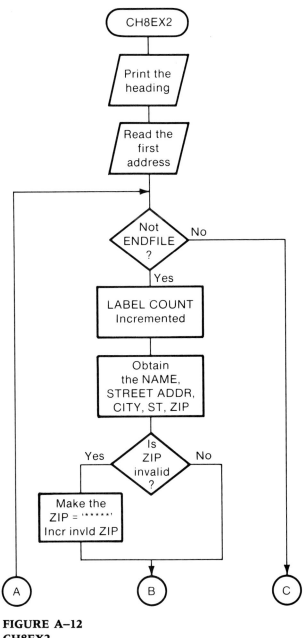

FIGURE A–12
CH8EX2

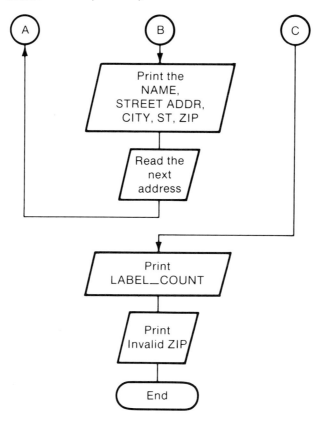

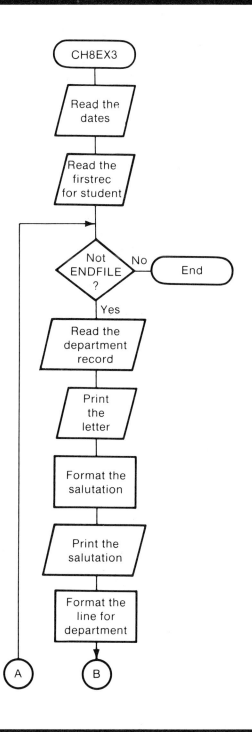

FIGURE A–13
CH8EX3

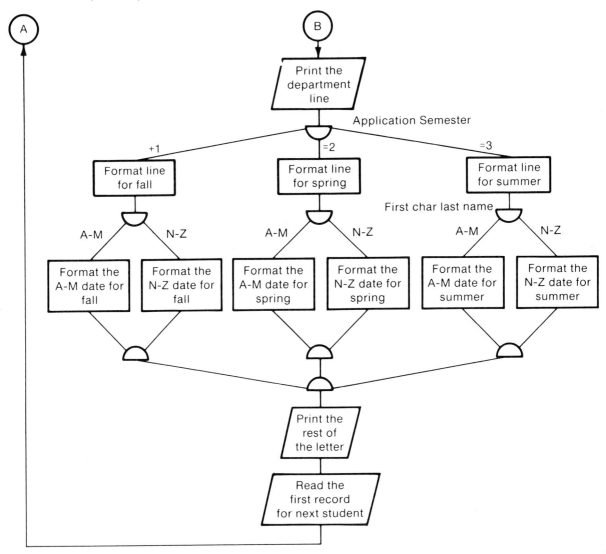

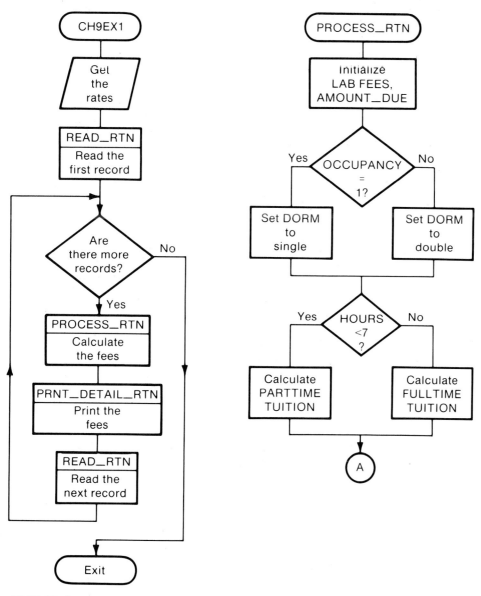

FIGURE A–14
CH9EX1

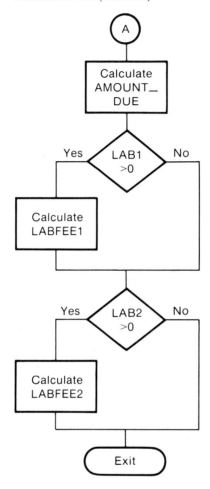

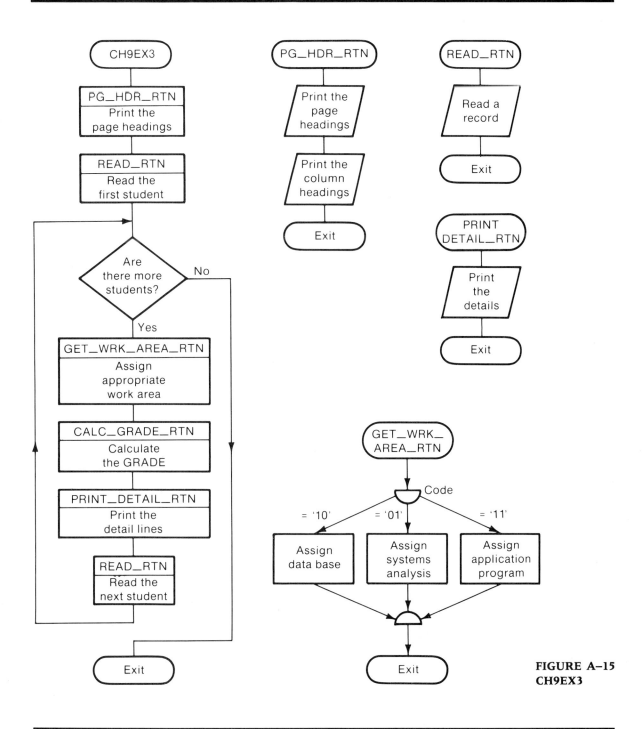

FIGURE A–15
CH9EX3

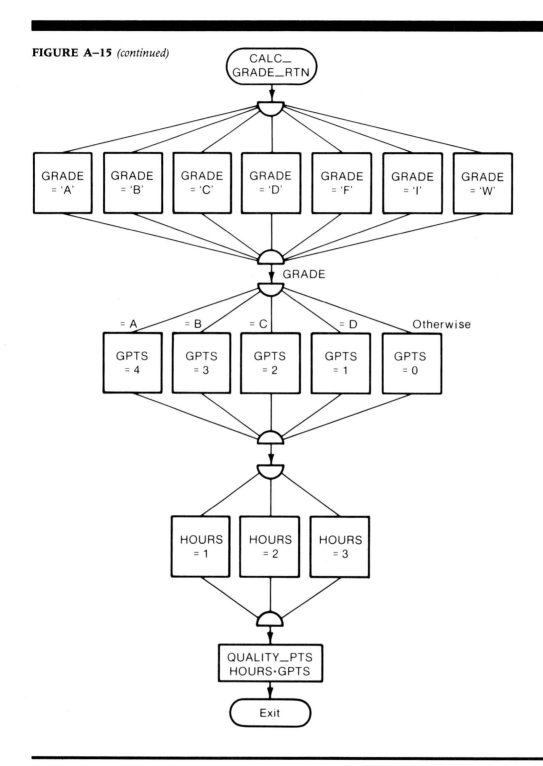

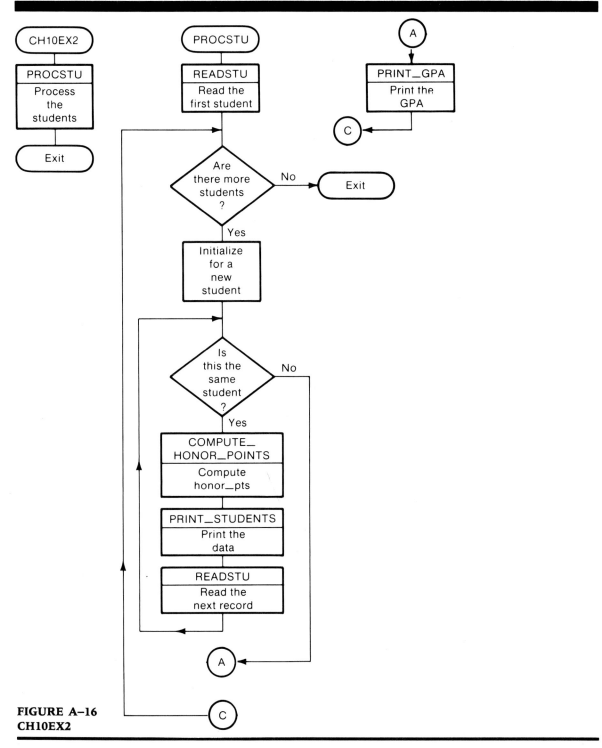

FIGURE A–16
CH10EX2

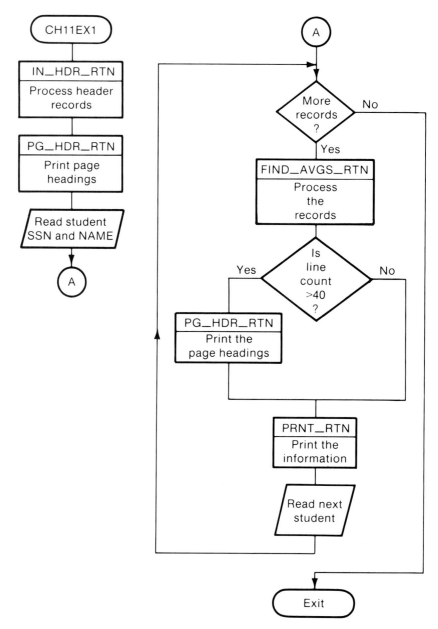

FIGURE A–17
CH11EX1

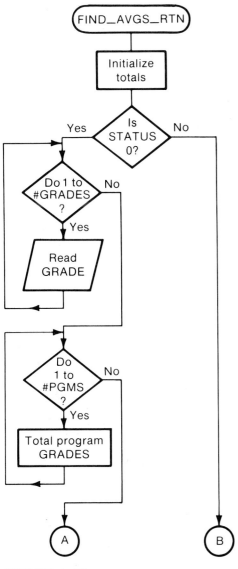

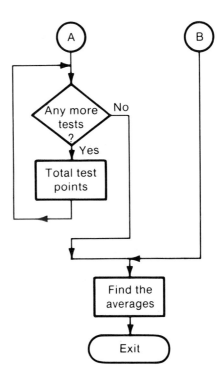

FIGURE A–18
CH11EX2

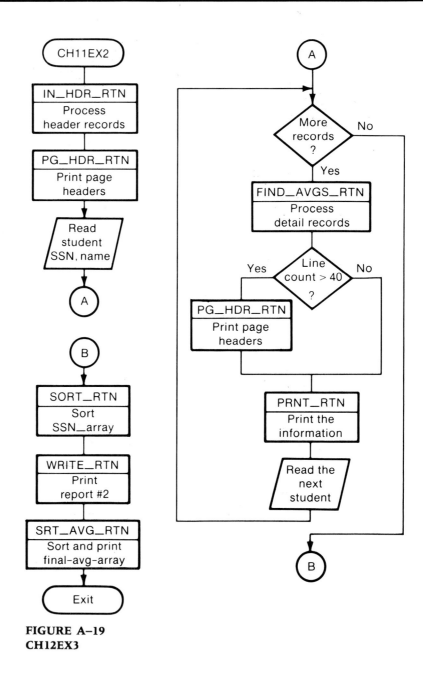

FIGURE A–19
CH12EX3

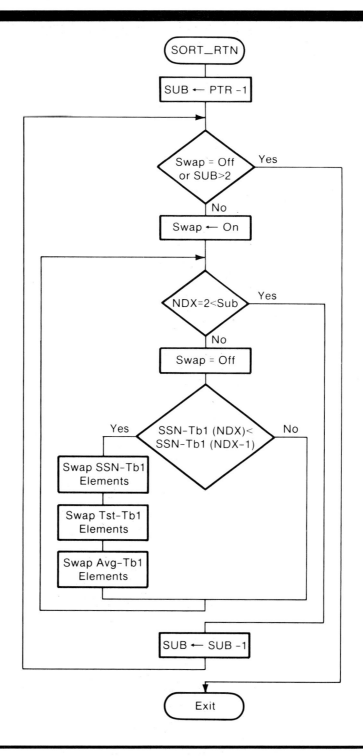

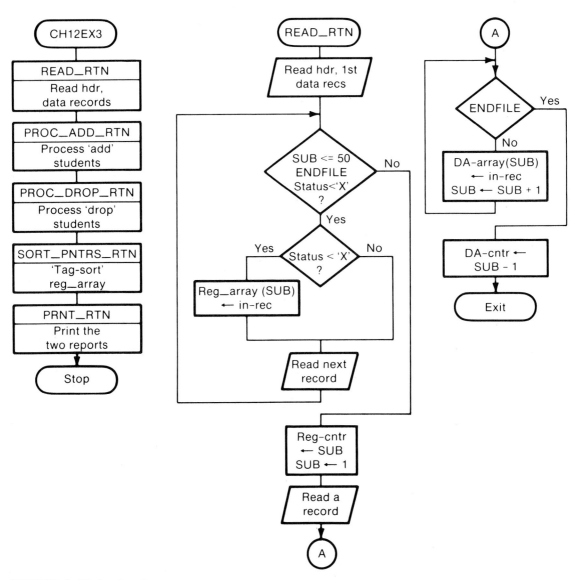

FIGURE A–19 *(continued)*

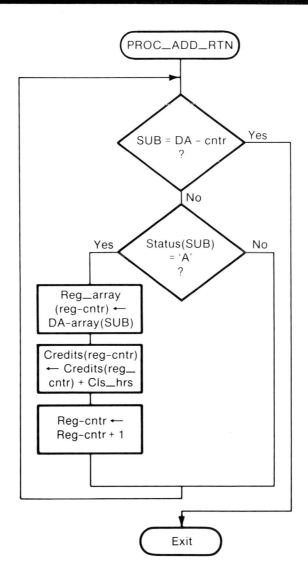

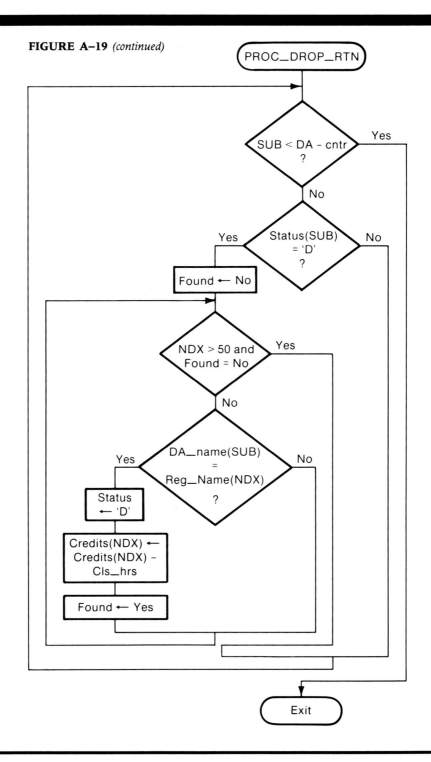

PROC_DROP_RTN

SUB < DA – cntr
?

Yes

No

Status(SUB)
= 'D'
?

Yes

No

Found ← No

NDX > 50 and
Found = No

Yes

No

DA_name(SUB)
=
Reg_Name(NDX)
?

Yes

No

Status
← 'D'

Credits(NDX) ←
Credits(NDX) –
Cls_hrs

Found ← Yes

Exit

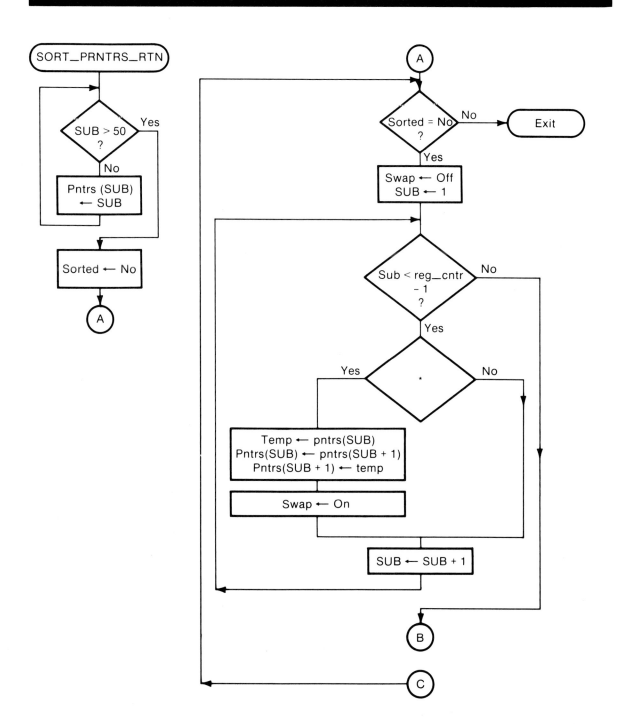

* Reg-name(pntrs(SUB)) >
Reg-Name(pntrs(SUB + 1))

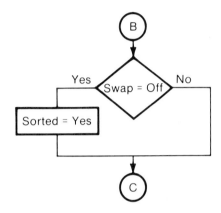

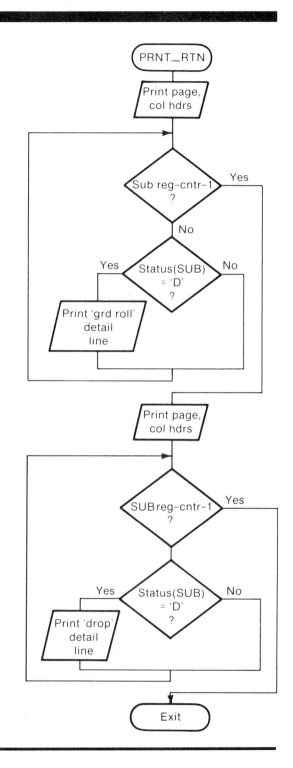

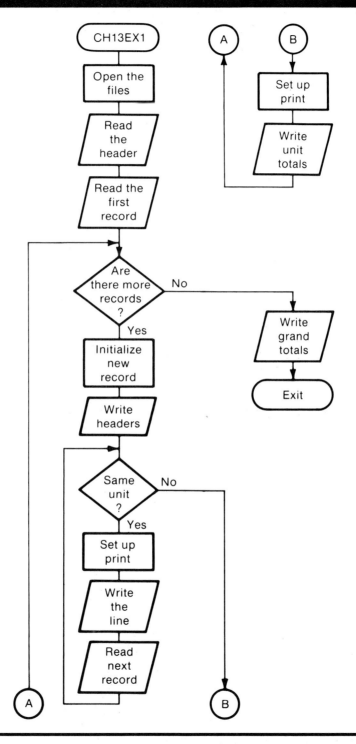

FIGURE A–20
CH13EX1

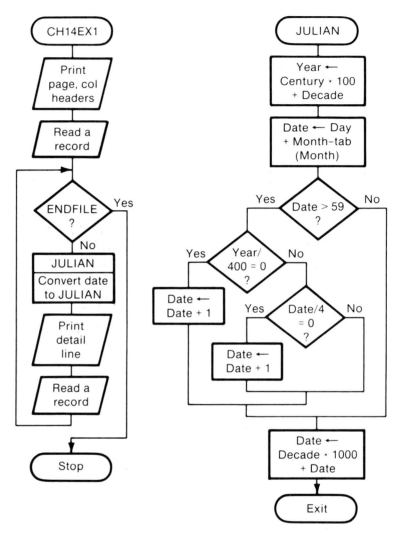

FIGURE A–21
CH14EX1

APPENDIX B

ANSWERS TO

SELECTED

EXERCISES

When the exercise has a single part, we provide answers for the odd-numbered exercises. When an exercise has more than one part, we provide answers for every other part (e.g., 1a, 1c, 1e). We provide no answers for the exercises that require a coded program.

Chapter 1

1. a. X = 19; X = 4; X = 22; X = 5; X = 2; X = 6; X = 9;
X = 11; X = 15; X = 3; X = 21; X = 45; X = 7; X = 34;
X = 13; X = 12; X = 40; X = 8
SUM = 306

   c. MAX = 99

2. a. START
     Read X
     N ← 1

```
                    Sum ← 0
                    DO WHILE (Sum < 20)
                        Print X
                        Sum ← Sum + X
                        Read X
                    ENDDO
                    Print Sum
                STOP
        c.  START
                Read X
                Counter ← 1
                Max ← 0
                DO WHILE (Counter < 5)
                    IF (X > Max)
                        Max ← X
                    ENDIF
                    Read X
                    Counter ← Counter + 1
                ENDDO
                Print Max
            STOP
```

3. Flowcharts are shown in Figure B-1.

4. a.
```
        START
            Read Number
            Sum ← 0
            Count ← 1
            DO WHILE (Count < 101)
                Sum ← Sum + Number
                Read Number
                Count ← Count + 1
            ENDDO
            Print Sum
        STOP
```

4. The flowchart is shown in Figure B-2.

```
        c.  START
                Read Number
                Max-num ← 0
                Count ← 1
                DO WHILE (Count < 6)
                    IF (Number > Max-num)
```

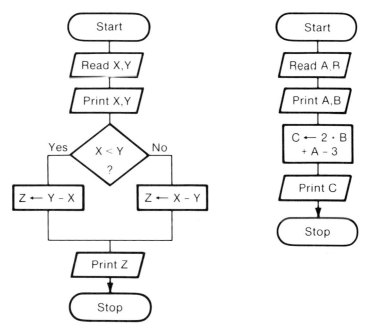

FIGURE B-1
Flowcharts for Exercise 3a (left) and 3c (right).

```
        Max-num ← Number
    ENDIF
    Read Number
    Count ← Count + 1
  ENDDO
  Print Max-num
STOP
```

The flowchart is shown in Figure B-3.

Chapter 2

1. 8 halfwords; 64 bytes; 512 bits.

3. To print binary digits would create strings of digits very difficult to read. One hexadecimal digit represents four binary digits (with necessary leading zeros) exactly and results in a number much easier to read.

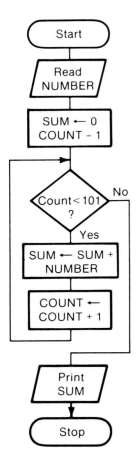

FIGURE B–2
Flowchart for Exercise 4a.

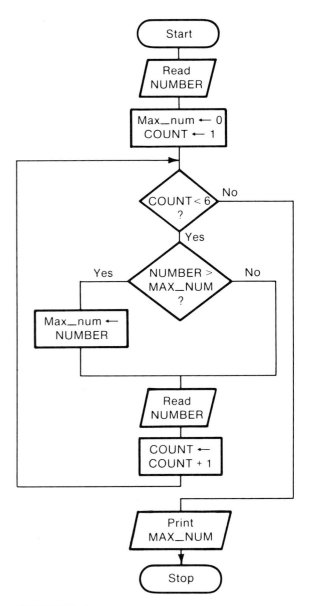

FIGURE B–3
Flowchart for Exercise 4c.

5. The compilation stage translates PL/I to machine language, checks for syntax errors, and prints a listing of the source code and messages for any errors detected. Execution carries out the instructions contained within the program. Logic errors might be detected during execution.

7. a. Valid
   c. Valid
   e. Invalid: too many characters
   g. Valid
   i. Invalid: begins with a number

8. a. The colon is the delimiter for a label.
   c. The semicolon is the delimiter for a word.

9. Comments in a PL/I program are intended to make the program easier for a fellow human being to read; they have no effect on the execution of the code.

11. a. AVERAGE = 79;
    c. CLASS = COURSE;

12. a. GET LIST (X);
    c. PUT LIST (GRD_1, GRD_2, GRD_3);

Chapter 3

1. a. DECLARE NUM_A FIXED DECIMAL (3);
   c. DECLARE NUM_C FIXED DECIMAL (6,3);
   e. DECLARE NUM_E FLOAT DECIMAL (7);
   g. DECLARE NUM_G FLOAT DECIMAL (7);
   i. DECLARE CHAR-I CHARACTER (11);

2. a. DECLARE NUM_A FIXED DECIMAL (3)   INITIAL (123);
   c. DECLARE NUM_C FIXED DECIMAL (6,3) INITIAL (123.123);
   e. DECLARE NUM_E FLOAT DECIMAL (7)   INITIAL (2652.398);
   g. DECLARE NUM_G FLOAT DECIMAL (7)   INITIAL (1.697531);
   i. DECLARE CHAR-I CHARACTER   (11) INITIAL
                                              ('DECLARATION');

3. a.

A					

   c.

B	R	O	W	N		C	O	W	

4. a. DECLARE IN_VALUE   FIXED BINARY (15,0);
   c. DECLARE J_NUM   FLOAT DECIMAL (6);
   e. DECLARE E_NUM   FIXED DECIMAL (5,0);

5. a. 64
c. 37
e. 5
g. 405
i. 72
k. 20

6. a. `LEAST = MIN (N1, N2, N3, N4);`
c. `WHOLE_NUM = TRUNC (K3);`
e. `SQROOT = SQRT (ROOT);`

Chapter 4

1. A *field* is the smallest division of data that is treated as a unit, the amount of data read into a single variable. It might contain a social security number. A *record* is one or more fields treated as a unit for purposes of I/O. A record might contain a social security number, a name, a major field of study and a grade for each semester the individual has been in school. A *file* is a collection of related records and might contain one record for each student in the College of Liberal Arts and Sciences.

3.

```
ABC        25        23+2+4    29
PRICE      AREA
ALL DONE
```

5. `'STUDY'   666   10.66`

Chapter 5

1. a. True
c. True

2. a. `IF W = 85 OR X = 5 OR Z = -9 THEN`
c. `IF Z = W OR Z = 0 AND 2 = 2 OR X = (W - 5) THEN`

3. a. `A=4; A=3; A=2; A=1; A=0;`
c. `C=0 Q=10; C=2 Q=9; C=4 Q=8; C=6 Q=7;`

Chapter 6

1. "Efficiency" in today's programming environment means *cost-effectiveness* and includes the cost of program development and maintenance, as well as computer costs. It is generally more cost-effective to write an easily maintainable program that uses somewhat more computer resources.

3. After a short period away from a program, its author is likely to forget the significance of a particular variable. A meaningful variable name indicates its significance.

5. See the discussion in Section 6.1.2 of the text.

7. Top-down development is the process whereby the programmer begins with the overall problem to be solved. The problem is then broken into components, which may themselves be further broken down until each routine carries out a single, well-defined function.

9. A DO UNTIL loop has its condition tested at the bottom of the loop; therefore, the loop is always entered once. A GOTO is an unconditional branch, which at worst produces "spaghetti" code and at best violates the one-entrance, one- exit rule of structured programming.

11. When each variable is declared in a separate statement, the compiler produces a statement number which includes the declaration of that single variable only. Therefore, a syntax error in the declaration is easier to find.

13. In general, CHECK and PUT DATA give the values of the variables in the variable list or, if no variable list is included, the value of each variable known to PL/I at the time the statement executes. A programmer can generally trace these values to find the source of the error.

Chapter 7

1.
```
IF  X = Y  THEN
     IF  A = Z  THEN
          R = 1;
     ELSE
          R = 2;
ELSE
     R = 3;
```

3. a. Adler   Rate 2
    c. Carson   Rate 2
    e. Eikes   High-Risk Pool
    g. Gray   Rate 1

4. a. A2=BC;   A5=EF;
    c. A4=AB;   A5=AB;

5.

a.				
N1=1;	N2=4;	N2=5;	N2=6;	N2=7;
N1=2;	N2=4;	N2=5;	N2=6;	N2=7;
N1=3;	N2=4;	N2=5;	N2=6;	N2=7;

c.				
N1=1;	N2=1;	N2=11;	N2=21;	N2=31;
N2=41;	N1=26;	N2=1;	N2=11;	N2=21;
N2=31;	N2=41;	N1=51;	N2=1;	N2=11;
N2=21;	N2=31;	N2=41;	N1=76;	N2=1;
N2=11;	N2=21;	N2=31;	N2=41;	

Chapter 8

1. a. [empty grid]

    c. `|3|4|5|6| |S|O|U|T|H| |S|T|A|T|`

2. a. `|2|3|8|7|-|9|2|9| |`

    c. `|9|2|8|3|8|3|`

    e. When assigned a null string, CUSTOMER-NAME has a length of zero.

3. a. `|W|I|T|C|H| |W|O|R|L|D| | | | | | | |`

    c. [empty grid]

4. a. False

    c. False

    e. True

    g. False

5. a. I = 4

    c. J = 4

    e. I = 0

    g. I = 11

6. a.
```
PUT LIST (SUBSTR (SSN, 1, 3) || '-' || SUBSTR
(SSN, 4, 2) || '-' || SUBSTR (6));
```

    c.
```
DCL VERIFY_CODE     FIXED BINARY (15);
VERIFY_CODE = VERIFY (ACCOUNT_NUMBER, '/0123456789');
```

Chapter 9

1. a. VAR_1 = bb567   VAR_3 = 890ABCDEFG

    c. VAR_1 = 34567   VAR_4 = HIbbb

2. a.
```
GET EDIT (NAME, CLASS_NM, CLASS_NUM, SEC)
         (A(10), A(4), F(3), F(2));
```

    c.
```
GET EDIT (NAME, CLASS_NM, CLASS_NUM, SEC)
         (COL(6), A(5), A(3), X(3), F(2), F(2));
```

3.
```
          1         2         3
123456789012345678901234567890
```

    a.   COMPUTERS ARE FAST

    c.   OMPUTERS A

4. a. COL(20)   Start printing in column 20

      A(10)   Print HOLY COWbb

      F(4,2)   Will cause an error because the field is too small

c. LINE(7)   Skip to absolute line 7
   A(4)   Print HOLY
   SKIP(3)   Skip to the third line from the present line
   F(7,2)   Print bb69.13
   LINE(15)   Sklp to absolute line 15
   F(8,4)   Print 136.9000

e. COL(10)   Start printing in column 10
   A(8)   Print HOLY COW
   COL(10)   Start printing in column 10 of the next page
   F(7,2)   Print 69.13
   COL(10)   Start printing in column 10 of the next page
   F(5,1)   Print 136.9

5. a. PUT EDIT (ZWEI, EIN, BETA) (F(4), F(9,3), A(4));
   c. PUT EDIT (BETA, ZWEI, ALPHA (A(9), X(3), F(4), A(5));
   e. PUT EDIT (BETA, EIN, ALPHA, ZWEI)
               (COL(3), A(12), COL(1), F(10,6), COL(4), A(5),
               COL(1), F(9,4));

6. a. PUT EDIT (SUBSTR (ALPHA, 7)) (A(4));
   c. PUT EDIT (MIN (GR_1, GR_2, GR_3)) (F(4));

7. a. PUT EDIT ((B_2 + B_3 + B_4) / Q_2) (F(5,2));
   c. PUT EDIT (SQRT (V_1 + V_2)) (F(3));

8. a. PUT EDIT (NAME, SSN)   (PAGE, LINE(3), A(25), A(15))
               (EMP_#, GROSS)   (A, F(9,2))
                   (TAX_RATE, TAX_$) (F(2,2), F(6,2))
                   (NET_$)   (F(9,2));
   c. PUT EDIT (ALPHA, BETA)   (COL(6), A(10), X(2), A(3))
               (F_A, F_B)   (COL(20), F(5), F(7,2))
               (ZETA, X_1)   (X(7), A(9), X(3), F(2))
               (X_2, X_3)   (F(3), F(4));

Chapter 10

1. Internal subroutines are compiled with the main program they are a part of and are available to that program only. Variables declared in the main routine (global variables) are available to the internal subroutines. Each external subroutine is compiled as a separate program and typically is available to any program at the installation. Only those variables declared within the external subprogram is available to it.

3. *PROCESS allows you to compile an external subprogram as part of a main program. The *PROCESS (starting in column 1) must precede the code of the external subprogram.

5. a.   SQRT = the value returned by the function
             DATE = the current date as returned by the function
        c.   MIN = the value returned by the function
             DATE = the function itself
             ABS = the value returned by the function

Chapter 11

1.

Dimension		Lower Bound	Upper Bound	Extent
a.	1	−6	0	7
	2	1	20	20
c.	1	1	20	20
	2	40	45	6
	3	1	30	30

2. a.   DCL TABLE_1 (60)   FIXED DEC (5)   INIT (60(0));
     c.   DCL TABLE_3 (7)   CHAR   (3)   INIT
                                (*, 'AB', *, *, 'CD', *, 'EF');

3. a.   DCL ARRAY_3A (12)   CHAR (5);
          DCL ARRAY_3A (1:12)   CHAR (5);
     c.   DCL ARRAY_3C (-10:0) FIXED DEC (2);

4. a.   GET LIST ((ARRAY_4A (SUB) DO SUB = 1 TO 12));
     c.   GET LIST ((ARRAY_4C (SUB) DO SUB = -10 TO 0));

5. a.   DO SUB = L_BND TO H_BND BY 5;
                ARRAY_5A (SUB) = ARRAY_5A (SUB) + 15
          END;
     c.   PUT LIST (ARRAY_5C (SUB) DO SUB = 1 TO N));

6. a.   SUB = 31
     c.
          SUB = N + 1

7. a.   12 times
     c.   2 times

9. a. ARRAY_9 = ARRAY_10;
    c. ARRAY_9 = ARRAY_9 + ARRAY_10;

11. a. VALUE = 7
    c. VALUE = -3

12. a. VALUE = 22

13. a.

After pass 1:	07	21	16	32	85
After pass 2:	07	16	21	32	85
After pass 3:	07	16	21	32	85

14. a.

After pass 1:	07	16	21	32	85
After pass 2:	07	16	21	32	85

15. a. 1 seek
    c. 2 seeks
    e. 3 seeks

Chapter 12

1. No level number on ASTRUC.
No comma after ASTRUC.
Y   is a group item but has attributes.
Z   level number is out of ascending order.
A   level number is out of ascending order.
B   is an elementary item but has no attributes.
No closing semicolon.

3.

```
DECLARE
    1 PROF_REC,
        2 PROF_SSN,
            3 PROF_SSN_1        CHAR        (3),
            3 PROF_SSN_2        CHAR        (2),
            3 PROF_SSN_3        CHAR        (4),
        2 PROF_NAME,
            3 PROF_L_NAME       CHAR        (10),
            3 PROF_F_NAME       CHAR        (10),
        2 PROF_STATUS,
            3 PROF_RANK         CHAR        (9),
            3 PROF_DEPT         CHAR        (4),
            3 PROF_TENURED      CHAR        (1),
            3 PROF_SALARY       FIXED DEC (7,2);
```

4. a. 100.01

c. +0.88

e. −621

g. $1,827.58

i. 01 02 88

k. 90000.96CR

5. a. P'•••9V.99'

c. P'$•••9V.99'

7.
```
PUT EDIT (PROF_SSN_1, PROF_SSN_2, PROF-SSN_3)
         (A3, X(5), A(2), X(5), A(4), X(10))
         (PROF_L_NAME, PROF_F_NAME)
         (A(10), X(5), A(10), X(10))
         (PROF_RANK, PROF_DEPT, PROF_TENURED,
         PROF_SALARY)
         (A(9), X(5), A(4), X(5), A(1), X(5),
         P'$$$$$9V.99');
```

9.
```
DECLARE
    1  WEATHER (12),
       2  TEMP,
          3  T_HIGH        FIXED DEC (4,1),
          3  T_LOW         FIXED DEC (4,1),
       2  REL_HUM,
          3  RH_HIGH       FIXED DEC (3),
          3  RH_LOW        FIXED DEC (3),
       2  WIND,
          3  WIND_DIR      CHAR      (3),
          3  WIND_VEL      FIXED DEC (3);
```

11.
```
WEATHER_DATA = STRING (WEATHER);
```

Chapter 13

1. Record I/O executes faster than stream I/O.

In record I/O a file contains discreet records; in stream I/O it does not.

In record I/O you must define every byte in the record.

In record I/O you cannot transmit either more or less than a single record with a single I/O statement.

In record I/O there is no automatic data conversion as there is in stream I/O.

3. a. DCL SYSIN FILE INPUT STREAM ENV (F BLKSIZE (40));

c. DCL STDNT FILE INPUT RECORD ENV (F BLKSIZE (75));

4. a.   OPEN FILE (SYSIN);
CLOSE FILE (SYSIN);
OPEN FILE (PRNTR);
CLOSE FILE (PRNTR);
OPEN FILE (STDNT);
CLOSE FILE (STDNT);
OPEN FILE (TCHR);
CLOSE FILE (TCHR);

5. a.   DCL SYS_AREA   CHAR (40);
DCL PTR_AREA   CHAR (133);
DCL STD_AREA   CHAR (75);
DCL TCH_AREA   CHAR (121);

7.   DECLARE
1   STDNT_OUT,
2   UNUSED_1   CHAR   (37) INIT (' '),
2   SSN   CHAR   (9),
2   UNUSED_2   CHAR   (5) INIT (' '),
2   NAME   CHAR   (20),
2   UNUSED_3   CHAR   (5) INIT (' '),
2   TUITION   P'$$$9V.99',
2   UNUSED_4   CHAR   (5) INIT (' '),
2   DORM_FEE   P'$$$$9V.99',
2   UNUSED_5   CHAR   (37) INIT (' ');

9.   READ FILE (STDNT) INTO (STDNT_IN);
WRITE FILE (PRNTR) FROM (COL_HDRS);
WRITE FILE (PRNTR) FROM (STDNT_OUT);

11.   DECLARE
1   OFF_CAMPUS   DEFINED STDNT_IN,
2   LAST_NAME CHAR   (10),
2   LCL_ADDR CHAR   (19),
2   TUITION   FIXED DEC (5,2));

Chapter 14

1. The three types of procedures are main procedures, subroutine procedures, and function procedures. Main procedures are distinguished by OPTIONS (MAIN) as part of the PROCEDURE statement. A subroutine procedure does not return any values, while a function procedure returns a single value.

3. Typically a function contains a block of code that is used by a very large percent of the programs at the installation. For the function to be available to all programs, it must be external.

5. A subroutine is never declared as such, but a function might have to be.
A function is invoked by coding its name; a subroutine is invoked through a CALL statement.
A function returns a single value; a subroutine returns no values.
If you want to reference the value returned by a function, you must supply a variable to receive it.

7. If a descriptor list is included, the dummy argument is assigned those attributes.
If no descriptor list is included, the attributes are derived from the literal itself.

8. a.
```
DCL S_LENGTH   FIXED DEC (6,2);
DCL S_WIDTH    FIXED DEC (5,2);
DCL R_AREA     FIXED DEC (9,2);
```
   c.
```
DCL S_LEG1     FIXED DEC (2);
DCL S_LEG2     FIXED DEC (2);
DCL R_HYPOT    FIXED DEC (5);
```

9. a.
```
DCL RECAREA ENTRY (FIXED DEC (6,2), FIXED DEC (5,2))
              RETURNS (FIXED DEC (9,2));
R_AREA = RECAREA (S_LENGTH, S_WIDTH);
```
   c.
```
DCL  PYTHAG ENTRY (FIXED DEC (2), FIXED DEC (2))
              RETURNS (FIXED DEC (5));
R_HYPOT = PYTHAG (S_LEG1, S_LEG2);
```

A P P E N D I X   C

PL/I AND

PL/C

DIFFERENCES

PL/C was designed to be a subset of PL/I. Any program that runs under PL/C should run under any version of PL/I. However, different diagnostic features have been added to PL/C to assist in learning PL/I. If you wish to run a PL/C program that contains these features under PL/I, then the features must be enclosed in pseudocomments.

The following chart summarizes the statements and keywords that are *not* available in PL/C.

ADDBUFF (x)	Option of the ENVIRONMENT attribute used to request additional memory for DIRECT files.
ADDR (x)	Built-in function that returns the address of the variable specified.
ALLOCATE	A statement that requests storage for CONTROLLED or BASED variables.

ALLOCATION (x)	A built-in function that is used to show the allocation of a CONTROLLED or BASED variable. The results of the function differs with the different compilers.
AREA	A condition that is used with an on-unit that gets control when a request for allocation exceeds the memory available to the program.
AREA [(size)]	An attribute used in a declare statement to specify the number of bytes (size) that can be used for allocation of a based variable.
ARGx	An option used in communications between different languages.
ASCII	An option of the ENVIRONMENT attribute used to specify that the data is in ASCII code.
ASM	An option of OPTIONS that says that an entry point is in assembler language.
ATTENTION	A condition used in a ON statement that is used during conversation with a terminal to show that the user has pushed the attention signal.
BACKWARD	An option used with the ENVIRONMENT attribute that says a VSAM file is to be processed backwards.
BACKWARDS	An attribute used when declaring a file that says the file is to be read backwards.
BASED[(pointer)]	An attribute that is used in the DECLARE statement to specify that a variable is to be used in list processing. If the pointer is specified it is associated with the variable.
BUFND(x)	An option of the ENVIRONMENT attribute that specifies x number of data buffers to be used with a VSAM data set.
BUFNI(x)	An option of the ENVIRONMENT attribute that specifies x number of index buffers to be used with a VSAM data set.
BUFOFF	An option of the ENVIRONMENT attribute that is used to specify the length of the prefix field on an ASCII block.
BUFSP(x)	An option of the ENVIRONMENT attribute that specifies the number of bytes (x) needed for the total buffer space for a VSAM data set.
CMDCHN	An option of the ENVIRONMENT attribute that is used to simulate blocked records with the 3540 diskette I/O device.
COBOL	An option of the ENVIRONMENT attribute that says that the COBOL algorithm is to be used to map records from the file specified.

COBOL	An option of the OPTIONS attribute that says the routine named is a COBOL program.
COMPILETIME	A built-in function that returns the date and time of compilation in a character-string.
COMPLETION(event)	A built-in function that specifies whether or not an event is completed.
CONDITION	An attribute that declares an identifier as a condition.
CONNECTED	An attribute used in the DECLARE statement that is used to specify that an array occupies a contiguous storage location.
CONTROLLED	An attribute used in the DECLARE statement that specifies the CONTROLLED storage class. The storage for the variable is obtained via an ALLOCATE statement and then freed with the FREE statement.
CURRENTSTORAGE(x)	A built-in function that returns the storage required by the variable x.
D	An option of the ENVIRONMENT attribute that says that an ASCII data set is unblocked and variable length.
DB	An option of the ENVIRONMENT attribute that says that an ASCII data set is blocked and variable length.
DEFAULT	A statement that gives the programmer control over the default rules.
DEFINED	An attribute that gives multiple names to the same area of memory.
DELAY(x)	A statement that is used to delay the execution of a program for 1000 milliseconds of real time.
DESCRIPTORS	An option of the DEFAULT statement that specifies the default attributes for parameters.
DISPLAY	A statement that sends a string to the operator's console for displaying.
EMPTY	A built-in function that is used to reset the storage of a variable to the empty status.
EVENT	An attribute/option of the CALL, DELETE, DISPLAY, READ, REWRITE, and WRITE statements that states that multitasking or asynchronous I/O is to be done.
EXCLUSIVE	An attribute that is used in declaring a file to prevent other tasks from accessing the file at the same time.
EXTENTNUMBER(x)	An option of the ENVIRONMENT attribute that is used for INDEXED and REGIONAL files to state the number of disk extents allowed.

FBS	An option of the ENVIRONMENT attribute that specifies that the file is in fixed-length blocked spanned format.
FETCH	A statement that is used to copy a procedure from disk into memory so that it can be executed.
FILSEC	An option of the ENVIRONMENT attribute that sets up security on any later access to the 3540 diskette.
FORTRAN	An option of the OPTIONS attribute that says the entry point is that of a FORTRAN program.
FREE	A statement that returns storage to the system that was previously allocated.
FS	An option of the ENVIRONMENT attribute that says a file has unblocked records and if on a DASD that every track will be filled.
FUNCTION	An option of the ENVIRONMENT attribute that specifies an operation to be performed on an IBM device.
G(max-size)	An option of the ENVIRONMENT attribute that indicates the maximum length of a teleprocessing file.
GENERIC	An attribute used in declaring a series of entry points that are to be chosen based upon the attributes of certain arguments.
HALT	A statement that passes control to a terminal thereby stopping the conversational mode of a program.
HIGHINDEX	An option of the ENVIRONMENT attribute that indicates the type of device that is being used for the high level index.
IN(variable)	Option of ALLOCATE and FREE statements that determines the area—variable—in which to allocate or free memory.
INDEXAREA[(x)]	An option of the ENVIRONMENT attribute that makes the high-level index resident in main memory.
INDEXMULTIPLE	An option of the ENVIRONMENT attribute that indicates that the index to be created for an indexed file is to have multiple levels.
INTER	An option of the OPTIONS attribute that tells PL/I not to handle interrupts in the specified entry point.
IRREDUCIBLE	An attribute used when declaring an entry point to specify that different values will be returned every time the routine runs.

LEAVE	An option of the ENVIRONMENT attribute that is used to allow a reread backwards by specifying that the tape be left at the end.
LEAVE	A statement that is used to exit a DO group to the first statement (executable) following the END.
LIKE	An attribute that specifies that the elements of two structures are to be identical.
LOCATE	A statement that moves previously located records to a file after allocating a buffer.
NCP(x)	An option of the ENVIRONMENT attribute that is used with asynchronous I/O.
NOFEED	An option of the ENVIRONMENT attribute that is used with the IBM 3540 diskette.
NOLOCK	An option of the READ statement used to show that exclusive use of a record is not needed.
NOMAP NOMAPIN NOMAPOUT	Options of OPTIONS used when interfacing with another language.
NOOPTIMIZE	A compliler option that specifies that the compiler is to do no optimizing of the code.
NOSTRINGSIZE	A condition prefix that makes the STRINGSIZE condition disabled for the statement.
NOTAPEMK	An option of the ENVIRONMENT attribute used to indicate that there is not a tape mark preceding the file located on magnetic tape.
NOWRITE	An option of the ENVIRONMENT attribute used with index files for space optimization.
NULL	A built-in function that returns the null address value.
OFFSET(n)	An attribute that is used to set a variable to an address that is relative to another variable.
OFFSET(x,n)	A built-in function that translates a pointer variable to an offset value from another variable.
OMR	An option of the ENVIRONMENT attribute that indicates a specific type of card input.
ONSYSLOG	An option of the OPTIONS option in the PROCEDURE statement that sends error messages to the operator's console.
OTHERWISE	A clause in the SELECT statement.
PARMSET	A built-in function used with the preprocessor.
PASSWORD(password)	An option of the ENVIRONMENT attribute that specifies the password of a VSAM file.
PENDING(file)	A condition use in the ON statement raised when there are no records in a TRANSIENT file for which a READ has been issued.

PLIRETV	A built-in function that returns the current value in the PL/I return code.
POINTER	An attribute indicating that the variable will be used to contain addresses.
POINTER(x,n)	A built-in function used with pointer variables.
POLY(x,n)	A built-in function that invokes the polynomial function on the vectors x and n.
POSITION(x)	An attribute used in the DECLARE statement to specify that two variables will occupy the same location.
PRIORITY(x)	An option of the CALL statement that assigns a priority to the task.
PRIORITY(task)	A built-in function that returns the priority of a task.
R(x)	An option of the ENVIRONMENT attribute used for teleprocessing files to state the maximum record size x.
RANGE	An option of the DEFAULT statement specifing a range of identifiers that will be affected by the DEFAULT.
RCE	An option of the ENVIRONMENT attribute used with an IBM 3525 or IBM 3505 to specify that only certain columns will be read.
RECSIZE(x)	An option of the ENVIRONMENT attribute indicating the logical record length.
REDUCIBLE	An attribute used in declaring an entry point to say that the compiler is allowed to optimize by reducing the number of references.
REENTRANT	An option of the OPTIONS in the PROCEDURE statement used to state that the routine can be reentrant.
REFER	An option of the BASED attribute used to declare varying length arrays or strings.
REGIONAL(1) REGIONAL(2) REGIONAL(3)	Options of the ENVIRONMENT attribute to specify the data set organization.
RELEASE entry	A statement used to release the memory occupied by the PL/I procedure (entry).
REPEAT	An option of the DO statement used with the WHILE and UNTIL option for specifing the end condition.
REPLY(x)	An option of the DISPLAY statement where x is the variable to which the operator response will be assigned.
REREAD	An option of the ENVIRONMENT attribute and CLOSE statement to rewind the tape for further reading.

RETCODE	An option of the OPTIONS attribute used when a non-PL/I program wants to pass a return code to PL/I.
REUSE	An option of the ENVIRONMENT attribute stating that the VSAM OUTPUT file is a workfile and is empty every time it is opened.
SAMEKEY(x)	A built-in function that is used with a VSAM file.
SCALARVARYING	An option of the ENVIRONMENT attribute that indicates the length field is included with the record that indicates the length of the varying string.
SELECT	A statement that allows the selection of alternatives based upon tests.
SET(p)	An option of ALLOCATE, LOCATE, and READ statements that sets a pointer to the start of the area specified.
SIS	An option of ENVIRONMENT attribute used with a VSAM file.
SKIP	An option of the ENVIRONMENT attribute used with a VSAM file.
STATUS(event)	A built-in function that returns the status of an EVENT variable (event).
STORAGE(x)	A built-in function that returns the number of bytes needed by the variable x.
STRINGSIZE	A condition that is raised when an assignment between strings requires truncation.
iSUB	A dummy variable of DEFINED attribute.
SYSTEM	An option of the DEFAULT statement.
TASK	An attribute of the PROCEDURE statement that specifies the name of a task.
TASK(x)	An option of the CALL statement that creates a task named x when invoking the entry point specified.
TOTAL	An option of the ENVIRONMENT attribute that provides in-line code for I/O operations.
TP(M $\mid$ R)	An option of the ENVIRONMENT attribute that says a teleprocessing file is to be transmitted in either record (R) or message (M) format.
TRANSIENT	An attribute that declares a teleprocessing file.
TRKOFL	An option of the ENVIRONMENT attribute that specifies that track overflow is allowed.
UNBUFFERED	An attribute that specifies that the records being used in I/O do not need to be buffered.
UNLOAD	An option of the ENVIRONMENT attribute that rewinds and unloads a tape file at CLOSE.
UNLOCK	A statement that release exclusive control of a file.
UNTIL	An option of the DO statement that specifies the end condition.

VALUE	An option of the DEFAULT statement.
VARIABLE	An attribute that says the specified ENTRY, FILE or LABEL is to be a variable rather than a constant.
VBS	An option of the ENVIRONMENT attribute that says the file is to be variable blocked spanned.
VOLSEQ	An option of the ENVIRONMENT attribute that specifies that it is used with the IBM 3540.
VS	An option of the ENVIRONMENT attribute that specifies that the file is variable spanned.
VSAM	An option of the ENVIRONMENT attribute that specifies that the data set is to be a VSAM one.
WAIT	A statement that says the task will wait until one of the events in the list completes.
WHEN	A clause of the SELECT statement.
WRTPROT	An option of the ENVIRONMENT attribute that is used with the IBM 3540.

We have excluded all the preprocessor statements from the above list because none of them is supported by PL/C.

APPENDIX D

BUILTIN

FUNCTIONS

The built-in functions of PL/I are divided into several classes and are presented below alphabetized within these classes.

Arithmetic   Mathematical
Array handling   Miscellaneous
Condition handling   Multitasking
Event   Storage
Input/output   String handling

The following is a *brief* discussion of all PL/I BUILTIN functions. Many of these functions are discussed in more detail earlier in the text; for a more complete discussion of the others, see the IBM *OS and DOS PL/I Reference Manual*.

Arithmetic

ABS (x)   x expression

Returns the absolute value of x in a REAL mode with the base scale, and precision of x.

ADD (x,y,p[,q])   x,y expressions; p integer specifying the number of digits to be maintained; q optionally signed integer specifying the scale factor of the result

Returns the sum of x and y with the precision specified by p and q; base, scale, and precision are determined by the rules for evaluating expressions.

BINARY (x[,p[,q]])   x expression; p integer specifying number of digits to be maintained; q optionally signed integer specifying the scale factor of the result

Returns the binary value of x with the precision specified by x. If p and q are omitted the result is determined by the rules for converting base.

CEIL (x)   x real expression

Returns the smallest integer value greater than or equal to x; result has the mode, base, scale, and precision of x.

COMPLEX (x,y)   x,y real expression

Returns the complex value x + yI; if x and y differ in base, conversion is to binary, if they differ in scale, conversion is to floating-point; result is in the common base and scale.

CONJG (x)   x expression

Returns the conjugate of x; if x is real it is converted to complex; the result has the base, scale, mode, and precision of x.

DECIMAL (x[,p[,q]])   x expression; p integer specifying the number of digits to be maintained; q optionally signed integer specifying the scale factor of the result

Returns the decimal value of x with the precision specified by p and q; the result has the mode and scale of x.

DIVIDE (x,y,p[,q])   x,y expressions; p integer specifying number of digits to be retained; q optionally signed integer specifying the scale factor of the result

Returns the quotient of x/y; the mode, base, and scale follow the rules for the evaluation of expressions.

FIXED (x[,p[,q]])   x expression; p integer specifying total number of digits in the result; q optionally signed integer specifying the scale factor

Returns the fixed-point value of x with the precision specified by p and q; result has the base and mode of x.

FLOAT (x[,p])   x expression; p integer specifying the minimum
                                       number of digits in the result
Returns the floating-point value of x with the precision specified by p; the result
has the base and mode of x.

FLOOR (x)   x real expression
Returns the largest integer value less than or equal to x; the result has the base,
scale, mode, and precision of x.

IMAG (x)   x expression (if real, it is converted to complex)
Returns the coefficient of the imaginary part of x; the mode of the result is real;
the result has the base, scale, and precision of x.

MAX (x1,x2,. . .,xn)   x1,x2,. . .,xn list of expressions; the maximum
                                       number of arguments is 64; all arguments must be
                                       real, and they are converted to a common base and
                                       scale
Returns the largest value from the set of arguments; the result is real with the
common base and scale of the arguments.

MIN (x1,x2,. . .,xn)   x1,x2,. . .,xn list of expressions; the maximum
                                       number of arguments is 64; all arguments must be
                                       real, and they are converted to a common base and
                                       scale
Returns the smallest value from the set of arguments; the result is real with the
common base and scale of the arguments.

MOD (x,y)   x,y real expressions; if y is zero, the ZERODIVIDE
                                       condition is raised
Returns the smallest nonnegative value that must be subtracted from x to make
it evenly divisible by y (i.e. the remainder expressed as a whole number); the
result has the common base and scale of the arguments.

MULTIPLY (x,y,p[,q])   x,y expressions; p integer specifying the number of
                                       digits to be maintained; q optionally signed integer
                                       specifying the scale factor of the result
Returns the product of x * y with the precision specified by p and q; the base
and scale are determined by the rules for the evaluation of expressions.

PRECISION (x,y,p[,q])   x,y expressions; p integer specifying the number of
                                       digits to be maintained; q optionally signed integer
                                       specifying the scale factor of the result
Returns the product of x * y with the precision specified by p and q; the base
and scale are determined by the rules for the evaluation of expressions.

REAL (x)   x expression (if real, it is converted to complex)

Returns the real portion of x with the base, scale, and precision of x.

ROUND (x,y)   x expression; y optionally signed integer at which rounding is to occur; if y is greater than 0, it specifies the digit position to the right of the decimal point; if it is negative or 0, the specified digit is y − 1 to the left of the decimal point.

Returns the value of x rounded at the position specified by y and with the mode, base, and scale of x.

SIGN (x)   x real expression

Returns a real, fixed-point binary value with a precision of (15,0) to indicate if x is positive (+1), zero (0), or negative (−1).

TRUNC (x)   x real expression

Returns an integer value that is the truncated value of x; the mode, base, scale, and precision are those of x.

Array Handling

ALL (x)   x array expression (converted to a bit string if it is not already one)

Returns a bit string in which each bit is a 1 if the corresponding bit in each element of x exists and is a 1; the result is as long as the longest element.

ANY (x)   x array expression (converted to a bit string if it is not already one)

Returns a bit string in which each bit is a 1 if the corresponding bit in any element exists and is a 1; the result is as long as the longest element.

DIM (x,y)   x array expression; x must not have fewer than y dimensions

Returns the extent of the y dimension of x as a real fixed-point binary value with a precision of (15,0).

HBOUND (x,y)   x array expression; y expression specifying a dimension of x; x must not have fewer than y dimensions

Returns the upper bound of dimension y of x as a real fixed-point binary value with a precision of (15,0).

LBOUND (x,y)   x array expression; y expression specifying a di-
                                       mension of x; x must not have fewer than y di-
                                       mensions
Returns the lower bound of dimension y of x as a real fixed-point binary value
with a precision of (15,0).

POLY (x,y)   x,y array expressions
Returns the polynomial formed from the two one-dimensional array expressions x and y; returned value has the precision of the longer argument.

PROD (x)   x array expression
Returns the product of all the elements in x with the precision of x.

SUM (x)   x array expression
Returns the sum of all the elements in x with the precision of x.

Condition Handling	DATAFIELD

DATAFIELD
Returns a character string whose value is the contents of the field that caused
the NAME condition to be raised.
ONCHAR
Returns a character string of one character containing the character that caused
the CONVERSION condition to be raised.
ONCODE
Returns the condition code as a fixed-point binary value with a precision of
(15,0).
ONCOUNT
Returns the number of conditions that remains to be handled when an on-unit
is entered; the value is a real fixed-point binary value with a precision of (15,0).
ONFILE
Returns a character string whose value is the name of the file for which an I/O
or CONVERSION condition is raised.
ONKEY
Returns a character string whose value is the key of the record that caused an
I/O condition to be raised.
ONLOC
Returns a character string whose value is the name of the entry-point for the
current call to the procedure in which a condition was raised.
ONSOURCE
Returns a character string whose value is the contents of the field being pro-
cessed when the CONVERSION condition was raised.

Event	COMPLETION (x)	x event reference

Returns a bit string of one bit specifying the completion value of x: incomplete ('0'B) or complete ('1'B).

STATUS (x)   x event reference; x must be inactive

Returns the status value of event reference x as a real fixed-point binary value with a precision of (15,0); if the event variable is normal, a '0'B is returned; if abnormal, a '1'B is returned.

Input/ Output   COUNT (x)   x file reference; file must be open and must be stream

Returns the number of data items transmitted during the last GET or PUT operation as a real fixed-point binary value with a precision of (15,0).

LINENO (x)   x file reference; the file must be open and have the PRINT attribute

Returns the current line number of x as a real fixed-point binary value with a precision of (15,0).

SAMEKEY (x)   x file reference; the file must have the RECORD attribute

Returns a one-bit value specifying if the record following the accessed record has the same key ('1'B) or not ('0'B).

Mathematical   ACOS (x)   x real expression

Returns the inverse cosine of the radians of x as a real floating-point value.

ASIN (x)   x real expression

Returns the inverse sine in the radians of x as a real floating-point value.

ATAN (x[,y])   x,y expressions; if both x and y are specified, each must be real; if x alone is specified, it can be real or complex

Returns the inverse tangent in radians of x or of a ratio x/y.

ATAND (x[,y])   x,y expressions; x and y must be real

Returns the inverse tangent in degrees of x or a ratio x/y; it is a real floating-point value.

ATANH (x)   x expression
      Returns the inverse hyperbolic tangent of x as a floating- point value with the
      mode, base, and precision of x.

COS (x)   x expression (value in radians)
      Returns the cosine of x as a floating-point value with the mode, base, and
      precision of x.

COSD (x)   x real expression (value in degrees)
      Returns the cosine of x as a real floating-point value with the base and precision
      of x.

COSH (x)   x expression
      Returns the hyperbolic cosine of x as a floating-point value with the mode,
      base, and precision of x.

ERF (x)   x real expression
      Returns the error function of x as a real floating-point value with the base and
      precision of x.

ERFC   x real expression
      Returns the complement of the error function of x as a real floating-point value
      with the base and precision of x.

EXP (x)   x expression
      Returns the base, e, of the natural logarithm system raised to the power of x as a
      floating-point value.

LOG (x)   x expression
      Returns the natural (i.e., to the base e) logarithm of x as a floating-point value
      with the mode, base, and precision of x.

LOG2 (x)   x real expression (greater than zero)
      Returns the binary logarithm of x as a floating-point value with the base and
      precision of x.

LOG10 (x)   x real expression (greater than zero)
      Returns the common (i.e., base 10) logarithm of x as a real floating-point value
      with the base and precision of x.

SIN (x)   x expression (in radians)
      Returns the sine of x as a floating-point value with the mode, base, and
      precision of x.

SIND (x)   x expression (in degrees)
 Returns the sine of x as a real floating-point value with the base and precision of x.

SINH (x)   x expression (in radians)
 Returns the hyperbolic sign of x as a floating-point value with the mode, base, and precision of x.

SQRT (x)   x expression (if real, must be greater than zero)
 Returns the positive square root of zero as a floating-point value with the mode, base, and precision of x.

TAN (x)   x expression (in radians)
 Returns the tangent of x as a floating-point value with the mode, base, and precision of x.

TAND (x)   x real expression (in degrees)
 Returns the tangent of x as a floating-point value with the mode, base, and precision of x.

TANH (x)   x expression (in radians)
 Returns the hyperbolic tangent of x as a floating-point value with the mode, base, and precision of x.

Miscellaneous   DATE
 Returns the current date in the form of yymmdd (yy—last two digits of current year; mm—current month; dd—current day).
PLIRETV
 Returns the PL/I return code as a fixed-point binary value with a precision of (15,0).
TIME
 Returns the current time in the form of hhmmssttt (hh—current hour; mm—current minute; ss—current second; ttt—current millisecond).

Multitasking   PRIORITY (x)   x task-reference
 Returns the priority associated with x relative to the priority of the current task; it is a real binary fixed-point value with a precision of (15,0).

Storage Control	ADDR (x)	x reference to a variable

Storage Control

ADDR (x)   x reference to a variable
Returns the pointer value that identifies the generation of x.

ALLOCATION (x)   x level-one unsubscripted controlled variable
Returns the number of generations of x that can be accessed in the current task; it is a real fixed-point binary value with a precision of (15,0).

CURRENT STORAGE (x)   x variable
Returns the implementation-defined storage (in bytes) required by x as a real fixed-point value with a precision of (31,0).

EMPTY
Returns an area of zero extent that can be used to free all allocations in an area; the value of the function is assigned to an area variable when the variable is allocated.

NULL
Returns the null pointer value.

OFFSET (x,y)   x pointer reference (within area y or a null pointer value); y area reference
Returns the offset value of pointer x relative to area y.

POINTER (x,y)   x offset reference; y array expression
Returns a pointer value identifying the generation specified by offset reference x.

STORAGE (x)   x a variable
Returns the implementation-defined storage (in bytes) allocated to the variable x; it is a real fixed-point binary value with a precision of (31,0).

String Handling

BIT (x[,y])   x,y expressions
Returns the bit value of x with the length specified by y.

BOOL (x,y,z)   x,y,z expressions
Returns a bit string that is the result of the boolean operation specified by z; the result has the length of the longer of x and y.

CHAR (x[,y])   x,y expressions
Returns the character value of x with the length specified by y.

HIGH (x)   x expression
    Returns a character string in which each character is the highest character in the
    collating sequence (hexadecimal FF); the result has the length specified by x.

INDEX (x,y)   x string-expression to be searched; y string-
                                      expression to be searched for
    Returns the starting position within x of a substring identical to y; the result is a
    real fixed-point binary value with a precision of (15,0); if y does not occur in x,
    the result is a zero.

LENGTH (x)   x string-expression
    Returns the current length of x as a real fixed-point binary value with a
    precision of (15,0).

LOW (x)   x expression
    Returns a character string in which each character is the lowest in the collating
    sequence (hexadecimal 00); the result has the length specified by x.

REPEAT (x,y)   x bit or character expression to be repeated; y
                                      expression
    Returns a bit or character string consisting of x concatenated with itself the
    number of times specified by y (i.e., y + 1 instances of x).

STRING (x)   x aggregate or element reference
    Returns an element bit or character string that is the concatenation of all the
    elements of x.

SUBSTR (x,y[,z])   x character expression; y,z expressions
    Returns a substring of x starting at the position specified by y and having the
    length specified by z; if z is omitted, the substring runs to the end of x.

TRANSLATE (x,y[,z])   x character expression to be searched; y character
                                      expression containing the translating values of
                                      characters; z character expression containing the
                                      characters that are to be translated
    Returns a character string the length of x; if the character in x is found in z, the
    corresponding character in y is copied to the result.

UNSPEC (x)   x expression
    Returns a bit string that is the internal coded form of x.

VERIFY (x,y)   x,y string expressions
    Returns the position in x of the leftmost character or bit that is not in the same
    position in y.

A P P E N D I X   E

PL/I DATA

CONVERSIONS

AND DEFAULTS

Appendix E summarizes the data conversions performed by PL/I and the default data types that PL/I assigns to variables.

DECLARE	Variable	MODE	BASE	SCALE	PRECISION
(1)	(2)	(3)	(4)	(5)	(6)

1. DECLARE is a PL/I keyword that may be abbreviated to DCL.

2. The variable name follows the rules defined for identifier names (see Chapter 2).

3. The MODE attribute of an arithmetic data item may be REAL or COMPLEX (CPLX).

4. The BASE attribute of an arithmetic data item may be DECIMAL (DEC) or BINARY (BIN).

5. The SCALE attribute of an arithmetic data item may be FIXED or FLOAT. A fixed-point data item (FIXED) is a number in which the position of the decimal (or binary) point is specified, either by (a) its appearance in a constant or by (b) a scale factor declared for a variable. A floating-point data item is a number followed by an optionally signed exponent. The exponent specifies the assumed position of the decimal (or binary) point, relative to the position in which it appears in the numbers.

6. The PRECISION attribute is used to specify the minimum number of significant digits to be maintained for the values of the data items. It may also be used with fixed-point data items to specify the position of the decimal (or binary) point. The general format is (p,q), where p is the length of the number of digits appearing in the number, q is the scale factor, the number of digits to the right of the decimal (or binary) point. The scale factor can be negative.

The following chart shows the default attributes that occur when certain attributes are not explicitly coded in a DECLARE statement.

Declared Attributes		Default Attributes
DCL (A, B, C)	FIXED DECIMAL	Precision of 5 digits
	FIXED BINARY	Precision of 15 bits (halfword)
	FIXED	DECIMAL (5,0)
	FLOAT DECIMAL	Precision of 6 digits
	FLOAT BINARY	Precision of 21 bits
	FLOAT	DECIMAL (6)
	DECIMAL	FLOAT (6)
	BINARY	FLOAT (21)

When a variable is not declared, the default attributes depend upon the first letter in the variable name as follows

First Letter	Default Attributes
I through N	FIXED BINARY (15)
A through H or O through Z or #, $, or @	FLOAT DECIMAL (6)

The following describes the storage requirements needed for the various data types in PL/I.

1. **FIXED DECIMAL** (p,q)
   Stored internally: ½ byte per digit plus ½ byte for the sign; (p + 1)/2 bytes required
   Default precision: p = 5 decimal digits
   Maximum precision: p = 15 decimal digits

2. **FIXED BINARY** (p,q)
   Stored internally as a binary integer as

   a. halfword when p <= 15
   b. fullword when 16 <= p <= 31

   Default precision: p = 15 binary digits (32,767 decimal)
   Maximum precision: p = 31 binary digits (21,147,483,647 decimal)

3. **FLOAT DECIMAL** (p)
   Stored internally: in floating-point representation as

   a. fullword (short floating-point representation) when p <= 6
   b. doubleword (long floating-point representation) when 7 <= p <= 16

   Default precision: p = 6 decimal digits
   Maximum precision: p = 16 decimal digits
   Range of exponent: (−10) to (10)

INDEX